Service Robot Applications

Edited by
Yoshihiko Takahashi

Service Robot Applications
Edited by Yoshihiko Takahashi

Published by InTech
Janeza Trdine 9, 51000 Rijeka, Croatia

Additional hard copies can be obtained from orders@intechopen.com

Service Robot Applications
Edited by Yoshihiko Takahashi
p. cm.
ISBN 978-953-7619-00-8

For EU product safety concerns:
IN TECH d.o.o., Prolaz Marije Krucifikse Kozulić 3, 51000 Rijeka, Croatia,
info@intechopen.com or visit our website at intechopen.com.

Meet the editor

Yoshihiko Takahashi is a Professor of Mechanical Systems Engineering at Kanagawa Institute of Technology. He has also been working for Toshiba Corporation since 1980. He has been teaching control systems, mechatronics, and robotics at the same school since 1996. His research interests are in human assist robotic systems, fuel cell electrical vehicle, and applications of sustainable energy. He is the author or coauthor of more than one hundred journal and international conference papers, and four books. Yoshihiko Takahashi received his Master's and Doctorate degrees in mechanical systems engineering from Tokyo University of Agriculture and Technology, Japan, in 1980 and 1995 respectively.

Preface

Service robotics is among the most promising technology dedicated to supporting the elderly since many countries are now facing aging populations coinciding with a decrease in the amount of the young working population. Service robots assist human beings, generally by performing dirty or tedious work, such as household chores. Service robots, in some cases, may replace human caretakers in their ability to care for elderly people.

Additionally, those being cared for maintain an increased level of dignity when receiving assistance with more personal activities such as using the restroom. A user may ask a robot system for service without embarrassment. It is therefore possible to receive potentially better service in comparison with that of a human caretaker by using intelligent service robots. Service robotics delivers an excellent research example of system engineering as it encompasses many scientific research fields including mechanical engineering, electrical engineering, computer science, human science, welfare science, and many more.

The aim of this book is to provide new ideas, original results and practical experiences regarding service robotics. This book provides only a small example of this research activity, but it covers a great deal of what has been done in the field recently. Furthermore, it works as a valuable resource for researchers interested in this field.

Editor

Yoshihiko Takahashi
Kanagawa Institute of Technology
Japan

Contents

1

Robotic Patient Lift and Transfer

Roger Bostelman and James Albus
National Institute of Standards and Technology
USA

1. Introduction

Pollack says "today, approximately 10 percent of the world's population is over 60; by 2050 this proportion will have more than doubled" and "the greatest rate of increase is amongst the oldest old, people aged 85 and older." [Pollack, 2004] She follows by adding that this group is therefore subject to both physical and cognitive impairments more than younger people. These facts have a profound impact on how the world will keep the elderly independent as long as possible from caregivers. Both physical and cognitive diminishing abilities address the body and the mental process of knowing, including aspects such as awareness, perception, reasoning, intuition and judgment. Assistive technology for the mobility impaired includes the wheelchair, lift aids and other devices, all of which have been around for centuries. However, the patient typically or eventually requires assistance to use the device - whether to: push the wheelchair, to lift themselves from the bed to a chair or to the toilet, or guide the patient through cluttered areas. With fewer caregivers and more elderly in the near future, there is a need for improving these devices to provide them independent assistance. As further background, the authors have included sections on wheelchairs and lift devices.

1.1 Wheelchairs

Wheelchairs have been around for four hundred years since the first dedicated wheelchair, called an "invalids' chair," was invented for Phillip II of Spain. Later, in 1932, engineer Harry Jennings, built the first folding, tubular steel wheelchair similar to what is in use today. That chair was built for a paraplegic friend of Jennings named Herbert Everest. Together they founded Everest & Jennings. [Bellis, 2005]

There has been an increasing need for wheelchairs over time. In [Van der Woude, 1999] they state: "Mobility is fundamental to health, social integration and individual well-being of the human being. Henceforth, mobility must be viewed as being essential to the outcome of the rehabilitation process of wheelchair dependent persons and to the successful (re)integration into society and to a productive and active life. Many lower limb disabled subjects depend upon a wheelchair for their mobility. Estimated numbers for the Netherlands, Europe and USA are respectively 80,000, 2.5 million and 1.25 million wheelchair dependent individuals. These groups are large enough to allow a special research focus and conference activity. Both the quality of the wheelchair, the individual work capacity, the functionality of the wheelchair/user combination, and the effectiveness of the rehabilitation program do indeed determine the freedom of mobility. Their optimization is highly dependent upon a continuous and high quality research effort, in combination with regular discussion and dissemination with practitioners."

There is also a need for smart wheelchairs as people are living longer than before, will typically become less mobile with time, and will have reduced cognitive abilities and yet will need and perhaps want to remain independent. With fewer, younger and more capable assistants available for these elders, it creates a need for personal robotic care. Standards are being set for these mobile devices including manual and powered devices. Intelligent powered chairs have not yet been standardized.

1.2 Patient lift

Just as important as wheelchairs are the lift devices and people who lift patients into wheelchairs and other seats, beds, automobiles, etc. The need for patient lift devices will also increase as generations get older. When considering if there is a need for patient lift devices, several references state the positive, for example:

- "The question is, what does it cost not to buy this equipmente A back injury can cost as much as f50,000, and that's not even including all the indirect costs. If a nursing home can buy these lifting devices for f1,000 to f2,000, and eliminate a back injury that costs tens of thousands of dollars, that's a good deal," [Marras, 1999]
- 1 in every 3 nurses become injured from the physical exertion put forth while moving non-ambulatory patients, costing their employers f35,000 per injured nurse. [Blevins, 2006]
- 1 in 2 non-ambulatory patients fall to the floor and become injured when being transferred from a bed to a wheelchair. - [US Bureau of Labor Statistics, 1994]
- "Nursing and personal care facilities are a growing industry where hazards are known and effective controls are available," said Occupational Safety and Health Administration (OSHA) Administrator John Henshaw. "The industry also ranks among the highest in terms of injuries and illnesses, with rates about 2 1/2 times that of all other general industries..." - [Henshaw, 2005]
- "Already today there are over 400,000 unfilled nursing positions causing healthcare providers across the country to close wings or risk negative outcomes. Over the coming years, the declining ratio of working age adults to elderly will further exacerbate the shortage. In 1950 there were 8 adults available to support each person who is sixty-five years of age and older, today the ratio is 5:1 and by 2020 the ratio will drop to 3 working age adults per elder person." [Wasatch Digital IQ, 2003]

1.3 Mobile patient lift devices

Toward full independence for wheelchair dependents (WCD's) and for elders, there is a need for patient lift devices to move them from one floor to the next, from the bed or chair to the toilet, to a wheelchair, to cars, and other places, etc. Combination lift devices and wheelchairs have become available within the past several years. They provide stair/curb climbing, lift to reach tall shelves, etc. Standards for lift wheelchairs have not yet become available.

Discussions with healthcare professionals and patients indicate that WCD's:

- want to be self-sufficient even in a typical home and remain at home (i.e., not in a medical care facility) throughout their life,
- and/or homeowners don't want the home changed due to costs and intrusive changes, or even radically exchanging homes (e.g., selling 2 level to buy a 1 level home),

- want to be mobile; pick from and place things on shelves and cabinets; be at eye level to others; sit in their favorite chair; use a standard toilet; perform household tasks (cook, clean, hobbies); and not rely on others for these tasks.

In our research, we also found that:

- wheelchairs/powered chairs mobilize but, typically cannot lift above 25 cm - 33 cm (10 in - 13 in).
- many wheelchairs/powered chairs cannot fit through standard bathroom doors.
- wheelchairs/powered chairs cannot typically place WCD's in favorite chairs, on toilets or on beds.
- gyro-stabilized lift chairs cannot lift typical patients to reach upper cabinet and shelf heights and are relatively expensive as compared to powered chairs,
- ceiling, patient, and other lifts do not have full user-controlled mobility and/or are usually location specific,
- there is a need for multiple devices for mobility and lift equating to more patient/insurance costs and providing cumbersome maneuverability for the patient,
- rehabilitation assistance is virtually non-existent when used with current wheelchairs aside from some basics in stander, walker, and rollator (rolling walker) technologies.
- Devices for wheelchair dependents and elderly:
  - Are need specific
  - Attempt to be very inexpensive
  - Are mostly available in care centers, hospitals
  - Typically require additional caregiver dependence.

Organizations studying wheelchair and patient lift devices are:

- Universities who perform intelligent wheelchair and wheelchair standards research
- Many companies who provide "need specific" devices

However, few devices if any, exist to provide generic or multi-purpose tools, for example: wheelchairs that lift AND navigate on their own AND support patient rehabilitation AND sense crosswalks AND sense negative obstacles (stairs) well before approaching them.

New to the patient mobility and lift device technologies are:

- Sensors that have advanced well beyond the still-used ultrasonic sensors.
- CSEM SR3000[1], Sick LMS, Canesta D200, PmdTech 3-dimensional imagers
- Computer systems that are much faster, smaller, and less-expensive than ever before
- Powered chairs that are off-the-shelf items now, where several companies sell them
- Intelligence algorithms that are just beginning to be integrated into powered chairs - e.g., through doorway navigation

Perhaps missing in current devices are improved standards that now exist only for manual and for some powered chairs. No standards exist for intelligent chairs that use advanced sensors, computers and actuation systems. We believe that before intelligent chairs are commercialized and sold to the general public, a target safety design standard should be in place. Advanced lift wheelchair devices could provide improved device capabilities such as:

- Safety: e.g., Powered chairs that don't require stair blocks or caregiver watch; guidance for blind riders

---

[1] The mention of specific products is for illustrative purposes only, and is not meant to imply that NIST recommends these products or to suggest that they are the best available.

- Obstacle detection and avoidance: e.g., know when operator is about to run into an obstacle; guidance for blind riders
- Reduced dependency on caregivers for the elderly and disabled

People that may benefit from lift-wheelchair standards, performance metrics and advanced technology are the:

- Elderly
- Wheelchair dependents
- Disabled persons
- Blind persons
- Nurses, Caregivers

This chapter will provide some examples of past and current wheelchair and patient lift technologies and provides an in-depth review of a new NIST concept called the "Home Lift, Position, and Rehabilitation (HLPR) Chair," including sections covering: the structure and mobility design, patient lift, placement on other seats, rehabilitation, and control. Modifications to the initial design towards ergonomics and manufacturability are discussed followed by conclusions, future research, and a list of references.

2. Current wheelchairs and patient lift devices

2.1 Wheelchair examples

Manual wheelchairs, like the example shown if Figure 1, are: portable as they can fold to a smaller size, are relatively lightweight as compared to powered wheelchairs, have been around for hundreds of years, and have become a pseudo-standard in hospitals, shopping malls, homes and many other facilities to assist immobile patients.
The concept of power assistance for a manual wheelchair is relatively new, and represents a viable alternative for individuals who are unable to generate sufficient propulsion force to use a manual wheelchair, but do not wish to use a traditional powered mobility device. In a power assisted manual wheelchair, the traditional rear wheel hubs are replaced with motorized hubs that serve to magnify or reduce (i.e., brake) the propulsive force applied to the rear wheels by the user. Power assistance is proposed as the basis for a Smart Power Assistance Module (SPAM) that provides independent mobility to non-ambulatory individuals with visual impairments. [Cooper, 2004]
Powered chairs have become readily available on the market today and are made by several companies. These devices allow the operator to control their mobility without exerting manual force to move them and the device. Using one's arms to push a manual wheelchair can result in injuries, the need for surgery, and potential loss of independent mobility. Powered chairs can help eliminate these issues. Powered chairs are also being designed for sit-to-stand capability as shown in Figure 1 (center). This allows the patient to exert forces on the legs if possible and to reach items on shelves that are difficult for non-lift wheelchairs to access. Figure 1 (right) shows a gyro-stabilized lift wheelchair that allows a patient to be rolled up and over curbs and lifted relative to some cabinet heights and eye level to average-heights of standing persons.
It is however, important to note the need for operator attention while driving powered scooters or chairs. One of the authors personally watched an elderly person, who recently suffered from a stroke, driving a scooter while being inattentive to his surroundings inside a store. As a result, several times the elderly driver crashed into store displays, shelving and

other people. This dangerous situation begs for assistance from a caregiver to manually push this person around using a manual wheelchair. Alternatively, the powered chair could be equipped with advanced operator attention control interlocked to the low level mobility power to cut-off power to the drive system when the operator is inattentive.

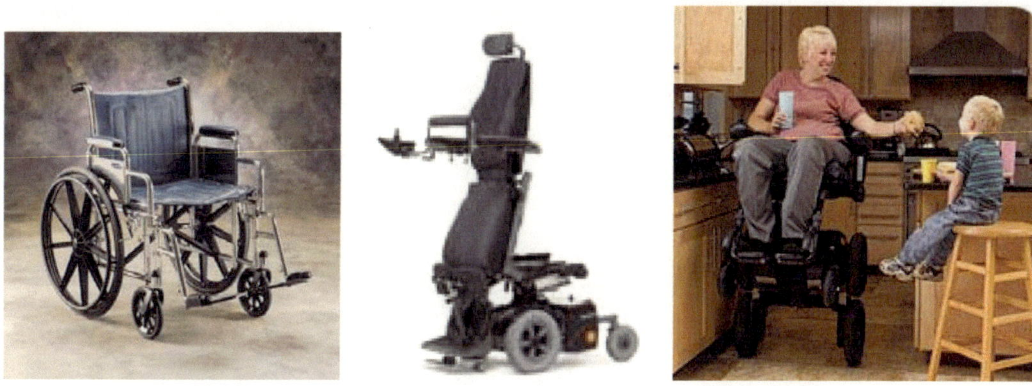

Fig. 1. (left to right) Manual Wheelchair, Powered Chair with patient sit-to-stand lift, Gyro-Stabilized Lift Wheelchair

2.2 Patient lift device examples

The Patient Pivot shown in Figure 2 allows a person to be lifted from a seated position, once strapped to the Pivot, and rotated forward to allow patient placement on another seat. The rotate mechanism uses levers allowing the caregiver to supply reduced force relative to lifting without the device.

Once a sling or hammock sheet is placed beneath the patient, lift devices such as the Hoyer and Patient lifts shown in Figure 2 can be used to power lift the patient from a seated or lying position. The device allows compliance for rotating the patient about the lift point and is manually mobilized by the caregiver. The Patient lift can reach the floor for patients who have fallen or can lift from seats or beds using straps attached to the device (not shown). This device also allows the device legs to be power separated for easy patient access to the device and/or to accommodate access to wide-base seats.

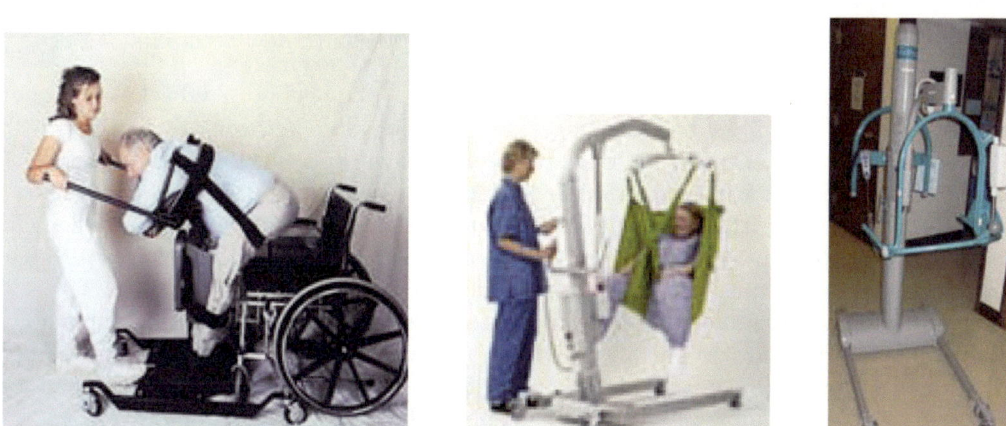

Fig. 2. (left to right) Patient Pivot, Hoyer lift with patient in a sling, and Patient lift

Wall or ceiling mounted patient lift devices (see Figure 3) provide lift and mobility from, for example a bed to a wheelchair. The sling or harness is required to surround the patient and therefore, must be initially placed beneath the patient and attached to the ceiling lift.
These lift devices have two degrees of freedom allowing lift of a harnessed patient in a sling like a crane and rotation about a horizontal pivot mounted to the wall or other support frame, such as a bed.

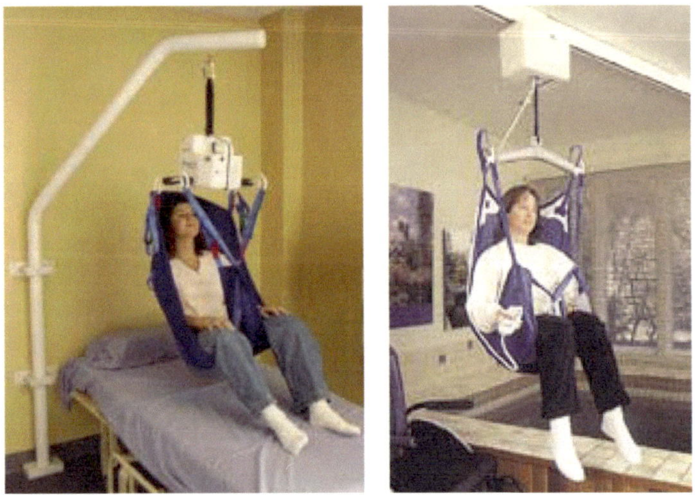

Fig. 3. Wall (left) and Ceiling (right) Mounted Patient Lift Devices

2.3 Patient transfer device examples

Some automobile and truck manufacturers have begun to develop transfer mechanisms built into the seat to allow access by the disabled. The chairs, made of actual car seats, are lifted and placed on the ground or into the vehicle.
Towards a home patient transfer concept, the Korea Advanced Institute of Science and Technology (KAIST) developed a system to transfer a person from a bed to a robotic wheelchair without assistance of another person in their Intelligent Sweet Home. [Park, et. al. 2007] The robotic wheelchair was equipped with autonomous mobility including: sensing, localization, obstacle detection, and motor control. The ultimate goal of a robotic wheelchair is to take the user automatically and safely to the destination.
For accomplishing the patient transfer objective, the robotic wheelchair moves autonomously to the predefined docking ready position and performs docking with the robotic transfer system. During the docking operation, bumpers of the robotic wheelchair detect collisions with the robotic transfer system and help make docking safe. Figure 4 shows the robotic wheelchair moving toward to the patient suspended in the patient lift. During autonomous moving, the robotic wheelchair performed localization, object detection, and motor control tasks.
The patient mobility, lift, and combinations of mobility and lift device examples cited above demonstrate that there are organizations who are or have recently been actively developing technology to transfer patients. What is missing is a single compact device to be used in a medical or caregiver facility and eventually the home that can independently:

- provide powered mobility for a patient,

- and lift them to reach from the floor to the ceiling or highest cabinets and shelves,
- and place the patient on a chair, bed or toilet
- and even provide some rehabilitation.

Built-in device intelligence is also required for patients who are not cognitively able to control the device for performing daily tasks.

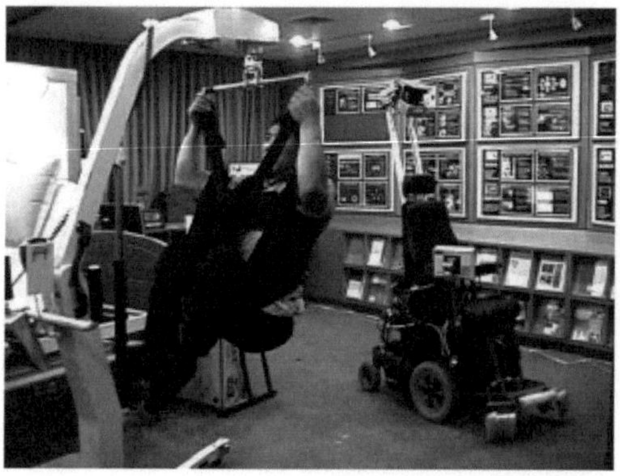

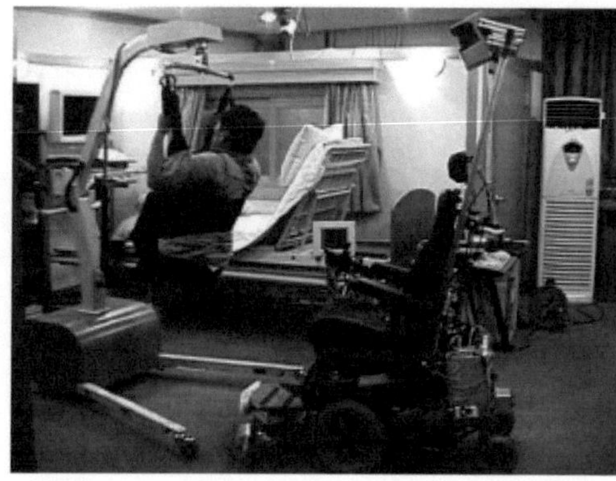

Fig. 4. (left) Autonomous moving to the predefined position, (right) Docking operation

3. Home lift, position, and rehabilitation (HLPR) chair

In 2005, the National Institute of Standards and Technology's (NIST) Intelligent Systems Division (ISD) began the Healthcare Mobility Project to address the staggering healthcare issue of patient lift and mobility. ISD researchers reviewed currently available technology through a survey of patient lift and mobility devices [Bostelman & Albus 2006-1]. The example cited above and many others are shown in this report. The report exposed a need for technology that includes mobility devices that can also lift and maneuver patients to other seats and technology that can provide for rehabilitation in the home to help the patient become independent of the wheelchair.

An additional area investigated in the survey was intelligent wheelchairs. NIST has been studying intelligent mobility for the military, transportation, and the manufacturing industry for at least 20 years through the Intelligent Control of Mobility Systems (ICMS) Program. [NIST, 2000] NIST is researching a standard control system architecture and advanced 3D imaging technologies within the ICMS Program. The NIST Healthcare Mobility Project is then applying them to intelligent wheelchairs where NIST has begun outfitting the HLPR Chair with computer controls. Although throughout the world there are or have been many research efforts in intelligent wheelchairs, including: [Kuno, et. al. 2000; Patel, et. al. 2002; Song et. al. 1999; Yanco, et. al. 1995] and many others, the authors could find no sources applying standard control methods nor application of the most advanced 3D imagers prototyped today to intelligent wheelchairs. Therefore, NIST began developing the HLPR Chair to investigate these specific areas of mobility, lift and rehabilitation, as well as advanced autonomous control.

3.1 HLPR chair design

The HLPR Chair [Bostelman & Albus 2006-2] prototype, shown in Figure 5, is based on a manual, steel, inexpensive, off-the-shelf, and sturdy forklift. The forklift includes a U-frame base with casters in the front and rear and a rectangular vertical frame. The lift and chair frame measures 58 cm (23 in) wide by 109 cm (43 in) long by 193 cm (76 in) high (when not in the lift position) making it small enough to pass through even the smallest, typically 61 cm (24 in) wide x 203 cm (80 in) high, residential bathroom doors. The HLPR Chair frame could be made lighter with aluminum instead of steel.

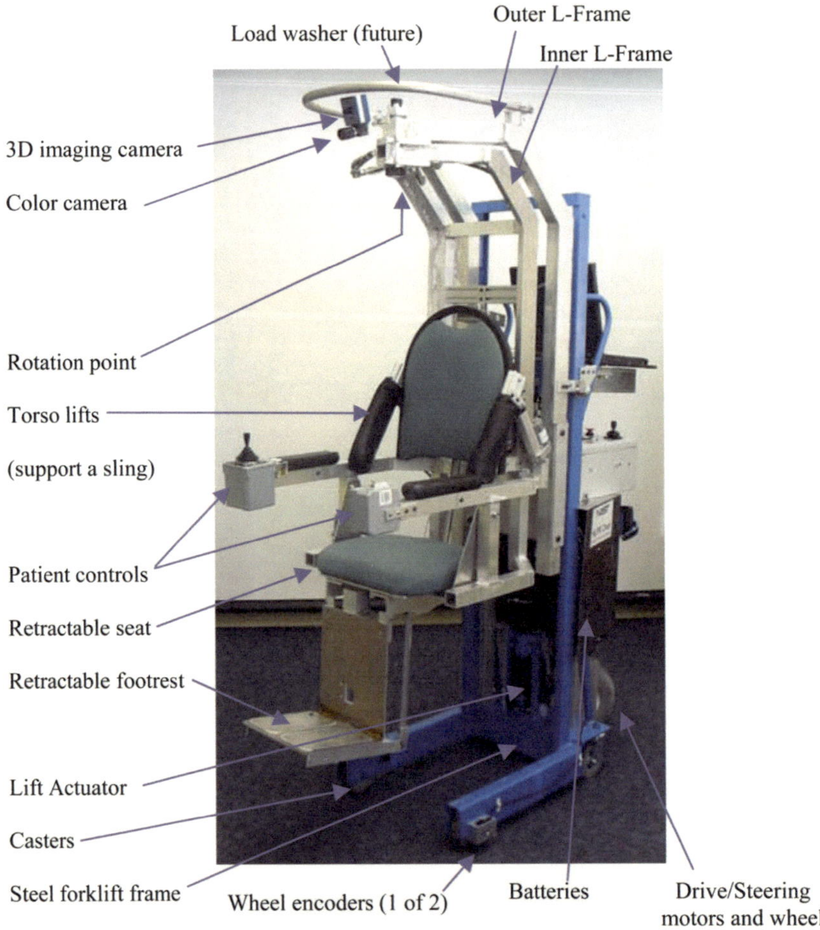

Fig. 5. The HLPR Chair prototype.

The patient seat/stand mechanism is a double, nested and inverted L-shape where the outer L is a seat base frame that provides a lift and rotation point for the inner L seat frame. The L frames are made of square, aluminum tubing welded as shown in the photograph. The outer L is bolted to the lift device while the inner L rotates with respect to the seat base frame at the end of the L as shown in Figure 5. The frames rotation point is above the casters at the very front of the HLPR Chair frame to allow access outside the wheelbase when the seat is rotated π rad (180°) and is the main reason access to other seats is available. Drive and steering motors, batteries and control electronics along with their aluminum support frame provide counterweight for the patient to rotate beyond the wheelbase. When not rotated, the center of gravity remains near the middle of the HLPR Chair. When rotated to π rad (180°) with a 136 kg (300 Lb) patient on board, the center of gravity remains within the wheelbase for safe seat access. Heavier patients would require additional counterweight.

The HLPR Chair is powered similarly to typical powered chairs on the market. Powered chairs include battery powered, drive and steer motors. However, the HLPR Chair has a tricycle design to simplify the need to provide steering and drive linkages and provide for a more vertical and compact drive system design. The drive motor is mounted perpendicular to the floor and above the drive wheel with chain drive to it. The steering motor is coupled to an end cap on the drive motor and provides approximately π rad (180°) rotation of the drive wheel to steer the HLPR Chair. The front of the robot has two casters mounted to a Ushaped frame.

The prototype high-speed drive motor is geared down through a chain-drive providing HLPR Chair speeds up to 0.7 m/s (27 in/s). Also, the drive amplifier gain has been adjusted to provide sufficient speed for typical eldercare needs and can be readjusted to allow for higher speeds as desired.

Steering is a novel single wheel design hard stopping the wheel at just beyond π rad (180°) for safety of the steering system. Steering is reverse Ackerman controlled as joystick left rotates the drive wheel counterclockwise and joystick right rotates the drive wheel clockwise. The steering rotation amount can be limited by reducing the amount of drive speed when large steering angles are commanded so as not to roll the frame during excessive speed. The navigation and control of the vehicle under this novel rear wheel steer and drive is currently under study and will be described in later publications.

For access to the HLPR Chair and for mobility, the HLPR Chair seat is lowered as shown in Figure 6. A seat belt or harness will be required for eldercare occupant safety. For access/exit to/from the HLPR Chair, the footrest can be retracted beneath the seat. For mobility, the footrest is deployed to carry the feet. Also, manually rotated feet pads can be deployed to provide a wider footrest. When retracted, the footrest pads automatically rotate within the footrest volume.

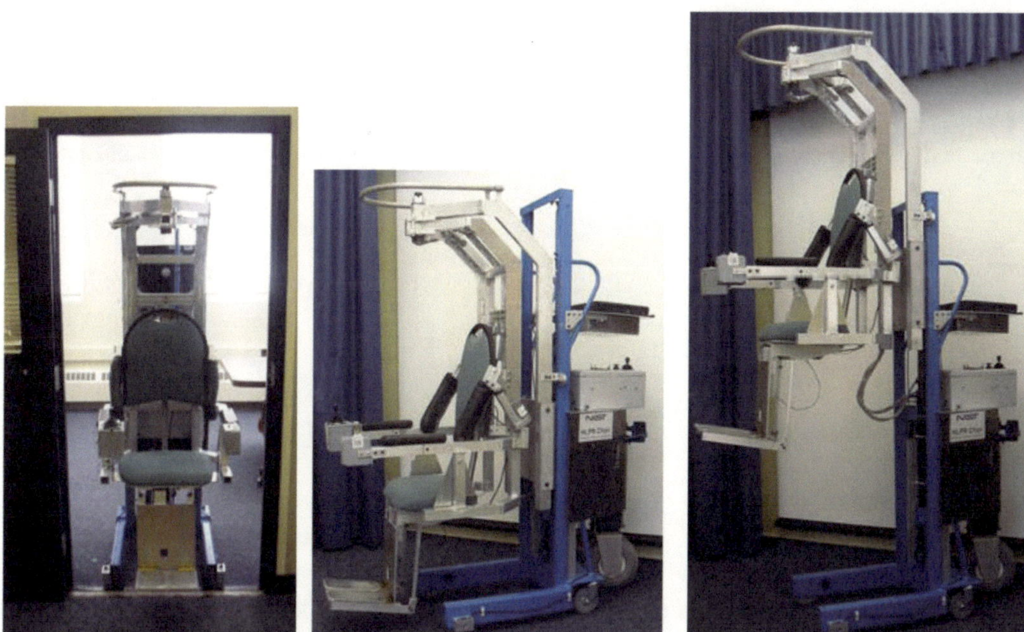

Fig. 6. The HLPR Chair in the mobility configuration showing the front view relative to a typical doorway (left), the side view (center) and the patient lift position (right).

3.2 Patient lift

Patient lift capability is designed into the HLPR Chair to allow user access to high shelves or other tall objects while seated. The HLPR Chairs' patient lift (see Figure 6 - right) is approximately 1 m (36 in) to reach what a typical, standing 2 m (6 ft) tall person could reach. This is a distinct advantage over marketed chairs and other concepts. [Bostelman & Albus 2006]. The additional height comes at no additional cost of frame and only minimally for actuator cost.

Lift is achieved by a 227 kg (500 Lbs) max. lift actuator that can support 681 (1500 Lbs) statically on the HLPR Chair prototype. The actuator can be replaced with a higher capacity unit if needed. The actuator connects to a lift plate with a steel chain that is fixed to one end of the HLPR Chair frame and to the lift plate at the other end. The actuator pushes up on a sprocket of which the chain rolls over providing 0.9 m (36 in) lift with only a 0.45 m (18 in) stroke actuator. The outer L-frame is then bolted to the lift plate. Rollers mounted to the lift plate roll inside the HLPR Chair vertical C-channel frame.

3.3 Placement on other seats

It is estimated that 1 in 3 nurses or caregivers will develop back injuries [9]. Most injuries occur because the patient is relatively heavy to lift and access to them is difficult when attempting to place the patient onto another seat. Wheelchair dependents have difficulty moving from a seat, to their wheelchair and back without a caregivers help or other lift mechanisms. The HLPR Chair was designed with the patient lift, as explained previously, to not only access tall objects, but to also pick up and place the patient in other chairs, on toilets, and on beds.

Figure 7 shows the concept of placing a patient onto a toilet. Figure 8 (left) shows the HLPR Chair prototype in the rotated position and Figure 8 (right) shows it in the torso support position similar to the Figure 7 (center and right) graphic.

To place a HLPR Chair user on another seat, they drive themselves to, for example, a toilet, seat, or bed. Once there, the HLPR Chair rotates the footrest up and beneath the seat and the

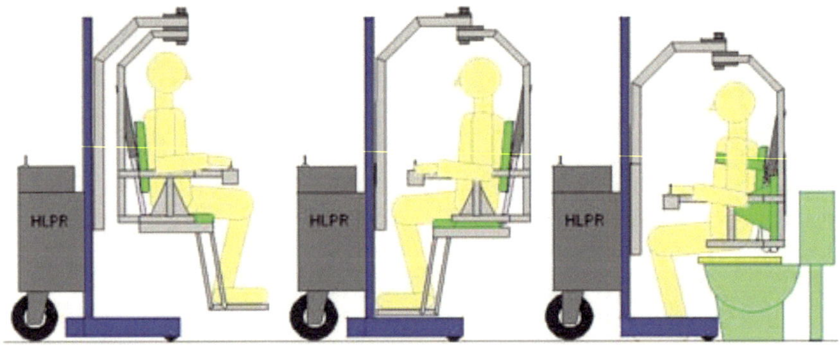

Fig. 7. Graphic showing the concept of placing a patient onto a toilet or chair with the HLPR Chair. The patient drives to the target seat (left), manually rotates near or over the seat (middle) while the torso lifts support the patient and the seat retracts, and then is lowered onto the seat - toilet, chair or bed (right).

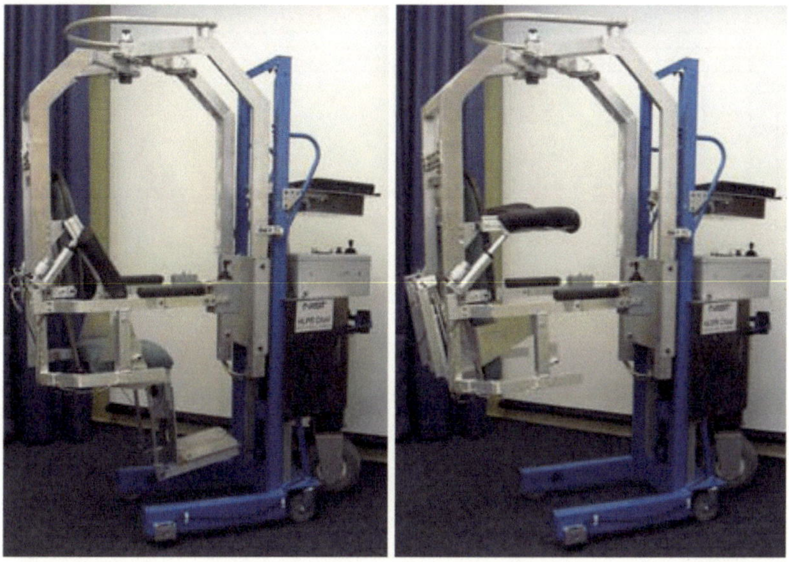

Fig. 8. The HLPR Chair in the same positions as in the center and right Figure 7 graphics placing a person on another seat.

patients feet are placed on the floor personally or by a caregiver. The HLPR Chair inner Lframe can then be rotated manually with respect to the chair frame allowing the patient to be above the toilet. Padded torso lifts, similar to crutches, then lift the patient from beneath his/her arm joints. The seat, with the footrest beneath, then rotates from horizontal to vertical behind the patients back clearing the area beneath the patient to be placed on the toilet, seat, bed, etc.

Once the person is placed on a toilet, the HLPR Chair can remain in the same position to continue supporting them from potential side, back or front fall. However, when placing a person onto a chair, the HLPR Chair must lift the patient and the patient manually rotates the chair from around the patient and out of the patients space. The HLPR Chair could then conceptually be driven from the seat location, using radio frequency or through voice commands, to a charging or waiting location and out of the patients view. When requesting to be picked up again, the patient could conceptually call the HLPR Chair remotely and have it return to the same pick up location and reverse the seat placement procedure.

For home use, the HLPR Chair is narrow enough to fit through typical doorways and openings. The turning radius of the HLPR Chair is approximately 76 cm (30 in). However, the HLPR Chair has a unique 'chair rotation within frame' design that, in many typical seat-access maneuvers, makes up for the longer turning radius. Figure 9 shows a CAD top view drawing of a typical bathroom in a home.

To place a patient onto, for example, a very soft sofa requires a simple, additional step not explained above. In order to rotate the chair about the patient once they have been seated on the sofa, it must first be lifted above the cushion and patients legs. Ideally, the patient is more comfortable on this type of seat than on the HLPR Chair or on a rigid chair. Once the person is placed on a , the HLPR Chair can remain in the same position to continue supporting the patient reducing potential for falling to the side, back or forward.

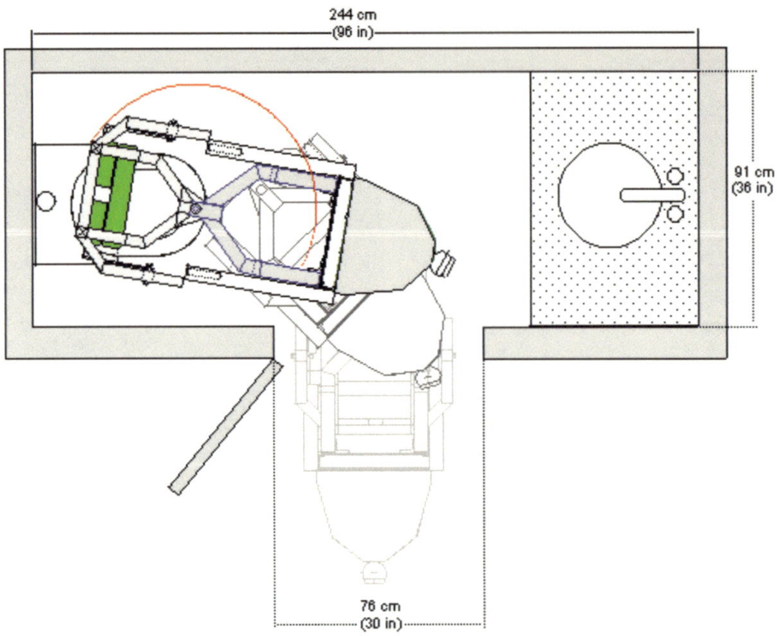

Fig. 9. CAD top view drawing of the HLPR Chair accessing a typical bathroom and toilet. The radius drawn is the area needed for the seat frame to rotate.

3.4 Manual control

The HLPR Chair controls include a joystick that sends drive controls to power amplifiers that control the drive and steering. The patient lift actuator is also controlled with the same type power amplifier through a rocker switch. A lever switch is used to control seat and footrest retraction or deployment.

The footrest, seat and torso lift actuators are direct powered switched forward and reverse from the battery through momentary rocker switches. Actuators for the footrest and each torso lift has 8cm (3 in) stroke while the seat includes a 31 cm (12 in) actuator to rotate it from seated position to behind the back and vice versa.

Behind the seat and frame and above the drive/steer wheel is the electronics box that houses the controls for the HLPR Chair while also providing a "Nurse" or caregiver control panel that duplicates the patient controls at the seat. The Nurse control panel (see Figure 10) includes all the control functions for a nurse or caregiver to drive or lift a dependent patient. Control redundancy is designed into the HLPR Chair to also allow a caregiver to quickly gain control of the device as needed. A "Nurse/Patient" switch on the Nurse control panel allows switching between the rear (Nurse) controls and the chair (Patient) controls.

3.5 Towards autonomous control

The HLPR Chair was recently modified (see Figure 10) to include encoders, attached between its' frame and front caster wheels, a computer and computer interface electronics. The encoder design included adapting a shaft to one side of each caster wheel, passing it through a bearing attached to the frame and to an encoder. Although the encoder and housing add an additional 2.5 cm (1 in) to each side of the base, the overall HLPR Chair base width is still within the chair-frame width and therefore, within the overall HLPR Chair width of 58 cm (23 in). The encoders provide 3600 pulses per revolution allowing relatively fine measurement over a 12.7 cm (5 in) diameter caster wheel or approximately 90 pulses/cm (230 pulses/in) of linear travel. The relatively high measurement accuracy of the wheels will support development of accurate path planning and control algorithms for the HLPR Chair.

Included in the Nurse control panel is a computer/manual switch. While switched in manual mode, all of the "Nurse" - labeled (rear) controls on the box or on the "Patient" - labeled (chair) can be used. While in computer control, drive and steer are controlled by an onboard computer. The computer is currently a personal computer (PC) laptop interfaced to off-the-shelf input/output (I/O) devices housed in the box beneath the PC and connected through a universal serial bus (USB) interface. This design was chosen as a simple developer interface to the HLPR Chair prototype knowing that the computer and its interfaces can be significantly reduced in size as future commercial versions are designed.

Software drivers for the HLPR Chair drive and steer control were written in C++ under the Linux operating system.

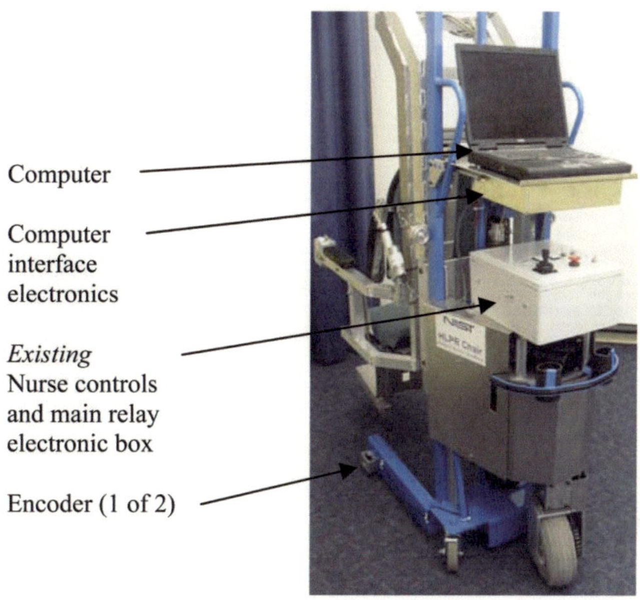

Fig. 10. The HLPR Chair with recently added front wheel encoders, development computer and interface electronics. Recently added advanced 3D imaging camera and color camera are shown in Figure 5.

This low level control is now ready to add planned HLPR Chair navigation and obstacle avoidance control. NIST and the University of Delaware (UD) are teaming to use the NIST standard software control architecture for intelligent machines called 4 Dimensional/ Real-time Control System (4D/RCS) and UD's robot behavior generation. [Sira-Ramirez & Agrawal, 2004] NIST has recently applied 4D/RCS to a Defense Advanced Research Project Agency (DARPA) Project called Learning Applied to Ground Robots (LAGR). [Albus, et al. 2006] The 4D/RCS structure developed for LAGR is shown in Figure 11. The basic premise of the 4D/RCS columns of boxes are to sense the environment around the robot (left column), to place the sensed information into a world model (middle column), then plan and generate appropriate navigational paths and input these paths into the robot actuators in real time (right column). The horizontal rows of 4D/RCS boxes stack from a servo level control (bottom row) to grouped pixels, a lower resolution map, and a higher level planner (top row).

The authors plan to adopt this standard control architecture on the HLPR Chair so that advanced 3D imagers, such as the ones shown in figure 5, and robust control algorithms can be "plug-and-played" to address the variety of patient mobility controls that may be needed. An earlier version (from the one pictured in figure 5), 3D imaging camera was mounted on an early version of the HLPR Chair and a control algorithm was developed and tested. Results of this test, as explained in [Bostelman, et. al. 2006], clearly show detected obstacles in the vehicle path and a planned path around the obstacles.

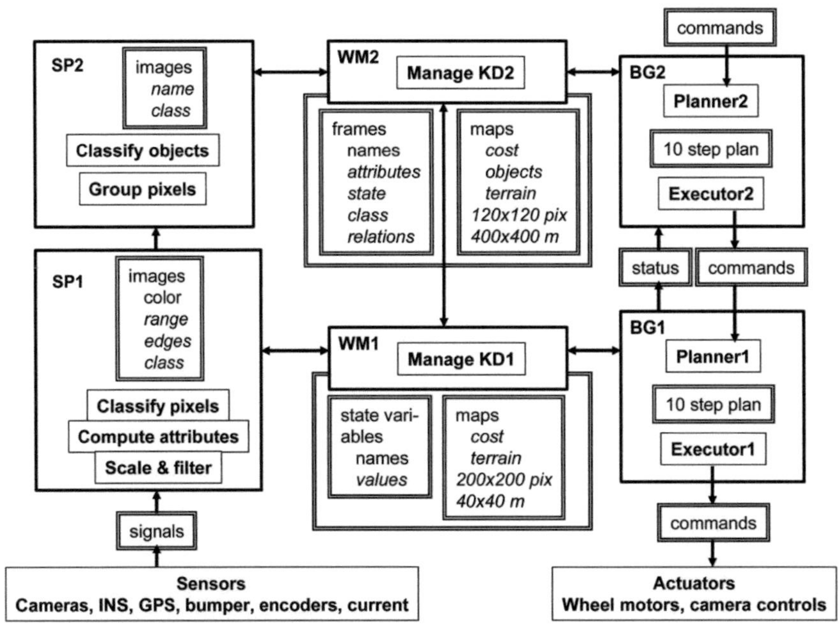

Fig. 11. NIST 4D/RCS 2-level, hierarchical control architecture developed for the DARPA LAGR Project and planned for implementation on the HLPR Chair.

3.6 Patient rehabilitation

HLPR Chair enhances patient rehabilitation through a load sensor and control on the lift actuator, as described in [Banala, et. al. 2007]. The authors designed rehabilitation into the HLPR Chair to allow, for example, stroke patients to keep their legs active without supporting the entire load of the patients body weight. The patient, once lifted, could walk while supported by the HLPR Chair driving at a slow walking pace towards regaining leg control and perhaps eliminating the need for a wheelchair.

To accomplish rehabilitation, the HLPR Chair includes, as explained in the Placement on Other Seats section, footrest and seat rotate behind the patient while he/she is lifted with torso lifts. However, instead of being placed low on a seat, the patient lift continues to mov up lifting the patient as they move their legs beneath them to standing position. The HLPR

Chair's open U-frame base allows access to the floor directly beneath the patient for standing. Figure 12 shows a photograph of the prototype in this configuration and a concept of how the HLPR Chair can be used for patient rehabilitation.

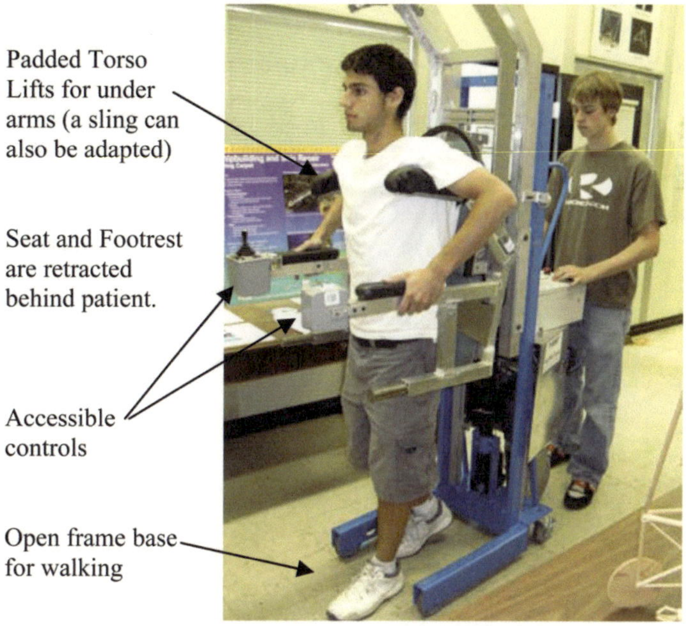

(a)

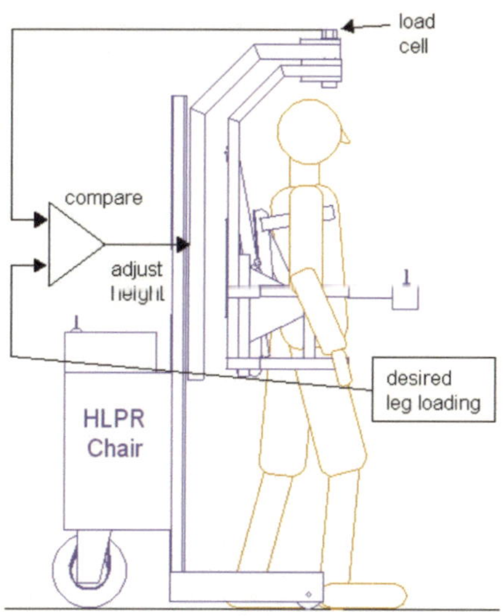

(b)

Fig. 12. (a) The HLPR Chair prototype in the rehabilitation/walking configuration. Summer Interns (Alex Page and Robert Vlacich) demonstrate the patient and nurse configuration as part of their official duties. (b) Graphic showing the concept of how the HLPR Chair can be used for patient rehabilitation and incorporate future legs load control.

Additionally, the patient can be continuously monitored with a load sensor at the L-frames rotation point. The patient could adjust the amount of load he/she wishes to place onto their legs and on the floor by rotating a dial on the controls from 0% to 100%. Load control is a future concept to be applied to the HLPR Chair prototype in the next several months.

4. Improved HLPR chair ergonomics and manufacturability

Modifications have continued on the HLPR Chair targeting more ergonomic designs, less expensive manufacturability, and a more load-distributed patient support while being transferred to other seats or in the standing position. The more ergonomic design and less expensive manufacturability were achieved by using a bent tubing design for the seat and base frames. Thin wall, 3.2 mm (0.125 in) wall thickness, tubing was bent into the curved shape as shown in Figure 13. Also shown in Figure 14 is the seat frame designed to be wider as compared to the first HLPR Chair design shown in figure 5, as well as being curved similar to the new base frame design. The wider frame allows standard, off the shelf seats to be used on the HLPR Chair. To allow very low friction and hollow rotary joint design for electrical cable harnessing, a thin, inexpensive bearing was used between the frames allowing a very smooth, low side-to-side torque design. These two designs were an order of magnitude decrease in manufacturability cost from the previous design.

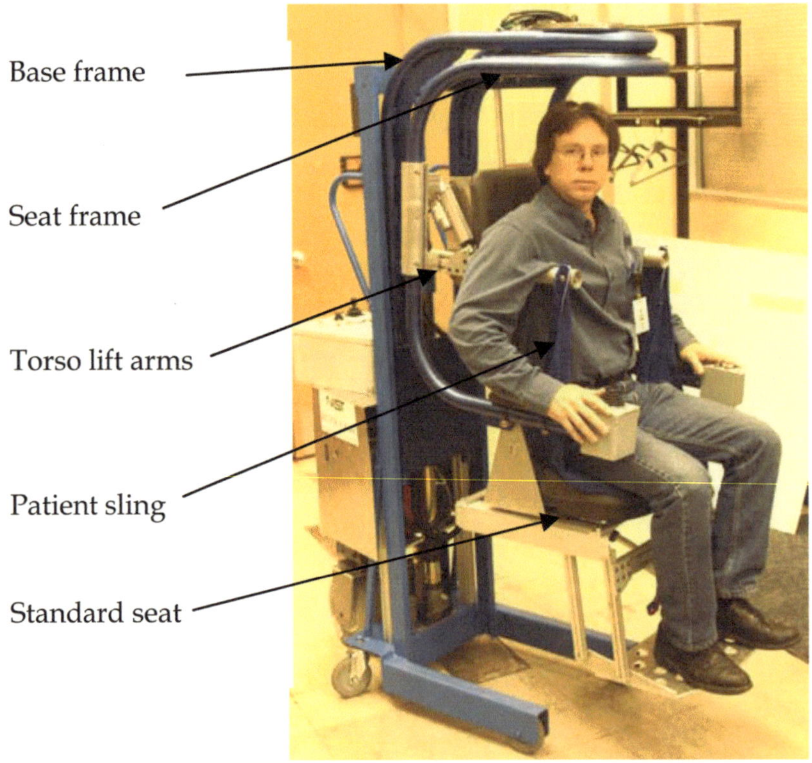

Fig. 13. The modified base and seat frames of the HLPR Chair from the previous welded design.

To address the more load-distributed patient support need when patients are transferred to other seats or in the standing position, a modified sling design was used (see Figure 14). The modified sling combines a typical medical sling and climbers harness but, uses much less material. The same torso lifts are used as in the previous design although they now do not provide the patient lift. They are simply used as a lift mechanism for the sling which now lifts the patients legs and buttocks. The original torso lift arms now only provide a sense of security to the patient without causing underarm injury as may have been experienced with only using torso lift arms as in the previous design shown in Figure 5. A standing position sling is also planned for integration with the HLPR Chair. The design will be slightly different from the seated sling design since it will include an off-the-shelf climbers or rehabilitation harness modified to fit the HLPR Chair and suspended from above the patient.

Fig. 14. Photographs of the HLPR Chair sling showing (left) the sling ready for the patient to be seated, (center) a patient with the sling ready to lift while seated on the HLPR Chair seat, and (right) the patient fully supported by the sling with the HLPR Chair seat rotated behind the patient.

5. Conclusions and future research

There have been many patient transfer devices designed and developed over many years. The HLPR Chair was designed to be a revolutionary patient lift and mobility system for wheelchair dependents, the elderly, stroke patients, and others desiring or even requiring personal mobility and lift access. The system shows promise for moving these groups of patients into the work force and removing the burden placed on the healthcare industry. The system has been prototyped to show the basic concept of such a patient lift and mobility system. The HLPR Chair was built to demonstrate its relatively inexpensive capabilities to the healthcare industry and to demonstrate potential, near-term capabilities with robust controls for mobility and rehabilitation.

Autonomous mobility control using the 4D/RCS standard control architecture and integration of advanced 3D imagers is planned for a next step through teaming with the University of Delaware under a federal grant. Force loading for rehabilitation of patient legs will also be studied in the near term.

Ergonomics and manufacturability of devices such as the HLPR Chair are critical for the general public and manufacturers to accept an appealing, safe and low cost patient transfer design.

Commercialization is now being considered by the healthcare industry. Figure 15 shows concept drawings developed by Innova Robotics and Automation, Inc. of a more ergonomic and commercialized version of the HLPR Chair. Collaborations for proving the service capabilities and evaluating performance of commercial versions of the HLPR Chair and also, setting safety and performance standards for this type of assist device are being pursued and expected in the near future.

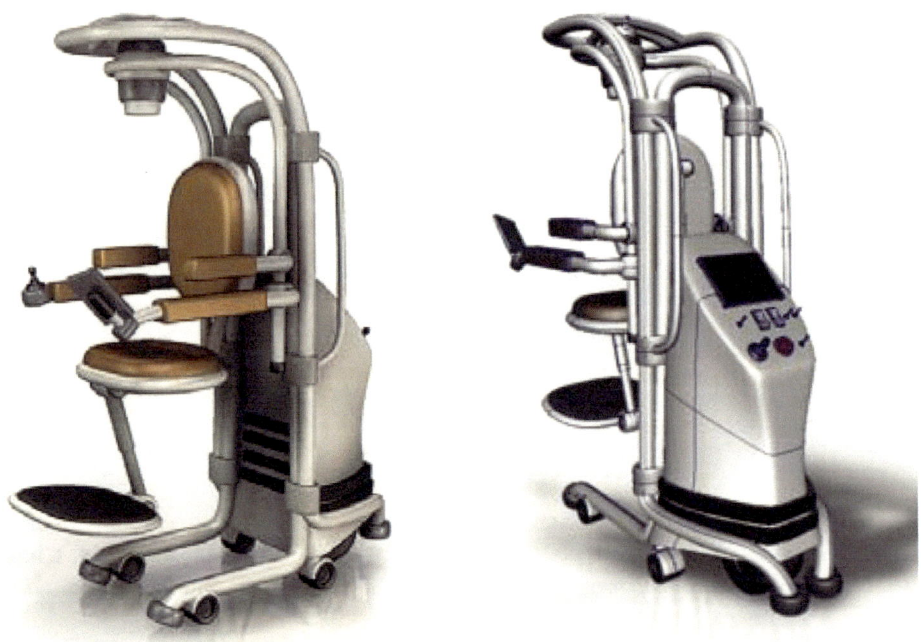

Fig. 15. Front (left) and rear (right) concept drawings of the HLPR Chair as a more ergonomic and commercialized version.

6. References

Albus, J., Bostelman, R., Hong, T., Chang, T., Shackleford, W. & Shneier, M. (2006). *Integrating Learning into a Hierarchical Vehicle Control System*, Integrated Computer-Aided Engineering Journal.

Banala, S., Agrawal, S., Fattah, A., Scholz, J., Krishnamoorthy, V., Rudolph, K., & Hsu, W-L. (2007). *Gravity Balancing Leg Orthosis and its Performance Evaluation*, IEEE Transactions of Robotics.

Bellis, M. (2005). "Wheelchair History," *http://inventors.about.com/library/inventors/* blwheelchair.htm

Blevins Medical, Inc. (2006). Healthcare Statistics: http://www.patientlift.net/ 282164.html.

Bostelman, R. & Albus, J. (2006-1). Survey of Patient Mobility and Lift Technologies Toward Advancements and Standards, NISTIR u7384.

Bostelman, R &, Albus, J. (2006-2). HLPR Chair - A Service Robot for the Healthcare Industry, 3rd International Workshop on Advances in Service Robotics, Vienna, Austria.

Bostelman, R., Russo, P., Albus, J., Hong., T., & Madhavan, R. (2006). Applications of a 3D Camera Towards Healthcare Mobility Aids, IEEE International Conference on Networking, Sensing and Control, Ft. Lauderdale, FL, April.

Cooper, R., Simpson, R., LoPresti, E., Guo, S. & Ding, D. (2004). NIH - Small Business Innovative Research Project, http://www.herlpitt.org/research.htm.

Dictionary.com (2007).

Henshaw, J. (2005). *http://www.osha.gov/SLTC/nursinghome/solutions.html*, Occupational Safety and Health Administration.

Kuno, Y., Murashima, T., Shimada, N. & Shirai, Y. (2000). Intelligent Wheelchair Remotely Controlled by Interactive Gestures, International Conference on Pattern Recognition, vol. 04, no. 4, p. 4672.

Marras, W. (1999). Lifting Patients Poses High Risk for Back Injuries, Ohio State University, http://researchnews.osu.edu/archive/resthome.htm.

NIST Intelligent Control of Mobility Systems Program website (2007). http://www.isd.mel.nist.gov/ researchvareas/mobility/index.htm.

Park, K., Bien, w., Lee, J., et. al. (2007). *Robotic Smart House to Assist People With Movement Disabilities*, Autonomous Robots Journal, V. 22, Iss. 2, Ppg. 183-198.

Patel, S., Jung, S-H., Ostrowski, J., Rao, R. & Taylor, C. (2002). Sensor based door navigation for a nonholonomic vehicle, GRASP Laboratory, University of Pennsylvania, Proceedings IEEE International Conference on Robotics and Automation, Washington, DC.

Pollack, M. (2004). Intelligent Technology for Adaptive Aging Presentation, AAAI-04 American Association for Artificial Intelligence Conference Keynote Address, 2004

Sira-Ramirez, H., Agrawal, S. (2004). Differentially Flat Systems, Marcel Dekker (Control Engineering Series), Hardbound, ISBN 0-8247-5470-0, 467 pages.

Song W., Lee, H., Bien, w. (1999). "KAIST - KARES: Intelligent wheelchair-mounted robotic arm system using vision and force sensor, Robotics and Autonomous Systems, vol. 28, no. 1, pp. 83-94(12), 31, Publisher: Elsevier Science.

U.S. Bureau of Labor Statistics (1994). from Blevins website: http://www.patientlift.net/ 282164.html.

Van der Woude, L., Hopman, M., & Van Kemenade C., (1999). *Biomedical Aspects of Manual Wheelchair Propulsion: The State of the Art)),* Volume 5, Assistive Technology Research Series, 392 pp., hardcover

Wasatch Digital iQ. (2003) "InTouch Healthxs Remote Presence Robot Used by Healthcare Experts," http://www.wasatchdigitaliq.com/parser.phpenavyarticle&articlevidy 43, Santa Barbara, CA & Salt Lake City --(Business Wire).

Yanco, H., Hazel, A., Peacock, A., Smith, S. & Wintermute, H. (1995). Initial Report on Wheelesley: A Robotic Wheelchair System, Department of Computer Science, Wellesley College.

2

Sensory-Motor Coupling in Rehabilitation Robotics

Alejandro Hernandez-Arieta, Konstantinos Dermitzakis, Dana Damian, Max Lungarella and Rolf Pfeifer
University of Zurich, Artificial Intelligence Laboratory, Switzerland

1. Introduction

The general well-being of people has always been a strong drive towards the improvement of available technologies and the development of new ones. Recently, a greater longevity and the consequent increase of physically challenged elder adults have increased the significance of research on assistive technologies such as rehabilitation robots, power-assist systems, and prosthetic devices. One important goal of these research endeavors is the restoration of lost motor function for people with disabilities (e.g. locomotion, manipulation, and prehension) with prostheses such as robot hands, arms, ankles, and legs (Lebedev et al., 2006; Kato et al., 2006). Although such prosthetic devices are increasingly more intuitive to use and yield better functionality, most users still fail to recognize the prosthesis as an integral part of their body – a problem akin to the one affecting people with lesions to their afferent nervous system (Fig. 1) (Tsakiris et al., 2007). Such lack of integration makes the control of the prosthetic device cumbersome, and thus leads to excessive and unwanted cognitive load (Hunter et al., 2003). To counter this problem, some prosthetic applications attempt to strengthen the user-environment coupling (Pylatiuk et al, 2006), for example by feeding back visual or vibro-tactile stimuli to the user.

Clearly, one important challenge of rehabilitation robotics is how to "trick" the user of the prosthesis to accept the device as an integral part of their body. There are at least three ways to tackle this challenge: (1) Exploit the trends in information technology such as more energy-efficient and powerful microcontrollers, allowing faster sensory feedback and better real-time performance at a smaller scale. (2) Use smart materials and adaptive mechanisms to reduce computational complexity and off-load more to the morphological properties of the prosthetic device (Dollar & Howe, 2007; Pfeifer et al., 2006; Pfeifer et al., 2007). (3) Improve human-machine interaction by developing "intelligent" machines that can guess the user's intention and are able to adapt to it, e.g. by employing artificial neural networks, pattern recognition and machine learning techniques (Yokoi et al., 2004). Experiments have shown, for instance, that the inclusion of the patient's intention in the control of electrical stimulation applied to the leg muscles can improve the performance of a paraplegic support system (Riener et al., 2000).

In the context of rehabilitation robotics, this idea was tested by implementing an artificial controller for a functional electrical stimulation support system for paraplegics that considers the whole-body dynamics involved in human walking to predict the next desired

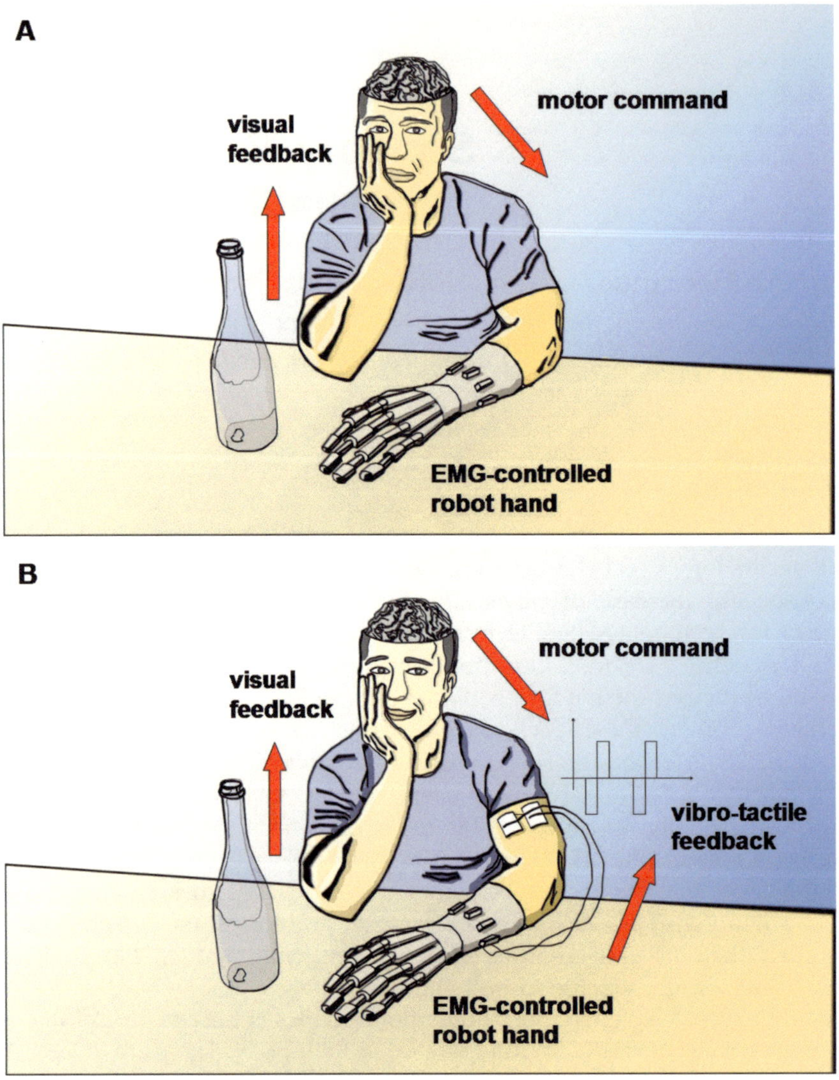

Fig. 1: Disruption of sensory-motor coupling. The disruption of the sensory-motor pathway has a direct effect on the interaction with the environment. (A) A lack of body awareness necessitates an increased cognitive effort to restore control of the body in an adapting environment. (B) The insertion of sensory feedback reduces cognitive effort and improves the interaction with prosthetic devices.

movement (Heliot et al., 2007). The coexistence of artificial and natural controllers allowed the patient some degree of control of otherwise paralyzed limbs. There are two implications from these studies. First, the use of "smart" machines can assist in improving the interaction, but the human brain must be included into the equation - being the most adaptive "machine" at our disposal. Second, in applications in which human and "smart" device interact directly, there is a need for an intuitive communication channel between man and machine. Several studies have addressed the realization of health-related robotic platforms that interact with humans. Ideally, such "human-centered" platforms (Riener et al., 2005; Riener, 2007) have to be safe, flexible, mechanically compliant, adaptive towards the user's needs, and easy to use. Moreover, they need to actively involve the patient in the rehabilitation process such that the recovery of lost motor function can be sped up. "Human-centered" strategies have to be contrasted with traditional "controller-centered" approaches in which the patient has to submit to the controller, e.g. a desired reference signal. Our long-term objective is to reduce system complexity by incorporating the plasticity of the human body and brain into the rehabilitation process. Towards achieving this goal, in this chapter, we present a case study based on an EMG-controlled prosthetic hand in which we study the sensory-motor patterns emerging from the interaction between the human, the robot hand, and the surrounding environment. An adaptive learning mechanism attempts to match the movements of the robot hand to those of the user's hand. To address the lack of integration, we apply an electrical stimulation to the user whenever the robot hand touches an object, eliciting a tactile sensation on the user's body. By employing a functional magnetic resonance imaging (fMRI) device, we then evaluate the human-machine interaction looking for the requirements behind the production of the illusion of "body ownership" illusion. We test two sensory modalities (visual and tactile) in the manipulation of the robot hand to gain insights into the mechanisms responsible for the extension of the user's body representation. Apart from looking at neural activity in the motor cortex, we also investigate related changes that occur in the sensory cortex.

2. Prosthetics applications

In what follows, we analyze the two principal characteristics for the interaction with prosthetic devices: human intention detection and sensory feedback.

2.1 Human intention detection

One of the biggest challenges for the field of human-machine interaction is the prediction of the intent of subjects to perform actions such as hand or finger movements. There exist many methods for predicting human movement intention, ranging from the real-time analysis of sensory data (Heliot et al., 2007) to the assessment of biological and physiological signals such as electroencephalograms (EEGs) or electromyograms (EMGs) (Kato et al. 2006, Bitzer & van der Smagt, 2007). Because of its relevance to the content of this chapter, we only review the use of EMG signals in the context of prosthetics applications.

EMGs are the electrical manifestation of the neuromuscular activity associated with a contracting muscle. Two properties make EMG signals particularly well suited for detecting the intention of movements: (1) EMG signals are directly linked to the desire of movement of a person, whether the movement is executed voluntarily or is initiated through a reflex response; and (2) EMG signals are emitted early, before the muscles contract, and hence can be used for prediction. The problem of intention detection has been tackled by a plethora of scientific work with varying degrees of satisfaction (Wang et al., 2006; Katsis et al. 2005; Nazarpour et al., 2005). Other efforts have focused on the discrimination of a number of hand movements through feature extraction and on improving the effect of a real-time learning for a prosthetic hand (Khezri et al., 2007). Although the accuracy of the recognition is high, these algorithms have not been yet applied to a system that tests their actual efficacy within the demand of a real amputee. In general, feature extraction raises concerns with respect to the techniques that select the meaningful data in the context of real world scenarios, where muscles can get tired, and EMG signals are non-stationary. Ways of dealing with the nature of EMG signals in the processing stage rely on the use of different filters and signal processing techniques. One method to transform such signals into an efficient representation is to use families of functions invariant to translation and scaling. By feeding the resulting signals to a neuro-fuzzy classifier, it becomes possible to infer the intention of a person to stand up or sit down (Hussein and Granat, 2002). Support Vector Machines have also been employed for categorizing finger movements. In conjunction with a maximum likeliness measure, the results were sufficiently robust to partition the finger movements in the case of arm pronation (Bitzer et al., 2006).

2.2 Sensory feedback

Neurological studies suggest that the self-attribution of body parts is mediated by correlated multisensory feedback (Armel, 2003; Ramachandran et al., 2000). Therefore, when provided with synchronous stimulation, the brain combines the stimuli and associates them to a unique perceptual event. For instance, the sight of brushing of a rubber hand at the same time as brushing of the person's own hand (but hidden from view) is sufficient to produce a feeling of ownership of the fake hand (Ehrson et al., 2005). This illusion of body ownership is called the "rubber-hand illusion" (Constantini & Haggard, 2007; Ehrson et al., 2005) and seems to originate from the ambiguous but correlated sensory information fed to the brain, which leads to the sensation of having the rubber hand incorporated into the body schema. By using this finding, one could think of fooling the brain into accepting a prosthetic device as an integral part of the body, hence reducing the cognitive effort required for its control. This idea is supported by an experiment conducted on monkeys, which shows that cortical motor neurons that fire when a hand is grasping an object, also fire when the object is grasped with a pair of pliers (Umilta et al., 2008). The outcome of this experiment strongly indicates that the tool is embedded in the monkey's bodily representation as if it would be the monkey's own hand. The aforementioned experiments seem also to suggest that the brain forms internal representations based on the sensory information fed back to it. In this sense, one major drawback of EMG-controlled devices is the minimal or non-existent biofeedback, that is, information on the prosthetic device in relation to the body.

The human body incorporates a robust and redundant sensory system, by which if a part fails, nearby ones are used in order to restore the lost function. The user of a prosthetic device usually needs to overcome the lack of tactile and proprioceptive data with visual feedback, which increases the cognitive effort required to control the device (Weir, 1998). This conscious effort is one of the main reasons that amputees abandon the use of current myoelectric devices (Biddiss, 2007). We conclude that prosthetic devices need to include a feedback source that enables the user to extend his physiological proprioception (Simpson, 1974). Such sensory feedback is of course a prerequisite not limited only to prosthetic applications; spinal-cord injury patients also share the same requirement.

Methods for providing feedback to the human body can be classified into two categories: invasive and non-invasive. Invasive methods directly stimulate nerve fibers to transmit sensations to the brain. For example, Shimojo et al. (2003) inserted electrodes in the nerve axons of a person that were used to transmit tactile information sampled from a robot. More recently, Dillon et al. (2005) implemented a controller for a robot hand that used both afferent and efferent neural paths for communication from and to the human body, i.e. for controlling the robot hand and for receiving tactile feedback from the robot platform. Non-invasive methods are more widespread in their implementations. The most common method to transmit information to the body is transcutaneous (surface-mounted) electrical stimulation (Back-y-Rita et al., 2003; Kaczmarek et al., 2006). There is also relevant research looking for appropriate stimulation areas in the human body for data transmission (Riso, 1999); as well as the modulation used to increase the amount of data transmitted into the body (Kaczmarek, 2000; Kim, 2005). Besides electrical stimulation, mechanic vibrators are an additional option for data transmission. Although they are typically used in haptic interfaces (Honma et al., 2004), they also find application in the context of prosthetics (Rios-Poveda, 2002).

3. Experimental setup

For our experiments, we used a prosthetic hand controlled through EMG signals (Yokoi et al., 2004). The raw EMG signals were processed producing a set of feature vectors that in turn are used to generate a database of intended motions. Feedback to the user of the hand is provided through transcutaneous functional electrical stimulation (Szeto, 1992; Kackzmareck, 2006). In what follows, we give an overview of the parts composing the experimental setup. First, we describe the prosthetic hand and the EMG-based intention detection system. Then, we present the visuo-tactile feedback system. Finally, we expose the fMRI scan setup and the related data analysis.

3.1 Prosthetic hand

The EMG-controlled prosthetic "humanoid" hand employed in this study is composed of five fingers and has 13 degrees-of-freedom (DOF) (Hernandez-Arieta et al., 2006a). Each finger has three joints and two DOFs, the distal-inter-phalangial joint (DIP) and the proximal–interphalangial joint (PIP) are actuated by the same tendon, and the metacarpal (MP) joint is actuated by a single tendon. The wrist and the MP joint of the thumb control used two motors for the actuation of pronation/supination and extention/flection movements. The robot hand had pressure sensors placed over the PIP joint of fingers, on the fingertips, and in the palm. Force sensing resistor (FSR) based pressure sensors - due to their flexibility and ease of installation - were used to detect the interaction with the environment. EMG signals were detected by surface electrodes placed at muscle sites on the residual limb. The raw signals were processed and eventually used to control the robot hand.

3.2 EMG signals and classification

For the EMG patterns classification, we used a feed-forward neural network with an automatic learning mechanism (Kato et al., 2006). The EMG pattern classification system is composed of three units (Fig. 3): (a) an analysis unit, (b) a classification unit, and (c) a supervision unit. The analysis unit is in charge of extracting feature vectors from the raw EMG signals.

The unit performs a Fast Fourier Transform of the acquired EMG signals producing a power density spectrum. We extract the feature vector from the power spectrum of all the EMG sensors. For this study, the feature vector extracts 8 samples from 3 EMG channels. The classification unit is in charge of generating clusters for the recognition of several hand movements. It consists of a supervised feed-forward artificial neural network with back propagation for the calculation of the weights. The supervision unit provides the system parameters for the evaluation of the feature vectors in the classification unit. Once a feature vector has been identified, the unit generates a control command to produce the desired robot hand movement. The supervision unit evaluates and updates the classification unit until the system achieves the expected motion, looking for the mapping function that denotes the relationship between the feature vectors and the expected robot hand motion. It receives 16 feature vectors for each motion that was included into the feature vectors' database.

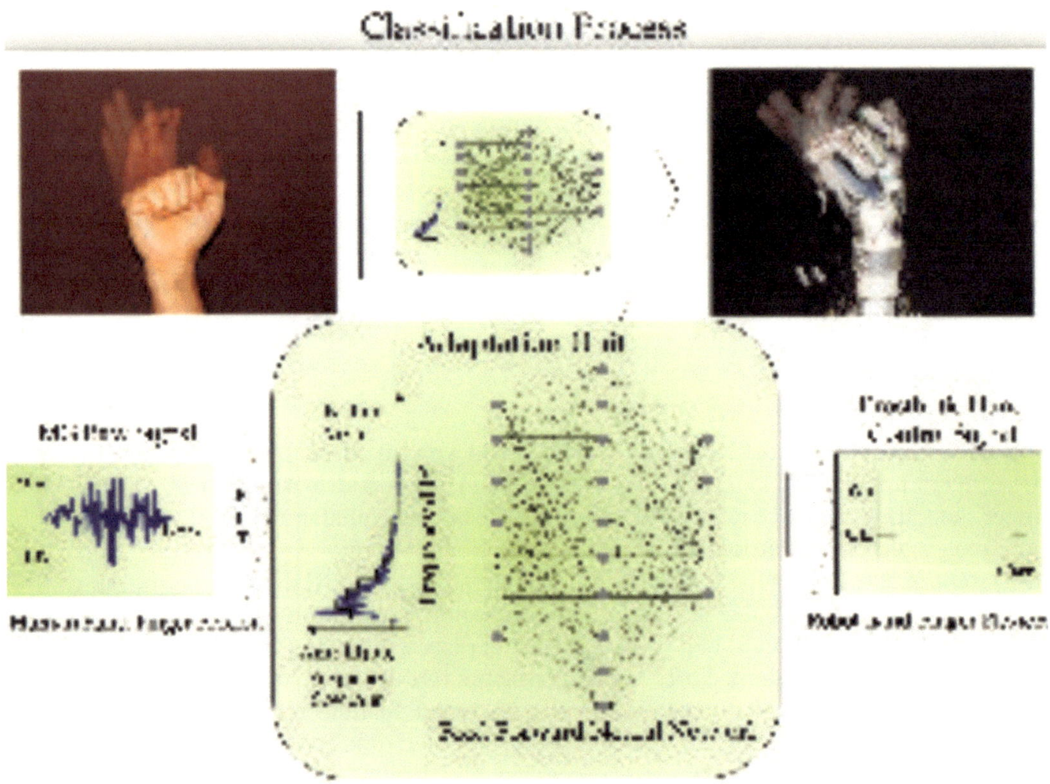

Fig. 2 EMG classification process. Raw EMG data is converted into a set of feature vectors that are fed into the ANN for classification. The ANN is trained to map each feature vector into a hand configuration.

3.3 Visuo-tactile feedback

We used a functional electrical stimulation device (Hernandez-Arieta, 2006b) to provide feedback by directly stimulating the skin's mechanoreceptors. The produced electrical signal follows the guidelines given from previous applications of electrical stimulation in prosthetic applications that define the required frequency, voltage and waveform shape (Pfeiffer, 1968; Melen & Meindl, 1971; Szeto, 1992). A high frequency, bi-phasic signal is efficient in interacting with the sensory system (Fig. 3). To regulate the intensity of the stimulation and to avoid damaging the skin (due to excessive charge accumulation), the duty rate of the positive and negative phases of the signal were changed simultaneously, while the frequency was kept constant (Grill & Mortimer, 1995).

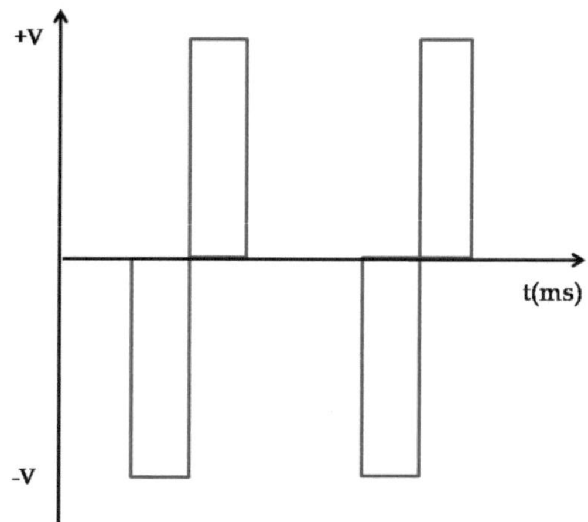

Fig. 3 Electrical stimulation waveform. The negative part of the signal depolarizes the nerves' axons, promoting vibro-tactile sensation. The positive charges avoid tissue damage by eliminating charge accumulation.

Like other sensory systems, mechanoreceptors habituate to constant stimuli. Hence, we set the stimulation intensity to be strong enough to be detected by the skin mechanoreceptors, while being weak enough to not stimulate noxious (pain) receptors and muscle fibers. Figure 4 presents the experimental concept. Whenever the robot hand touches an object, the pressure sensors located in the hand are used to electrically stimulate the user, providing tactile feedback. Because the person can in addition see the robot hand's movements, there is also visual feedback.

3.4 fMRI scan setup

To accommodate visual feedback, the robotic hand has to be within the line of sight of the user. However, the fMRI apparatus involves strong magnetic forces, prohibiting the placement of the robot hand within the user's visible area. To overcome this difficulty, a video camera, in conjunction with a set of mirrors, was used to project the prosthetic hand and its surrounding environment within the fMRI room (Figure 4). The EMG sensors and the stimulation electrodes were coated with aluminum foil to shield them from magnetic fields. The EMG sensors were then placed on the right forearm, while the electrical stimulation electrodes were placed on the upper left arm of the subjects. This was done to reduce the effects of the electric stimulation over the EMG acquisition process. The volume acquisition was done with a 1.5T MAGNETOM Vision plus MR scanner (Siemens, Erlangen, Germany) using the standard head coil. We used foam padding around the subjects' head to minimize head motion and discarded the first five volumes of each fMRI scan because of non-steady magnetization; the analysis was performed using the remaining 54 scans. The fMRI protocol was a block design with one epoch for the task and rest conditions. Each epoch lasted 24 [s] which is equivalent to three whole-brain fMRI volume acquisitions. We used the Statistical Parametric Mapping software 2 (Holmes, 1994) for the analysis of the data. The duration of the first type of experiment (see Section 4) was 5 [s], with a scan time of 3 [s] and a rest time of 2 [s] between scans and we acquired 4 sessions of 35 scans. In the second type of experiment (Section 4) the duration for one scan was 7 [s], with a scan time of 3 [s] and a rest time of 4 [s] between scans; we performed two sessions of 35 scans each. We used an echo-planar imaging (EPI) template to realign the scans and transform them into the standard stereotactic space of Talairach (Talairach, 1988). Data was then smoothed in a spatial domain (full width at half-maximum = 8 x 8 x 8 [mm]) to improve the signal-to-noise ratio. After specifying the appropriate design matrix, the delayed box-car function as a reference waveform, the condition, and the slow hemodynamic fluctuation unrelated to the task, the subject effects were estimated according to a general linear model taking temporal smoothness into account. We then applied a proportional scaling for the global normalization. Once this process was completed, we compared the estimates using the linear contrasts of rest and task period to test the hypotheses about regionally specific condition effects. The resulting set of voxel values for each contrast constituted a statistical parametric map of the t statistic, SPM(t). For the analysis of each session, we assigned a threshold of $P<0.001$ to the voxels and significant clusters, which were not corrected for multiple comparisons. The minimum threshold for comparison was set to $T=2.9$ for passive application of electrical stimulation, and $T=3.26$ for the active manipulation of the robot hand.

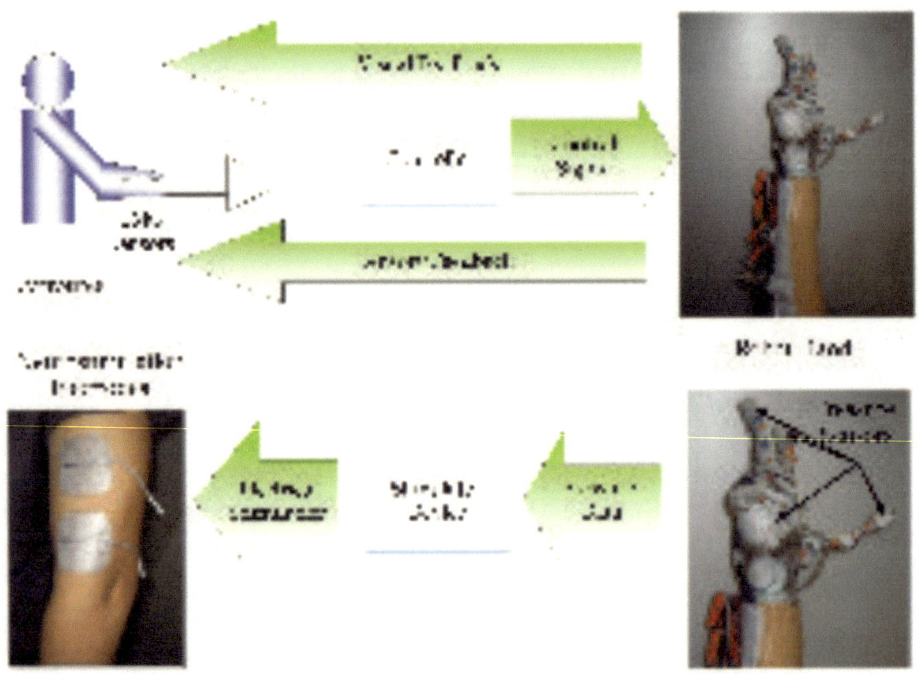

Fig. 4 Experimental setup. Top: the participant receives both visual and tactile feedback from the robotic platform. Bottom: tactile information derived from the pressure sensors is used to provide electrical stimulation to the left upper arm of the user of the device.

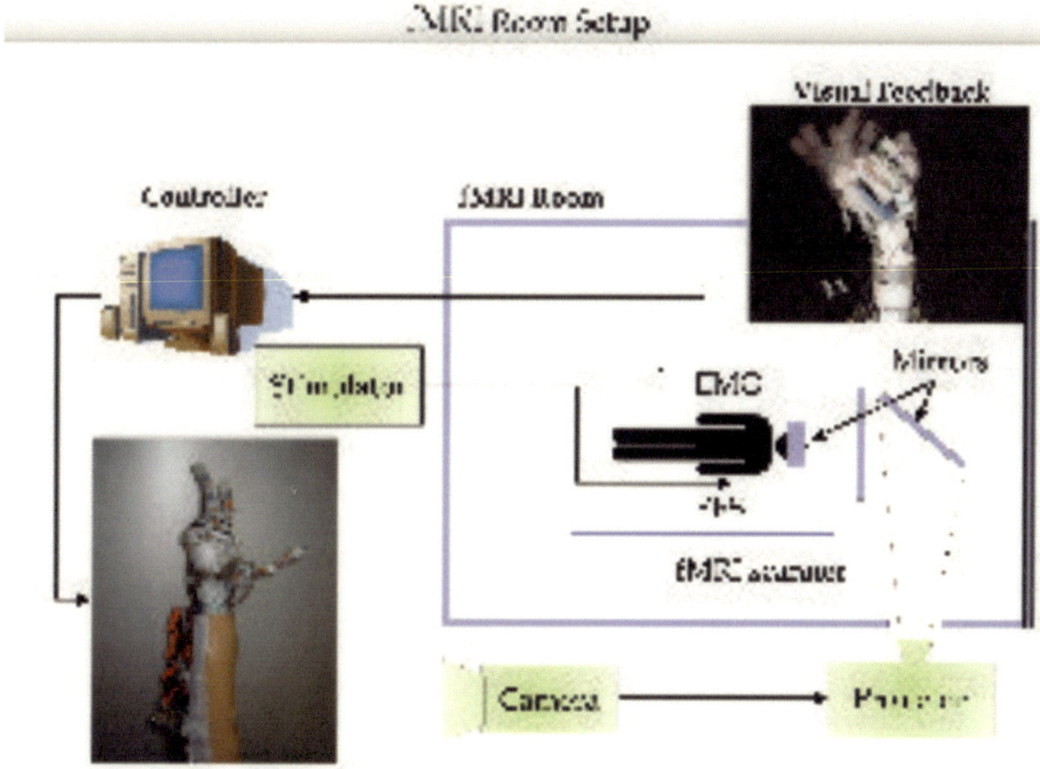

Fig. 5 fMRI room setup. The robot hand is placed outside the fMRI scanner room. A video camera records the robot hand whose movement is then projected through a set of mirrors inside the fMRI scanner.

4. Experiments

In order to evaluate the effects of the interaction between the sensory and motor modalities during the function recovery, we conducted "passive" and "active" experiments. In the passive experiments, we applied an electrical stimulation to the subjects irrespectively of the configuration of the robot hand. Concurrently, we measured the subjects' cortical activation response.

We conducted two types of active experiments, one in which there was only tactile feedback and one in which there was both visual and tactile feedback. The first type aimed at providing insight into the relationship between the subjects' sensory and motor areas without the use of visual feedback. We asked the subjects to open and close the fingers of the robot hand. The neural network processing the EMG signals was trained individually for each participant using a finger flexion/extension motion. A ball was periodically placed inside the robot hand. Every time the subjects grabbed the ball with the hand, the pressure sensors located at the fingertip triggered the electrical feedback process, stimulating the subject's upper left arm. To evaluate the multimodal sensory-motor relationship, visual feedback was added. The subjects were shown an object and were asked to grasp it using

the robot hand. The movements of the hand as well as its surrounding environment were projected to the visual field of the user using the method described in Section 3.4.
Three participants took part in our experiments: one test subject and a control group composed of two healthy subjects. The test subject was a woman in her 50's with a right arm amputation performed 5 years before this study. As her amputation was above the wrist level, the majority of the forearm was intact. The control group consisted of two healthy men in their 20's, with no visible physical abnormalities. To avoid biased results, for all participants three EMG sensors were placed in locations focusing on the major muscle groups on the forearm: extensor digitorums, flexor digitorum superficialis, and flexor digitorum profundus. To measure the effect of the continuous use of the robot hand, the amputee patient was asked to use the robot hand on a daily basis over a period of three months. fMRI scans were taken at the beginning of the study, one month after, and at the end of the study. After each session, the subjects had to fill in a questionnaire regarding the sensations perceived during the tasks.

5. Results

At the beginning of each experimental session, in order to establish a common ground for the active experimentation that followed, all subjects were subjected to passive electrical stimulation. The results are presented in Figure 6, using a base T value of 2.9 for comparison. Passive electrical stimulation without manipulating the control hand showed no activation of the sensory and motor cortices (Brodmann areas 3 & 4 respectively) in either hemisphere. However, the parietal area of both hemispheres, responsible for processing sensory data, did present mild activation. At the end of the three-month experimentation period and using the robot hand on a daily basis, an apparent reversion of the cortical reorganization process in the amputee's brain can be observed (Figure 5). The results obtained from the fMRI portray a major reduction of cortical activity in the general area of the brain. The parietal area and motor cortex present a significant activation reduction. The answers both the amputee and the control subjects provided in the questionnaire clearly indicate the "rubber-hand illusion"; they were feeling as if their actual right hand was touching the presented object. The results from the control subjects however show a more specific activation of the motor cortex (Brodmann Area 4) than that in the amputee's case. This is speculated to relate to the amputees' unused lost limbs, not utilized in daily life activities. In addition, the lack of sensory input from a lost limb results in a cortical reorganization process occurring after an amputation, essentially recycling the no longer active cortical regions for using on other tasks. Using a base T value of 3.6, the control subjects presented activation of Brodmann area 4 and Brodmann area 3 in the left hemisphere of the brain for the manipulation of the robot hand with tactile and visual feedback. There was no visible activation of Brodmann area 3 in the right hemisphere. To identify the influence of the visual sensory modality on the generation of tactile illusions, visual feedback was suppressed. Figure 6 shows the results from applying electrical stimulation in the left upper arm of the patient without visual feedback. The comparison between the usual stimulation with visual feedback shows an increase in the somatosensory area (Brodmann Area 3). The motor cortex exhibits the same level of activation as with the included visual feedback experiments. The activation for the case with visual feedback had a T value of 4.49 for MNI coordinates x=-30, y=-26, y=52. In absence of visual feedback, the T value was 6.48 for MNI coordinates x=-24, y=-26, z=52. In the blind case, the right hemisphere presented activation of Brodmann area 3 at MNI coordinates x=24, y=-30, z=51. The continuous use of the robot hand with tactile feedback during a period of three months led to a reduction of the cortical reorganization that the patient suffered after amputation.

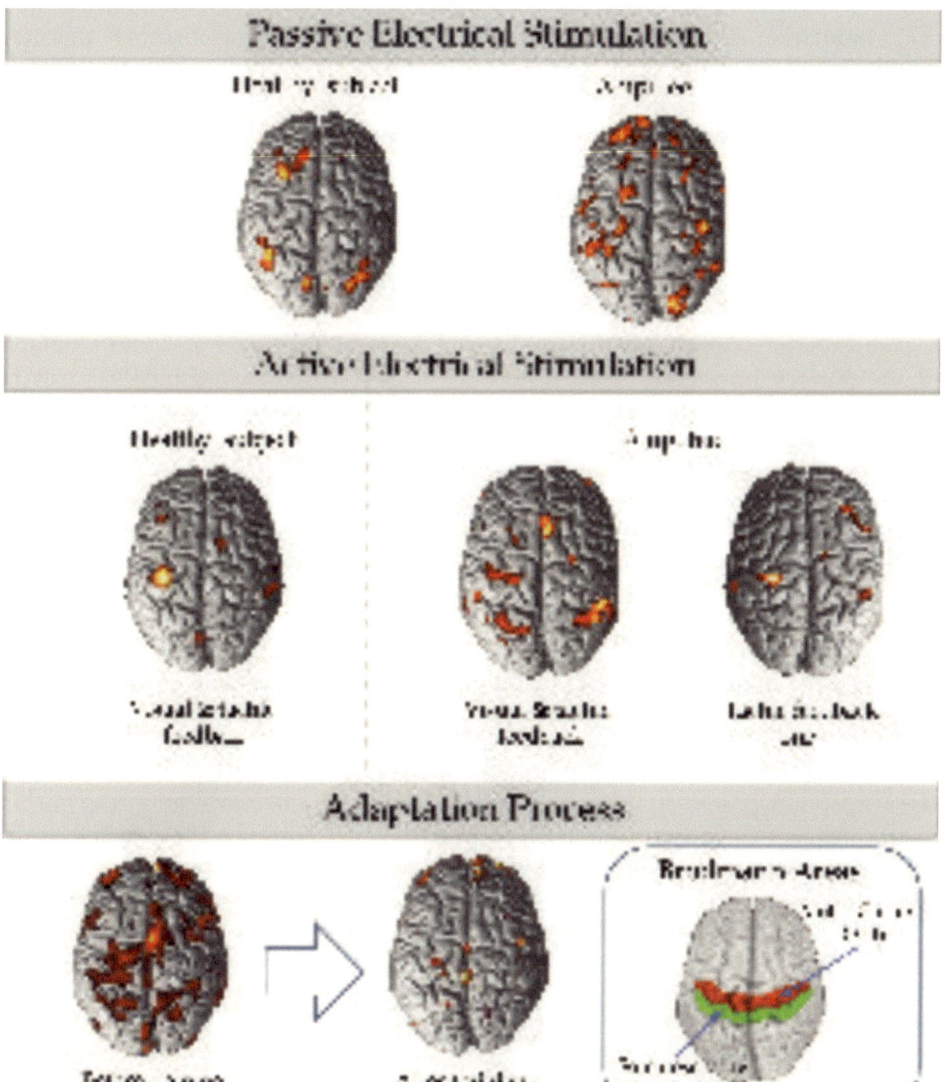

Fig. 6: Experimental results. Top: Passive electrical stimulation shows no activation of Brodmann area 3 for the healthy subject. The amputee presents cortical reorganization with light broad activation. Center: The healthy subject presents no activation of the right hemisphere for active manipulation of the robot hand with multisensory feedback. The amputee presents no activation of the right hemisphere for both multisensory feedback and tactile feedback. Bottom: After three months of use, the amputee's neural activity displays reversion of cortical reorganization. The amputee's brain presents less activation in Brodmann areas 3 & 4.

6. Discussion

It is interesting to observe how the brain reacts when stimulation occurs as an isolated event during the passive experiments. In such a case, the parietal lobe "perceives" the stimulation as a source of pain or as an uncorrelated stimulus presented to the body. However, when electrical stimulation is used for grasping an object, there are three channels that work in synergy; the intention of the subject, the visual feedback, and the stimulus provided by the electrical stimulation. This sequence of events takes place several times during the scanning process, allowing the brain to combine the various pieces of information as a single, simultaneous and repetitive event. The fMRI results provide an indication of how the brain adapts to accommodate the perception induced from electrical stimulation applied on the left arm. During active experimentation, the primary area in the somatosensory cortex (Brodmann Area 3) which is related with the left arm points to an increased activation, sufficient to register changes in the fMRI data. The brain clearly identifies the stimulation as taking part on the left arm. However, comparing these results to the ones obtained during the passive experimentation, we find additional areas of increased activation, not present during the passive tests. The sensory area associated with the right hand is activated, suggesting that the brain is correlating the multi-sensory input as a single event, localizing it in the right hand (in this case, the prosthetic hand). The subjects do not interact directly with the object in question; rather, they do so through the robot hand. Such results clearly indicate the role of brain plasticity in the creation of new communication channels between the user of a robotic device, and the device itself. fMRI measurements proved to be useful and reliable for objectively measuring changes in cortical activation while using the robotic platform, allowing for detailed feedback on the workings of the subjects' brain. The removal of visual feedback as a sensory channel during experimentation leads to two interesting observations: (1) the influence of visual feedback on the illusion of ownership decreased over extended periods of use; and (2) tactile habituation alone is enough for generating the illusion of ownership of the robot hand. The use of myoelectric prostheses acts against the cortical reorganization process that takes place after an amputation. The simultaneous application of electrical stimulation along with an adaptive prosthetic system that adheres to the user's intention promotes the generation of the illusion of ownership; the amputee is able to reconstruct their impaired body image by incorporating the prosthesis. In this respect, a particular emphasis should be given to the nature of the artificial signals provided to the human body. They should be shaped in such fashion that they promote the inclusion of the prosthetic system in the user's body schema. Electrical stimulation has proven to be a feasible way to accomplish this goal, reinforcing the vigor of the involved muscles and, in consequence, shaping the body schema. All these results open the possibility to develop novel man-machine interfaces that allow for the subconscious control of an external (e.g. prosthetic) device.

7. Conclusions

To answer the question as to what extent the sensory and motor function are inter-operating during rehabilitation, we presented fMRI data on apparent brain activation during the use of both motor and sensory modalities. An EMG-controlled prosthetic hand provided the necessary motor function recovery, and an electrical signal was utilized for tactile feedback.
The inclusion of tactile feedback while actively controlling the EMG prosthetic device produced a tactile illusion in the left hemisphere of the patient's brain. This result is surprising given that we applied the stimulation to the opposite arm. The production of the

tactile illusion points to a correlative behavior of the brain in processing active motor commands and their corresponding feedback. To test the conditions for the generation of the tactile illusion, we performed both passive and active administrations of electrical stimulation, where the interaction of the robot hand with an object triggered the active administration of electrical stimulation. We tested the influence of the different sensory modalities by removing visual feedback from the setup.

We confirmed the presence of the tactile illusion in all tests except for the passive administration of electrical stimulation. These experiments serve to demonstrate how the active involvement of disabled individuals improves the sensory-motor function recovery by effectively promoting the correlation process in the brain. Of equal importance, the illusion of ownership is only dependant on tactile feedback with or without the inclusion of visual feedback. Our findings point to a flexible structure that can be modified if we provide adequate feedback within an adequate period.

Future work will aim at contributing more to the human-centered approach to rehabilitation. The results concerning the interaction of motor and sensory modalities for function recovery are already promising; nevertheless, it is imperative to test a larger number of patients. Finally, intercepting the sensory-motor interaction between human and prosthetic devices allows an investigation into ways of improving and shaping it. Linked with the need of an amputee, such an approach bears the importance of relieving the user from a steady conscious mental activity in order to manipulate the prosthetic device.

8. References

Adee, S. (2008). Dean Kamen's "Luke Arm" Prosthesis Readies for Clinical Trials, *IEEE Spectrum*, February.

Atkins, D.J., Heard. D.C.Y., & Donovan, W.H. (2001). Epidemiologic overview: Individuals with upper limb loss, and their reported research priorities. *Int. J. of Prosthetics and Orthotics*, 25: 228–227.

Au, S.K., Bonato, P., & Herr, H. (2005). An EMG-position controlled system for an active ankle-foot prosthesis: an initial experimental study. *Proc. of 9th Int. Conf. on Rehabilitation Robotics*, pp. 375–379.

Bach-y-Rita, P., Tyler, M.E., & Kaczmarek, K.A. (2003). Seeing with the brain. *Int. J. of Human-Computer Interaction*, 15(2): 285–295.

Biddiss, E., Beaton, D., & Chau, T. (2007). Consumer design priorities for upper limb prosthetics. *Disability and Rehabilitation: Assistive Technology*, 2(6): 346–357.

Bitzer, S. & van der Smagt, P. (2006). Learning EMG Control of a robotic hand: Towards Active Prostheses. *Proc. of 2006 IEEE Int. Conf. on Robotics and Automation*, pp. 2819–2823.

de Castro, M. C. & Cliquet, A. (2000). Artificial sensorimotor integration in spinal cord injured subjects through neuromuscular and electrotactile stimulation. *Artificial Organs*, 24 (9): 710–717.

Dhillon, G.S., Horch, K.W., & DeBakey, M. E. (2005). Direct neural sensory feedback and control of a prosthetic arm. Neural Systems and Rehabilitation Engineering, ISSN-1534-4320, 13 (14): 468–472.

Dollar, A.M. & Howe, R.D. (2007). The SDM hand as a prosthetic terminal device: a feasibility study. *Proc. of the 2007 IEEE Int. Conf. on Rehabilitation Robotics*..

Grill, W. M. & Mortimer, T. (1995). Stimulus waveforms for selective neural stimulation. *IEEE Engineering in Medicine and Biology*, 14(4): 375–385.

Hacking, H.G.A, Berg, J.P., Van Der Dahmen, K.T., & Post, M.W.M. (1997). Long term outcome upper limb prosthetic use in the Netherlands. *Eur. J. of Physical Medicine and Rehabilitation*, 7: 179–181.

Heliot, R., Azebedo, C., & Espiau, B. (2007). *Rehabilitation Robotics*, Sashi S. Kommu (ed.), I-tech Education and Publishing, Vienna, Austria, ISBN 978-3-902613-04-2, pp: 163–186.

Henrik-Ehrsson, H., Holmes, N., & Passigham, R. (2005). Touching a rubber hand: feeling of body ownership is associated with activity in multisensory brain areas. *J. of Neuroscience*, 25(45): 10564–10573.

Henrik-Ehrsson, H., Wiech, K., Weiskopf, N., Raymond, J. D., & Passingham, R. E. (2007). Threatening a rubber hand that you feel is yours elicits a cortical anxiety response. *Proc. Nat. Acad. of Sciences USA*, 104(23): 9828–9833.

Hernandez-Arieta, A., Kato, R., Yokoi, H., Yu, W., (2006a). Development of a multi-DOF electromyography prosthetic system using the adaptive joint mechanism. *Applied Bionics and Biomechanics*, 3(2): 101–111.

Hernandez Arieta, A., Yu, W., Yokoi, H., Arai, A. (2006b). fMRI study of the cross-modal interaction in the brain with an adaptable EMG prosthetic hand with biofeedback, *Proc. of the 28th Ann. Int. Conf. of IEEE Eng. In Med. and Biol. Soc.*, pp: 1280–1284.

Holmes, A.P. (1994). Statistical Issues in Functional Brain Mapping. *PhD Thesis*, University of Glasgow.

Hunter, J.P., Katz, J., & Davis, K.D. (2003). The effect of tactile and visual sensory inputs on phantom limb awareness. *Brain*, 126(3): 579–589.

Kaczmarek, K.A., Nammi, K., Agarwal, A.K., Tyler, M.E., Haase, S.J., & Beebe, D.J. (2006). Polarity effect in electrovibration for tactile display. *IEEE Trans. on Biomed. Eng.*, 53(10): 2047–2054.

Kaczmarek, K. A.. (2000). Electrotactile adaptation on the abdomen: preliminary results. *IEEE Trans. Rehabil. Eng.*, 8(4): 499–505.

Kato, R., Yokoi, H., & Arai, T. (2006). Competitive learning method for robust EMG-to-Motion Classifier. *Proc. of Conf. on Intelligent Autonomous Systems 9*, IOS Press, ISBN 1-58603-595-9

Kim, G., Asakura, Y., Okuno, R., & Akazawa, K. (2005). Tactile substitution system for transmitting a few words to a prosthetic hand user, *Conf. Proc. IEEE Eng. Med. Biol. Soc.*, 7,pp: 6908-6911.

Lancaster, J. L., Woldor, M. G., Parsons, L. M., Liotti, M., Freitas, C. S., Rainey, L., Kochunov, P. V., Nickerson, D., Mikiten, S. A., & Fox, P. T. (2000). Automated talairach atlas labels for functional brain mapping. *Human Brain Map*, 10(3):120-131

Kuyper, M.A., Breedijk, M., Muldres, A.H.M, Post, M.W.M, & Prevo, A.J.H. (2001). Prosthetic management of children in the Netherlands with upper limb deficiencies. *Prosthetics and Orthotics International*, 25: 228-237

Lebedev, M. A., & Nicolelis, M.A.L. (2006). Brain-machine interfaces: past, present and future, *Trends Neuroscience*, 29(9): 536-546

Lotze, M., Grodd, W., Birbaumer, N., Erb, M., Huse, E., & Flor. H. (1999). Does use of a myoelectric prosthesis prevent cortical reorganization and phantom limb pain? *Nature Neuroscience*, 2(6): 501-502

Lotze, M., Montoya, P., Erb, M., Hulsmann, E., Flor, H., Klose, U., Birbaumer, N., & Grodd W. (1999). Activation of cortical and cerebellar motor areas during executed and imagined hand movements: an fmri study. *J Cognitive Neuroscience*, 11(5)pp: 491-501

Lotze, M., Flor, H., Grodd, W., Larbig, W., & Birbaumer, N. (2001). Phantom movements and pain. An fmri study in upper limb amputees. *Brain*, 124(11): 2268-2277

Logothetis, N. K. (2002). The neural basis of the blood-oxygen-level-dependent functional magnetic resonance imaging signal. *Phil. Trans. Real Soc. London B | Biological Sciences*, 357(1424): 1003-1037

Melen, R. D. & Meindl, J. D. (1971). Electrocutaneous stimulation in a reading aid for the blind. *IEEE Trans Biomed Eng*, 18(1) :1-3

Pappas, I. P., Popovic, M. R., Keller, T., Dietz, V., & Morari, M. (2001). A reliable gait phase detection system. *IEEE Trans. Neural Sys. Rehabil. Eng.*, 9 (2): 113-125

Pfeifer, R., & Bongard, J. C. (2006).*How the Body Shapes the Way We Think: a New View of Intelligence (Bradford Books)*. The MIT Press, ISBN:0262162393

Pfeifer, R., Lungarella, M. and Iida, F. (2007). Self-organization, embodiment, and biologically inspired robotics. Science, 318: 1088–1093.

Pfeiffer, E. A. (1968). Electrical stimulation of sensory nerves with skin electrodes for research, diagnosis, communication and behavioral conditioning: a survey. *Med. Biol. Eng.*, 6(6):637-651, November

Plettenburg, D.H. (1998). Basic requirements for upper extremity prostheses: the wilmer approach. In H.K. Chang and Y.T. Zhang, editors, *Proceedings of the 20th Annual International Conference of the IEEE Engineering in Medicine and Biology Society*, ISBN 0-7803-5164-9, volume Vol. 5/6, pp: 2276-2281

Pylatiuk, C., & Doderlein, L. (2006). "Bionic" arm prostheses. State of the art in research and development, *Orthopade*, 35: 1169-70, 1172-5

Ramachandran, V.S., & Rogers-Ramachandran, D. (2000). Phantom limbs and neural plasticity. *In Archives Neurology*, 57: 317-320

Rios-Poveda, A. (2002). Myoelectric prosthesis with sensorial feedback. *In University of New Brunswick MyoElectric Controls/Powered Prosthetics Symposium*

Riso, R. R., Ignagni, A. R., & Keith, M. W. (1991). Cognitive feedback for use with fes upper extremity neuroprostheses. *IEEE Transactions Biomedical Engineering*, 38(1):29

Riso, R. R.(1999). Strategies for providing upper extremity amputees with tactile and hand position feedback--moving closer to the bionic arm. *Technol Health Care*, 7:401-409

Riener, R., Ferrarin, M., Pavan, E. E., & Frigo, C. A. (2000). Patient-driven control of fes-supported standing up and sitting down: experimental results. *IEEE Transactions in Rehabilitation Engineering*, 8(4):523-529

Riener, R., Frey, M., Bernhardt, M., Nef, T., & Colombo, G. (2005). Human-centered rehabilitation robotics. *Proc. of the 2005 IEEE 9th Int. Conf. on Rehabilitation Robotics*

Riener, R. (2007). Robot-aided rehabilitation of neural function in the upper extremities, *Acta Neurochir Suppl*, 97 (1): 465-471

Rorden, C., & Brett, M. (2000). Stereotaxic display of brain lesions. *Behav Neurol*, 12(4):191-200

Sabolich, J.A., & Ortega, G.M. (1994). Sense of feel for lower-limb amputees: A phase one study. *Journal of Prosthetics and Orthotics*, 6: 36-41,

Szeto, A. Y. & Farrenkopf, G. R. (1992). Optimization of single electrode tactile codes. *Annals of Biomedical Engineering*, 20 (6): 647-665

Simpson, D.C. (1974). The choice of control system for the multimovement prosthesis: Extended physiological proprioception (e.p.p.). In P. Herberts, R. Kadefors, and R. Magnussonand I. Petersen, editors, *The Control of Upper-Extremity Prostheses and Orthoses*. Springfield, Charles Thomas

Scott, R. N. (1990). Feedback in myoelectric prostheses. *Clin Orthop Relat Res*, (256):58-63

Shimojo, M., Suzuki, T., Namiki, A., Saito, T., Kunimoto, M., Makino, R., Ogawa, Hi., Ishikawa, M., & Mabuchi, K. (2003). Development of a system for experiencing tactile sensation from a robot hand by electrically stimulating sensory nerve fiber. *In 2003 IEEE Int. Con. on Robotics and Automation*, pp: 1264-1270

Strother, S. C. (2006). Evaluating fmri preprocessing pipelines. *IEEE Eng. in Med. and Bio. Magazine*, 25(2): 27-41

Talairach, J. & Tournoux, P. (1988). Co-Planar Stereotaxic Atlas of the Human Brain: 3-dimensional Proportional System: an Approach to Cerebral Imaging. *Thieme Medical Publishers*, Inc., ISBN 0865772932

Tsakiris, M., Bosbach S. & Gallagher, S. (2007). On agency and body-ownership: Phenomenological and neuroscientific reflections. *Consciousness and Cognition*. 16 (3): 645-60.

Veltman, D., & Hutton, C. (2000). Spm99 manual. Technical report, *Wellcome Department of Imaging Neuroscience*

Weir, R.F., Childress, D.S., & Licameli, J.N. (1998). Motor Prostheses and the Lack of Physiologically Appropriate Feedback: Their function and connections with Manipulators, *Robots and other Human-Machine Systems*, chapter Section 3.21. MIT Press

Yokoi, H., Hernandez Arieta, A., Katoh, R., Yu, W., Watanabe, I., & Maruishi, M. (2004). Mutual Adaptation in a Prosthetics Application, *Embodied Artificial Intelligence edited by Fumiya Iida, Rolf Pfeifer, Luc Steels, and Yasuo Kuniyoshi*, LNCS/LNAI series of Springer

3

Mobility Assistive Robots for Disabled Patients

Chung-Hsien Kuo
Department of Electrical Engineering
National Taiwan University of Science and Technology
Taiwan

1. Introduction

2.1 Background

Wheelchairs are one of the most important mobility aids for disabled patients and elders. In general, wheelchairs are categorized as manual and powered types. From the viewpoints of movement behaviours, the powered wheelchair can be regarded as a mobile robot (Asai, 2001) if additional environmental sensing components are built and an intelligent motion controller is included. Therefore, robotic wheelchairs (Luo et al., 1999) are proposed integrating novel sensing and intelligent control technologies to improve the convenience and comfort of mobility assistance.

In general, robotic wheelchairs integrate the sensing, intelligent computing, and communication technologies to improve the autonomy when compared to the conventional powered wheelchairs. The assistive technologies resolve the problems of conventional powered wheelchairs of:

1. Stable velocity control of wheels in terms of feedback motion control system.
2. Comfortable wheelchair driving in terms of velocity adjustment of wheels for various degrees of turning.
3. Collision avoidance of wheelchairs for approaching still and moving objects or person.

2.2 State of the art

In recent years, several robotic wheelchair projects and literatures were proposed. Their system focused on the autonomy, intelligence, safety, and navigation of robotic wheelchairs. The VAHM (French acronym for Autonomous Vehicle for people with Motor Disabilities) project proposed an autonomous wheelchair to assist the disabled people who are unpractical to drive a conventional powered wheelchair (Bourhis et al., 2001). The software architecture of VAHM robotic wheelchair is categorized as the physical, local and global levels. In the local level, the freespace detection, wall detection, and localization are implemented as the perception controller; the freespace search, direction following, wall following, path planning, motion control, obstacle avoidance are implemented as the navigation controller; and finally, the man-machine interface and command interpretation are implemented as the communication controller. The VAHM robotic wheelchair photo is shown in Fig. 1.

The SENARIO (Sensor Aided Intelligent Wheelchair Navigation System) project developed an intelligent wheelchair navigation system (Bourhis et al., 1997), and it provided fully autonomous and semi-autonomous control modes. In the fully autonomous mode, the control system is capable of accepting the user's command. When the control system receives a "go to goal" command, it locates its current position and the target position from the map. Consequently, a path is generated, and the wheelchair executes by following the desired path. In addition, the wheelchair can avoid obstacles in terms of the ultrasonic and infrared sensors. In the semi-autonomous mode, the user's command can override the action generating from the autonomous control mode, and the users can drive the wheelchair using a specified path in a special environment. The SENARIO robotic wheelchair photo is shown in Fig. 1.

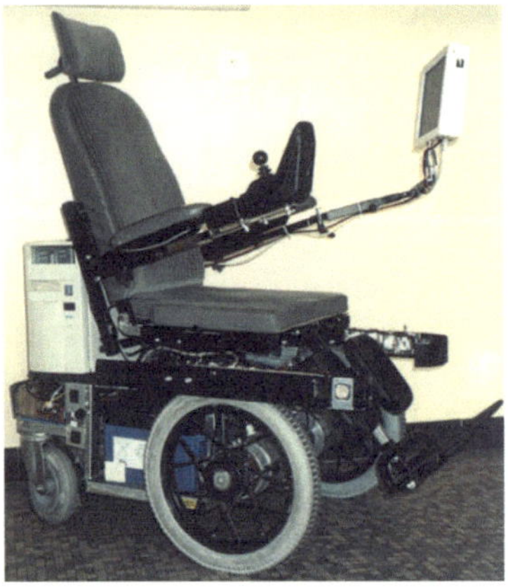

Fig. 1. Photo of robotic wheelchair for the VAHM project (Bourhis et al., 2001)

Fig. 2. Photo of robotic wheelchair for the SENARIO project (Bourhis et al., 1997)

The Bremen Autonomous Wheelchair project (Lankenau et al., 2001) developed a "Rolland" wheelchair, and it is implemented by following the skills of the driving and route assistants. The "Rolland" is capable of smooth speed control, turning around in an open space, turning around in a corner, obstacle avoidance, and doorway passage. It provides the mobility assistance for the handicapped people. The robotic wheelchair photo of 'Rolland' is shown in Fig. 3.

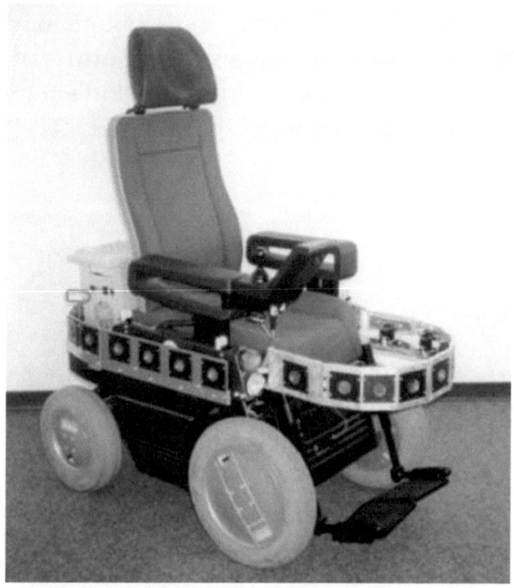

Fig. 3. Photo of Bremen autonomous wheelchair (Lankenau et al., 2001)

At the same time, the SIAMO (Spanish acronym for Integral System for Assisted Mobility) project (Mazo et al., 2001) developed an electronic system to guide the autonomous wheelchairs for the disabled and elder people. The proposed autonomous wheelchair system integrates an innovative user-machine interface, a complete sensory subsystem (ultrasonic, infrared, vision, etc), and an advanced strategy of control and navigation system to guide the wheelchair with safety and comfort. Especially, the face and mouth positions can be traced properly. The SIAMO robotic wheelchair photo is shown in Fig. 4.

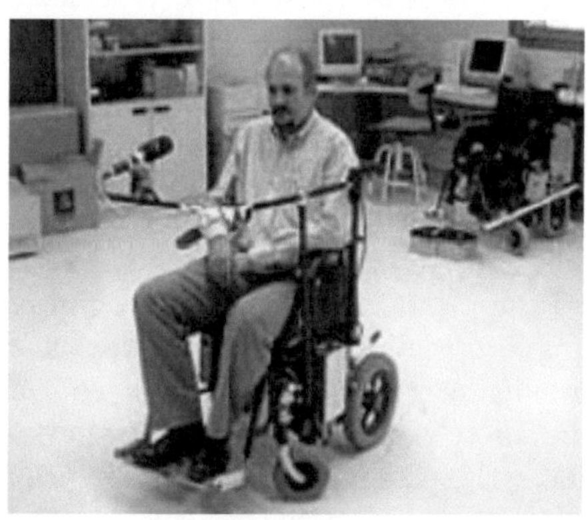

Fig. 4. Photo of robotic wheelchair for the SIAMO project (Mazo et al., 2001)

The autonomous wheelchair can be also developed as the omni-wheeled platform. Such a driving platform is not suitable for the narrow space with non-straight paths of indoor environments, such as patient rooms in hospitals. In that paper, an omni-wheeled platform (Kuo et al., 2006) is proposed as the robotic wheelchair platform. Due to the property of free movements in orientations, the robotic wheelchair can easily drive in a narrow space with non-straight paths. In addition to the omni-wheeled platform, the robotic wheelchair also integrates obstacle detection sensors to perform autonomous navigations using the fuzzy control system. However, such an omni-wheeled platform can be used in the indoor environment. A heavy-duty omni-wheel with larger size and a four-wheel configuration can be further used to improve the mechanical reliabilities. The omni-wheeled robotic wheelchair photo is shown in Fig. 5.

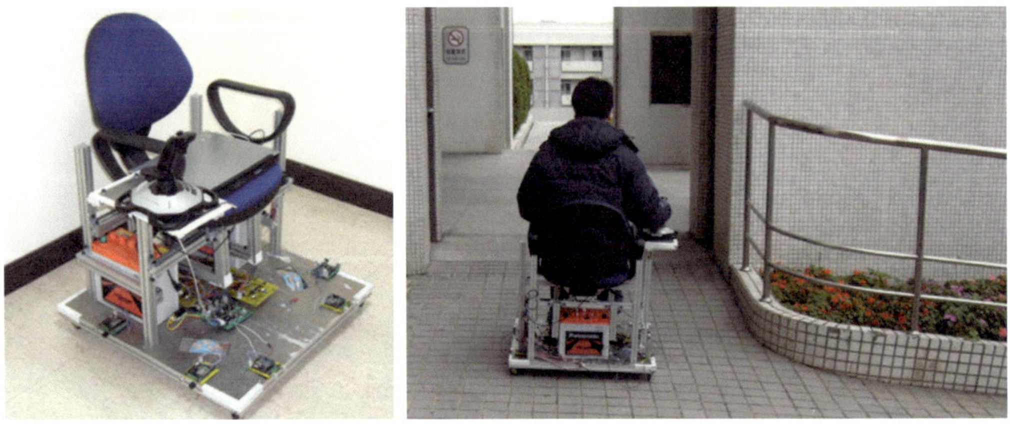

Fig. 5. Photo of omni-wheeled robotic wheelchair (Kuo et al., 2006)

2.3 Practical considerations of mobility assistive robots

Development of autonomous navigation system for robotic wheelchairs seems to be an important issue. However, most of navigation systems are developed based on personal computers such as desktops or laptops. Such a computational architecture increases the weight, cost and power consumption of the robotic wheelchair. Therefore, a chip based solution is discussed in this chapter to improve the feasibility of the robotic wheelchairs. Two important issues are proposed:

1. Human-centered design for mobility assistive robots: the conventional robotic wheelchair constructs the autonomous navigation technologies, and the user may drive the robotic wheelchair without using his (her) hand. Such an autonomous operation is not feasible when complicated environment presents. Therefore, the user joystick input should be reserved, and the joystick input can be regarded as a 'virtual' goal for the autonomous navigations. In this manner, the user is capable of controlling the wheelchair, and the intelligent robotic wheelchair controller is used to justify the joystick command for the safety and comfort considerations.
2. Low cost implementations: The cost for developing an autonomous wheelchair is quite expensive when compared to conventional powered wheelchair. The additional costs are ultrasonic sensors and personal computers. To reduce the cost, a distributed chip based robotic wheelchair supervisory controller is discussed in this work to improve the reliability, and to reduce the cost as well.

Finally, this chapter is organized as follows: section 3 introduces the fuzzy based navigations; section 4 describes the human-centered navigations; section 5 illustrates the distributed chip based implementations; section 6 illustrates the experiments and verifications; and finally the conclusions and future works are discussed in section 7.

3. Fuzzy based navigations

3.1 Fuzzy control system

The fuzzy control system developed for the robotic wheelchair is to emulate a skilled powered wheelchair user. The skilled powered wheelchair user drives the wheelchair depending on the following situations:

1. Goal direction and distance: the user drives the wheelchair in a higher velocity when the distance to the goal is long. The user reduces the velocity when the wheelchair rotates in a small radius for the safety and comfort concerns.
2. The user changes the wheelchair direction smoothly when obstacles appear in front of the wheelchair.
3. The user keeps a distance to the wall when a wall is beside the wheelchair.

According to the previous situations, the driving behaviors of a skilled robotic wheelchair user can be modeled as an intelligent controller. The intelligent controller percepts distance information of obstacles, and makes decision for the wheelchair steering and velocity. Therefore, the intelligent controller is responsible of three navigation functions, including wall-following, goal-seeking, and obstacle avoidance.

The fuzzy control system (Negnevitsky, 2005) uses the fuzzy sets and the membership functions to describe the facts with uncertainty. The inference engine is executed according to the input status and knowledgebase to determine the control policy. In this chapter, the robotic wheelchair uses the fuzzy logics as the kernel of the intelligent controller.

The case study in this chapter uses 7 ultrasonic sensors, and they are used as the inputs of the fuzzy controller, as shown in Fig. 6. The operation of the navigation fusion is indicated as in Fig. 7. In this figure, three navigation functions are dynamically switching according to the distributions of close obstacles. The switching reference table is shown in Table 1. In this table, the u1 to u7 indicate the signals collected form the ultrasonic sensors. Note that 'X' indicates no obstacle detected; 'O' indicates an obstacle is detected; and '-' indicates 'don't care'. In addition, the steering (i.e., velocity allocations of two wheels) control policy can be inferred from the results of the wall-following, goal-seeking, and obstacle avoidance fuzzy functions.

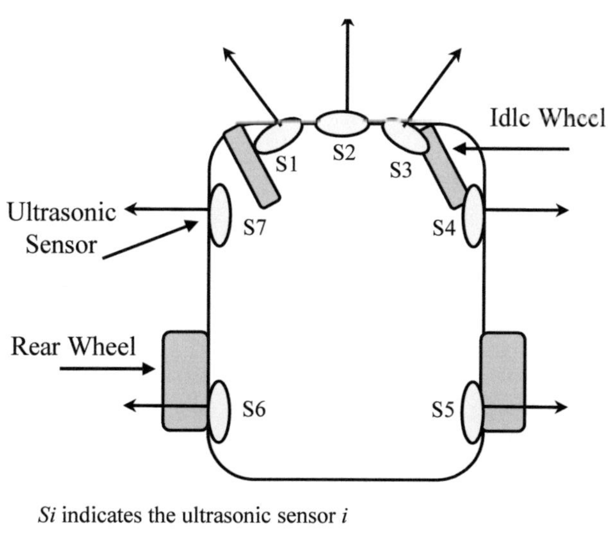

Si indicates the ultrasonic sensor *i*

Fig. 6. Sensor allocations of case study

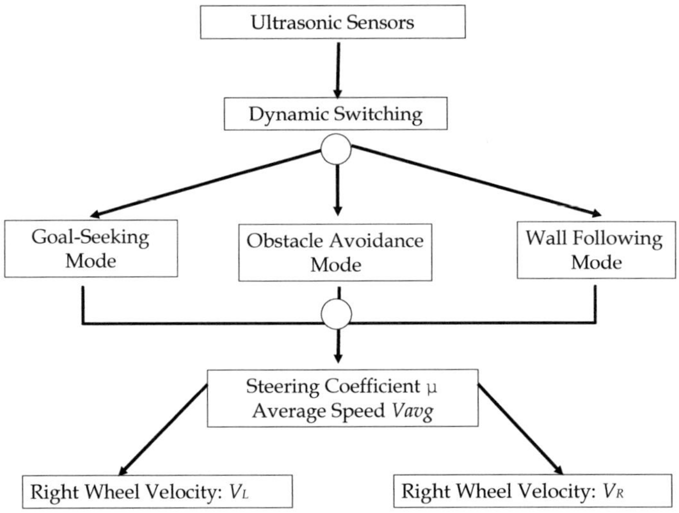

Fig. 7. Fusions of navigation functions

u1	u2	u3	u4	u5	u6	u7	Execution Function
×	×	×	×	×	×	×	Goal-Seeking
×	×	×	×	○	○	×	Goal-Seeking
×	×	×	×	○	×	×	Goal-Seeking
×	×	×	×	×	○	×	Goal-Seeking
×	×	×	○	○	×	×	Wall-Following
×	×	×	×	×	○	○	Wall-Following
×	×	×	○	○	—	○	Wall-Following
×	×	×	○	○	○	—	Wall-Following
×	×	×	—	○	○	○	Wall-Following
×	×	×	○	—	○	○	Wall-Following
○	—	—	—	—	—	—	Obstacle Avoidance
—	○	—	—	—	—	—	Obstacle Avoidance
—	—	○	—	—	—	—	Obstacle Avoidance
○	○	—	—	—	—	—	Obstacle Avoidance
○	—	○	—	—	—	—	Obstacle Avoidance
—	○	○	—	—	—	—	Obstacle Avoidance
○	○	○	—	—	—	—	Obstacle Avoidance

Table 1. Navigation fusion table

3.2 Goal-seeking mode

The goal-seeking mode is defined as none of the ultrasonic sensor detected obstacles, and the sensors in the same side of the robot ('S4 and S5' or 'S6 and S7') do not detect any obstacles simultaneously (as indicated in Table 1). For this navigation mode, the fuzzy rule table is shown in Table 2. Where θ is the angle between the direction of the wheelchair (α) and the direction of from two rear wheels' centre to the target (β), as shown in Fig. 8; μ is the steering coefficient. The linguistic parameters of fuzzy control are also presented: NB indicates negative big; NM indicates negative medium; NS indicate negative small; ZE indicates zero; PS indicates positive small; PM indicates positive medium; and PB indicates positive big. The input membership function of θ and the output membership function of μ are shown in Fig. 9(a) and Fig. 9(b), respectively.

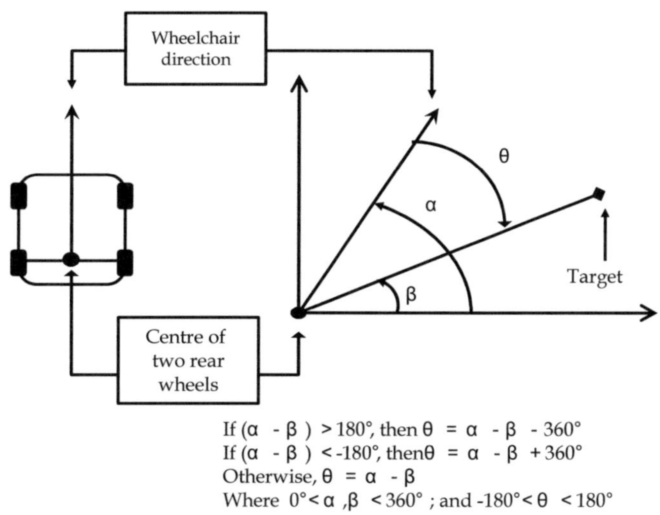

Fig. 8. Angle parameter definition for fuzzy controller

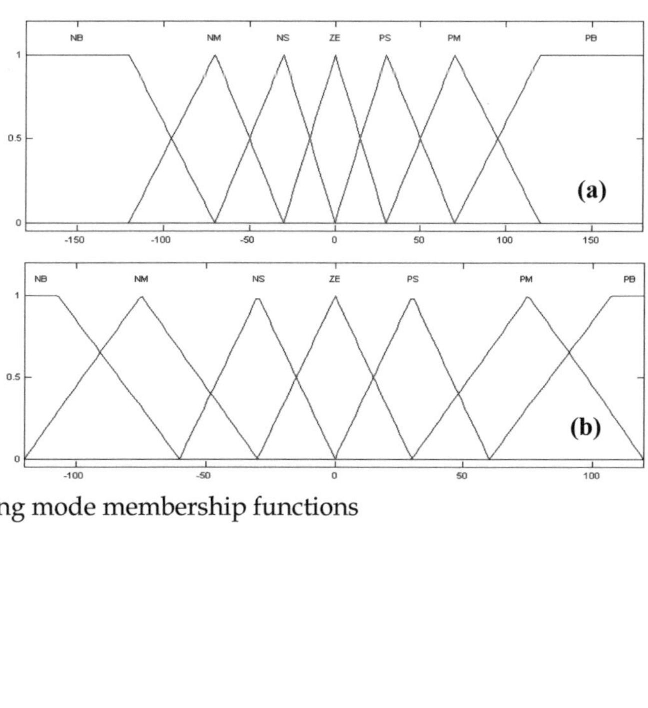

Fig. 9. Goal-seeking mode membership functions

θ	NB	NM	NS	ZE	PS	PM	PB
μ	NB	NM	NS	ZE	PS	PM	PB

Table 2. Goal-seeking mode fuzzy rule table

3.3 Wall-following mode

The wall-following mode is defined as S1, S2, and S3 which do not detect any obstacles and two ultrasonic sensor pair in the same side (S4, S5 pair or S6, S7 pair) which detects obstacles. In this mode, the fuzzy rule table is shown in Table 3. Their membership functions are shown in Fig. 10.

Condition 1: following the right wall

μ		*Ds4*			
		S	M	B	L
Ds5	S	NS	PS	PM	PM
	M	NS	ZE	PS	PM
	B	NM	NS	PS	PM
	L	NM	NM	ZE	PM

Condition 2: following the left wall

μ		*Ds7*			
		S	M	B	L
Ds6	S	PS	NS	NM	NM
	M	PS	ZE	NS	NM
	B	PM	PS	NS	NM
	L	PM	PM	ZE	NM

Table 3. Wall-following mode fuzzy rule table

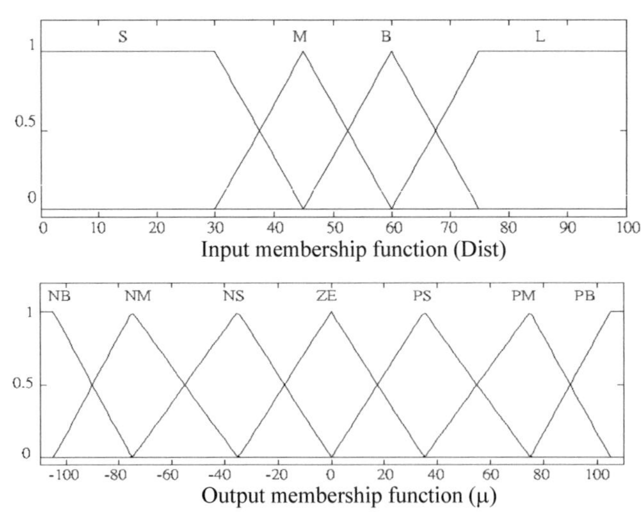

Fig. 10. Wall-following mode membership functions

3.4 Obstacle avoidance mode

The obstacle avoidance mode is defined as at least one of the front ultrasonic sensors (S1, S2, or S3) which detect obstacles. In this mode, the fuzzy rule table is defined in Table 3. Their membership functions are shown in Fig. 11. Where Dsi is the distance measured from Si.

Condition 1: if Ds1 is Small (S)

μ		*Ds3*			
		S	M	B	L
Ds2	S	PB	PB	PB	PB
	M	PB	PB	PB	PB
	B	PB	PB	PB	PB
	L	PB	PB	PB	PB

Condition 2: if Ds1 is Medium (M)

μ		*Ds3*			
		S	M	B	L
Ds2	S	PB	PB	PB	PB
	M	NB	PB	PB	PB
	B	NB	PB	PM	PM
	L	NB	PB	PM	PM

Condition 3: if Ds1 is Big (B)

μ		*Ds3*			
		S	M	B	L
Ds2	S	NB	NB	PB	PB
	M	NB	NB	PB	PB
	B	NB	NM	PM	PM
	L	NB	NM	PM	PS

Condition 4: if Ds1 is Large (L)

μ		*Ds3*			
		S	M	B	L
Ds2	S	NB	NB	NB	PB
	M	NB	NB	NB	PB
	B	NB	NM	NM	PM
	L	NB	NM	NS	ZE

Table 4. Obstacle avoidance mode fuzzy rule table

3.5 Velocity justification model

The navigation modes are executed in a mutual exclusion manner. The output of navigation mode is the steering coefficient, and it determines the degree of turning of the robotic wheelchair. Subsequently, the steering coefficient and the minimal distance of from wheelchair to desired target or nearest obstacle (denoted as dist) are further used to determine the average driving velocity (denoted as Vavg) of the robotic wheelchair. Such a design aims to increase the safety of steering, and it is quite important to avoid the turnover of wheelchairs from their sides. In this manner, increasing steering angle will result in the decreasing in speed of the wheelchair. Such a design meets the behaviors of skill wheelchair user mentioned at the beginning of this section. Consequently, the fuzzy rule table is shown in Table 5. Their membership functions are shown in Fig. 12.

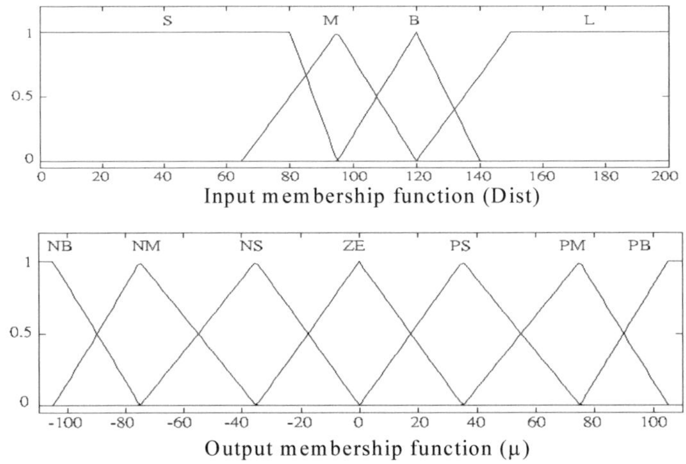

Fig. 11. Obstacle avoidance mode membership functions

$V_{avg.}$		dist			
		S	M	B	L
μ	NB	ZE	ZE	PS	PM
	NM	ZE	ZE	PS	PM
	NS	ZE	PS	PM	PB
	ZE	ZE	PS	PM	PB
	PS	ZE	PS	PM	PB
	PM	ZE	ZE	PS	PM
	PB	ZE	ZE	PS	PM

Table 5. Velocity justification fuzzy rule table

4. Human-centered navigations

Most of conventional robotic wheelchairs construct their intelligent navigation system using autonomous manners. However, such an autonomous operation is not feasible when in complicated environments. Therefore, the user joystick input should also be the best way to control the robotic wheelchair. Nevertheless, the joystick may be sensitive to user wrist input. Improper inputs may result in the dangers of users. The proposed fuzzy based navigation functions should be included to avoid dangerous situations and to improve safety and comfort before the wheel control command is made.

From previous discussions, the fuzzy navigation model can be properly executed only when the target position is determined. For most of users, they have no idea about how to set the target positions. It is also inconvenient to force users to set the target positions for each segmental path. More seriously, the global localization and map of environment is frequently not available for the robotic wheelchairs for most situations. Consequently, the autonomous navigation technology is not feasible for mobility assistances.

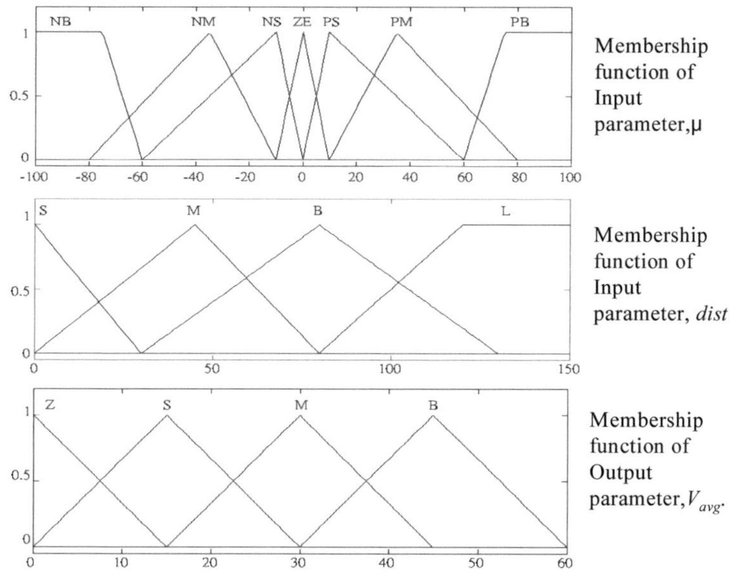

Fig. 12. Velocity justification model membership functions

In this work, the intelligent navigation technology will be constructed without the global positioning and electric environment map. More specially, no target position is required for the robotic wheelchair. The wheelchair users can use their eyes to understand where they are and to use the joystick to determine where to go. The navigation functions are used to assist the user to control the robotic wheelchair in more safe and comfortable.

Detailed description of the human-centered navigation is shown in Fig. 13. The joystick command will be normalized as (pos_x, pos_y). To determine the target of the navigation functions, the normalized joystick command will be further transformed as the virtual target position. In order to improve the comfortability and to reduce the sensitivity of the joystick, the joystick command will not be converted linearly.

Instead, a hybrid function is proposed by combining exponential and parabolic functions, as shown in Fig. 14 . The hybrid function Y is defined as:

$$Y = e^{mw} - 1 \qquad (w <= 0.5\,Jmax) \tag{1}$$

$$Y = aw^2 + bw \qquad (w > 0.5\,Jmax) \tag{2}$$

Where Y is the output of the hybrid function; w is smoothed signal of the joystick; a, b and m are constants. Note that Y can be used to determine target_x and target_y independently. Jmax is the maximum value of normalized pos_x or pos_y (i.e., unity). Consequently,

$$a = (2Y_{\max} - 4Y_{set}) / J_{\max}^{2} \tag{3}$$

$$b = (4Y_{set} - Y_{\max}) / J_{\max} \tag{4}$$

$$m = 2\ell n(Y_{set} + 1) / J_{\max} \quad (5)$$

Where Ymax is the maximum meaningful distance to the target for x-and y-coordinates of the robotic wheelchair; and Yset is an adjustable parameter, and it is determined in terms of the health condition of the user. For example, the Ymax is set as 400 cm, and this setting indicates that target position in a distance (either x or y direction) larger than 400 cm should conduct similar results. Such a value can be regarded as the limit position of joystick in each direction.

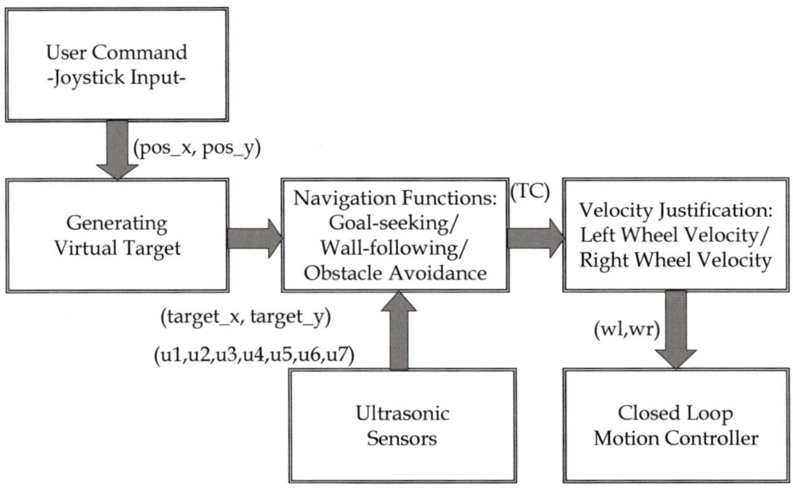

Fig. 13. Detailed description of the human-centered navigation

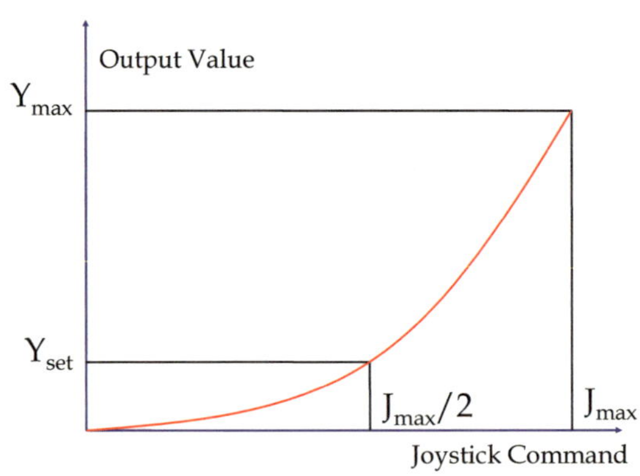

Fig. 14. Hybrid function conversion for joystick input

The target position cooperating with the ultrasonic sensor data will be used for the fuzzy navigation function to generate the steering coefficient (TC). Finally, the TC and the minimum distance of obstacle or the target are used to infer the angular velocities of two wheels.

5. Distributed chip based implementations

As mention before, most of conventional robotic wheelchair control systems are constructed using the personal computers, and they are less reliable. At the same time, the cost and weight are also drawbacks. In this work, a distributed chip based robotic wheelchair supervisory controller is discussed. For this purpose, a low cost computing processor of the programmable system-on-chip (PSoC) (http://www.cypress.com, 2008) is selected. In general, the performance of low cost PSoC is limited.

To extend computational capability of the PSoC, a distributed computing architecture is proposed. The individual navigation function are implemented in a task based PSoC. Consequently, the operations of the whole system achieve acceptable performance in real-time. The proposed distributed computing architecture is shown in Fig. 15. The intra-communications are established in terms of the I2C and the serial interfaces. There are 10 PSoC modules used in this work, and they are marked using the gray rectangles. These modules are further elaborated as follows:

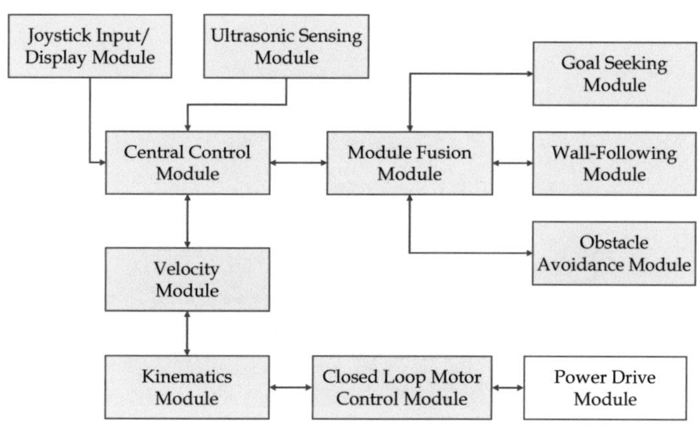

Fig. 15. Distributed PSoC based computing architecture

1. Joystick input/ display module: this module is responsible of collecting user's joystick command. The hybrid function conversion for joystick input is implemented in this module. In addition, the display module shows the wheelchair status such as battery status, obstacle information, driving distance, etc. Especially, this module also provides the serial communication interface to connect with a personal computer or PDA.
2. Ultrasonic sensing module: this module collects seven ultrasonic sensor data.
3. Goal-seeking module: this module implements the fuzzy logic based goal-seeking navigation function
4. Wall-following module: this module realizes the fuzzy logic based wall-following navigation function
5. Obstacle avoidance module: this module implements the fuzzy logic based obstacle avoidance navigation function.
6. Module fusion module: this module realizes the navigation fusions as shown in Fig. 7 and Table 1.

7. Central control module: this module is responsible of coordinating the user joystick command, ultrasonic sensor data and autonomous navigation functions.
8. Velocity justification module: this module implements the fuzzy logic based velocity justification function.
9. Kinematics module: this module converts the steering coefficient and the average velocity as angular velocities of two motors.
10. Closed loop motor control module: this module implements the closed loop angular velocity controller in terms of proportional-integral-differential (PID) control algorithm.

Finally, the power drive module is responsible of driving the wheelchair motors. In addition, based on the proposed architecture, the user may drive the intelligent robotic wheelchair like as a conventional powered wheelchair; however, the goal-seeking, wall-following, obstacle avoidance, and velocity control functions are continuously executing at the backend to perform more comfortable and safe driving.

Based on the proposed distributed computing architecture, a prototype is fabricated in laboratory. Fig. 16 shows the control board. The size of this controller is 10 cm * 15 cm. Finally, a robotic wheelchair is constructed as shown in Fig. 17.

Fig. 16. Photo of the proposed distributed PSoC based controller.

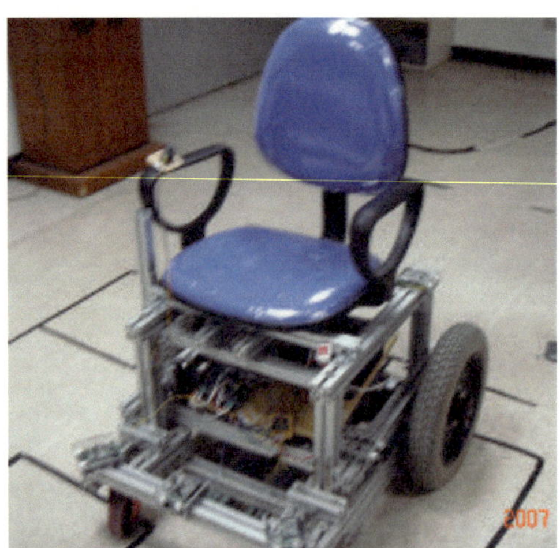

Fig. 17. Photo of an assembled robotic wheelchair.

6. Experiments and verifications

The experiments focuses on the validations of the proposed PSoC based distributed computing architecture. Initially, the PC based simulation environment is constructed to generate correct navigation results, as shown in Fig. 18. These results are used to valid individual navigations functions. The PC based navigation program is developed based on the navigation fusion architecture shown in Fig. 7 and Table 1.

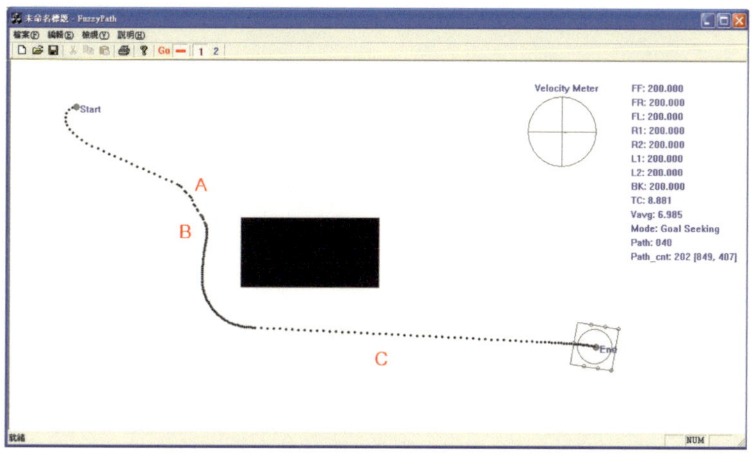

Fig. 18. PC based simulation program for autonomous navigations.

This program is coded using the Microsoft Visual C++ (Kruglinski, 1996). A black rectangle emulates the obstacles. The start and end points are indicated on the screen. The robotic wheelchair executes the goal-seeking mode initially.

When the robot closes to the obstacle, the obstacle avoidance mode takes over (point A). Subsequently, the wall-following mode executes (point B) to follow the left and bottom sides of the obstacle. Finally, the goal-seeking mode executes again (point C) to real the end point. Note that the path history of the robotic wheelchair is represented using the "dot". These dots are recorded in constant time. Therefore, the effects of determining average velocity can be obviously observed from the dots' density on path. Such an effect validates the fuzzy model of the average velocity function. In addition, the robotic wheelchair avoids collisions all the way.

The next step is to validate the correctness of individual PSoC based navigation functions. In this experiment, the PC based simulation program and the PSoC based computing architecture are executing in parallel, and they use identical simulation conditions with the same start and end points. The inference results of the PSoC based computing modules are transmitted to the PC system. Therefore, the screen displays the results of two wheelchairs and paths: blue path is plotted for PC based navigation results; and red path is plotted for PSoC based navigation results.

Results of PSoC based inference are compared with PC based solution. Fig. 19 shows the comparison results of the goal-seeking navigation function. In this figure, the block located at the left-upper corner indicates the operated joystick coordinates. The list box located at the left-lower corner shows the replied steering coefficient of the PSoC controller. The circle located at the right-upper corner represents the velocity meter that contains the direction and value to the wheelchair moving velocity.

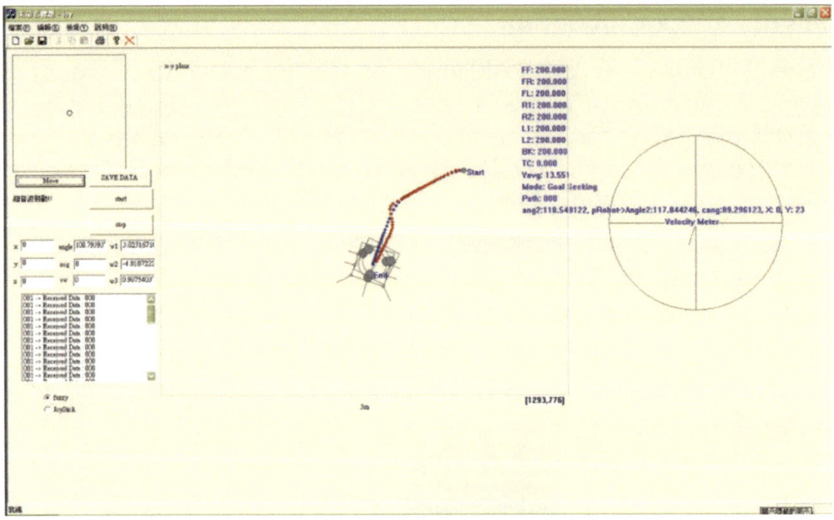

Fig. 19. Comparisons of goal-seeking navigation function.

In addition to the goal-seeking function, the wall-following and obstacle avoidance functions are also validated as shown in Fig. 20 and Fig. 21 respectively. Apparently, the blue and red path can all reach the desired goal; however, the PC based solution is more accurate than the PSoC based solution since the limited digital resolutions of the 8-bits processors. Nevertheless, the PSoC solution is also applicable in real applications.

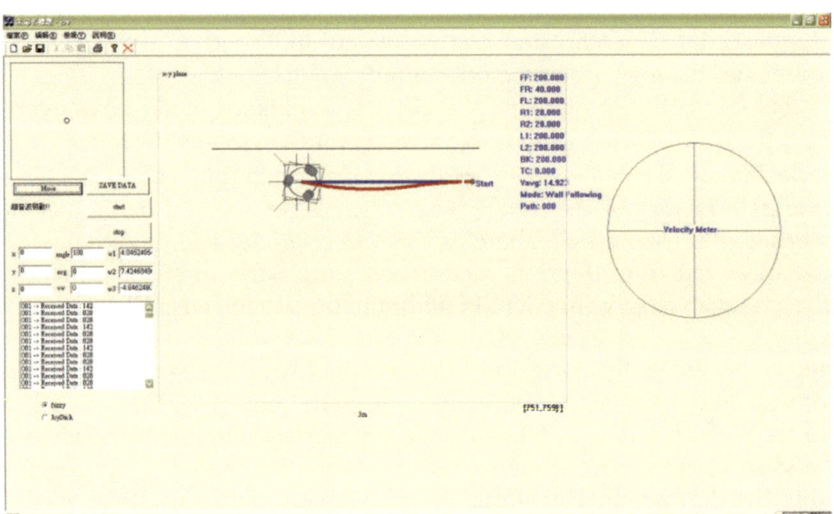

Fig. 20. Comparisons of wall-following navigation function.

In addition to the accuracy of navigation functions, the real-time property is also evaluated. Table 6 shows the computational time of inferring the fuzzy logic based navigation functions. Note that the time is measured in second, and it includes the communication traffic time. The computation time of a navigation function can be finished within 0.14 second. It is reasonable to the wheelchair users because of the driving speed of the wheelchair being not fast for the user who needs the navigation assisted functions.

Fig. 21. Comparisons of obstacle avoidance navigation function.

Navigation Function	Computation Time	Error %
Goal-Seeking	0.125 s	8 %
Wall-Following	0.11 s	7 %
Obstacle Avoidance	0.14 s	8 %

Table 6. Computation time evaluations

Finally, a practical test result of plotting the path of the wheelchair on the ground is evaluated in terms of different maximum wheelchair velocities (20%, 40%, 60% and 80%). The blue line indicates the wall of this experiment. Form the results of experiments, the wall-following fuzzy navigation function can work properly. Due to computational time limitations of the chip, large velocity results in a delay in sensor response and larger error.

7. Conclusions and future works

This chapter presents the navigation technologies of mobility assistive robots for disabled patients. The autonomous navigations of conventional robotic wheelchairs are not easy to be implemented because the inconvenience of setting target positions for each segmental path, and unavailable global localization and map of environment. Therefore, a feasible robotic wheelchair is achieved by controlling the wheelchair using the joystick, and the hybrid function converts joystick command as the virtual target for the fuzzy based navigation functions. In this manner, the proposed navigation functions become important assistive function to justify user's command and to perform more safe and comfortable wheelchair driving mechanism. On the other hand, a PSoC based distributed computing architecture is implemented to control the robotic wheelchairs. The proposed architecture is proposed to reduce the costs, size, and power consumptions and to increase the reliability of the popular PC based navigation system. Experiment results validated the PSoC based navigation functions. Based on practical experiments, the computational error is within 8 % and the time elapsed for computation and communication traffic is within 0.14 second. More specially, the proposed architecture is more feasible than the PC based navigation system. In the future, the navigation algorithm will be optimized to increase the computational efficiency and accuracy as well. At the same, in-hospital test will be planned to verify the clinical effectiveness.

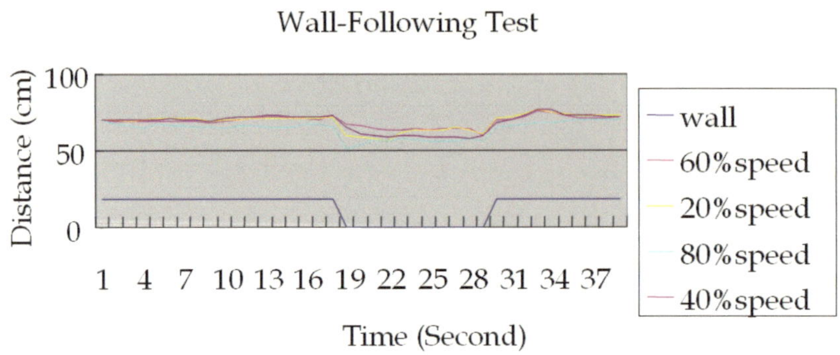

Fig. 21. Comparisons of obstacle avoidance navigation function.

8. References

Asai, H. (2001). Motion Planning of Obstacle Avoidance Based on Guide Style for Autonomous Mobile Robot, *Proceedings of the Seventh International Conference on Virtual Systems and Multimedia*, pp. 694 -700, Berkeley, California, U.S.A.

Bourhis, G.; Horn, O.; Habert, O.; & Pruski, A. (1997). The Autonomous Mobile Robot SENARIO: A sensor Aided Intelligent Navigation System for Powered Wheelchairs. *IEEE Robotics & Automation Magazine,* Vol. 4, 60-70.

Bourhis, G.; Horn, O.; Habert, O.; & Pruski, A. (2001). An Autonomous Vehicle for People with Motor Disabilities. *IEEE Robotics & Automation Magazine,* Vol. 8, 20-28.

Kruglinski, D.J. (1996). *Inside Visual C++ - The Standard Reference for Programming with Microsoft Visual C++ Version 4,* Microsoft, Washington.

Kuo, C.H.; & Chen, H.W. (2006). Human-Oriented Design of Autonomous Navigation Assisted Robotic Wheelchair for Indoor Environments, *Proceedings of IEEE International Conference on Mechatronics*, pp. 230-235, Budapest, Hungary.

Lankenau, A.; & Rofer, T. (2001). A Versatile and Safe Mobility Assistant. *IEEE Robotics & Automation Magazine,* Vol. 8, 29-37.

Luo, R.C.; Chen, T.M.; & Lin, M.H.. (1999). Automatic Guided Intelligent Wheelchair System Using Hierarchical Grey-fuzzy Motion Decision-making Algorithms, *Proceedings of the 1999 IEEE/RSJ International Conference on Intelligent Robots and Systems*, pp. 900-905, Kyongju, Korea.

Mazo, M. (2001). An Integral System for Assisted Mobility. *IEEE Robotics & Automation Magazine,* Vol. 8, 46-56.

Negnevitsky, M. (2005). *Artificial Intelligence – A Guide to Intelligent Systems,* Addison Wesley, New York.

4

Intelligent Wheelchair Robot "TAO Aicle"

Osamu Matsumoto[1], Kiyoshi Komoriya[1],
Tsutomu Hatase[2], Tadao Yuki[3] and Shigeki Goto[3]
[1]National Institute of Advanced Industrial Science and Technology,
[2]FUJITSU LIMITED,
[3]AISIN SEIKI Co.Ltd.
Japan

1. Introduction

Non-motorized manually operated wheelchairs are widely utilized in hospitals, welfare facilities, etc., to transport physically handicapped and elderly patients. Typically, these wheelchairs require the presence of healthcare staff who are burdened with actually moving the wheelchairs. To reduce the burden on staff in such facilities, we have developed a motorized wheelchair with autonomous traveling capability by applying robotics and mechatronics technologies. We believe that our research results will be useful for reducing the physical labor of the care workers and also promote the independence of the wheelchair users.

In previous research, intelligent wheelchairs with autonomous traveling capability guided by a vision system, laser range finder, etc. have been developed and demonstrated to a limited extent (Mori, 2005) (Adachi, 2003). However, such wheelchairs are not practical due to problems of safety, reliability, high cost, etc.

In this chapter we introduce an intelligent wheelchair with autonomous traveling capability called the "TAO Aicle". This wheelchair can accurately determine its position and direction using internal sensors and assistance from the surrounding environment including the Global Positioning System (GPS) and Radio Frequency Identification (RFID). To assure safe travel, we utilize assistance from environment, especially RFID, which is a distinctive characteristic in this research. Regarding autonomous navigation of mobile robots using RFID information in the environment, some research studies have been reported (Hahnel, 2004) (Tsukiyama, 2003) (Yamano, 2004). In those studies RFIDs were attached to walls at specific points such as destinations. When the mobile robot approaches the RFIDs, it detects them and knows that it has reached its destination. This use seems to focus on identifying a place of arrival. In our system the position information is provided by RFID and the wheelchair uses it directly for its trajectory control. As the similar research, in indoor environment, the mobile robot can travel using the position information from passive RFID tags in the floor and active RFID tags on the ceiling (Kim, 2006). In this research the robot needs RFID tags arranged in the high density for precise positioning.

On the other hand, in our research, we propose a position compensation method using the position information including the RFID tags which are locally arranged in the surface of a road. We also confirm the effectiveness of this compensation method by the high performance of the trajectory following (Matsumoto, 2006).

To confirm the effectiveness of our research results, we demonstrated "TAO Aicle" by offering trial rides to visitors at the Aichi Expo 2005 for six months (from 25th Mar. to 25th Sep. in 2005) (Hatase, 2006).
In this chapter the total system of "TAO Aicle" and the positioning method using GPS and RFID are introduced. An autonomous traveling control method based on the estimation of position and direction using a Kalman filter is proposed. The characteristics of trajectory following are discussed and the demonstration experimental results at the Aichi Expo are reported.

2. Total system of autonomous travelling service using intelligent wheelchair

Figure 1 shows the system configuration of the experiment system that was constructed at the EXPO site. Wheelchairs and servers are continuously connected via wireless LAN. Positioning data and wheelchair status data are sent from TAO Aicle to servers, and map data, route data, and position-related information data are sent from the server to each TAO Aicle. On the server display the position of each wheelchair can be monitored and important status information, such as battery power remaining, can also be monitored. The intelligent functions of TAO Aicle are thus realized not by the robot itself but through interaction with the infrastructure and servers. Figure 2 shows a part of outdoor traveling course of TAO Aicle constructed beside the building at the EXPO site.

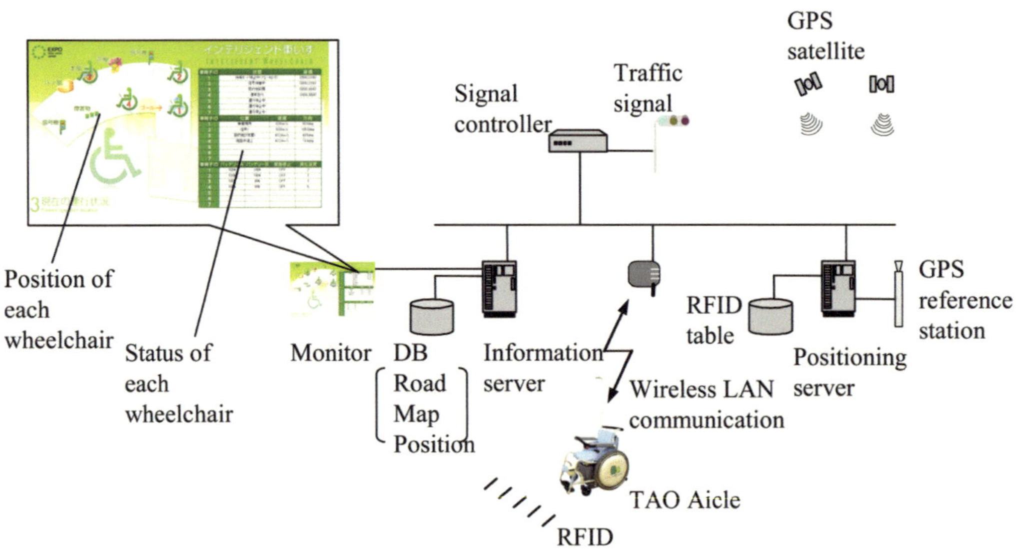

Fig. 1. Total System of autonomous traveling service using intelligent wheelchair

3. Intelligent wheelchair robot "TAO Aicle"

TAO Aicle is an improved version of the currently commercially available "TAO Light II" electric motorized wheelchair manufactured by Aisin Seiki Co., Ltd. Figure 3 is an overview of TAO Aicle showing the wheelchair hardware, internal and external sensors, computers, etc. to allow autonomous traveling. TAO Aicle was designed and manufactured using lightweight frames and driving units (motors, encoders, magnetic brakes and batteries). The frames are reinforced versions of those used in "TAO Light II" to withstand the frequent use of the trial rides. Also the capacity of the batteries was increased by assigning one battery to computer power and the other one to motor driving power. The long stick-like object installed at the back of the wheelchair is a GPS antenna, with which the robot can accurately determine its position using GPS. A laser range sensor is installed at the front-right, with which the robot can detect obstacles in front of it. User interface equipment (Safety Information Provision Terminal) is installed at the front end of the right armrest. For the safety of users, emergency stop button, wheelcovers, anti-tipover and seatbelt are installed.

Fig. 2. Course panorama

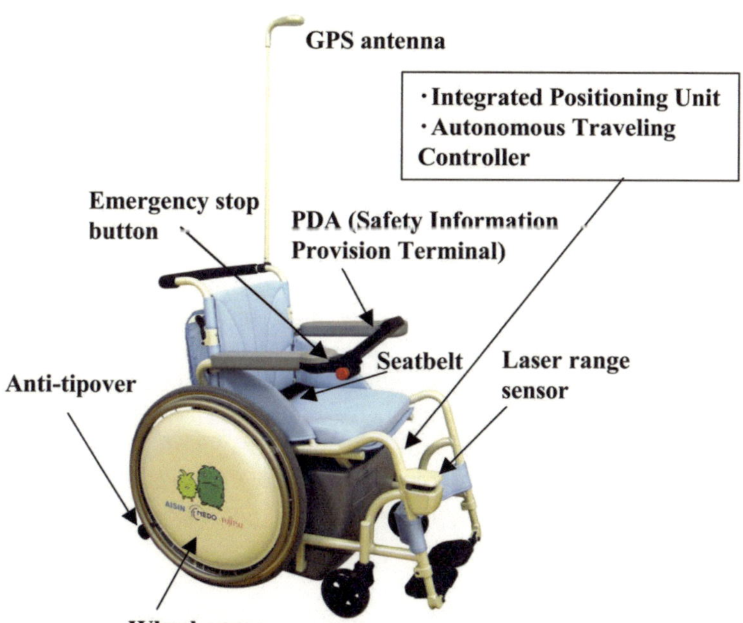

Fig. 3. Intelligent wheelchair robot TAO Aicle

Table 1 shows the specification of TAO Aicle. The right and left wheels are driven by DC servo motors with reduction gears, which is the same drive system as used in "TAO Light II". Each wheel is controlled independently so that the steering is controlled by the difference of the rotation angle between the right and left wheels. The overall wheelchair size is approximately the same as "TAO Light II" exclusive of the GPS antenna. The total weight is approximately 40 kg which is twice as heavy as "TAO Light II". The power source is a lithium-ion battery and the operating time with a full charge is approximately 3 hours. The traveling speed is potentially more than 6 km/h, however we limited the maximum speed to less than 2 km/h for safety demonstrations at the Aichi Expo site.
TAO Aicle consists of the following sub-systems. The assistance systems from the environment are shown in the next section.

Driving System	DC servo motors (90[W])x 2
Size (Length x Width x Height)	940 [mm] x 640 [mm] x 870[mm] (Height of GPS antenna: 1500 [mm])
Sensors	Rotary encoder x 2, GPS receiver, RFID reader, Directional compass, Laser range sensor
Total weight	40 [kg]
Battery	Lithium-ion Battery (24 [V], 6 [Ah]) x 2
Traveling speed	2 km/h (at Aichi Expo site)

Table 1. Specification of TAO Aicle

3.1 Autonomous traveling controller

The Autonomous Traveling Controller is a control unit driving the right and left wheel motors and includes a board computer (CPU, AD/DA converters and counter) and motor amplifiers. Using communication with the Integrated Positioning Unit, it recognizes the wheelchair's absolute position and direction based on GPS, RFID tags and a directional compass. Combining the above absolute information and the calculated position and direction using the integration of rotary encoder signals attached to the right and left wheels, the Autonomous Traveling Controller can estimate the accurate position and direction of TAO Aicle. The estimation method is described in section 5. The trajectory following control system and the speed control system to ensure safe stops and smooth starts are also installed. Moreover, various functions, such as obstacle detection using a laser range sensor, trajectory generation for obstacle avoidance, and the control algorithm for detecting and avoiding a fixed obstacle and stopping when confronted by moving obstacles, are also installed on the Autonomous Traveling Controller. With these features, TAO Aicle can follow the designed path accurately and achieve safe and comfortable traveling.

3.2 Integrated positioning unit

The Integrated Positioning Unit integrates the positional information of GPS and an RFID reader, and the directional information of a directional compass. It provides the information to the Autonomous Traveling Controller. To function with other units, it includes interface software to the Autonomous Traveling Controller and Safety Information Provision Terminal, and wireless communication software to an access point as well as a coordinate function with a positioning server.

3.3 Safety information provision terminal

The Safety Information Provision Terminal is a user interface attached to an armrest on TAO Aicle. The basic hardware is Fujitsu's PDA (Pocket LOOX), but it has been customized to suit the operation of this wheelchair. Not only can the user select destinations with the PDA's touch panel, but also access a range of information provided in the form of images and spoken words. It also includes terminal software for the interface to the integrated positioning unit and wireless communication to an access point.

4. Assistance system for intelligent wheelchair robot

For the demonstration at the Aichi Expo site, we prepared assistance system to allow safety traveling of TAO Aicle. In this section, we introduce typical asistance systems, GPS, RFID and information provision server.

4.1 GPS

In order to travel accurately along the planned route, and in order to provide useful and timely information to users, accurate positioning is indispensable. TAO Aicle can determine its position and direction based upon the amount of wheel rotation. This is referred to as the "dead reckoning method." However, this method can lead to measurement errors accumulating in the integrated value. This means it is necessary to frequently receive absolute values (position and direction). One method is using GPS positioning. TAO Aicle has an accurate GPS receiver and can establish its own absolute position precisely. At the EXPO site, we installed a GPS reference station near the demonstration course in order to improve the positioning accuracy using the DGPS method.

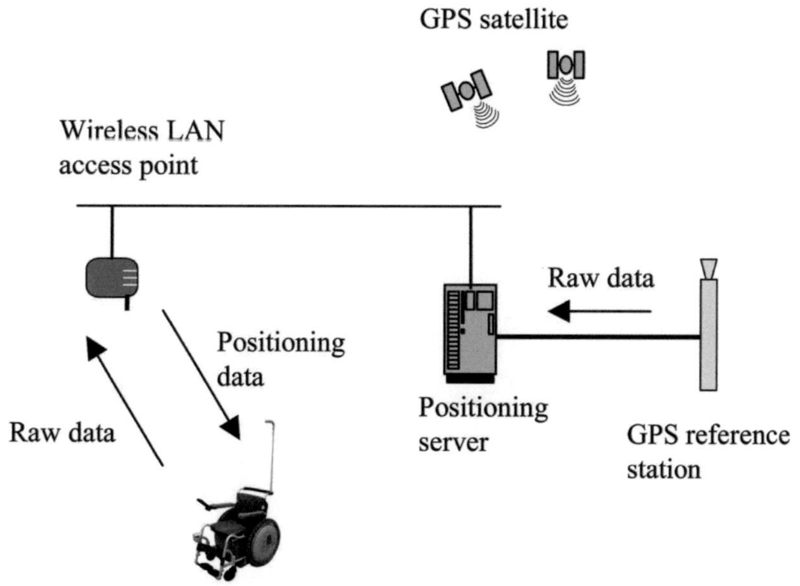

Fig. 4. Positioning mechanism using GPS

Figure 4 shows the system for accurate positioning using GPS. A rover receiver installed on TAO Aicle is connected to a positioning server via a wireless LAN access point. The rover receiver sends raw data to the positioning server every second. The positioning server receives the raw data from rover receivers and also receives data from a reference station which is installed nearby. The server calculates all receivers' accurate positions in real time and sends the positioning data to TAO Aicle. Accurate positioning is achieved by processing not only L1 code data but also carrier phase data. DGPS and RTK calculations are performed on the positioning server.

4.2 RFID

It is not possible to use GPS for indoor positioning. Even when outdoors, if the GPS sensor is too close to buildings, it is difficult to get accurate position information. For positioning, we propose the use of RFID (IC tags), which are being used recently to identify products in stores.

TAO Aicle has an RFID reader with an antenna installed under the seat. During operation it can read the RFIDs that are installed under the flooring panels. Each RFID has a unique code and the code can be converted into position data by referring to a database positioning table. Figure 5 shows the RFID that was actually used in the demonstration course at the EXPO site. The floor of the course was made of rubber blocks (Eco-rock) made from waste rubber material, each sized 10 x 20 x 5 cm. We installed a number of RFIDs under the rubber blocks along the course. An image of the actual landmark is shown in Fig. 5.

Fig. 5. Structure of RFID landmark

The antenna and RFIDs are shown in Fig. 6. This passive type RFID is a product of Fujitsu Limited, and it has a frequency of 13.56 MHz (ISO15693 Standard). The RFID antenna is also from Fujitsu Limited and it is a middle range type antenna. Figure 7 shows cross sections of the antenna's detection area in orthogonal vertical planes. These areas were experimentally measured by changing the relative position between the RFID and the antenna three dimensionally while keeping their faces parallel. There is a central detection space whose shape is an ellipsoid. The main detection space is surrounded by a sub-space as shown in the figure. When the distance between the RFID and the antenna is short, for example 5 cm, the detection area is relatively large. For the stable detection, the distance between the RFID and the antenna should be between 10 to 25 cm, and in this case the size of the detection area is about 20 x 30 cm. If one RFID is used, the positioning resolution is approximately the size of the antenna. When multiple RFIDs are detected simultaneously, the resolution of the positioning can be decreased by averaging the detected position data. As the maximum number of simultaneous RFID detections by the antenna is 16, it is possible to refine the resolution by arranging the RFIDs on the floor. In our localization experiment in which the antenna moves over arrayed RFIDs that are spaced at 90 mm intervals, the antenna detects one to three RFIDs by the antenna size (300 x 214mm) simultaneously and the deviation of the positioning is about 40 mm — less than half of the 90 mm spacing interval. This result indicates that the detection of multiple RFIDs helps to refine the positioning resolution. The RFID detection rate is about 5 Hz. While the antenna is moving attached to the wheelchair, it will not miss detecting an RFID inside the detection area.

We designed a RFID landmark for TAO Aicle to move over and get the positioning information, as well as to be able to change the course easily. The RFID landmark consisted of multiple RFIDs set perpendicular to the course at regularly spaced intervals. From a comparison of the antenna size and the size of the Eco-rock, the maximum number of simultaneous detections of RFIDs is about four and the positioning error is about 5 to 10 cm depending on the direction of measurement.

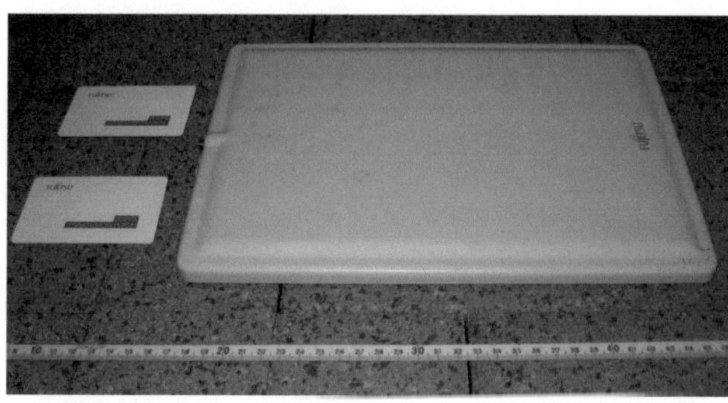

Fig. 6. RFIDs and antenna

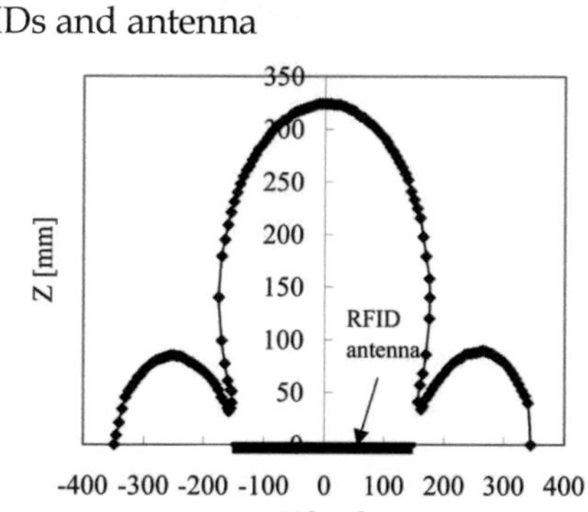

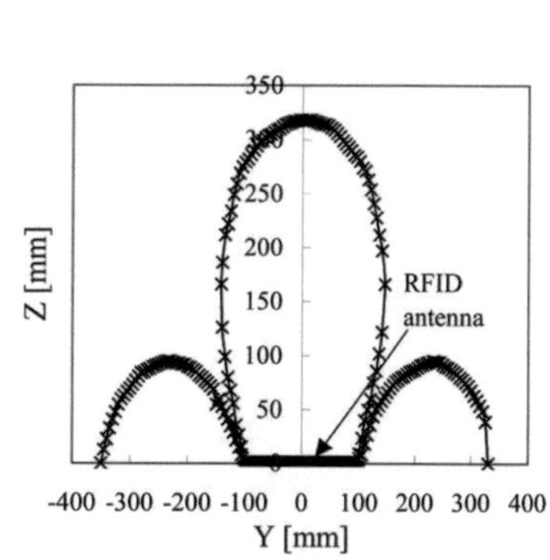

Fig. 7. Detection area of RFID antenna

4.3 Information provision server

For users of an intelligent wheelchair, information provision is indispensable. It starts, stops, and turns automatically according to various conditions. Other than in an emergency, users do not generally control the wheelchair, and if there is insufficient information, users may become uneasy. Examples of information that is currently provided is as follows:
- "The wheelchair is crossing a road."
- "The wheelchair will make a left turn soon."
- "You have arrived at the destination."

Each piece of information is provided at the moment the wheelchair reaches a specified position. LBS (Location-Based Service) technology can be used to achieve this. Fujitsu Limited had already developed a tourist information system and demonstrated this at a famous tourist resort in Kyoto in 2003 (Hatase, 2003). We applied the same technology used in the tourist information system to TAO Aicle.

Figure 8 shows a block diagram of the information provision system, and Figure 9 shows the terminal device screen. Information for users is provided by images and voice, displayed on the terminal device and transmitted by a speaker. When TAO Aicle is moving normally, the information on the screen shows mainly wheelchair conditions, such as speed, direction, and battery status (Figure 9(a)). On the other hand when it performs a particular action, the information on the screen shows the present situation or reason for the action (Figure 9(b)). The voice information tells users what the wheelchair is doing or what it will do next. The terminal device continuously receives accurate positioning data from the integrated positioning unit, enabling the terminal to play the appropriate voice data promptly when the wheelchair comes to a predetermined position.

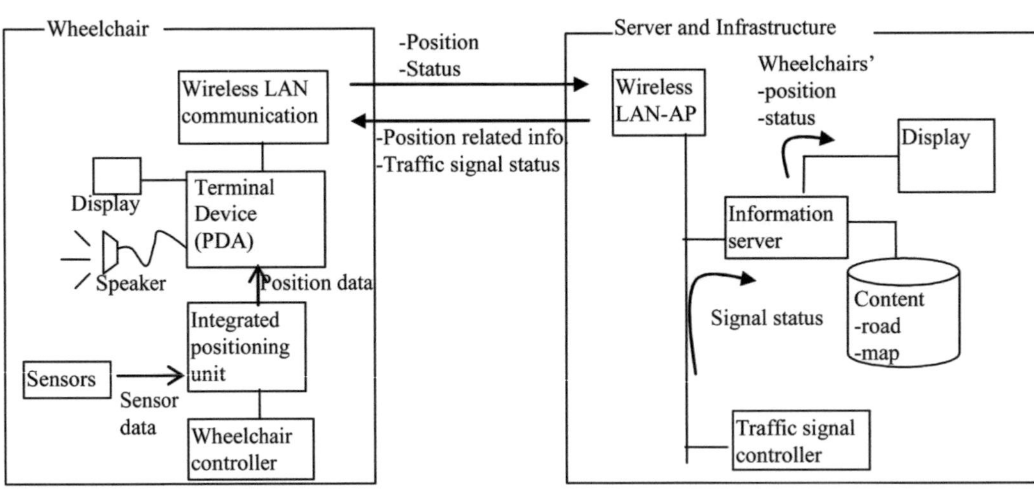

Fig. 8. Block diagram of information provision

The primary role of the information server is to store as much information as is necessary for controlling the wheelchair and to provide information to the user. The server retains information on the condition of relevant roads and shows each road's availability. At the request of the wheelchair controller, the server transmits this availability information to the wheelchair controller via the terminal device. The controller determines the route taking each road's availability into account. The server also returns the position-related information that is necessary for each wheelchair. The wheelchair terminal provides this information to the user according to the wheelchair's position. The traffic signal status is sent from the signal controller to the server and is then immediately sent to the wheelchair controller. Using the signal status data, the wheelchair can stop when the signal is red. The secondary role of the server is to monitor each wheelchair's position and status. It receives each wheelchair's status data continuously and displays this on an area map. Monitoring personnel can easily ascertain each wheelchair's situation and are immediately made aware of any emergency, should one occur.

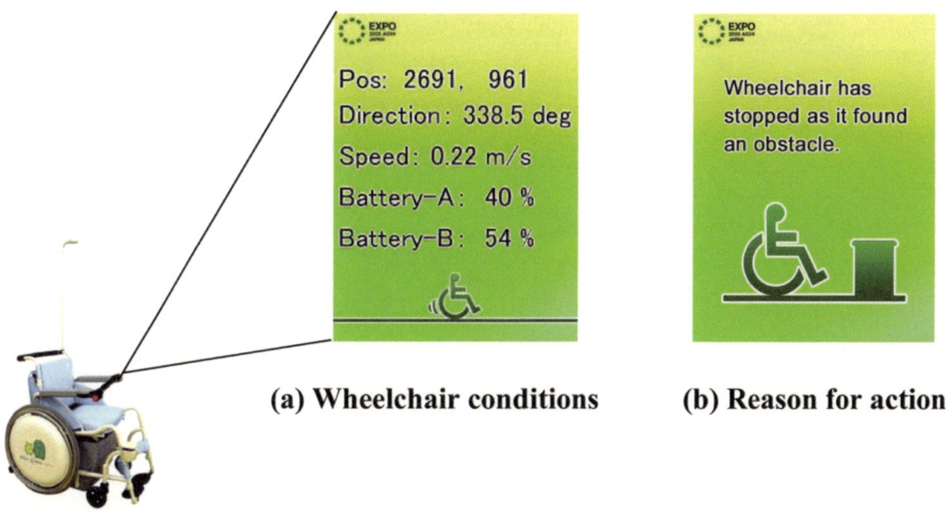

Fig. 9. Example of safety information provision terminal screen

5. Positioning method using GPS, RFID and directional compass

To achieve accurate trajectory following, the accurate present position and direction of TAO Aicle wheelchair are necessary. In our experimental setup, the position information from GPS was sent to the controller every second, while the position information from the RFIDs was sent to the controller when TAO Aicle passed over the RFIDs. To integrate the position (x, y) and the direction θ by dead reckoning using the rotary encoders of the right and left wheels, the direction θ from the directional compass and the position (x, y) information from GPS and the RFIDs, we constructed an estimating system for (x, y, θ) using a Kalman filter. Equation (1) shows the calculation of (x, y, θ).

$$\boldsymbol{x}_{fk} = \tilde{\boldsymbol{x}}_k + \left(\boldsymbol{X}_k^{-1} + \boldsymbol{Q}'^{-1}_s\right)\boldsymbol{Q}'^{-1}_s \hat{\boldsymbol{x}}_s \tag{1}$$

$\boldsymbol{x}_{fk}$: Calculated state vector (x, y, θ) using a Kalman filter

$\tilde{\boldsymbol{x}}_k$: State vector calculated from dead reckoning

$\boldsymbol{X}_k$: Error variance matrix of dead reckoning

$\boldsymbol{Q}'_s$: Error variance matrix of external sensors (RFID, GPS and directional compass)

$\hat{\boldsymbol{x}}_s$: State vector of the difference of (x, y, θ) between dead reckoning and external sensors

6. Autonomous travelling control

Our trajectory following control includes a Kalman filtering method for compensating position and direction using position information from GPS and RFIDs as described in the previous section. We adopted the conventional trajectory following method which is shown in Fig. 10 using eqs. (2)(3)(4). If the closest point on the planned path to the wheelchair is determined considering the present position and direction of the wheelchair, the reference velocities of the right and left wheels are calculated using eqs. (2)(3)(4). A velocity feedback control system is installed in the low-level controller to achieve the trajectory following control.

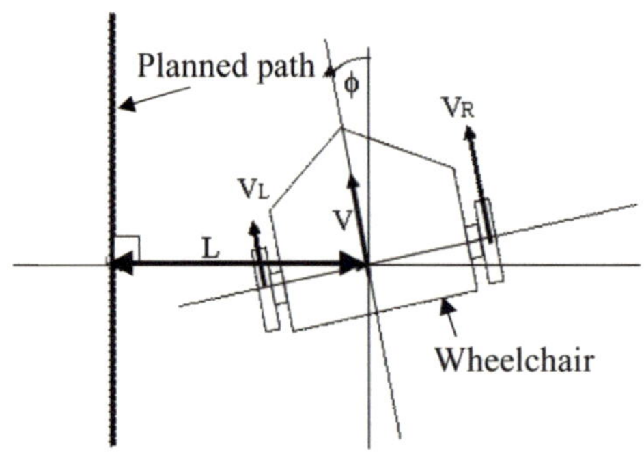

Fig.10. Trajectory following control method

Figure 11 shows a block diagram of this autonomous traveling control system including accurate positioning method using Kalman filter.

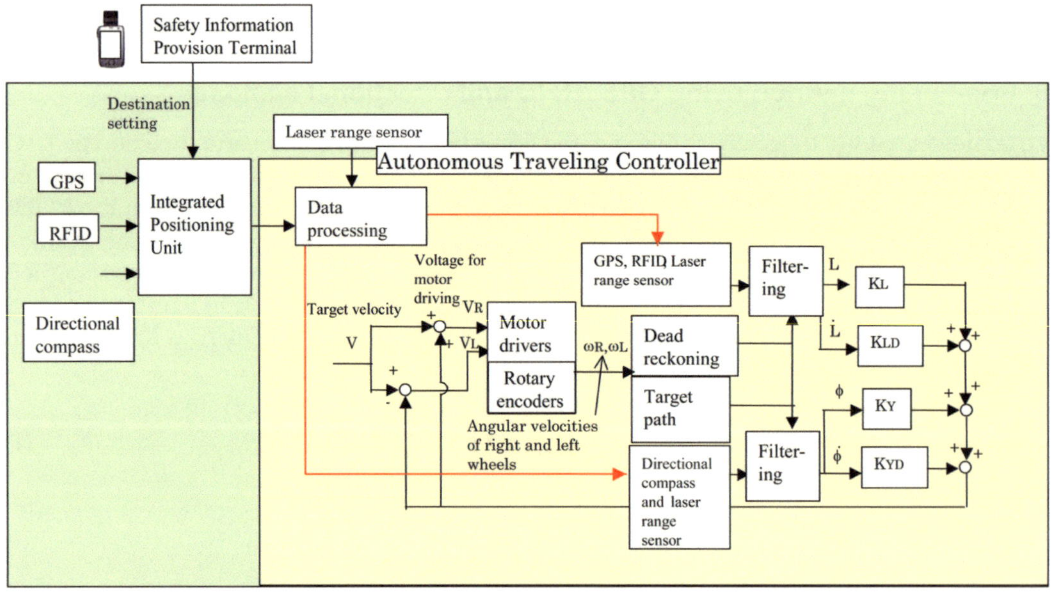

Fig.11. Block diagram of autonomous traveling control

$$V_R = V + \Delta V \tag{2}$$

$$V_L = V - \Delta V \quad (3)$$

$$\Delta V = K_L L + K_{LD}\dot{L} + K_Y \Phi + K_{YD}\dot{\Phi} \quad (4)$$

V: Target traveling velocity of wheelchair
V_R: Target velocity of right wheel
V_L: Target velocity of left wheel
L: Minimum distance between the wheelchair center and target path
Φ : Angle between the wheelchair yaw-direction and tangent direction of the closest point on the target path
K_L,K_{LD}: Feedback gains for position and velocity of minimum distance between the planned path and the wheelchair position
$K_Y K_{YD}$: Feedback gains for angle and angular velocity between the direction of planned path at the closest point to wheelchair and the wheelchair direction

7. Evaluation result of GPS positioning in Aichi Expo. Site

Figure 12 shows the GPS positioning result obtained during the test drive experiments. The solid line shows the position of TAO Aicle calculated in the controller and the dots show the calculated position based on GPS. In this test drive, the wheelchair left the starting point for the destination and after stopping at the destination it returned to the starting point. It crossed the crosswalk twice and avoided a fixed obstacle on the way. In Fig. 12 all dots appear to be plotted on the actual trajectory that the wheelchair passed, and the maximum deviation was approximately 40 cm. (This was better than average accuracy.) The effectiveness of the GPS positioning was confirmed through this experiment.

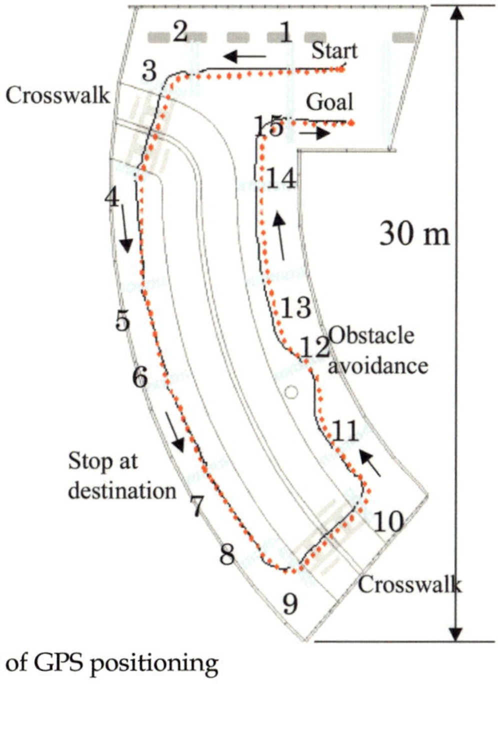

Fig.12. Evaluation result of GPS positioning

8. Experimental result of trajectory following in Aichi Expo. Site

The speed of the wheelchair was set relatively low at 1.4 km/h because various individuals were expected to try the demonstration. At this speed the number of RFIDs detected in the landmarks was measured along the demonstration course. The result is shown in Fig. 13. From this figure, the set of RFIDs in each landmark was measured at least twice.

The performance of the trajectory following control was tested by measuring the position of the wheelchair using surveying equipment (Total Station GTS-820A, Topcon Corp.). The resolution of the position measurement was 1 cm and the cycle time was 2Hz. For the measurement, an omnidirectional prism was set above the center of the wheel axis of the wheelchair.

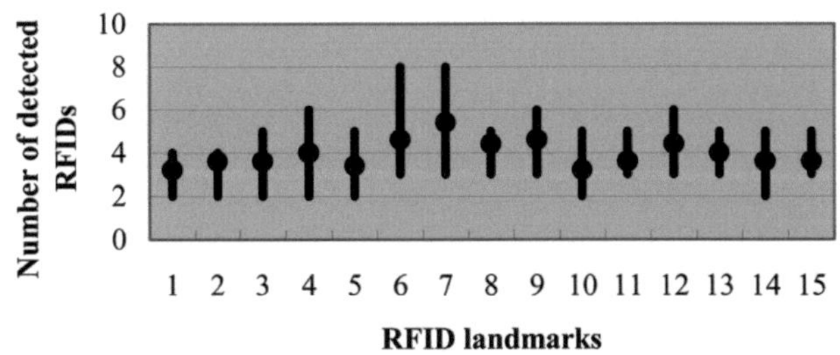

Fig.13. Number of detected RFIDs

The measured trajectory of the wheelchair was compared with the actual course as well as the course data calculated in the wheelchair controller. Figure 14 shows an example of autonomous traveling. After moving through one third of the total course, the wheelchair's deviation from the course was fairly large. The figure shows that at each landmark position the wheelchair recognized the deviation by detecting the RFIDs of the landmark and that it controlled its direction to follow the course. From these experimental results, the maximum deviation from the course was about 40 cm, and stable trajectory following control was achieved along the overall course. There was no dropout from the course during the total demonstration period of 6 months at the moving speed described above.

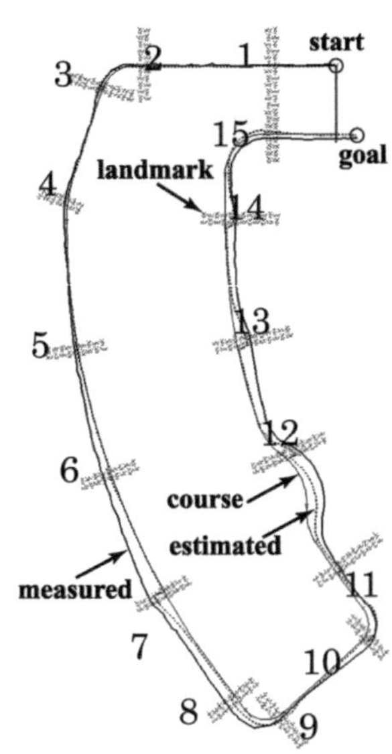

Fig.14. Results of trajectory following control

9. Evaluation result of information provision in Aichi Expo. Site

In the demonstration, the following information was provided as position-related information:
1. "You are coming to "Flower Shop" ("Bookshop", "Bakery") soon."
2. "The wheelchair is crossing the road."
3. "The wheelchair will make a left (right) turn soon."
4. "There is an upward (downward) slope ahead."
5. "You are coming to the terminal soon."
These informative statements should be provided at the optimal timing in order not to make users feel uneasy. They should neither be too early, nor too late. This was achieved without problem because TAO Aicle always knows its own position accurately in order to realize autonomous driving.
The following non-position-related information was also provided:
1. "The wheelchair will stop, as the next light is red."
2. "As the system found an obstacle ahead, the wheelchair will bypass it."
3. "Emergency! Push the emergency button!"
These voice statements were triggered by the wheelchair's action, however the timing was adjusted beforehand. Emergency information must be provided quickly after an emergency situation arises. Avoidance action should be announced not just before the actual action but sufficiently prior to the action because the user may feel uneasy when approaching an obstacle. This was possible as the laser range sensor can detect obstacles 4 m in front of the wheelchair. We evaluated that this information provision using voices and images performed optimally. Thanks to such information provision, visitors could enjoy autonomous traveling without worry.

10. Overall evaluation of demonstration experiment in Aichi Expo. Site

The feasibility of TAO Aicle was confirmed by the trial ride demonstrations at the Aichi Expo site. On a dedicated outdoor course, whose length was approximately 50 m with features similar to a downtown environment as shown in Fig. 15, TAO Aicle autonomously traveled along the course and stopped at the destination that the visitor selected then reached the goal position while detecting and avoiding obstacles on the way.
We developed seven TAO Aicles for the demonstration at Aichi Expo. site. Table 2 shows round (from start point to goal point) number of each TAO Aicle including test traveling before visitor's trial ride. Totally, TAO Aicles rounded the course at Aichi Expo. site 41,316 times. Table 3 shows travel distance of each TAO Aicle. Totally, TAO Aicles traveled approximately 2,107km. Figure 16 shows number of visitors at Aichi Expo. taking a test-ride on TAO Aicle. The total number of the visitors was 36,578 and the average number of visiters per a day is approximately 200. No severe accidents (e.g., collisions with the environment, the wheelchair tipping over) occurred during the period of the Aichi Expo (185 days).
Moreover, we received many positive comments from the visitors. Some typical comments were as follows:
- It is useful because we can go to a desired destination just by simple touch panel operations.
- Voice announcements help the user to feel comfortable.
- Early commercialization is desirable.

Fig. 15. Demonstration at Aichi Expo. Site

Through the experiment we were able to confirm the usability and reliability of this newlydeveloped intelligent wheelchair robot.

round

	Wheelchair (No.1)	Wheelchair (No.2)	Wheelchair (No.3)	Wheelchair (No.4)	Wheelchair (No.5)	Wheelchair (No.6)	Wheelchair (No.7)
March	311	228	251	344	161	0	85
April	671	1,400	1,533	807	1,565	320	1,090
May	776	1,400	545	668	1,150	1,511	766
June	1,014	514	0	864	1,626	905	825
July	878	554	586	1,204	1,199	899	493
August	550	1,082	1,042	1,869	1,121	1,196	995
September	649	835	969	986	885	495	1,499
Total	4,849	6,013	4,926	6,742	7,707	5,326	5,753
Total of all wheelchair	41,316						

Table 2. Round number of TAO Aicles

[m]

	Wheelchair (No.1)	Wheelchair (No.2)	Wheelchair (No.3)	Wheelchair (No.4)	Wheelchair (No.5)	Wheelchair (No.6)	Wheelchair (No.7)
March	15,824	11,721	12,852	17,497	8,201	0	4,350
April	33,888	71,571	78,259	40,744	80,338	16,389	55,480
May	39,595	71,526	27,719	34,054	58,653	77,403	39,045
June	51,653	26,253	0	43,965	83,135	46,297	42,066
July	44,580	27,837	29,976	61,248	61,326	45,621	24,944
August	27,844	55,505	53,234	94,781	57,022	60,456	50,889
September	33,155	42,756	49,695	50,353	45,283	25,265	76,799
Total	246,539	307,169	251,735	342,642	393,958	271,431	293,573
Total of all wheelchair	2,107,047						

Table 3. Travel Distance of TAO Aicles

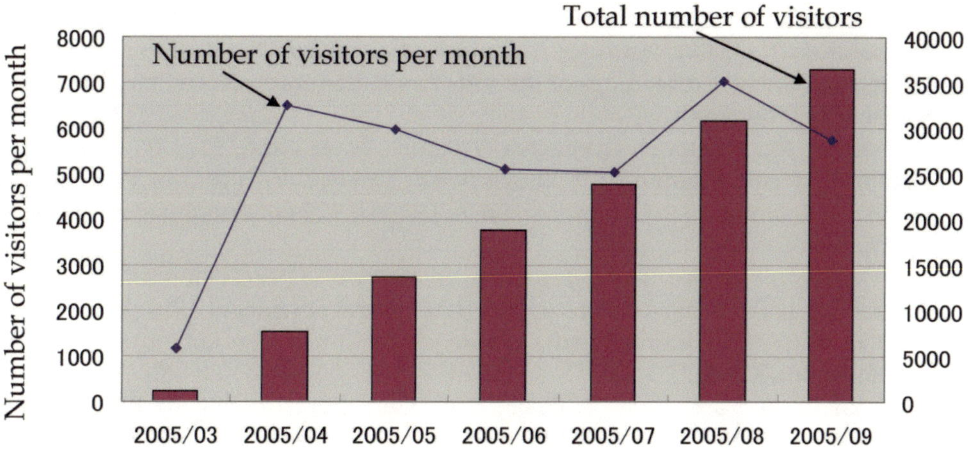

Fig. 16. Number of visitors taking a test-ride on TAO Aicle

11. Conclusion

In this chapter, we introduced an autonomous traveling control system for the "TAO Aicle" wheelchair with a view to increasing the quality of life of wheelchair users as well as their caregivers. Autonomous and safe traveling control was achieved by assistance from the infrastructure, such as GPS and RFID tags which provide accurate positioning. Our demonstration experiments at Aichi Expo confirmed the safety and reliability of the developed "TAO Aicle" system. By further analyzing minor problems that occurred (e.g., failure to avoid an obstacle, deviating from the desired path), we will seek to achieve improvements that enable the practical use of the "TAO Aicle" wheelchair.

12. Acknowledgement

The "TAO Aicle" wheelchair was developed as part of the "Project for the Practical Application of Next-Generation Robots (FY2004-2005)" of the New Energy and Industrial Technology Development Organization (NEDO) of Japan. We wish to express our gratitude to everyone supporting this project.

13. References

Adachi, Y.; Goto, K.; Khiat, A.; Matsumoto, Y. & Ogasawara, T. (2003). Estimation of User's Attention based on Gaze and Environment Measurements for Robotic Wheelchair, Proceedings of the 12th International IEEE Workshop on Robot and Human Interactive Communication (RO-MAN2003), pp.97-103.

Hahnel, D. et al. (2004). Mapping and Localization with RFID Technology, Proc. of Int. Conf. on Robotics and Automation (ICRA2004), pp.1015-1020.

Hatase, T.; Sato, K.; Wakamatsu, Y. & Nishimura, H. (2003). Development of a New Location-based Navigation/Information Service for Pedestrians, Proceedings of the 10th World Congress on Intelligent Transport Systems and Services.

Hatase, T.; Wakamatsu, Y.; Nishimura, H.; Yamamoto, H.; Toda,K.; Goto, S.; Matsumoto, O. & Komoriya, K. (2006) . An Application of Location-based Services to an Intelligent Wheelchair Robot, Proceedings of the 13th World Congress on Intelligent Transport Systems and Services.

Kim, B.; Tomokuni, M.; Ohara, K.; Tanikawa, T.; Ohba, K. & Hirai, S. (2006). Ubiquitous Localization and Mapping for Robots with Ambient Intelligence, Proc. of 2006 IEEE/RSJ International Conference on Intelligent Robots and Systems (IROS06), pp.4809-4814.

Matsumoto, O.; Komoriya, K.; Toda, K.; Goto, S.; Hatase, T. & Nishimura, H. (2006). Autonomous Traveling Control of the "TAO Aicle" Intelligent Wheelchair", Proc. Of 2006 IEEE/RSJ International Conference on Intelligent Robots and Systems (IROS06), pp.4322-4327.

Mori, H.; Nakata, T. & Kamiyama, Y. (2005). Next-Generation Intelligent Wheelchair as developed by ROTA, Proc. of 36th Int. Symposium on Robotics (ISR2005), pp.39

Tsukiyama, T. (2003). Navigation System for Mobile Robots Using RFID Tags, Proc. of Int. Conf. on Advanced Robotics (ICAR2003), pp.1130-1135.

Yamano, K. et al. (2004). Self-localization of Mobile Robots with RFID System by using Support Vector Machine, Proc. of Int. Conf. on Intelligent Robots and Systems (IROS2004), pp.3756-3761.

5

Perceptual Navigation for Semi-Autonomous Wheelchair Operations

H. Uchiyama, W. D. Potter, M. A. Covington, J. Tarver and R. Eunice
Artificial Intelligence Center, The University of Georgia
United States of America

1. Introduction

Studies have shown that individuals with disabilities, regardless of their age, benefit substantially from access to independent mobility, including electrically powered wheelchairs (Douglass & Ryan, 1987; Paulsson & Christoffersen, 1989; Verburg et al., 1991). Independent mobility increases vocational or educational opportunities, reduces dependence on caregivers, and promotes a feeling of self-reliance and self-esteem. Unfortunately, some individuals are unable to independently operate a powered wheelchair due to some combination of motor, sensory, or perceptual impairments.

With the aim of assisting those individuals, a number of studies have been conducted in the field of assistive technology which combines robotics and artificial intelligence to develop an intelligent wheelchair system. In robotics, the intelligent wheelchair system is classified as a robotized wheelchair, a subcategory of service robots. The World Robotics 2005 survey predicts that service robots will routinely assist elderly and physically challenged people by the end of 2010, and have large potential for future applications not only within the service industries, but also in our rapidly aging society as future caregivers (United Nations, 2005).

However, almost no robotized wheelchairs are currently on the market, and few are expected in the near future (United Nations, 2005). This may mean that the market for robotized wheelchairs is practically in the stage of infancy; most of the applications mainly serve research-based purposes rather than end customers.

This chapter presents our semi-autonomous wheelchair prototype, which aims to increase independent mobility for individuals with multiple disabilities, specifically those who are wheelchair-bound and severely vision impaired, by providing perceptual and navigational capabilities. The scope of our project focuses on developing a perceptual navigation system to enable the user to maintain the maximum control over the wheelchair movement.

2. Background

2.1 Previous work

The platforms of robotized wheelchairs in research are mostly power wheelchairs; some try to achieve autonomous control over the wheelchairs, and others aim for semi-autonomous control. We briefly survey some selected robotized wheelchair systems.

Applied AI Systems (TAO Project): Applied AI Systems, Inc. developed the robotic wheelchair prototype, TAO series, conducted by Gomi & Griffith (1998). The goal of the

TAO project is to create an add-on system for any standard powered wheelchair that will provide a high level of autonomy. The current model presented to the market is TAO-7 and is mostly used for research and development purposes.[1]

Kiss Institute of Practical Robotics (Tin Man Project): The Tin Man project is aimed at the development of a low-cost robotic wheelchair to aid people with impaired mobility while focusing on a more user-friendly interface and cost issues (Miller & Slack, 1995). The latest version, Tin Man II, is a robotized wheelchair which exhibits manual navigation with obstacle avoidance override, autonomous control, and manual mode. The Tin Man system was also used as a base platform for the Wheelesley project at MIT (see below).

MIT Artificial Intelligence Lab (Wheelesley): The goal of the Wheelesley project is the creation of a complete robotic wheelchair system to be used by users unable to manipulate a standard motorized wheelchair. This project aims to establish the system for both indoor and outdoor navigation with automatically switching navigation modes. Wheelesley was developed based on the Tin Man model and interacts with the seated user via a graphical user interface (Yanco, 1998).

Northeastern University, Deictically Controlled Wheelchair: The goal of this project is to create a "gopher" robot that can be given commands easily and accurately for the handicapped (Crisman & Cleary, 1998). The robot will retrieve objects in the environment autonomously via commands from users, which are communicated deictically.

University of Edinburgh (Smart Wheelchair): The CALL Centre of the University of Edinburgh has developed a wheelchair intended to be used by children who do not have the physical, perceptual or cognitive abilities to control an ordinary powered mobility aid (Odor & Watson, 1994).[2] The Smart Wheelchair exhibits collision detection followed by a maneuvering action, follows line tracks laid along the floor, and communicates with the user via a speech synthesizer or other feedback system.

University of Michigan (NavChair): The NavChair system intends to provide mobility support to users who are unable to drive standard motorized wheelchairs, and this is one of the most successful systems in the '90s (Levine et al., 1999). The NavChair uses several navigation modes and switches among them automatically. The navigation modes are as follows: general obstacle avoidance, door passage, and wall following.

While there has been much research to develop robotized wheelchairs, very few are sold commercially, such as those from Applied AI and CALL Centre, and none of those systems are intended to be used outside of a research lab or a training facility.

2.2 Semi-autonomous wheelchairs

With autonomous control, the system probes the environment, makes decisions, and fully controls the mobility of the wheelchair, which leaves the user totally dependent upon the equipment. While some users may be comfortable with an autonomous wheelchair transportation system, others want to be more involved with the process. It is essential for

[1] Information is available at http://www.aai.ca

[2] Information is available at http://callcentre.education.ed.ac.uk/

them to feel in control, while being responsible for both decision-making and motion rather than being a passenger.

A semi-autonomous (SA) wheelchair system comprises hardware equipment similar to autonomous wheelchairs, but the purpose of the system development focuses on maximizing the level of autonomy of users. Thus, the magnitude of assistance is designed to provide just as much as the user really needs. Generally, the type of assistance for a user varies from one to another; therefore, the specification of an SA wheelchair system needs to be determined based on the user's physical capabilities.

A number of research projects have developed various types of semi-autonomous wheelchairs. Borgolte et al. (1998) developed a robotized wheelchair featuring omnidirectional maneuverability to provide an intuitive semi-autonomous control system to assist people with severe or multiple disabilities. Argyros et al. (2002) presented a robotic wheelchair with a semi-autonomous navigation system comprising a panoramic vision system to support the users who have limited motor control of the upper extremities. The Mobile Internet Connected Assistant (MICA) project (Rönnbäck et al., 2006) designed a semi-autonomous wheelchair system in which the user either locally or remotely controls the chair by means of head-held sensors. Although using a slightly different platform, a power-assisted manual wheelchair,[3] Simpson et al. (2005) demonstrated the first prototype which attempts to provide the users who have visual impairments with some basic navigation tasks such as collision avoidance.

However, none of these systems provide or enhance perceptual capabilities for the seated user who has severe visual impairment.

2.3 Motivation

As stated in Section 1, the goal of our project is to increase the independent mobility of power wheelchair users with severe visual impairment by providing perceptual and navigational capabilities. We assume our typical SA wheelchair user is tactilely and auditorily competent with fine motor control of the upper extremities. In fact, our research efforts have been influenced by a former student who has exactly the disabilities we are mentioning. A collaborative discussion with the student at the initial stage of the project enabled us to elucidate such problems of interest as: collision avoidance, human and obstacle detection, drop-off avoidance, portal navigation, textual-information acquisition, and indoor navigation.

Many large universities provide on-campus curb-to-curb van transportation service to persons with mobility, visual, and other health-related impairments; however, typically no official care attendant service is provided inside a building. Therefore our focus is primarily on the situations in which the wheelchair users need to be on their own, independently maneuvering in indoor circumstances.

3. Design

In this section, we present an overview of our behavior-based architecture with *Behavior Cell* and *Behavior Network* components, and hardware design which consists of a sensor system and feedback module.

---

[3] Based on a manual wheelchair, the rear wheel hubs are replaced with motorized ones that magnify and/or adjust the propulsive force provided by the user.

3.1 Behavior based approach

Since the SA wheelchair system needs to be individualized to the user, our design principle is based upon modularity and flexibility. Further, due to the nature of the problems described in Section 2.3 which must be resolved quickly and handled concurrently, the *behavior-based control (BBC)* architecture (Arkin, 1998; Brooks, 1991; Matarić, 1992) is suitable for the base architecture of our design. Utilizing a customized BBC architecture, we define *behaviors* which emerge from an agent-environment interaction based on a number of loosely tied processes that run asynchronously and respond to problems in a parallel manner.

In our project, those behaviors are realized to accomplish their tasks through user-machine cooperation. We call the robotic wheelchair a *wheelchair-navigation agent*. The wheelchair-navigation agent consists of two cooperating sub-agents: the perceptual agent and the navigational agent. The perceptual agent acquires the state of the environment through sensors (*Perceptual Behaviors*), and the navigational agent interprets the state and passes the appropriate information to the user (*Navigational Behaviors*). Finally the user manipulates the wheelchair in combination with his/her own perception and judgment (*Manipulative Behaviors*).

Since we intend to leave maximum control on the user side, the degree of mobility assistance from the wheelchair-navigation agent is set to minimal; that is, the system does not directly intervene in controlling the wheelchair at all. We also assume that our target users are able to generate an internal representation (*cognitive map*) of the familiar environment; that is, they have some knowledge of the starting or current location and the destination as well as some landmarks in between. Based on such assumptions, assisting local navigation, including obstacle avoidance, and landmark matching are more immediately needed tasks than more complicated tasks, such as planning the whole path and navigating autonomously. The main focus of our project is on pursuing perceptual and navigational assistance, thus *semi-autonomous control*.

3.1.1 Behavior cell

We present a unit of the behavioral structure, a *Behavior Cell*, which is based upon the BBC architecture with an extended input/output feature. A Behavior Cell consists of an input/output (I/O) component, a behavioral function component, and an internal storage component (Figure 1). It structurally resembles an artificial neuron; however, it has a logical gate in addition to widely extended functions such that the innervation link between cells can run by both Boolean and numeric means.

A Behavior Cell does not have to employ all components; it may or may not contain the behavioral function and the internal storage components depending upon what features it needs. Behavior cells communicate with other Behavior Cells, sensors, and effectors through their I/O components.

The I/O component consists of a subset of the following eight I/O ports: Port-EB, excitatory inputs; Port-IB, inhibitory inputs; Port-DI, sensory/behavioral inputs; Port-RS, a reset signal; Port-IS, an innervation signal; Port-AB, an activation output; Port-EO, an effect output; and Port-AO, actuator outputs. The excitatory and inhibitory inputs are linked to the corresponding behaviors' activation output ports. When any activation (inhibition) conditions are met, the behavior is activated (deactivated). Our architecture allows both Port-EB and Port-IB to specify activation (inhibition) conditions by using logical expressions (Boolean algebraic functions), such as $Activation = (Activation1 \wedge Activation2) \vee Activation3$.

Port-DI takes various types of data inputs from sensors or other behaviors (effect outputs). When Port-IS receives an innervation signal from outside or from Port-RS, the behavior checks or sends its inputs and outputs. If Port-RS receives a reset signal from the system, the behavior will clear all dynamic contents of the storage component and activate Port-IS. Port-AB contains an activation value (binary) that is linked to the value of Port-EB. Port-EO contains an effect value that is derived from the behavioral function. If the behavior is connected to its effector(s), Port-AO sends actuator outputs to them. The generic features of each port are summarized in Table 1.

Port ID	Name	Feature
Port-EB	Excitatory Inputs	Binary value is derived from a predefined logical expression and the connected input values.
Port-IB	Inhibitory Inputs	Binary value is derived from a predefined logical expression and the connected inputs values.
Port-DI	Sensory/Behavioral Inputs	Input data from sensors and/or other behaviors. A various types of the data (including arrays) are allowed.
Port-IS	Innervation Signal	Innervates the Behavior Cell.
Port-RS	Reset Signal	Clear temporal memory
Port-AB	Activation Output	Binary value linked to Port-EB and Port- IB
Port-EO	Effect Outputs	Outputs (the result of computation)
Port-AO	Action Outputs	Action control to effectors.

Table 1. The generic features of input/output ports of a Behavior Cell.

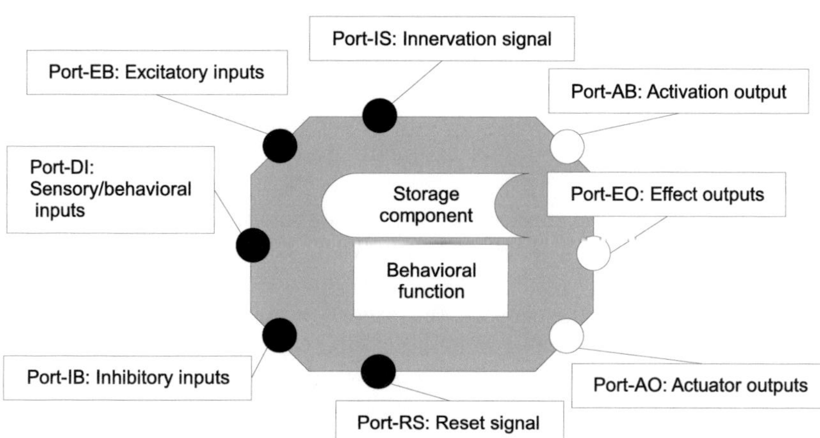

Fig. 1. The basic structure of a Behavior Cell.

The behavioral function component provides flexible activation/computation functionalities, such as algebraic sum, sigmoid, Gaussian, and logical expressions, as well as a simple by-pass function (e.g. a direct link between an input and output). More complicated functionalities, such as fuzzy logic inference operators or artificial neural networks, can also be implemented.

The storage component provides a storing capability of the current state onto its dynamic data, which enables the behavior to achieve goals that contain temporal sequences. It may also contain internal static data which all instantiated behaviors can share and refer to, as well as individual constant data that a behavior utilizes as permanent reference information, such as threshold values or look-up tables.

The activation/computation process performed by a Behavior Cell is as follows:

1. When Initialization/Reset input (Port-RS) is activated, it refreshes the internal dynamic memory and innerves Innervation Input (Port-IS).
2. When Innervation Input receives a signal, check the value of Effect Inputs (Port-EB). If true, set Activation Output (Port-AB) value to 1 (true) and go to the next step, otherwise return.
3. Check the value of Inhibitory Inputs (Port-IB) to see whether the behavior is inhibited. If false, go to the next step, otherwise set Activation Output (Port-AB) to 0 (false) and return.
4. In case of using Port-EO: Using the information from Sensors/Behavior Inputs (Port-DI), derive the return value from the behavioral function and write this value to Effect Output (Port-EO) and return. Store the necessary data in the internal memory if so designed.
5. In case of using Port-AO: Similar to (4), derive the return action commands from the behavioral function and send the commands to the effectors via Action Outputs (Port-AO) and return. Store the necessary data in the internal memory if so designed.

3.1.2 Behavior network

Similar to other Behavior-Based architectures (for instance, Nicolescu & Matarić, 2002), our approach also enables behaviors to consist of other behaviors. Such behaviors are implemented based on a subset of corresponding behaviors, thus represented as a *Behavior Network*. The types of relationships between a Behavior Network and its corresponding behaviors may vary; they can be hierarchical or interdependent. In a Behavior Network, behaviors communicate with each other through their port-to-port links, and precondition dependence characterizes the links; thus, the activation of a behavior is dependent on its pre-conditional links.

An individual Behavior Cell can connect to multiple Behavior Cells, and similarly, multiple Behavior Cells can be linked to a single Behavior Cell. This multiple-connectivity allows a Behavior Cell to be a member of multiple Behavior Networks, which makes component-like behaviors, such as interface to the sensors, reusable. Containing multiple task-oriented and reactive behaviors (*functional behaviors*) enables a Behavior Network to accomplish various tasks, such as command arbitration, learning, and planning, while asynchronously performing tasks within the distributed architecture. The bottom line of the Behavior-Based philosophy, the distributed architecture, is preserved such that the behaviors are relatively simple, organized into modules, and performed in a parallel fashion.

A Behavior Network must also behave as structurally equivalent as a Behavior Cell when observed from outside. In order to do so, each Behavior Network must contain a specific type of Behavior Cell (*I/O cell*), which accomplishes the tasks related to input/output communication and activation sequence inside of the Behavior Network. Figure 2 depicts a generic Behavior Network that consists of I/O cells and functional behaviors.

The I/O cells are categorized in two types: Boolean I/O cells that exchange Boolean signals and activation I/O cells that control sequential activation in the Behavior Network. Figure 3a illustrates a generic Boolean I/O cell that consists of Port-EB (excitatory inputs), Port-RS (reset signal), Port-IS (innervation signal), Port-AB (activation output), and a Boolean algebraic function. The Boolean algebraic function can employ various symbolic logic expressions, and the result from the function is directly connected to the activation output (Port-AB).

Activation I/O cells are responsible for the sequence of innervation/reset functionalities in the Behavior Network. An activation I/O cell consists of an excitatory input, an inhibitory input, an innervation input, an activation output, action outputs, and a storage component that contains a predefined activation sequence of the behaviors (Figure 3b). The activation I/O cell innerves the reset/innervation ports of the behaviors that belong to the Behavior Network according to the sequence stored on the storage component.

The functional behaviors deal with data and/or actuator communication. Connections between functional behaviors consist of excitatory links (between Port-AB and Port-EB), inhibitory links (between Port-AB and Port-IB), (sensory and/or behavioral) data links (between Port-EO and Port-DI), and actuator links (between Port-AO and effectors). A functional behavior in a Behavior Network can also be another Behavior Network.

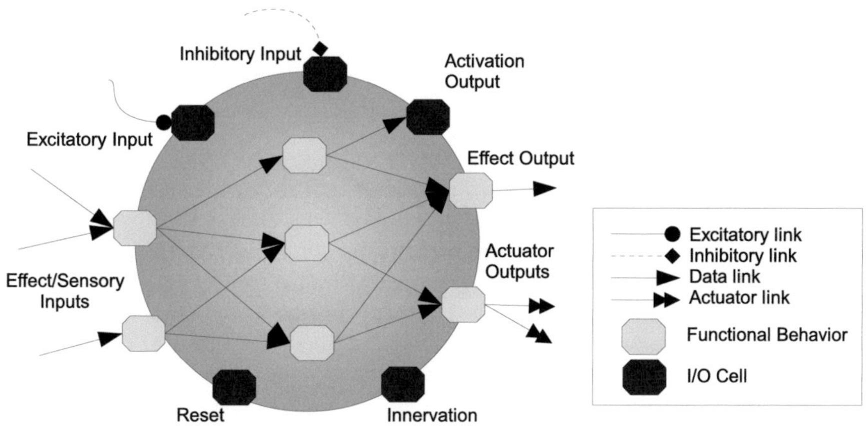

Fig. 2. Schematic diagram of a Behavior Network.

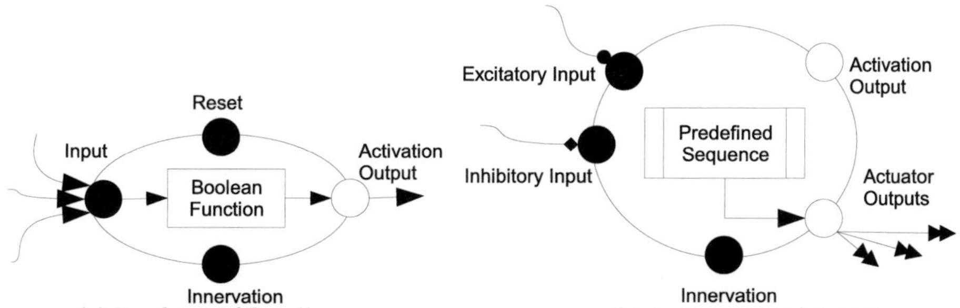

(a) Boolean I/O cell.

(b) Activation I/O Cell.

Fig. 3. Schematic diagrams of I/O cells.

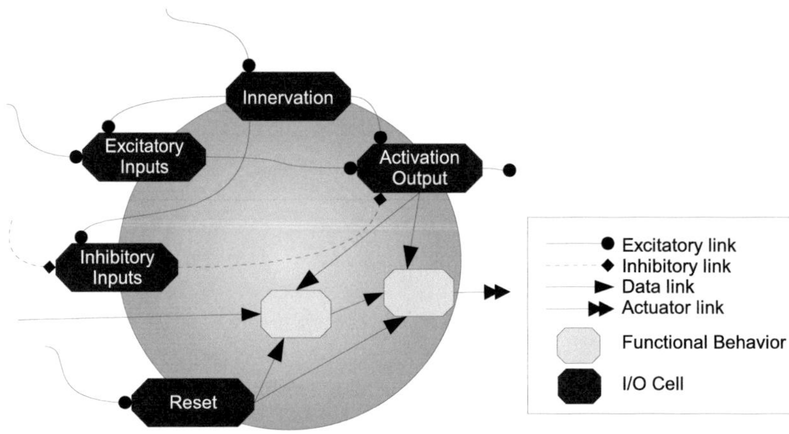

Fig. 4. Schematic diagram of I/O cell links in a Behavior Network.

However, the activation process of a Behavior Network differs from the one of a Behavior Cell. An Innervation cell is first innerved and it innervates the Excitatory Inputs cell, Inhibitory Inputs cell, and Activation Output cell in this order. If the value of Port-AO (activation output) of an Activation Output cell is true, it will innervate the functional behaviors in a predefined order otherwise the whole process will return; thus the Behavior Network will be deactivated (Figure 4).

3.2 Sensor system

In this research, we design and employ a sensor system which consists of three types of modules: a stationary ranging module, a motorized vision module, and a motion sensing module. The stationary ranging module is attached to the wheelchair and constantly measures distances to any objects that are within the scope of the sensors. The motorized vision module is designed to have the ability to pan/tilt and to acquire depth data as well as scenery images, so that the perceptual behaviors can utilize more detailed information. The motion sensing module consists of a gyroscope and accelerometers and continuously tracks the position, orientation, and velocity of the wheelchair.

3.2.1 Stationary ranging module

Employing ultrasonic range sensors (sonar sensors) for the stationary ranging module appears appropriate because sonar sensors commonly used in robotics research have an effective range (typically between a few centimeters and 6 m) that is suitable for indoor navigation. Sonar sensors are also inexpensive and easy to install. We chose to employ Devantech SRF08 because of its cost-perfomance and size efficiency. The specification of the SRF08 and its beam pattern are shown in Figure 5.[4] According to the specification, the ultrasound wave propagation profile (beam width) of an SRF08 is roughly 45°.

It is common to employ an array of sensors in order to obtain a wide coverage area around a mobile robot (specifically for omni-directional robots). However, the wheelchair we employed is kinetically non-holonomic; that is, although it could physically move laterally, it is not equipped with the mechanism to control such movement. Based on the kinetic

[4] Retrieved from http://www.acroname.com

property of the wheelchair and the beam pattern of an SRF08, we use six sonar sensors attached to the front, and two sonar sensors are placed to the rear, as illustrated in Figure 6a such that we have more coverage on the front side of the wheelchair. Figure 6b depicts the sonar installation at Front Right and Front-Side Right location.

The eight sonar sensors circumferentially attached to the wheelchair are coupled into four opposing pairs: Front Left/Rear Right, Front Right/Rear Left, Front-Side Right/Side Left, and Side Right/Front-Side Left. The sensor controller is implemented to fire pairs at an appropriate interval to minimize the crosstalk problem.

Sonar reading is a default task on the sensor controller, therefore the sensor reading cycle runs continuously while the power is on. On each reading cycle, the sonar readings are bundled into a data packet and sent to the host computer via a USB cable.

SRF08 main specification

Voltage	5 V
Current	15 mA (typical)
Frequency	40 kHz
Maximum Range	6 m
Minimum Range	3 cm

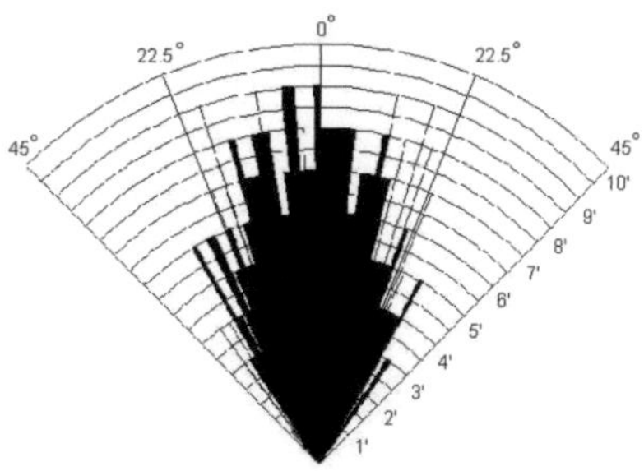

Fig. 5. Devantech SRF08 main specification (left) and its beam pattern (right).

3.2.2 Motorized vision module

Since typical ultrasonic sensors only measure the distance to the closest object within range, certain information (e.g. the precise location and orientation of a door-opening) is difficult to obtain from them. In contrast, a vision based ranging system, specifically a laser rangefinder system, is suitable to acquire such geometric information because of its accuracy in determining distance within a reasonable amount of time.

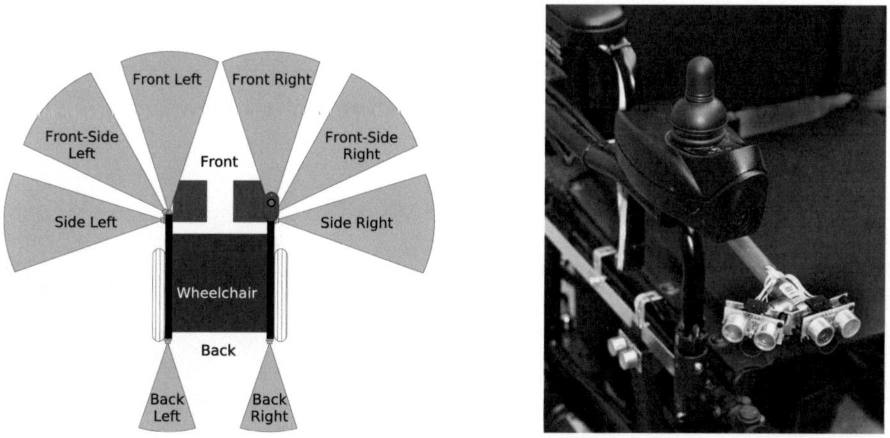

(a) Schematic diagram of installation of the stationary ranging modules.

(b) Sonar sensors installation on Front Right/Front-Side Right.

Fig. 6. Installation of the stationary ranging module.

A variety of commercial laser rangefinders are availble with ease of installation; however, there are a few disadvantages: Some have limited scanning range or area (viewing angles), some are too large to mount on wheelchairs, and all of the pre-made laser rangefinders are currently quite expensive.

An alternate way to resolve the problem is to create a custom laser rangefinder. With an inexpensive camera and a laser pointer, a simple laser rangefinder can be configured by utilizing the triangulation principle and image processing methodology (range imaging using structured lighting). Figure 7a illustrates that the coordinate (x, y) of the laser reflecting point P can be calculated from observed point a in the acquired image, the focal length (f), the baseline distance (b) between the camera center and the laser light source, and the projection angle θ by Formula 1. The minimum and maximum distances (depths) that can be measured by this method are illustrated as S_{min} and S_{max} respectively.

$$(x, y) = \frac{b}{f\cos\theta - a}(f, a) \tag{1}$$

Our approach is to employ a CCD camera (a webcam) as a low cost optical sensor in combination with a laser line generator emitting a horizontal laser line such that depth mapping is extended from a single point to a horizontal cross-sectional image in a view.

An advantage of using a custom-made laser rangefinder is that we are able to use the camera for other purposes as well. A few examples of computer vision techniques that can be useful for wheelchair operations include door frame detection, human-face detection, and textual information extraction. In this project, we introduce the doorway detection technique as a part of the perceptual behaviors (Section 4.1.6).

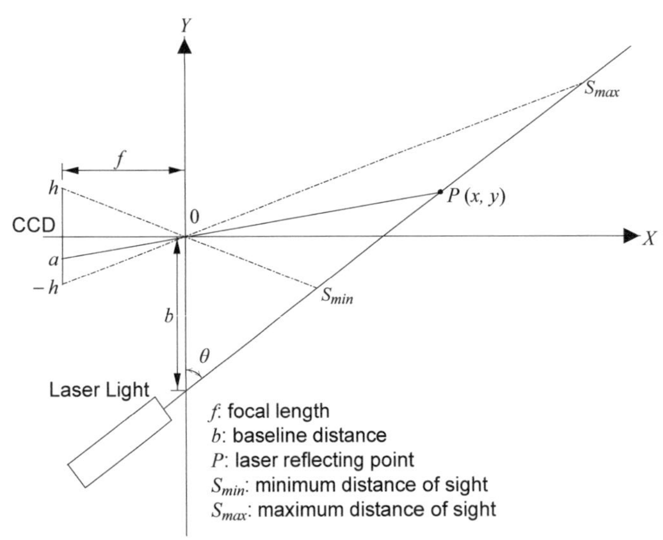

(a) Camera-centered triangulation geometry (adapted from Jain et al., 1995).

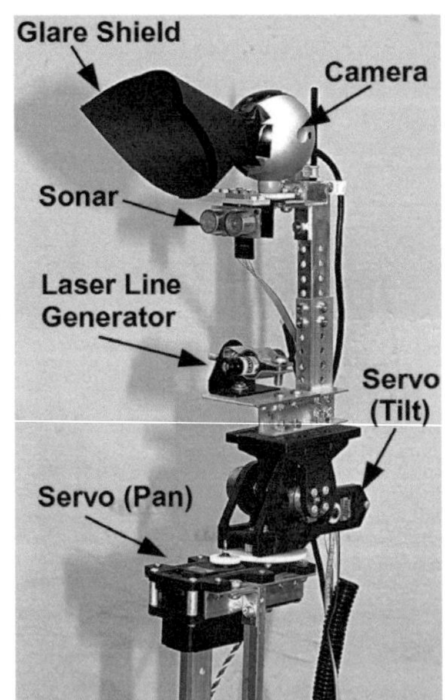

(b) The motorized vision module prototype.

Fig. 7. Camera geometry and implementation.

We tried a variety of optical filters in front of the camera lens to enhance response to the reflected 635 nm laser beam and rejection of unwanted ambient red light, while allowing the scenery to be visible to the blue and green pixels in the camera. We settled on a broadband nebula filter designed for astronomers (Celestron™ LPR-94126A, which is no longer made). This filter transmits two bands, 450–530 nm and 630 to beyond 700 nm. Other broadband nebula filters, seemingly similar, do not work because they do not transmit at 635 nm. Plans are being made to test a custom-made interference filter. One drawback of all interference filters is poor performance when light does not strike them orthogonally; thus the edges of the field of the wide-angle camera lens are bright and pinkish.

A typical inexpensive CCD camera generally has a limited field of view which is not enough to cover the surrounding environment of the wheelchair. To overcome this drawback, we designed the laser range finder to be motorized by mounting it on a pan/tilt servo device.

Figure 7b illustrates the prototype of the motorized vision module. It consists of an off-the-shelf webcam (Logitech QuickCam® Pro 4000), a laser line generator module (distributed by World Technology Inc.), a pan/tilt module (ServoCity™ SPG785 and SPT400), and a microcontroller (Rowley Associates Ltd. CrossFire® LPC-2138) that accepts signals from the host computer (PC) and controls the servo motors and the laser module. We also add a sonar sensor (Devantech SRF08) to make the module incorporate more redundancy by acquiring information from different sensory channels. A hardware specification list is shown in Table 2.

Hardware Equipment	Product name: Specification
Camera	Logitech Quickcam® Pro 4000 Optical sensor type: CCD Image capture resolution: 640 x 480 (Video) Frame rate: 15 fps (640 x 480) Effective focal length: 4.5 mm Field of view: 42° Interface: USB
Optical filter	Celestron™ LPR-94126A
Laser module	Laser line generator (Class II, output power: 10 mW, wavelength: 635 nm, fan angle: 45°)
Pan/tilt system	(Pan) SPG785 Gearbox: Hitech HS-785HB (0–420°) (Tilt) SPT400 Tilt System: Hitech HS-425BB (0–135°)
Microcontroller	Rowley Associates Ltd. CrossFire® LPC-2138 Evaluation kit

Table 2. Hardware specification of the motorized vision module.

3.2.3 Motion sensing module

An integrated dual-axis gyroscope (InvenSense, Inc. IDG-300) and an accelerometer breakout board (SparkFun™ Electronics) containing an ADXL-320 accelerometer (Analog Devices, Inc.) were employed for the motion sensing modules. These modules are connected to the microcontroller (LPC-2138) and obtain the acceleration along X and Y axes and the yaw-axis angular velocity of the wheelchair.

3.3 Tactile feedback module

Communication between the user and the SA wheelchair via a versatile and robust man-machine interface plays an important role in achieving successful transportation activities. For those who have visual impairment, the sensory channels available for communication are limited to audible or tactile perception.

At first glance, auditory communication such as voice navigation or warning beepers may seem to be the easiest way to relay information to users. However, making noise might be socially inappropriate in certain situations. The auditory signals could also be overwhelmed by strong background noise in the environment (such as a fire alarm). Furthermore, additional sound sources are undesirable with respect to preserving the user's resource of attention since people with severe visual impairment rely extensively on their auditory system to collect environmental information.

Human tactile perception is robust and suitable for multimodal sensing (van Erp & van Veen, 2004). While people with blindness already use their haptic sensory system (e.g. using a cane), the somatic sensory system of another area of skin can still be used. Since the user passively receives most information, an effective form of tactual information should comprise vibrations (*vibrotactile signals*).

The outputs from the perceptual behaviors should consist of various kinds of vibrotactile signals conveying information via human tactile perception. Previous work evaluating a variety of tactile feedback systems has shown substantial potential (Geldard, 1975; Gemperle et al., 2001; Tan & Pentland, 1997; Tan et al., 2000; Traylor & Tan, 2002; Zelek et al., 2003). For example, Tan et al. (2000) have derived several benefits by studying cutaneous sensory saltation in order to develop a haptic interface. The sensory saltation may provide directional information that requires an otherwise highly intuitive grasp. A vibrotactile display can be established by relatively simple hardware configurations, such as an array of tactors, and can be perceived on many body sites, such as the fingertips and the back (Cholewiak & Collins, 2000).

3.3.1 Vibrotactile glove

In our design, we conclude that the vibrotactile display should reasonably be located on the hand because we have defined that our users can operate the SA wheelchair with little or no problems using their hand. In contrast, the fingertips would not be a convenient place for using a vibrotactile display because many individuals with severe visual impairment already use their hands to control a joystick, hold a cane, or read Braille signs using their fingertips. Further, a display mounted on the back seat would not be ideal because the users would frequently need to adjust themselves to put their back against the seat.

Although the palm side of the hand is much more sensitive in tactile perception than the back side of the hand, volar flexion during the joystick operations makes it difficult to design a robust vibrotactile display which guarantees consistent pressure of a vibrotactor to the skin. On the other hand, the back side of the hand maintains its flatness better during joystick operation. Therefore, we chose the back side of the hand as a locus of the vibrotactor.

Since our purpose is to convey translated visuospatial information to the user, an array of vibrotactors should obviously be employed rather than a single vibrotactor. The choice of an appropriate size of array depends upon the area of the locus, the size of each vibrotactor, and the tactile sensitivity of the corresponding cutaneous area.

Naturally, a suitable form of carrying the vibrotactile display is a glove (*Vibrotactile Glove*). The Vibrotactile Glove consists of the vibrotactile array inside the glove so as to face the back side of the hand and a motor-array controller, which generates and sends signals to the vibrotactors.

3.3.2 Vibrotactile array

Figure 8a shows the vibrotactile array which consists of vibrotactors, each of which is directly mapped to correspond to the ultrasonic sensors as shown in Figure 6a. We used an ordinary baseball glove and cut it on the little-finger side to make it easier to install the vibrotactor array. Each vibrotactor contains a miniature vibrating electric disk motor attached with a connector pin (Figure 8b) and is attached inside of the glove using Velcro®. The vibrating motor is thin, light-weight, and characterized by an off-centered weight causing lateral vibration when the specified voltage (3 V) is applied. It is also fully self-contained in a housing shell, thus no moving parts are visible. A vibrotactor generates a vibrotactile stimulus consisting of a rapid vibration lateral to the skin surface. Although the terminals of the vibrotactors are fully sealed by epoxy adhesive coating, finger cots are attached on the surface that make contact with human skin in order to maintain constant contact pressure between the vibrotactors and the skin surface and to minimize the effects of perspiration, body oils, hand cream, and so forth.

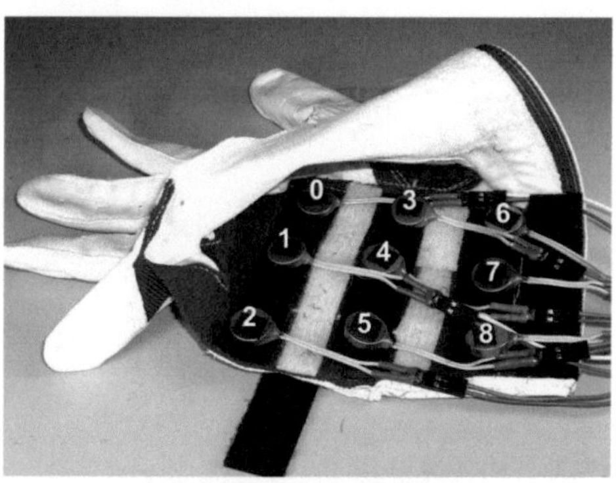

(a) The layout of the vibrotactor array on the back side of the glove.

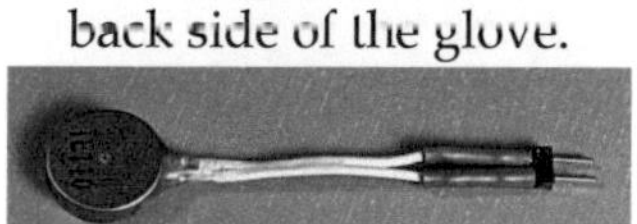

(b) The vibrotactor consisting of a vibrating disk motor and connector pin.

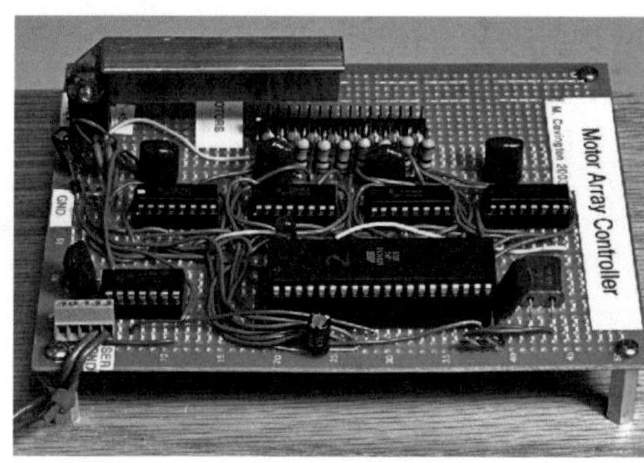

(c) The prototype of the motor-array controller.

Fig. 8. The Vibrotactile Glove.

3.3.3 Motor-array controller

We implemented a motor-array controller on a single-board computer to control an array of vibrating disk motors. The controller receives control signals from a PC by a serial port and executes various patterns of vibration in an array of up to 16 motors. Each time the motor-array controller receives a control signal from the host computer, a vibration pattern is generated by an assembly-language program running on a CPU of the controller.

The prototype of the motor-array controller is shown in Figure 8c. The CPU is an Atmel AT89C52, a flash-memory derivative of the Intel 8051 (MCS/51) with the same instruction set. Serial input is received through an MC1489A or equivalent RS-232 line receiver. Two of the CPU's output ports (P0 and P1) are coupled to the motors through SN754410 motor control ICs. The prototype controller is much bulkier than necessary; it was built large to facilitate experimentation. A much smaller version has been designed.

3.4 System architecture

An overview of the system architecture is depicted in Figure 9a. It illusrates a layered structure which constructs the hardware layer (camera, servos, laser module, and the Vibrotactile Glove), the microcontroller unit (MCU) layer for controlling sensors and the Vibrotactile motor array, the communication layer, and the application layer. Both the camera and the sensor controller are connected to the host computer with USB cables. The motor-array controller receives control codes from the host computer via a serial port. In the diagram, the thin arrows represent the flow of data, and the thick arrows represent the flow of actuator commands. The overview of our SA wheelchair prototype is shown in Figure 9b.

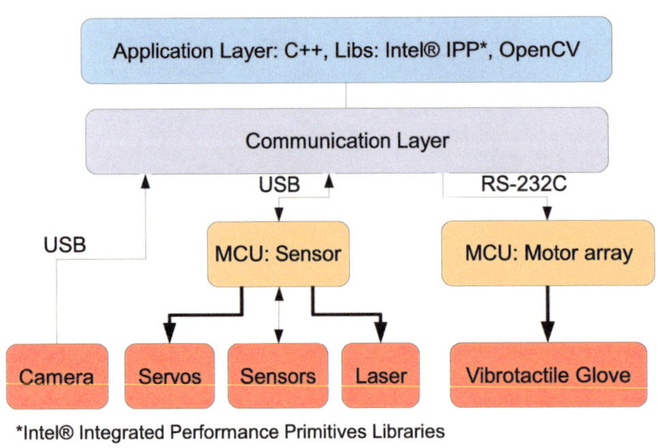

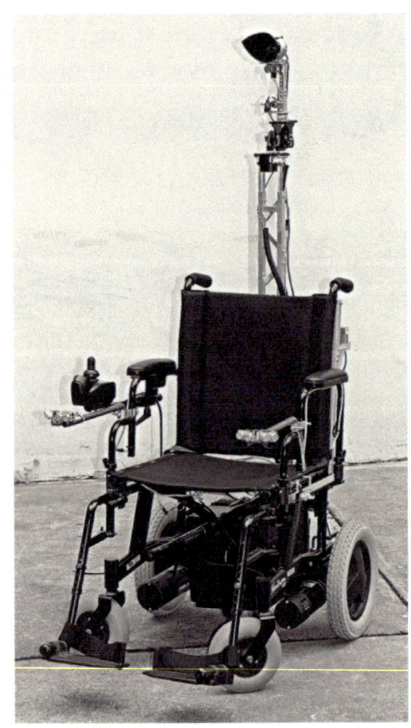

Fig. 9. System architecture (left) and the SA wheelchair prototype (right).

4. Implementation

4.1 Behavior implementation

The essential goal of the perceptual behaviors is to provide navigational and/or warning guidance to the user through a tactile interface. The perceptual behaviors are designed for indoor usage with partial domain knowledge, such as destination and portal information. A set of typical specifications of the goals/tasks that should be performed by the wheelchair-navigation agent are as follows:

> While roaming without any specific instructions from the user, the wheelchair-navigation agent should mainly (1) notify of obstacles; however, under crowded circumstances, (2) navigate the orientation of an available free space to the user. Simultaneously (3) search for a landmark (e.g., a doorway) from the image, and if a doorway is found, (4) assist the user to pass through it.

From the user's perspective, the first specification, notifying obstacles, simply constructs a perceptual support, almost direct translation from ranging information into a vibrotactile representation. This specification, however, will almost certainly fail to serve the purpose when the number of obstacles increases because of the *masking effect*[5] (Verrillo, 1983). In addition, it is far more effective to navigate the user to an available space (orientation) when in a crowd or a narrow corridor.

The second specification, navigating a free-space, serves in a complementary manner to notifying of obstacles since if there are many free areas available, it is also difficult for the wheelchair-navigation agent to determine the best orientation to navigate. By these specifications, the user will be either notified of obstacles or navigated to the recommended orientation one way or another.

The third specification, searching for a landmark, although it needs to be concurrently performed with other specifications, is not brought to the user's attention until a landmark is found (e.g., a doorway). This specification mainly relies on sensory input from the motorized vision module and machine/computer vision techniques performed on the host computer in order to accomplish more complicated tasks.

Once the doorway is confirmed as a valid portal to go through, the fourth specification is designed to perform a plan-like navigation task from the current location to the destination (the other side of the door) based on the acquired geometric information and the result of the user output (wheelchair movement) while taking into account dynamic changes in the environment such as obstalces.

To accomplish these tasks, the perceptual/navigational behaviors of interest in our project are constucted: namely, two reactive behaviors, `Obstacle notification` and `Free-space finding`; a task-oriented behavior, `Doorway navigation`; a command arbitration behavior, `Navigation command manager`; and `Sensor command manager` which manages the commands for the sensor-related effectors (servo motors and the laser line generator).

Figure 10 illustrates the schematic diagram of the perceptual/navigational behaviors accompanied with the sensor-effector systems. The thin arrows represent data links, and the thick arrows represent actuator links. Other port-to-port links are not shown in this diagram.

4.1.1 Obstacle notification

The `Obstacle notification` behavior is a reactive behavior which takes sensor readings from the ultrasonic sensors and directly maps into vibrotactile representations. It consists of two subordinate behaviors: `Sonar reader` and `Obstacle detection`. The `Sonar`

---

[5] When multiple vibrotactile stimuli simultaneously present to the skin, one stimulus decreases the detectability of another.

`reader` behavior is a component-like behavior which reads sensor data from the micro-controller. The `Obstacle detection` behavior interprets the sensor data into a vibrotactile representation which expresses the distance of an obstacle within the scope of each sensor.

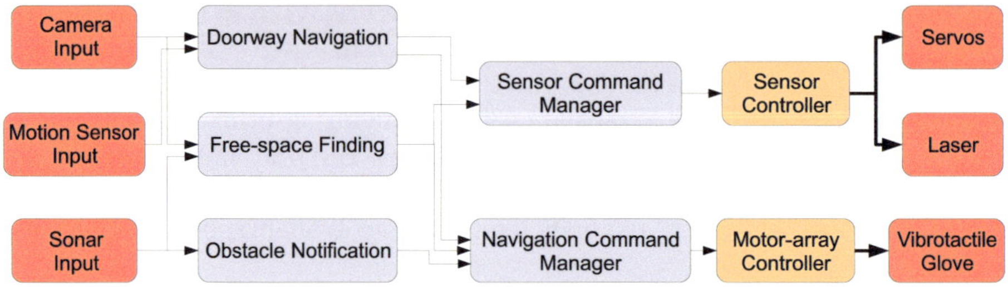

Fig. 10. Schematic diagram of the perceptual and navigational behaviors (thin arrows: data links, thick arrows: actuator links).

4.1.2 Free-space finding

The `Free-space` finding behavior is, by default, another reactive behavior relating to the sensor readings, but its tasks is different from the `Obstacle notification` behavior. Instead of notifying of obstacles, it reports a band of orientation in which no immediate obstacles are found. This behavior is particularly effective to find a way out when the SA wheelchair is surrounded by many obstacles. It consists of `Sonar reader`, `Way finder`, and `Range finder` behaviors. It primarily relies on sonar readings; however, if the sonar readings do not provide enough information, it invokes the `Range finder` behavior to acquire more accurate depth information.

4.1.3 Doorway navigation

The task of passing through a doorway, in general, comprises several subtasks addressing dynamic determination of a maneuvering trajectory and decision making which relies on high level cognitive processes. Some previous studies demonstrate a creation of a local map in order to generate a maneuvering trajectory for an autonomous vehicle (Patel et al., 2002; Surmann et al., 2003). Such maneuvering trajectories may shape a complicated curvature which may not be appropriate to represent to the user via a sequence of tactile signals. We propose a simple guidance approach that utilizes the *Pivotal Point/Zone*, an intermediate goal point (area) from which the wheelchair can straightforwardly move toward the doorway.

As illustrated in Figure 11, the pivotal trajectory is composed of a series of simplified trajectories (the solid arrows) and a sharp turning point instead of a continuous curved trajectory (the dotted arrow) representing a maneuvering trajectory for manual driving. Even though the pivotal trajectory is not an optimal path, the navigational signals for the pivotal trajectory remain simple and intuitive, such as "go front-left," "swivel clockwise," and "go forward."

Figure 12 illustrates the `Doorway navigation` behavior which comprises the following behaviors: `Doorway detection`, `Range finder`, `Door validation`, `Object tracker`, and `Path planner`. The `Doorway detection` behavior searches for a doorframe from the image stream, and if it finds a candidate, the `Range finder` and `Door validation` behaviors are invoked to confirm the validation of the candidate. Once the doorframe is confirmed, the `Object tracker` behavior traces the door in order to fixate the camera to the doorframe. In the mean time, the `Path planner` behavior is invoked to undertake a repetitive process of localizing the wheelchair, sending a signal indicating the desired orientation to the user, and adjusting the path plan, until the wheelchair reaches the sub-goal (pivotal point). At the pivotal point, the user is guided to swivel until the wheelchair straightforwardly faces the doorway and then to move forward.

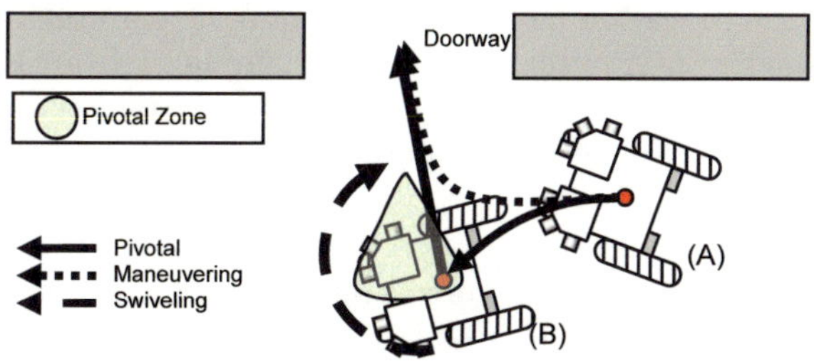

Fig. 11. When the wheelchair-navigation agent detects a doorway at position (A), it first navigates the user to the Pivotal Zone, position (B). At the Pivotal Zone, the agent alerts the user to swivel (clockwise in this example) until the correct orientation appears toward the center of the entrance, and then to move toward the entrance.

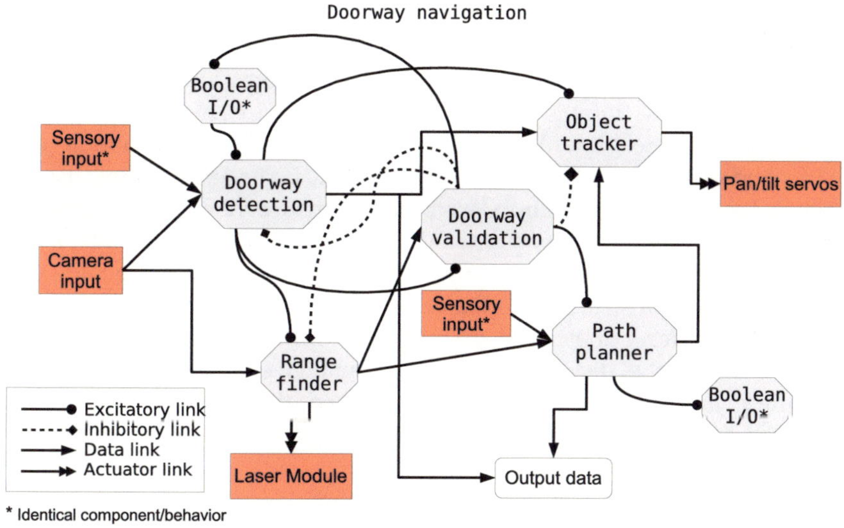

Fig. 12. Schematic diagram of `Doorway navigation` Behavior Network.

4.1.4 Sensor command manager

The `Sensor command manager` behavior accepts input commands from the `Doorway navigation` and `Free-space finding` behaviors, more specifically, the `Object tracker` and `Range finder` behaviors in the `Doorway navigation` behavior, and the `Way finder` and `Range finder` behaviors in the `Free-space finding` behavior. When the `Object tracker` and `Range finder` behaviors in the `Doorway navigation` behavior are active, a request of the `Range finder` behavior in the `Free-space finding` behavior to move servo motors is suppressed.

4.1.5 Navigation command manager

The `Navigation command manager` behavior acceptes navigation commands from the `Doorway navigation`, `Free-space finding`, and `Obstacle notification` behaviors, and arbitrates the commmands. By default, commands from the `Obstacle notification` behavior are sent to the motor-array controller. When more than two immediate obstacles are detected, the commands from the `Free-space finding` behavior will be sent. If the `Doorway navigation` behavior issues a command, a command the from `Obstacle notification` behavior is suppressed and the command from `Free-space finding` will be fused with the command from the `Doorway navigation` behavior. The details of the command fusion strategy are not discussed in this chapter.

4.1.6 Doorway detection

The `Doorway detection` behavior constitutes a computer-vision based algorithm to classify doorways. Doorways are classified based on the presence of the door jambs and the lintel. These are represented in the image as two strong vertical lines connected by a horizontal line. First, the image from the camera is converted to greyscale and reduced to half its original size. Edges are extracted by using a custom convolution filter which combines the accurate edge localization of the Laplacian kernel with the directional bias of a Sobel filter to improve detection of vertical lines.

The detection of vertical lines is achieved by applying the Probabilistic Hough Transform (Kiryati et al., 1991) to the edge image. The technique searches a proportion α ($5\ \% < \alpha < 15\ \%$) of randomly selected edge pixels in the image to detect line segments. The vertical line segments are passed to an algorithm that merges line segments deemed close enough. The search for the lintel is limited to the region between the top of the image and the top of the lowest vertical line. Within this sub-region, edge extraction is performed by means of a second-order partial derivative. Horizontal line segments are also detected by the Probabilistic Hough Transformation, examined to ensure they are sufficiently horizontal, and then passed to the merging algorithm. After both sets of line segments are collected, a matching algorithm is performed to determine combinations of two vertical lines connected by a horizontal line that construct a doorframe.

If the matching criteria for a doorframe are met, the `Range finder` behavior (Section 4.1.7) is invoked to measure the distance to the door which enables a validation process to increase accuracy and reduce false positives. Once the doorframe is validated, the dimensions of the door candidate are obtained and sent to the linked behaviors.

4.1.7 Range finder

The `Range finder` behavior also constitutes a computer-vision based algorithm to detect a laser line. The laser detection algorithm seeks each laser point per column while eliminating noise and erroneous detection. First, an acquired image is decomposed into its constituent RGB color channels, and an intermediate image is created by subtracting the green channel from the red. The algorithm then seeks a candidate pixel that represents the laser point by means of a second-order differential for each column.

The 8 neighboring pixels of each candidate are examined to eliminate single outlier pixels. After outlier points are eliminated, the resulting points are collapsed into a one-dimensional array representing the height in the original image where each point was found. Based on camera calibrations, these points are then mapped into distances and sent to the linked behaviors.

In both `Doorway detection` and `Range Finder` behaviors, we utilized an open source computer vision library, OpenCV, and Intel® Integrated Performance Primitives for real-time image processing.

4.2 Vibrotactile signal pattern

All the necessary information to operate the SA wheelchair must be symbolized as a set of signals, so that the wheelchair-navigaiton agent is able to transmit the signals to the vibrotactile display to generate vibrotactile signal patterns. The vibrotactile signal patterns for the SA wheelchair operation are categorized into three groups: warning signal, spatial representation, and directional guidance.

Warning Signal When an emergency situation occurs in which the SA wheelchair determines that it is too dangerous to drive the wheelchair or finds any sort of system malfunctioning, the warning signal is invoked, and the user must stop the wheelchair movement.

Spatial Representation The spatial representation is translated from visuospatial information to help the user to "sense" the orientation and distance of an obstacle.

Directional Guidance The directional guidances are designed to navigate the wheelchair user in the desired direction (e.g., "go-forward", "turn-left").

The vibrotactile display consists of nine vibrating motors. A vibrating motor generates an individual pulse and its characteristic is determined by controlling the duration of pulses and the interpulse interval. A sequential pulse consists of a sequence of individual pulses generated by more than one vibrating motor that indicates the desired direction.

4.2.1 Warning signal

The warning message does not need to contain any visuospatial information, thus, we designed all vibrotactors to repeat a long duration pulse simultaneously.

4.2.2 Spatial representation

The orientation and distance of an obstacle needs to be discretely mapped into the vibrotactile display due to its limited spatial resolution. The orientation of an object is mapped into the locus of the corresponding peripheral vibrotactors, each of which represents approximately 45 degrees of the surrounding area of the chair. The degree of distance is set into three levels: "near" is the range from the chair to approximately 0.6 m; "medium," 0.6 to approximately 1.2 m; and "far," 1.2 to approximately 2 m. These thresholds are primarily experimental, and will be adjusted during the field test.

The magnitude of each level is implemented by the density of the vibration pulse. Repetition of a single short vibration pulse represents the "far" range; three short pulses, the "medium" range; and a long pulse, the "near" range (Figure 13a).

As briefly mentioned in Section 4.1, the presence of multiple stimuli in vibrotactile perception can cause the masking effect which decreases the detectability of stimuli. In our primarily experiment, this phenomenon also occured. When two stimuli are simultaneously presented, one cannot distinguish the loci of stimuli if they are either different patterns or presented in nearby loci. When more than two stimuli are simultaneously presented, one cannot distinguish the loci of stimuli regardless of their closeness or patterns.

However, limiting to only a single obstacle to be presented is undesirable. To solve this dilemma by compromise, we configured the signal generation mechanism such that one repetitive cycle contains multiple phases, and a unique stimulus is permitted to present in each phase. Figure 13b illustrates an example of a signal pattern to notify of two obstacles at the north-east and west of the wheelchair. More phases per cycle are possible; however, the duration of one cycle increases as more phases are included, which would make the display "slow" (typically one phase takes approximately one second).

4.2.3 Directional guidance

The directional guidance behavior is designed to generate successively activating vibrotactors with a short interstimulus period. The center vibrotactor is first invoked with two short pulses followed by another two short pulses in the desired orientation. We set the number of short pulses to two in order to avoid confusion with spatial representation.
Figure 14 shows a sample pattern of the direction "go-forward-right" (or "north-east") by sing two tactors: tactor 4 and tactor 2, in sequence. Applying the carefully chosen interstimuli intervals between the tactors (20 to 200 ms (Geldard, 1985)), it is suggested that the user would feel on their hand a single sensation in the designated direction rather than two separate stimuli. In our preminarly experiment, we found that the appropriate interstimulus interval for the Vibrotactile Glove is between 10 and 30 ms.

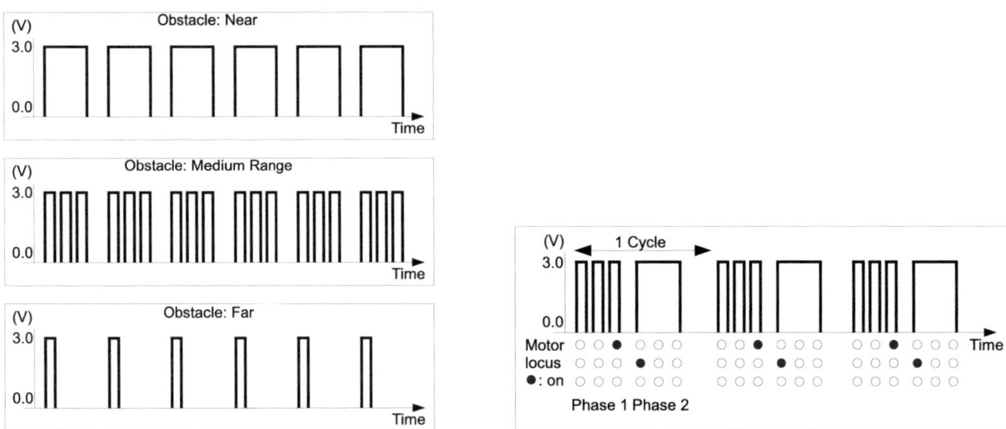

(a) Examples of signal patterns for notifying of an obstacle in different ranges (top: obstacle is near; middle: obstacle is in medium range; bottom: obstacle is far).

(b) An example of a signal pattern for notifying of two obstacles (north-east: medium range, west: near).

Fig. 13. Schematic diagrams of obstacle signal patterns.

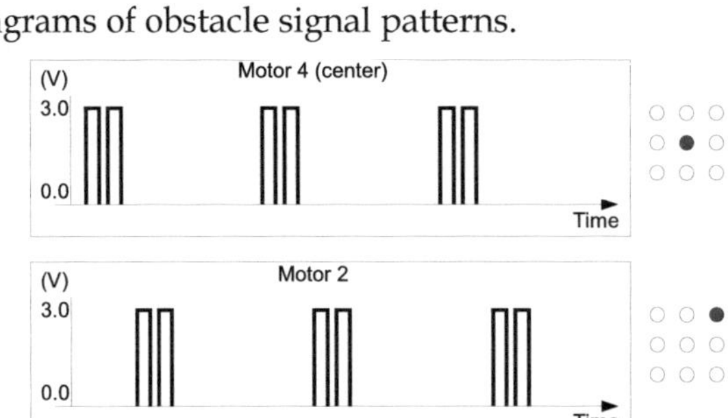

Fig. 14. An example of the pulse pattern to indicate direction ("north-east"). The dark circle in the array represents the activated tactor.

5. Results and current status

Currently we are in the middle of completing the implementation of the behaviors, and a field test is underway. Some results of the behaviors utilizing the computer-vision techniques (`Doorway detection` and `Range finder`) are presented.

As shown in Figure 15, the `Door detection` behavior successfully detected the two door frames from a captured image in the left pane. The top-right pane shows the result of the line detection algorithm; the number of vertical lines is 10; and horizontal lines, 7. In the bottom-right pane, two candidate door frames are identified by the line matching algorithm and marked as solid rectangles.

The results of the `Range finder` behavior are shown in Figure 16 and Figure 17. Figure 16a is a captured image at the same site where the `Doorway detection` behavior detected the two door frames. The laser reflecting points are successfully identified as long as the laser beam is not disturbed by the strong ambient light sources (Figure 16b). Since the door is closed, only a small amount of discrepancies of the laser reflecting points are observed on the candidate door (the right door).

Fig. 15. A sample result of the `Doorway detection` behavior : (left) a captured image; (top-right) detected vertical lines (n = 10) and horizontal lines (n = 7); (bottom-right) candidate door frames (solid rectangles).

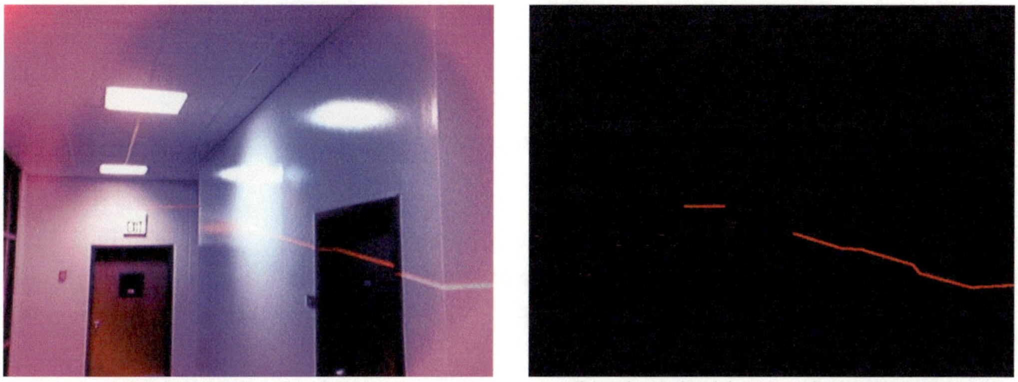

(a) A captured image. (b) Identified laser reflecting points.

Fig. 16. A result of the `Range finder` behavior (door closed).

Figure 17a is a captured image at the same site, but with the door open. In this occasion, a large amount of discrepancies on the laser reflecting points are observed (Figure 17b); thus, the right door frame in Figure 15 is identified as a door opening. Those laser reflecting points in the image will also be mapped into a set of distance and orientation data which will be used to construct a geometric profile around the wheelchair by the following `Path planner` behavior.

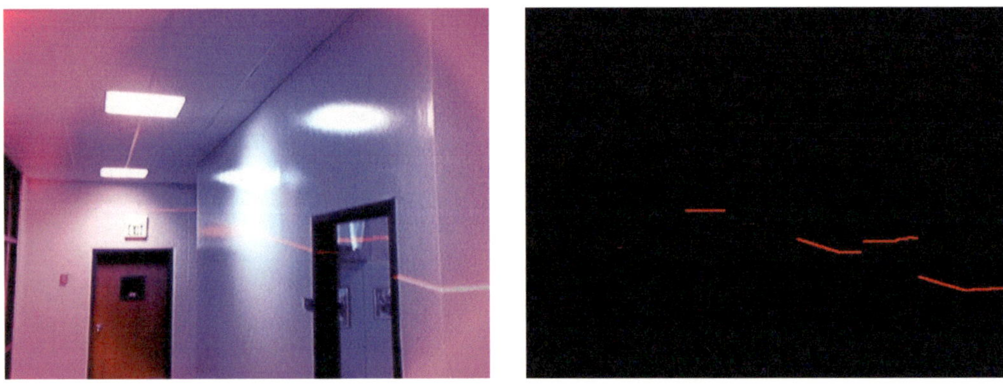

(a) A captured image. (b) Identified laser reflecting points.

Fig. 17. A result of the `Range finder` behavior (door opened).

6. Conclusion

In this chapter, we have presented our SA wheelchair prototype which consists of a stationary ranging module, a motorized vision module, and a tactile feedback module. The goal of our SA wheelchair project is to provide perceptual and navigational capabilities to wheelchairbound users with severe vision impairment.

Our SA wheelchair is particularly characterized by the two modules: the motorized vision module and the tactile feedback module. The motorized vision module consists of lowcost components but has exhibited versatile capabilities including laser ranging and other computer-vision techniques such as doorway detection. We have presented a novel tactile feedback module, a Vibrotactile Glove, which consists of an array of vibrotactors and conveys various information such as presence of obstacles and recommended orientation to the users without interference of their auditory perception.

We have proposed a custom BBC architecture consisting of Behavior Cells and Behavior Networks components to address the problems when the users with vision impairment need to operate wheelchairs. In this chapter, we have presented our schemes for the user to avoid obstacles and to go through a narrow doorway. However, these are only a small subset of the problems that exist, and there are a great number of problems challenging researchers. Particularly, drop-off avoidance, distinction between humans and stationary objects, and local mapping capability can be the next topics of interest.

Finally, we are hoping to continue to study this challenging and interesting research toward the goal of improving the quality of life of the users in the blind community.

7. References

Argyros, A. A., Georgiadis, P., Trahanias, P., & Tsakiris, D. P. (2002). Semi-autonomous navigation of a robotic wheelchair. *Journal of Intelligent and Robotic Systems*, 34, 315–329.

Arkin, R. C. (1998). *Behavior-Based Robots*. Cambridge, Massachusetts: The MIT Press.

Borgolte, U., Hoyer, H., Bühler, C., Heck, H., & Hoelper, R. (1998). Architectural concepts of a semi-autonomous wheelchair. *Journal of Intelligent and Robotic Systems*, 22(3), 233–253.

Brooks, R. A. (1991). Integrated Systems Based on Behaviors. *SIGART Bulletin*, 2(4), 46–50.

Cholewiak, R. W., & Collins, A. A. (2000). The generation of vibrotactile patterns on a linear array: Influences of body site, space, and time. *Perception & Psychophysics*, 62(6), 1220–1235.

Crisman, J. D., & Cleary, M. E. (1998). Progress on the deictically controlled wheelchair. In V. O. Mittal, H. A. Yanco, J. M. Aronis, & R. C. Simpson (Eds.) *Assistive Technology and Artificial Intelligence, Applications in Robotics, User Interfaces and Natural Language Processing*, vol. 1458 of *Lecture Notes in Computer Science*, (pp. 137–149). New York: Springer.

Douglass, J., & Ryan, M. (1987). A pre-school severely disabled boy and his powered wheelchair: A case study. *Child Care, Health Development*, 13, 303–309.

Geldard, F. A. (1975). *Sensory Saltation: Metastability in the Perceptual World*. Hillsdale, New Jersey: Lawrence Erlbaum Associates.

Geldard, F. A. (1985). The mutability of time and space on the skin. *Journal of the Acoustical Society of America*, 77, 233–237.

Gemperle, F., Ota, N., & Siewiorek, D. P. (2001). Design of a wearable tactile display. In *Fifth International Symposium on Wearable Computers* (ISWC '01), (pp. 5–12). IEEE Computer Society.

Gomi, T., & Griffith, A. (1998). Developing intelligent wheelchairs for the handicapped. In V. O. Mittal, H. A. Yanco, J. M. Aronis, & R. C. Simpson (Eds.) *Assistive Technology and Artificial Intelligence, Applications in Robotics, User Interfaces and Natural Language Processing*, vol. 1458 of *Lecture Notes in Computer Science*, (pp. 150–178). New York: Springer.

Jain, R. C., Kasturi, R., & Schunck, B. G. (1995). *Machine Vision*. McGraw-Hill.

Kiryati, N., Eldar, Y., & Bruckstein, A. (1991). A probabilistic Hough transform. *Pattern Recognition*, 24, 303–316.

Levine, S. P., Bell, D. A., Jaros, L. A., Simpson, R. C., Koren, Y., Member, S., & Borenstein, J. (1999). The NavChair assistive wheelchair navigation system. *IEEE Transactions on Rehabilitation Engineering*, 7, 443–451.

Matarić, M. J. (1992). Behavior-based control: Main properties and implications. In *IEEE International Conference on Robotics and Automation, Workshop on Architectures for Intelligent Control Systems*, (pp. 46–54). Nice, France.

Miller, D. P., & Slack, M. G. (1995). Design and testing of a low-cost robotic wheelchair prototype. *Autonomous Robots*, 2(1), 77–88.

Nicolescu, M. N., & Matarić, M. J. (2002). A hierarchical architecture for behavior-based robots. In M. Gini, T. Ishida, C. Castelfranchi, & W. L. Johnson (Eds.) *Proceedings of the First International Joint Conference on Autonomous Agents and Multiagent Systems (AAMAS'02)*. ACM Press.

Odor, J., & Watson, M. (1994). Learning through smart wheelchairs: A formative evaluation of the effective use of the call centre's smart wheelchairs as part of children's emerging mobility, communication, education and personal development. In *Final Report to The Nuffield Foundation and the Scottish Office Education Department.*

Patel, S., Jung, S.-H., Ostrowski, J. P., Rao, R., & Taylor, C. J. (2002). Sensor based door navigation for a nonholonomic vehicle. In *IEEE International Conference on Robotics & Automation*, (pp. 3081-3086). Washington, DC.

Paulsson, K., & Christoffersen, M. (1989). Psychosocial aspects of technical aids - how does independent mobility affect the psychosocial and intellectual development of children with physical difficulties? In *Proc. 2nd Int. Conf. of the Rehabilitation Engineering Society of North America, Ottawa*, (pp. 282–285).

Rönnbäck, S., Piekkari, J., Hyyppä, K., Berglund, T., & Koskinen, S. (2006). A semi-autonomous wheelchair towards user-centered design. In K. Miesenberger, J. Klaus, W. L. Zagler, & A. I. Karshmer (Eds.) *ICCHP*, vol. 4061 of *Lecture Notes in Computer Science*, (pp. 701–708). Springer.

Simpson, R., LoPresti, E., Hayashi, S., Guo, S., Ding, D., Ammer,W., Sharma, V., & Cooper, R. (2005). A prototype power assist wheelchair that provides for obstacle detection and avoidance for those with visual impairments. *Journal of NeuroEngineering and Rehabilitation*, 2(1), 30.

Surmann, H., Nüchter, A., & Hertzberg, J. (2003). An autonomous mobile robot with a 3D laser range finder for 3D exploration and digitalization of indoor environments. *Robotics and Autonomous Systems*, 45(3-4), 181–198.

Tan, H. Z., Lim, A., & Traylor, R. (2000). A psychophysical study of sensory saltation with an open response paradigm. In *Ninth (9th) International Symposium on Haptic Interfaces for Virtual Environment and Teleoperator Systems.*

Tan, H. Z., & Pentland, A. (1997). Tactual displays for wearable computing. In *The International Symposium on Wearable Computers*, (pp. 84–89).

Traylor, R., & Tan, H. Z. (2002). Development of a wearable haptic display for situation awareness in altered-gravity environment: Some initial findings. In *Symposium on Haptic Interfaces for Virtual Environment and Teleoperator Systems*, (pp. 159–164).

United Nations (2005). *World Robotics 2005*. Geneva: United Nations.

van Erp, J. B., & van Veen, H. A. (2004). Vibrotactile in-vehicle navigation system. *Transportation Research Part F: Traffic Psychology and Behaviour*, 7(Issues 4-5), 247–256.

Verburg, G., Balfour, L., Snell, E., & Naumann, S. (1991). Mobility training in the home and school environment for persons with developmental delay. In *Final Report to Ontario Mental Health Foundation and Ministry of Community and Social Services' Research and Program Evaluation Unit.*

Verrillo, R. T. (1983). Vibrotactile masking: Effects of one- and two-site stimulation. *Perception and Psychophysics*, 33(4), 379–387.

Yanco, H. A. (1998). Wheelesley: A robotic wheelchair system: Indoor navigation and user interface. *Lecture Notes in Computer Science*, 1458, 256.

Zelek, J. S., Bromley, S., Asmar, D., & Thompson, D. (2003). A haptic glove as a tactilevision sensory substitution for wayfinding. *Journal of Visual Impairment & Blindness*, 97(10), 1–24.

6

Intelligent Robot Human Interface using Cheek Movement for Severely Handicapped Persons

Yoshihiko Takahashi and Shinichiro Suzukawa
Department of System Design Engineering / Vehicle System Engineering
Kanagawa Institute of Technology
Japan

1. Introduction

Independent support is a major concern for individuals with severe handicaps. Welfare assist robot technology is a popular solution to this concern (Hashino, 1996; Takahashi, 2004). Assist robots offer a future substitute for human nurses and helpers. A significant advantage to using such robots is that patients are able to give commands to robots without the same concern of inconveniencing their human helpers. It is therefore very important to develop interfaces from which severely handicapped individuals may operate highly intelligent robots.

A number of human to robot interfaces have already been developed for severely handicapped individuals. These include an interface which uses voice commands to assist in eating (Tateno et al., 1993), an interface to move a user's upper-limbs (Homma and Arai, 1997), an interface with a joystick also for an eating assist robot (O'Connel et al., 1993; M.T.BA Cert. Ed. et al., 1993), an interface using a laser pointer and a joy stick was developed for bedridden users (Takahashi et al., 2002A), and a head space pointer, called the HeadMasterPlus(TM), developed by Prentke Romich International Ltd.

The HeadMasterPlus(TM) allows users to control a mouse pointer on a computer monitor by moving their head and click the mouse by using a puffing switch. This interface was utilized for quadriplegic individuals (Ito, 2002). More recently, an eating assist robot was developed using this head space pointer design (Takahashi et al., 2002B).

The objective of this paper is to provide an easy human interface to operate an eating assist robot for severely handicapped users, namely quadriplegic users. An interface which uses a joystick creates difficulty in operating an eating assist robot because a user is required to move the joystick with his/her chin while eating. An interface using the voice commands has similar difficulties. It is also difficult to conduct more sensitive motion control of an assist robot when using chin movements to control. A head space pointer greatly reduces these difficulties, however still requires a puffing switch to click the mouse pointer.

This paper proposes a human to robot interface which utilizes the head space pointer concept to move the mouse pointer, however instead of the puff switch, a newly developed switch using a fiber sensor to click the mouse pointer. The fiber sensor detects the user's cheek movements, so that a user is able to click simply by swelling his/her cheek. It is more accessible for a user to swell his/her cheek during eating than to utilize a puffer switch.

This paper will present at first the concept of the proposed human to robot interface, and follow with the detailed construction, experimental results, and application of this eating assist robot. The eating assist robot used in our experiments is comprised of a three dimensional manipulator with a spoon hand, and computer display. The experiments were conducted in cooperation with a quadriplegic patient.

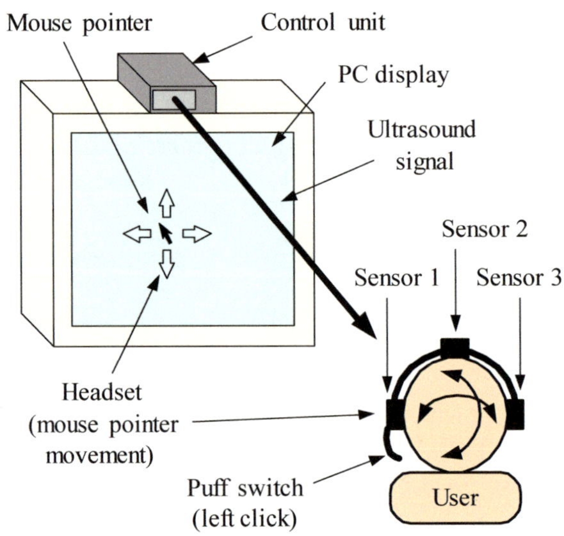

Fig.1. Head space pointer

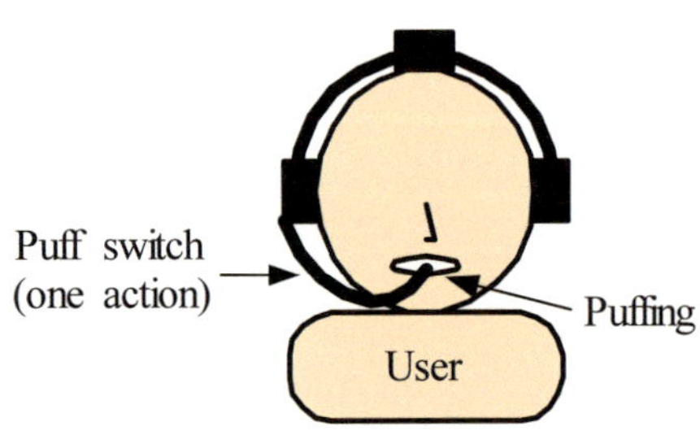

Fig.2. Puff switch of head space pointer

2. Concept of proposed human interface

Quadriplegic patients are assumed to be the target users. Fig.1 illustrates the configuration of a purchased head space pointer, where the control unit of the head space pointer is set on and connected to a computer. The user wears the controller headset. The control unit outputs an ultrasonic signal, and detects the movement of the headset which allows a user to move a mouse pointer on the computer display. The head space pointer uses a tube, or a puff switch (shown in Fig.2) to puff. Puffing on the tube allows a user to click operation switches (software control buttons) on the computer display. It is however strenuous for a user to puff while using his/her mouth for eating.

The proposed human interface resolves this matter with a newly designed input device. The proposed input device uses two fiber units to detect a user's cheek movements as shown in

Fig.3. A user clicks by swelling his/her cheek. The proposed cheek swelling switch can input two actions using left and right cheek movements. Figs. 4 and 5 demonstrate an example of clicking assignments. In Fig. 4 swelling the left cheek centers the mouse pointer, while swelling right cheek (Fig.5) clicks the left mouse button. Clicking assignments may be changed in the control software.

3. System design of cheek swelling switch

Fig.6 shows the mechanical structure of the proposed cheek swelling switch. Two fibers are attached at the both sides of the frame to detect the user's cheek movements. A user can put on the cheek swelling switch easily as shown in Fig.7. The distance between user's cheek and the fiber sensor can be adjusted. Fig.8 shows the hardware construction of the cheek swelling switch. The headset of the head space pointer is connected to a computer USB socket via a control unit. The output signals of the two fiber sensors are inputted into the A/D converter of the computer via amplifier units. Based on the output signals, control program determines the sensor voltage situation. The control program was written in C++.

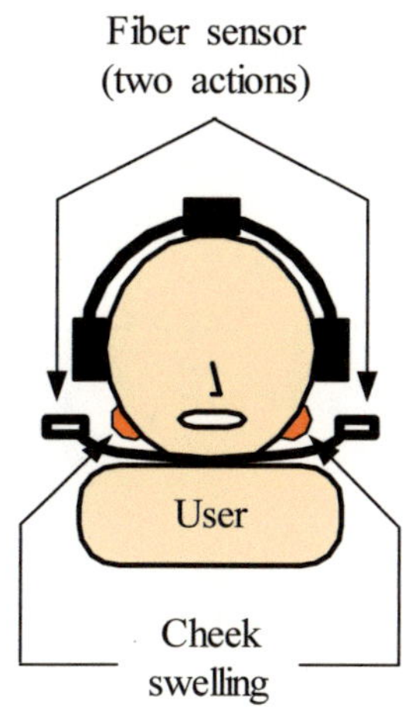

Fig.3. Cheek swelling switch

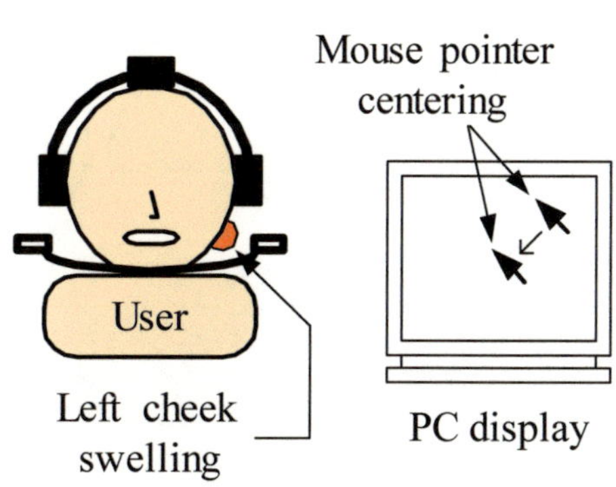

Fig.4. Example of clicking definitions (mouse pointer centering)

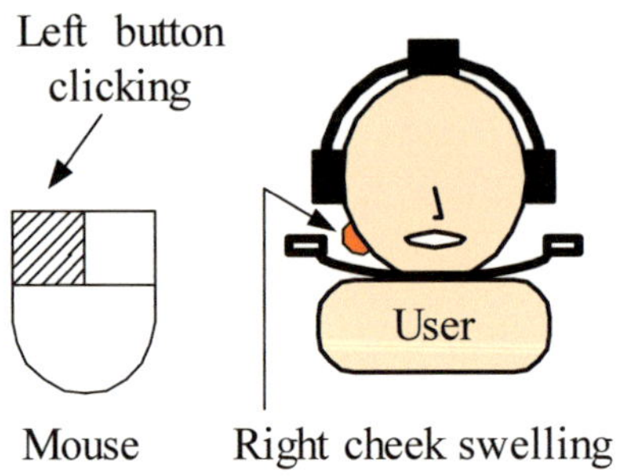

Fig.5. Example of clicking definitions (left button clicking)

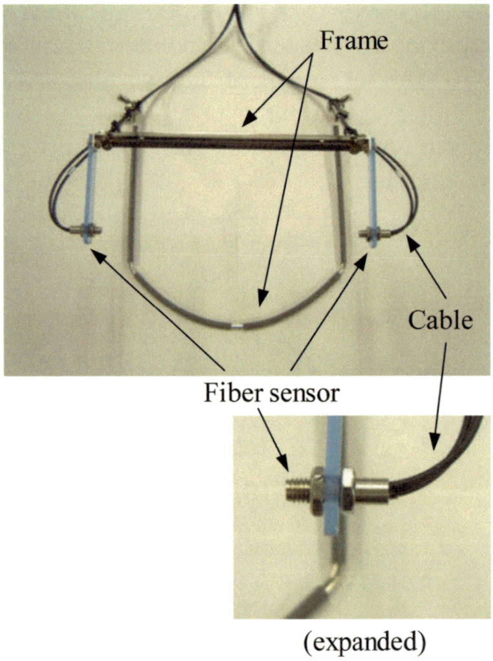

Fig.6. Mechanical structure of cheek swelling switch

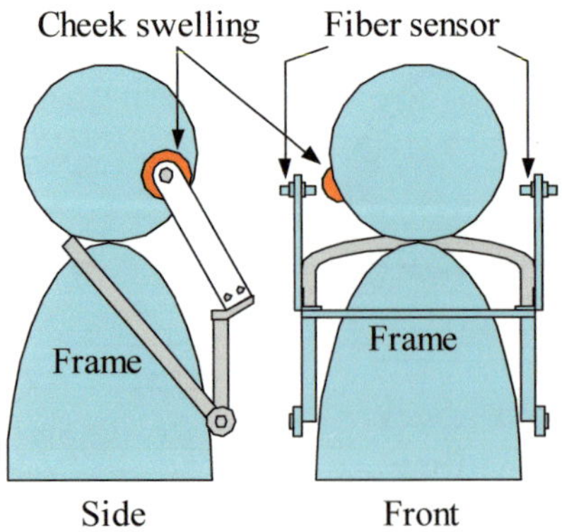

Fig.7. How to wear cheek swelling switch

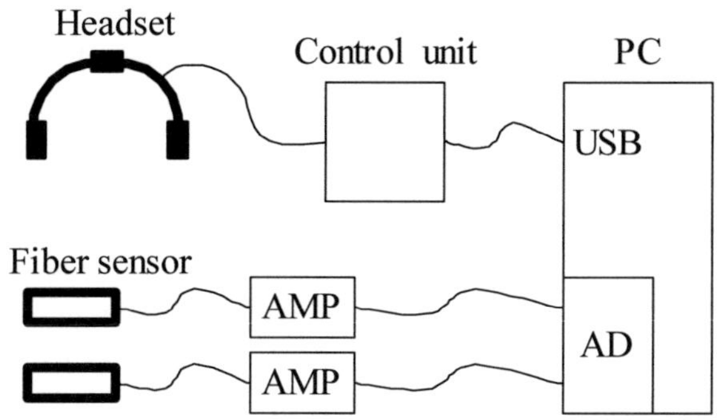

Fig.8. Electrical hardware of cheek swelling switch

4. Basic experiments of fiber sensor and voltage definition

Initially, the basic characteristics of the fiber sensor were investigated. Fig.9 exhibits the experimental results on the basic characteristics of the fiber sensor where the fiber sensor and a sheet of white paper were set on Aluminum blocks. The voltage output varied from approximately 1 to 9 volts in relation to a distance from approximately 140 mm to 8 mm. The measured distance corresponded to the voltage output one to one in the detecting range. The changing rate of the output voltage was higher in the detecting range under 50 mm. It is therefore confirmed that the fiber sensor is able to detect a distance of approximately 25 mm with a high voltage changing rate. Considering the experimental results, the distance between the sensor and user's cheek is designed to be approximately 25 mm.

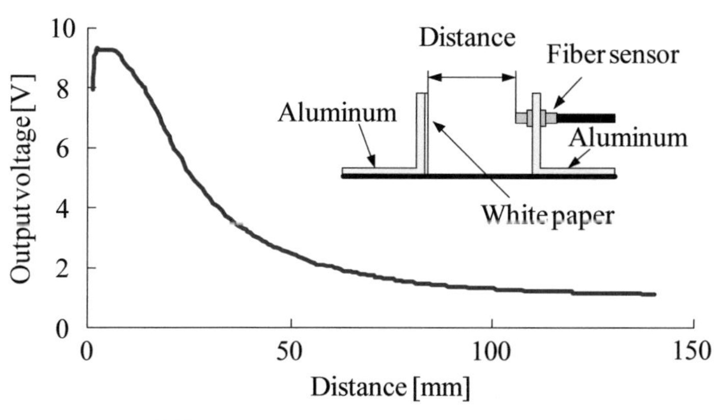

Fig.9. Basic characteristics of fiber sensor

Fig.10 presents the results of the output of the cheek swelling switch where human cheek surface was actually detected. The test subject was a 22-year old male. The test conditions were (1) normal, where a test subject does not eat nor swell his cheek, (2) eating food, where the test subject chewed his food five times, (3) swelling cheeks, where the test subject swells his right and left cheek two times each. In the normal conditions of (1), the output voltage change was very small because the test subject did not swell his cheeks. In the food eating conditions of (2), the output voltage changed approximately +/- 0.1 volts. In the right cheek swelling condition of (3), the right hand side voltage changed approximately 2.0 volts, however the left hand side voltage changed approximately 0.3 volts. In the left cheek swelling condition of (3), the left hand side voltage changed approximately 0.9 volts, however the left hand side voltage changed approximately 0.3 volts.

The following is confirmed by the experimental results; (a) the voltage change was very small at the conditions of normal and food eating, (b) the voltage change was bigger at the condition of cheek swelling, (c) one side cheek swelling influenced the other side cheek movement, however the influence was small, (d) the output voltage will not change largely during eating, the output voltage will change largely only when the test subject swells his cheek intentionally.

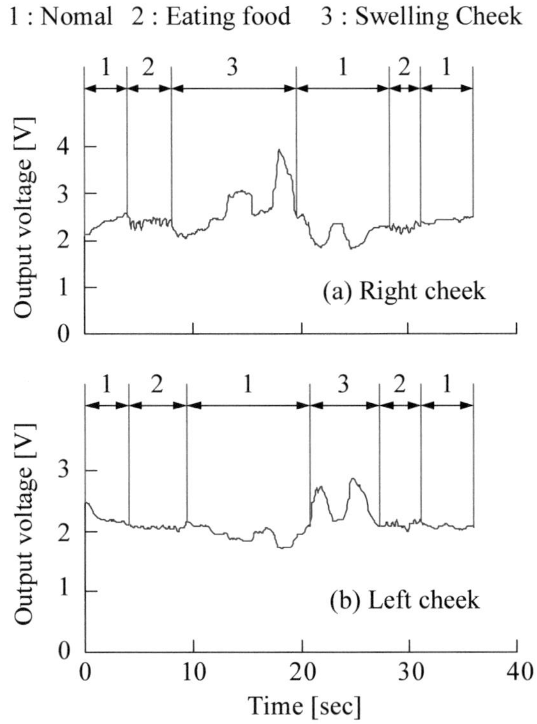

Fig.10. Explanation of output of cheek swelling switch

The program for the cheek swelling switch was designed with the consideration that the output voltage will change largely only when the test subject swell his cheek intentionally. Fig.11 shows the expanded voltage output showing the measurement region of 0 to 35 mm. We define the preset voltage value to click the mouse pointer from the output of the cheek swelling switch. When the output voltage exceeds the preset voltage, the control program sends the command to click the mouse. The preset voltage values can be changed in the control program. The distance between the user's cheek and the tip of the fiber sensor is adjusted so that the output voltage shows less than 4 volts. Three time periods; 1 second, 2 seconds, and 3 seconds are provided to confirm the user's response.

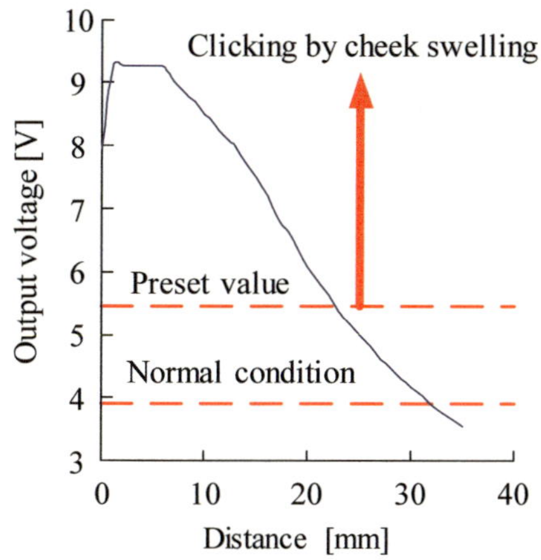

Fig.11. Voltage definition of cheek swelling switch

5. Experiments of cheek swelling switch

Fig.12 shows the experimental results of cheek swelling switch. The experiments were carried out using the following nine test subjects.

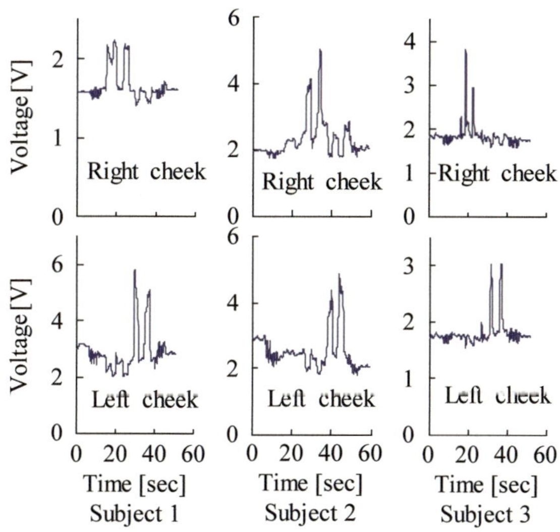

Fig.12.1 Experimental results of cheek swelling switch (subject 1 to 3)

Test subject 1: Male, 22 years old
Test subject 2: Male, 22 years old
Test subject 3: Male, 22 years old
Test subject 4: Male, 25 years old
Test subject 5: Female, 22 years old
Test subject 6: Male, 22 years old
Test subject 7: Male, 21 years old
Test subject 8: Male, 22 years old
Test subject 9: Male, 22 years old

The test conditions were the same as the conditions in Fig.10; (1) normal where a test subject does not eat nor swell his/her cheek, (2) food eating where the test subject chewed his/her food five times, (3) cheek swelling where the test subject swell his/her right and left cheek two times each. Similar results were observed as the experimental results shown in Fig.12. It is apparent that the output voltage does not change largely during eating, the output voltage changes largely only when a user swells his/her cheek intentionally

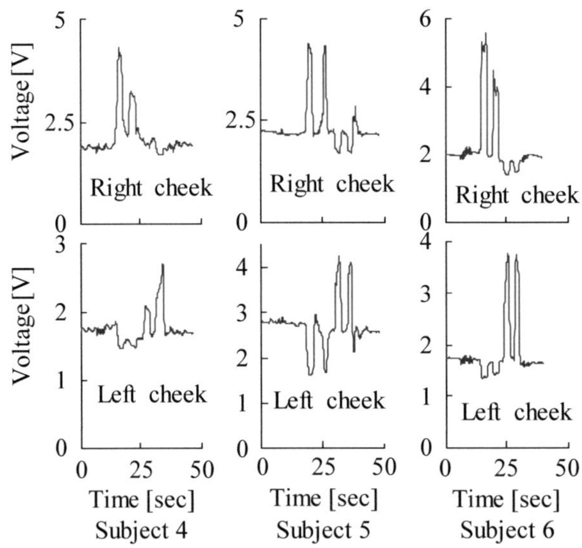

Fig.12.2 Experimental results of cheek swelling switch (subject 4 to 5)

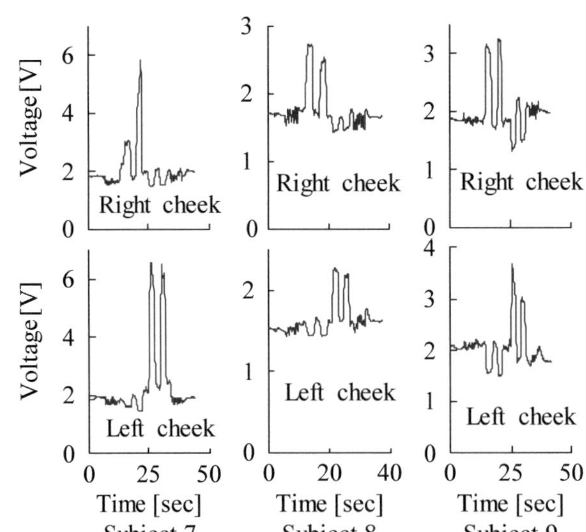

Fig.12.3 Experimental results of cheek swelling switch (subject 7 to 9)

6. Food eating assist robot with proposed human interface

Fig.13 shows the mechanical construction and the movement of the eating assist robot with the proposed human interface of the cheek swelling switch. Fig.14 shows the fabricated eating assist robot. The robot hand can be moved linearly in the three orthogonal directions. The eating assist robot is the computer controlled three dimensional robot arms with a robot hand and a dish rotation table. A spoon is attached at the tip of the robot hand, and can be changed easily to a fork. The spoon can be rotated and be withdrawn in order to dip into food easily. Three dishes are set on the dish rotation table where a user can only pick food up from the dish in front of him/her. The mechanism of the robot arm, spoon, and dish table can be moved by operating a personal computer. A CCD camera is mounted at the tip of the robot hand, and a user can monitor his/her face and mouth on the personal computer display during eating.

Fig.15 displays the hardware of the control system in which a DC motor and a potentiometer are respectively utilized as an actuator and a sensor. The computer inputs the potentiometer output, and outputs the control signal in order to create the feedback control loops. The desired values are calculated by the control program depending on the inputted operation commands.

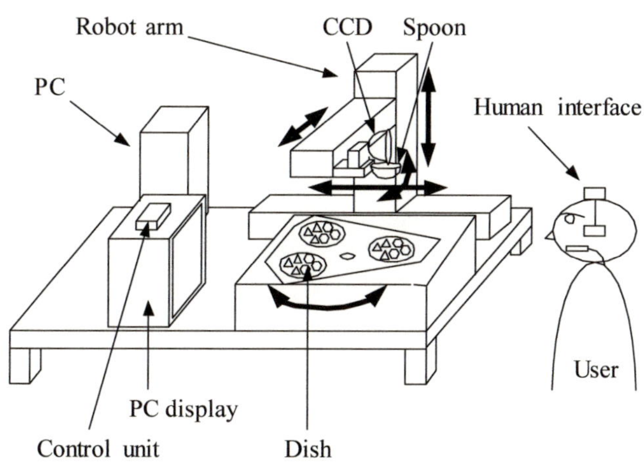

Fig.13. Eating assist robot with cheek swelling switch

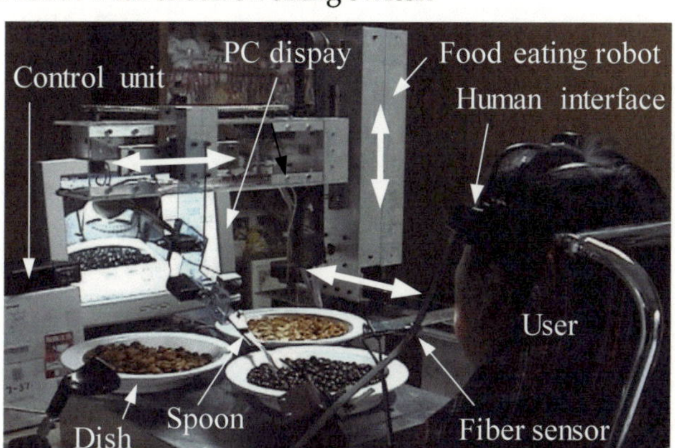

Fig.14. Photograph of eating assist robot with cheek swelling switch

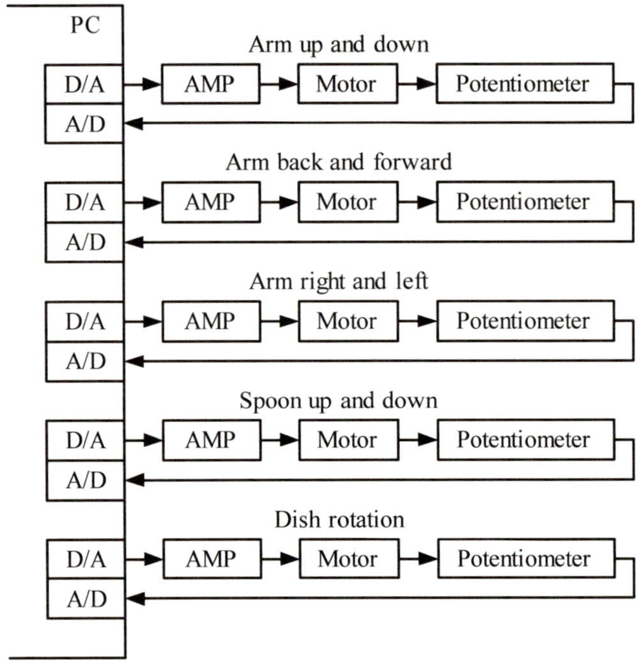

Fig.15. Electrical hardware of eating assist robot

Fig.16 is an example of the computer display while eating. Many operation commands are provided on the display, which a user can select and click. The proposed human interface of the cheek swelling switch is used to click the operation switches on the computer display. Examples of the operation command list are as follows:

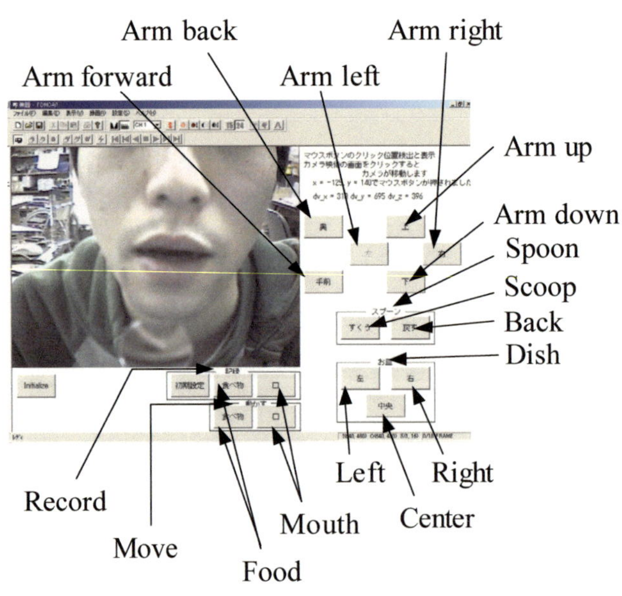

Fig.16. Example of control commands on PC display

Arm control commands; Left, Right, Up, Down
Spoon control commands; Scoop, Back
Dish control commands; Left, Center, Right
Record commands; Initialize, Food, Mouth
Move commands; Food, Mouth

We are also improving the operation commands based on the user's feedback comments. The record commands are used to record the preset two positions of food and user's mouth. After initializing the preset positions, a user can record the two positions. By clicking the preset commands, a user is able move the robot hand more easily without clicking many arm commands.

7. Evaluation experiments in cooperation with a quadriplegic patient

Eating assist experiments using the proposed human interface were conducted in cooperation with a quadriplegic patient. The experiments were executed at the patient's house (Ayase city, Kanagawa prefecture, Japan). The user is a wheelchair bound quadriplegic patient in his forties. The eating assist robot was placed on a specially designed table. The height of the table was adjusted to the height of the user's wheelchair. The user approached the table using his electric wheelchair. The evaluation experiments proved that a truly quadriplegic user is able to manipulate the spoon and eat the food using the proposed interface.

8. Conclusions

A new interface using cheek movement to trigger a cheek swelling switch was proposed to control an intelligent robot for severely handicapped individuals. Two fiber sensors detect the user's left and right cheeks independently. When the user swells his/her cheek, the fiber sensor detects the cheek movement and a control system inputs a command signal. The proposed interface uses the command signal to click the control commands on a computer display. A head space pointer and the proposed cheek swelling switch are combined to control the intelligent robot.

An eating assist robot with the proposed human to robot interface was designed, and evaluation experiments were conducted. The computer has many control commands on the display. A quadriplegic patient is able to manipulate the eating assist robot by using the interface simply by moving his/her head and swelling his/her cheeks.

Combination with the Robot Face projects (Hayashi et al., 2006, Takahashi et al., 2006) is to be addressed in future research.

9. Acknowledgments

The authors would like to express their deepest gratitude to Mr. Hisashi Kaneko for his cooperation with the experiments, and to Professor Takuro Hatakeyama, Waseda University, for his kind advice.

10. References

Hashino, H. (1996); Daily Life Support Robot, *Journal of Robotics Society of Japan,* Vol.14, No.5, pp.614-618

Hayashi, K., Onishi, Y., Itoh, K., Miwa, H. and Takanishi, A. (2006); Development and Evaluation of Face Robot to Express Various Face Shape, *Proceedings of the 2006 IEEE International Conference on Robotics and Automation,* pp.481-486

Homma, K. and Arai, T. (1997); Upper Limb Motion Assist System with Parallel Mechanism, *Journal of Robotics Society of Japan,* Vol.15, No.1, pp.90-96

Ito, E. (2002); Head Tracking and Voice Control Device for Quadriplegics to Use Multiple Home Applications, *Transactions of Human Interface Society (Japan),* Vol.4, No.1, pp.25-31

M.T.BA Cert. Ed. and J.S.BA Hons (1999); The Development of Handy 1, A Robotic System to Assist the Severely Disabled, *Proceedings International Conference on Rehabilitation Robotics,* pp.244-249

O'Connell, S. and M.T.BA Cert. Ed. (1999), A User's Perspective on the Handy 1 System, *Proceedings International Conference on Rehabilitation Robotics,* pp.115-121

Operator's Manual for the HeadMasterPlus(TM) (1994), Prentke Romich International, Ltd.

Takahashi, Y. (2004); Conceptual Design of Care-Given Robot, *Journal of Japan Society for Design Engineering,* Vol.39, No.1, pp.7-12

Takahashi, Y., Yashige, M., Nakamura, M. and Hirata, (2002A); Human Interface and Vibration Control of Robotic Manipulator Suspended from Ceiling, *Transactions of Institute of Measurement and Control,* Vol.24, No.5, pp.387-401

Takahashi, Y., Hasegawa, N., Takahashi, K. and Hatakeyama, T. (2002B); Experimental Evaluations of Human Interface Using Head Space Pointer for Eating Assist Robot and Human's Response to Robot Movement, *Journal of Japan Society for Design Engineering,* Vol.37, No.3, pp.122-128

Takahashi, Y., Goto, H. and Shigemoto, H. (2006); Facial Expressions of Simplified Robot Face for Health Care System, *Transactions of Institute of Measurement and Control,* Vol.28, No.2, pp.163-176

Tateno, M., Tomita, H., Hatakeyama, S., Miyashita, O., Maeda, A. and Ishigami, S. (1993); Development of Powered Upper-Limb Orthoses, *Journal of Society of Life Support Technology,* Vol.5, No.5, pp.156-164

7

Service Robotics: Robot-Assisted Training for Stroke Rehabilitation

Raymond Kai-yu Tong and Xiaoling Hu
Department of Health Technology and Informatics, the Hong Kong Polytechnic University, Hong Kong

1. Introduction

Stroke, a cerebrovascular accident, is a leading cause of permanent disability in adults, with clinical symptoms such as weakness, spasticity, contracture, loss of dexterity, and pain at the paretic side. Approximately 70% to 80% of people who sustain a stroke have limb impairment and require continuous long-term medical care to reduce their physical impairment [1, 2].

In the absence of direct repair on the damaged brain tissues after stroke, neurological rehabilitation is an arduous process, because poststroke rehabilitation programs are usually time-consuming and labor-intensive for both the therapist and the patient in one-to-one manual interaction. Recent technologies have made it possible to use robotic devices as assistance by the therapist, providing safe and intensive rehabilitation with repeated motions to persons after stroke [3, 4]. The most commonly reported motion types provided by developed rehabilitation robots are: 1) continuous passive motion, 2) active-assisted movement, and 3) active-resisted movement [5]. During treatment with continuous passive motion, the movements of the patient's limb(s) on the paretic side are guided by the robot system while the patient stays in a relaxed condition. This type of intervention was found to be effective in temporarily reducing hypertonia in chronic stroke [6], and in maintaining joint flexibility and stability for persons after stroke in the early stage (within 3 weeks of onset) [5]. However, passive movement did not significantly benefit motor improvement. In active-assisted robotic treatment (or interactive robotic treatment), the rehabilitation robot would provide external assisting forces when the patient could not complete a desired movement independently. For chronic stroke (>1 year after stroke), it has been found that with voluntary attempts by the patients, the interactive robotic treatments were more effective for motor functional improvement than treatments using continuous passive motions [5, 7]. Robotic treatment with active-resisted motion involved voluntarily completing movements against programmed resistance [8]. It has been found that repetitive practice of hand and finger movements against loads resulted in greater improvements in motor performance and functional scales than Bobath-based treatment [9], transcutaneous electric nerve stimulation, and suprathreshold electric stimulation on hand and wrist muscles [10]. In robotic therapy, repeated practice against opposing force mainly improved the muscle force from the elbow and shoulder in a reaching task, and also benefited the functional improvements in the wrist and hand for chronic stroke [8].

In recent years, robotic systems developed for stroke rehabilitation have been widely reported in the literatures. Most of these robotic systems can provide all the three motion types mentioned above for poststroke rehabilitation. For example, MIT-MANUS is one of the robotic systems for poststroke upper limb rehabilitation [11, 12]. The key feature of MIT-MANUS is its impedance control which could keep a compliant trajectory under perturbation. ARM Guide is a robotic system designed both for training and evaluation of upper limb reaching functions in a linear trajectory [13, 14]. Colombo et al. also designed a wrist manipulator with one-degree of freedom and an elbow-shoulder manipulator with two-degrees of freedom for the rehabilitation of upper limb movements [3]. They used admittance control to reduce the inertia and facilitate the movement.

Due to the effectiveness in motor improvement by active-assisted robotic treatment, the recent developments involving rehabilitation robots has been worked towards the interactive control, which allows the robotic system to react to patients' voluntary intention. However, most of the interactive controls used in rehabilitation robot are still preliminary and oversimplified on the voluntary inputs from stroke patients. The easiest and most commonly used algorithm for the interactive control is the "on/off" strategy, i.e., once the voluntary indicator from a subject is captured or disappears, the assistive function of robotic system will be triggered to help the subject finish the programmed motion [15]. Usually the subject only needs to stay passively and have a sensory experience after the assistive function is triggered. The voluntary indicator could be represented by limb torques, limb acceleration, electromyographic (EMG) signal amplitude, etc [5, 11, 16]. According to the sensorimotor integration theory, the voluntary motor efferent and the afferent sensor experiences together are important and helpful to promote the reorganization of the brain [17].

A more advanced design in the interactive control requires continuous voluntary motor inputs from a stroke patient to the robot, and the robotic system gives support, or even encourages the output from the motor system of the stroke patient but not to override, during the training. In this chapter, we introduced an interactive robotic system with continuous voluntary intention control developed in our laboratory for poststroke rehabilitation training, and its associated training effects.

2. Continuous EMG(cEMG)-driven robot for poststroke rehabilitation

EMG is the electricity generated in muscles under the control of the nervous system. When an action potential (AP) transmitted from a motor neuron axon to the muscle fibers it innervates, a motor unit action potential (MUAP) is evoked. EMG signal of a muscle is the summation of MUAPs evoked at the same time. Therefore, EMG activities could be used to reflect the nervous instructions on a muscle. In comparison with the amplitude of APs in nerve fibers stimulating a muscle, the EMG amplitude of the muscle is higher (usually in millivolt), if detected non-invasively by surface electrodes from skin surface. Thus, EMG signals have been used as indicators of subjects' voluntary intention in some orthotic and prosthetic devices for paralyzed people [18-21], and the trigger signals in some interactive robotic system for stroke rehabilitation training [15]. In our works, we developed an interactive robotic system with continuous EMG control [22] for poststroke rehabilitation training on elbow joint with the attempt to encourage the participants to involve more voluntary efforts during the training [23, 24]. This developed robotic system was also named as PolyJbot.

i. The Design of the Robot

The experimental setup for the elbow training with the robotic system is shown in Fig 1. During the training, a subject was seated, and the affected upper limb was mounted horizontally on a cEMG-driven robotic system. The subject would conduct elbow flexion and extension horizontally by tracking a target cursor on a computer screen, moving with an angular velocity at 10°/sec in both flexion and extension directions from 0° (full extension) to 90° of the elbow angle. During the tracking, continuous interactive torque would be provided by the robot to help the subject finish the task.

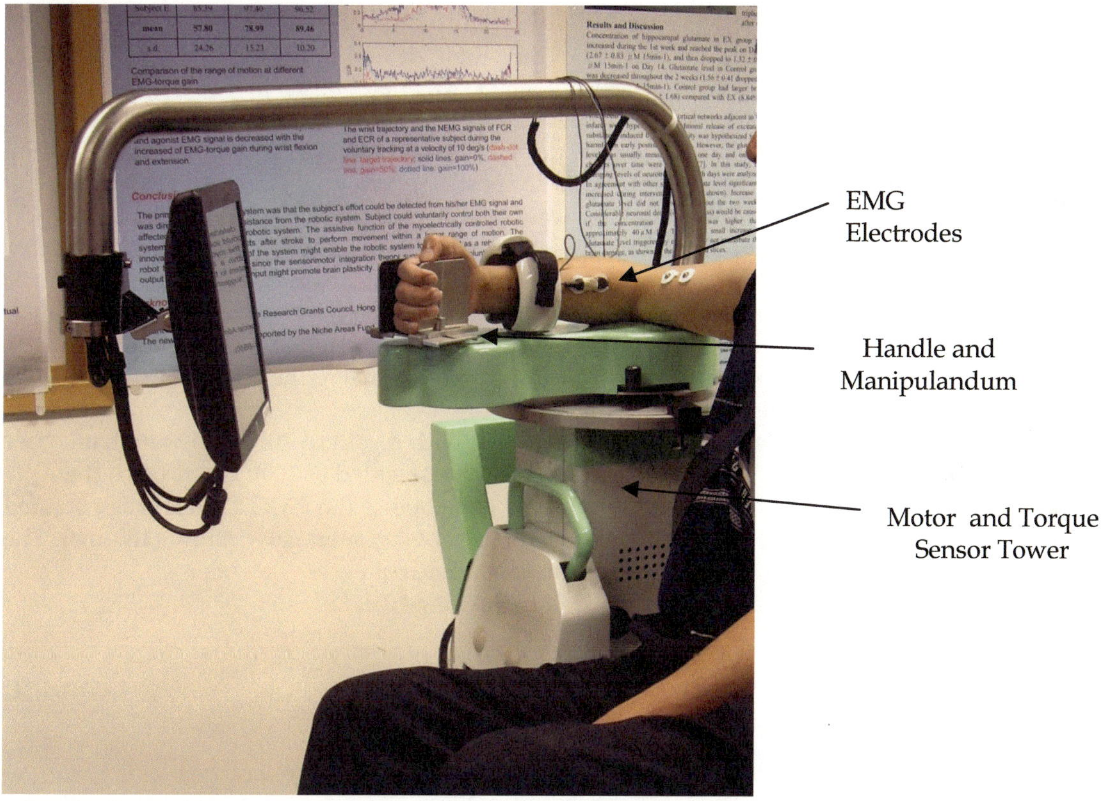

Fig. 1. Diagram of the continuous electromyography driven robotic system.

The detailed structure of the cEMG-controlled robotic system for poststroke elbow training is illustrated in Fig 2, which consists of a personal computer (PC), a PC based data acquisition device, an actuated mechanical part, and an EMG amplifier. The captured EMG signals together with the torque and the angle signals can be inputted through the data acquisition (DAQ) card into the computer. The PC-based program has three functions: 1) to provide a programmed tracking task; 2) to generate a control signal and control the motor to provide interactive torque; and 3) to guide the movement of a subject with visual feedback, which displays both the target and the actual elbow joint angle in a computer screen in front of the subject. The mechanical part of the robotic platform with one-degree of freedom (DOF), i.e. elbow flexion and extension on the horizontal plane, was designed and fabricated for assisting the movement of elbow flexion and extension. A two-layer of aluminium plates were connected by four aluminium pillars inside of a motor and torque sensor tower. The lower plate was fixed to the base of the tower. The direct drive (DDR) brushless AC servo motor was fixed to the lower plate. The motor was connected to a torque sensor. The other end of the torque sensor was connected to a manipulandum with an orthosis to fix the forearm. During the training, the forearm of the paretic side of a subject was placed into the orthosis on the manipulandum which could rotate with the motor; and the elbow angle signals were measured via readings of the positions of the manipulandum. A belt was used to fasten the shoulder joint in order to keep the joint position still during elbow flexion and extension.

In the design of the interactive control algorithm of the robotic system, EMG signals from the main antagonist muscle pairs related to the elbow flexion and extension was detected , i.e. the biceps brachii (BIC) and the triceps brachii (TRI, lateral head) muscles. Clinically, it is observed that most of persons after stroke have difficulties in elbow extension rather than flexion in the affected side, and the paretic upper limbs are always flexed even in the resting state [25]. One of the major reasons is the muscle spasticity commonly observed in flexors, e.g. the biceps brachii (BIC) muscle, and the weakness in extensors, e.g. the TRI muscle [26, 27]. Reduced nervous control on muscle coordination in achieving a target movement is another factor that affects the elbow movements in persons after stroke; and significant muscle cocontractions, i.e. co-shortening, may be found during elbow extension in the affected limb [28, 29]. Therefore, in our robotic system, the assistive torques provided by the motor was only in the extension phase of a tracking task, and was proportional to the amplitude of TRI EMG signal, described as follows [22, 24].

1. The raw TRI EMG signals were first amplified with a gain of 1000 and were band-pass filtered in 10-400 Hz band. Then, the signals were sampled at 1000 Hz through the A/D card and captured by the PC-program. The envelope of the EMG signals was obtained by full-wave rectification and filtering with a moving average window (100 ms). The processed TRI EMG at this step was named as, EMG_{TRI}.
2. TRI EMG, EMG_{TRI}, was then normalized to the range 0-1, i.e.,

$$M_t = \frac{EMG_{TRI} - EMG_{trest}}{EMG_{tIMVE} - EMG_{trest}} \tag{1}$$

   where EMG_{trest} was the amplitude of processed TRI EMG signal at rest, and EMG_{tIMVE} was the maximal amplitude of the processed TRI EMG signal during the maximum isometric voluntary extension (MIVE) at 90 deg of the elbow angle, which was detected before the training session.
3. The assistive torque during the extension was formulated as:

$$T_a = G \cdot T_{IMVE} \cdot M_t \tag{2}$$

   where, T_{IMVE} was the maximum value of the extension torque at the elbow angle of 90°, and G is a constant gain used to adjust the magnitude of the assistive torque for a training program [22, 24].

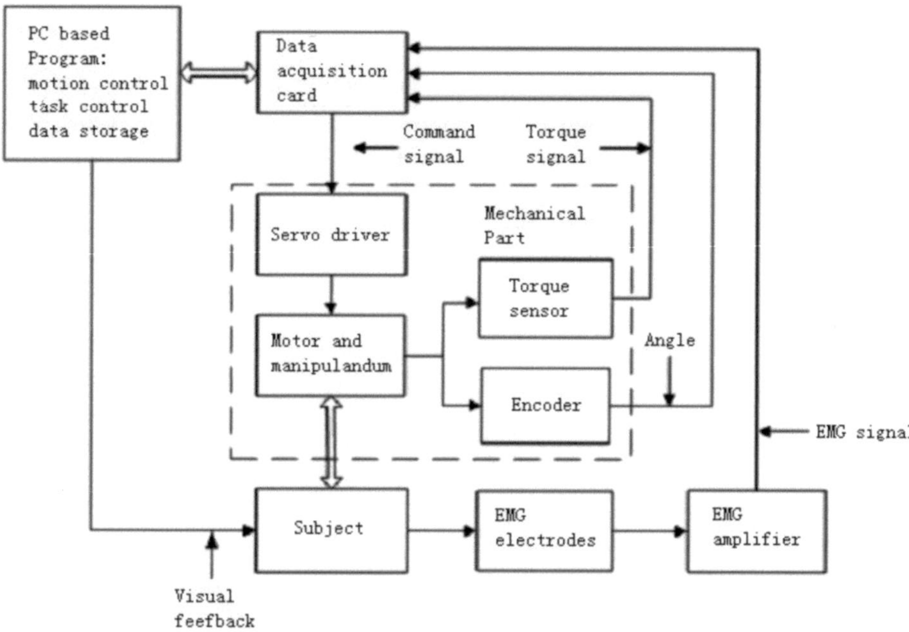

Fig 2. Architecture of the continuous electromyography driven robotic system

The interactive robotic system we developed not only provided the assistive torque to the affected limb during elbow extension, but also could generate resistive torque for muscle strength training in the tracking tasks. The resistive torque could be applied in both flexion and extension directions, that is,

$$T_r = a \cdot T_{MVC}, \tag{3}$$

where T_r was the resistive torque, T_{MVC} that included two parts, the torque of isometric maximum voluntary flexion (IMVF, detected before the training), T_{IMVF}, (applied in the flexion phase only) and the torque of IMVE, T_{IMVE}, (applied in the extension phase only), and a was a percentage value used to adjust the range of resistive torques applied for a training. The selection on the values of G in Eq 2 and a in Eq 3 is related to specific training program, during which subjects should not feel too difficult or too easy to follow. The net torque provided by the robot during the training should be the combination of the assistive torque and the resistive torque, that is,

$$T_n = T_a - T_r, \tag{4}$$

where T_a was the assistive torque, and T_r was the resistive torque. The resistive torques provided by the robotic system were highly related to the torque values of isometric voluntary contractions before a training session. In a multi-session training program, the resistive torques provided by the robot were adaptive to the variation of muscle force generation during the training course, i.e., to keep the effective muscular effort at a level associated with a possible increase in muscle force during the training. Although T_a and T_r in Eq 4 would tend to cancel, the two torques were directly related to the personal effort of the subjects during the training. Therefore, the net torque provided by the robot was interactive to the motor ability of subjects.

ii. Robot as an Evaluation System

Rehabilitation robot could share the large portion of the repeated labor work in a long-term physical training program, with the proper administration by a physical therapist. Another advantage of using robot-assisted poststroke rehabilitation training is that the robot could be a platform for quantitative monitoring on the motor recovery process during rehabilitation training, due to the standardized experimental setup and the high repeatability of training motions compared to the modes manually offered. However, in many works on rehabilitation training, only pre- and post- training evaluations were conducted to assess the training effects, even in most of studies related to robot-assisted rehabilitation [5]. The commonly used clinical scores include the Fugl-Meyer Assessment (FMA)[30], the FIM instrument [31], the Motor Status Score (MSS) test [32], and the Modified Ashworth Scale (MAS) [33].

In our previous work on the poststroke elbow training assisted by the interactive robot, several robotic parameters were used to monitor the motor recovery process during a training program [22, 24]. They were:

1. EMG activation levels of muscles: EMG activation level mainly reflects signal amplitude, and is related to the muscle power, or force, generated in a task. Besides the EMG signals of BIC and TRI muscles, another two muscles related to the shoulder joint, i.e. the anterior deltoid (AD) and posterior deltoid (PD) muscles were also studied in the training program. The raw EMG signals of these four muscles during elbow tracking tasks were recorded and stored in a computer. Then, the linear envelope of the recorded EMG signals was obtained by i) full-wave rectification, ii) low-pass filtering (10Hz cutoff frequency with forth-order, zero-phase forward and reverse Butterworth filter), iii) subtraction of the average EMG activity during the resting state, and iv) normalized to the maximum value of EMG activation in each muscle during either a training session or the IMVE and IMVE of each session. The EMG activation level of a muscle in a tracking trial was calculated by averaging the EMG envelope of the trial.
2. Muscle cocontraction indexes (CI) of muscle pairs: After obtaining the EMG envelope of the muscles of interest, we also calculated the coactivations between different muscle pairs during the tracking task, which could be expressed by the cocontraction index as introduced in Frost's study [34], that is,

$$CI = \frac{1}{T}\int_T A_{ij}(t)dt \tag{5}$$

where $A_{ij}(t)$ was the overlapping activity of EMG linear envelopes for muscles i and j, and T was the length of the signal. The value of a CI for a muscle pair varied from 0 (no overlapping at all in the signal trial) to 1 (total overlapping of the 2 muscles with both EMG levels kept at 1 during the trial). The representative segments of EMG envelopes obtained in the tracking experiment are shown in Fig 3. The CI value of a muscle pair represents the coordination pattern in a motion task.

3. The root-mean-squared-error (RMSE) of tracking trajectories: The root-mean-squared-error of a trial calculates the difference between the target and the actual elbow angles during the tracking task. This measure represents the performance of a tracking task,

i.e. the small the RMSE value, the better the accuracy of performance. An example of the measured actual elbow angle, in comparison with the target angle, is shown in Fig 3.

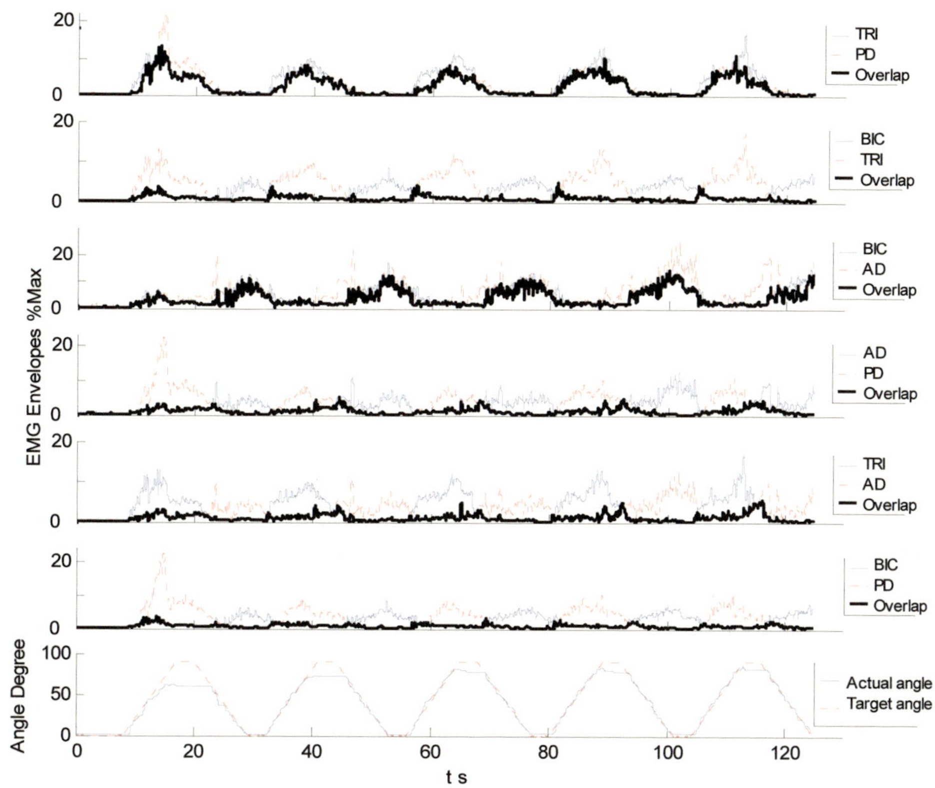

Fig 3. Representative signal envelopes of electromyography captured in a tracking trial, as well as the angle signals during the tracking.

3. Elbow training with cEMG-driven robot for chronic stroke

The developed cEMG-driven robot has been used in the elbow training for people with chronic stroke; and the training effects were evaluated by both traditional clinical assessments and robotic parameters during the training with the purpose of quantitatively revealing the recovery process during the training [22, 24].

i. Training Protocol

After obtaining approval from the Human Subjects Ethics Sub-Committee of the Hong Kong Polytechnic University, we recruited 6 hemiplegic subjects after stroke for the study. All of the subjects were in the chronic stage (at least 1y postonset of stroke; 5 men, 1 woman; age, 51.1±9.7y) All subjects received a robot-assisted elbow training program consisting of 20 sessions, with at least 3 sessions a week and at most 5 sessions a week, and finished in 7 consecutive weeks. Each training session was completed in 1.5 hours. Before and after the training, we adopted 2 clinical scores to evaluate the voluntary motor function of the paretic upper limb (the elbow and shoulder) of the subjects: the FMA (for elbow and shoulder; maximum score, 42) and MSS (shoulder/elbow; maximum score, 40). Spasticity of the paretic elbow of each subject before and after the training was assessed by the MAS score. The clinical assessments of this study were conducted by a blinded therapist.

During each training session, each subject was seated with the paretic arm mounted on the robotic system as specified in Fig 1. EMG signals were detected from the muscles of TRI, BIC, AD, and PD muscles. Before the tracking tasks in a session, each subject was asked to conduct the IMVF (duration, 5s) and IMVE (duration, 5s) of the elbow at a 90° elbow angle at a repetition of 3 times respective, with a 5-minute rest break after each contraction to avoid muscle fatigue. The EMG signal trials with maximum torque values of IMVE and IMVF were used for control algorithm calculation. In each training session, there were 18 trials of the tracking tasks; and in each trial there were 5 cycles of elbow flexion and extension, following the target cursor moving with 10°/Sec on the computer screen. The assistive torques during the extension phases were provided according to the formula of Eq 2, where G, the gain, used in this study was equal to 0%, 50%, and 100% alternatively applied to the tracking trials in a session. The resistive torques in the tracking trials in a training session were provided according to Eq 3, where a, the percentage value, used in this work was 10% and 20% alternatively applied to the tracking trials in a session.The robotic parameters of the EMG activation levels, CI values, and RMSE of the tracking trajectories were also calculated for each trial of all even sessions. The averaged values of the CIs and RMSEs of all trials in a session for subject were used as the experimental readings for statistical analyses. Statistical analyses on the variation of the overall cocontraction indexes, overall EMG activation levels, and the overall RMSEs across the sessions, summarizing the performance of all subjects, were carried out by analyses of variance (ANOVAs) with Bonferroni post hoc test. A paired t test was used for comparison of the clinical scores before and after the training. The statistical significant level was 0.05 in the study.

ii. Training Effects

After the 20 sessions of robot training of the elbow extension/flexion, motor improvements could be observed in all subjects, associated with the improved clinical scores of FMA, MSS, and MAS, as shown in Table 1. It was found that the mean values of scores for FMA and MSS had increased significantly ($P<0.05$) and that of the MAS score also decreased significantly ($P<0.05$) after the training. For each subjects, the clinical scales indicated improvements in at least 2 items.

Improvements also were observed in the tracking performance evaluated by the RMSE values, shown in Fig 4. The overall RMSE values decreased significantly across the training sessions ($P<0.05$). The statistically significant decrease occurred at the 10th session compared with the values of the 2nd, 4th, and 6th sessions. The decreasing process of RMSE during the training represented a motor re-learning procedure in a physical task [8], which is characterized by a plateau of little or no change in performance [35]. Therefore, in this work, after the 10th session, the tracking skill had been stably developed or learned by the subjects after stroke, because the RMSE reached its steady state after the 10th session.

The EMG activation levels of the individual muscles (see Fig 5) and the cocontraction indexes of the muscle pairs (see Fig 6) were used to investigate the muscle coactivation patterns during the task-oriented and robot-assisted training. Before the 10th session, the overall RMSE values during tracking were relatively high. This was associated with the higher overall EMG levels of the BIC, TRI, and AD muscles, during which the cocontractions, i.e., the overall cocontraction indexes, between the different muscle pairs were also high. This was mainly due to the over-activation of muscles during the initial period of motor learning for a skillful task, and the spasticity after stroke, which could cause extra muscle activities. Significant decreases were found in EMG levels of BIC and AD muscles in later training sessions ($P<0.05$, after the 10th session), which was mainly due to the reduction of muscle spasticity, since the tracking skill of the subjects was stable after session 10.

Subject	Pre-Training			Post-Training		
	MAS	FMA	MSS	MAS	FMA	MSS
1	1	15	15	1	17	19.6
2	2	14	16.4	0	17	18.6
3	2	14	12	1+	15	16
4	2	12	18.6	1	15	18.6
5	2	14	12.6	1	14	14.4
6	2	10	14.4	1+	14	21
Mean±SD	1.8±0.4	13.2±1.8	14.8±2.4	0.96±0.5	15.3±1.3	18.0±2.4

Table 1. The clinical scales measured before and after the training.

The cocontraction between the antagonist muscle pair around a joint in subjects without impairment could contribute to stabilizing the joint in a static motion [36], and to movement accuracy in a dynamic motion [37]. However, excessive cocontractions are energetically expensive [37], and abnormal muscle coactivation patterns, mainly cocontractions, in the paretic limb after stroke even degrade the accuracy and efficiency of limb movements [38]. Significant decreases in the cocontraction index values of all muscle pairs (see fig 5) were observed during the training ($P<0.05$), which suggested overall improvement in the coordination of the individual muscles in the tracking task. There are two major reasons that could explain the decreased cocontraction index of a muscle pair: the reduction in the EMG level of the muscles, and the reduced cocontracting phase of the two muscles. The decrease in the overall CI of TRI and PD was mainly due to the reduction of the EMG level of TRI, since the variation of this CI was consistent with the change in the overall TRI EMG levels. However, for the overall CI between BIC and TRI, further decrease after 10th sessions were observed, when the overall EMG levels of the two muscles were almost unchanged. It suggests that this decrease was not related to the reduction in the excessive EMG activation of the antagonist muscle pairs, but the reduced cocontracting phase of the BIC and TRI. It also indicated an improved contracting/relaxing phasic pattern of the BIC and TRI during the elbow extension and flexion in the later sessions of the training. The decrease of the overall CI for BIC and TRI appeared later with a longer time course than the decreases in EMG levels of the respective BIC and TRI muscles. There was no steady state reached by the decreasing CI for BIC and TRI within the 20 sessions. It implied that further improvement in the coordination between the BIC and TRI possibly could be obtained by providing additional sessions. The decreasing in the overall CIs of BIC & AD, TRI & AD, and BIC &PD in later training sessions indicated the better isolation in movements related to the two adjacent joints, i.e. the shoulder and elbow, during the elbow tracking task. This improvement was related to the reduction in excessive activation of shoulder muscles, e.g. the AD muscle. The decease in the overall CIs of BIC&AD, AD&PD, and TRI&AD till the 20th session also suggests that further reduction in these CIs was possible with more training sessions.

4. Conclusion

Significant motor improvements in the poststroke subjects were obtained after the elbow training assisted by the cEMG-driven robot. Robotic parameters also could be used to quantitatively monitor the process of the motor recovery, which provided more information on the extent of the motor recovery than the clinical scores assessed before and after the training. In the future works, comparison should be carried out between the training effects by the cEMG-driven robot and those by the other rehabilitation robots with control strategies, such as, the EMG-triggered control [15], impedance control [11], etc, by large-scale randomized control trials. In the robot design aspect, the cEMG-driven interactive control algorithm can be applied to other joints, such as, the wrist, the knee, and the ankle. The related robot-assisted training effects on the different parts of the body also need to be evaluated in future works. It is possible that the training effects may be elevated by integrating more vigorous visual sensory feedbacks, such as virtual reality programs.

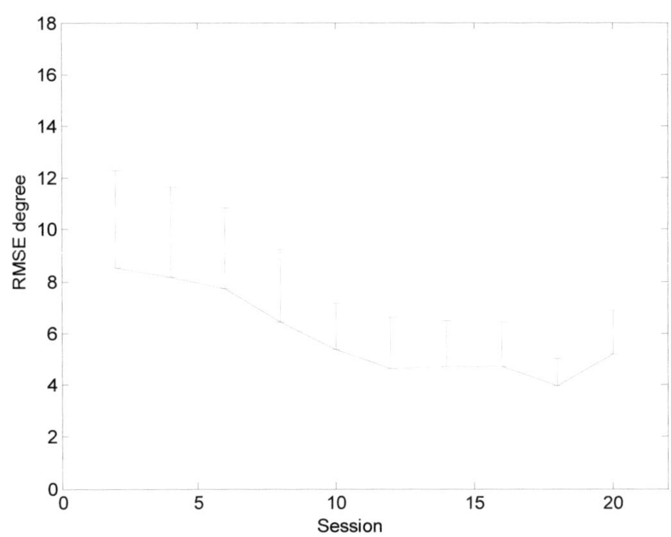

Fig 4. The variation of the overall RMSE values across the training sessions, represented by the values of mean and standard deviation (SD) (error bars).

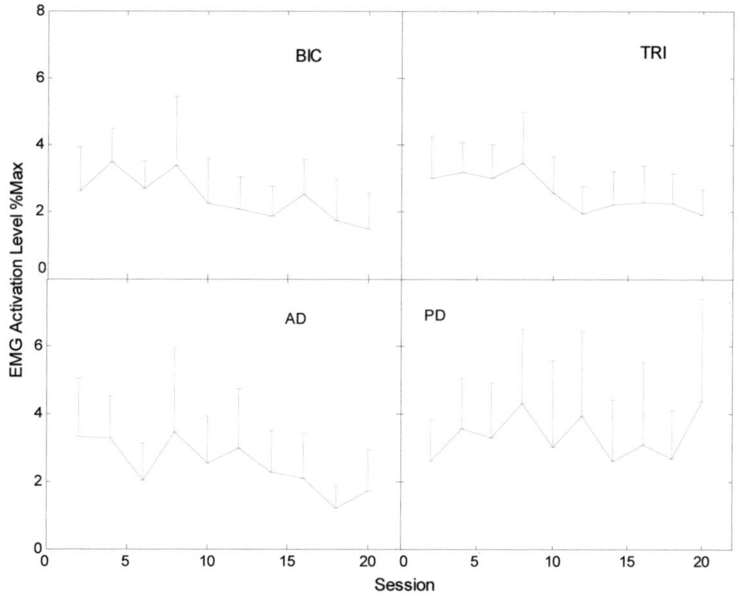

Fig 5. The variations of the overall electromyographic activation level for the biceps brachii, triceps brachii, anterior deltoid, and posterior deltoid muscles, represented by values of mean and SD (error bar).

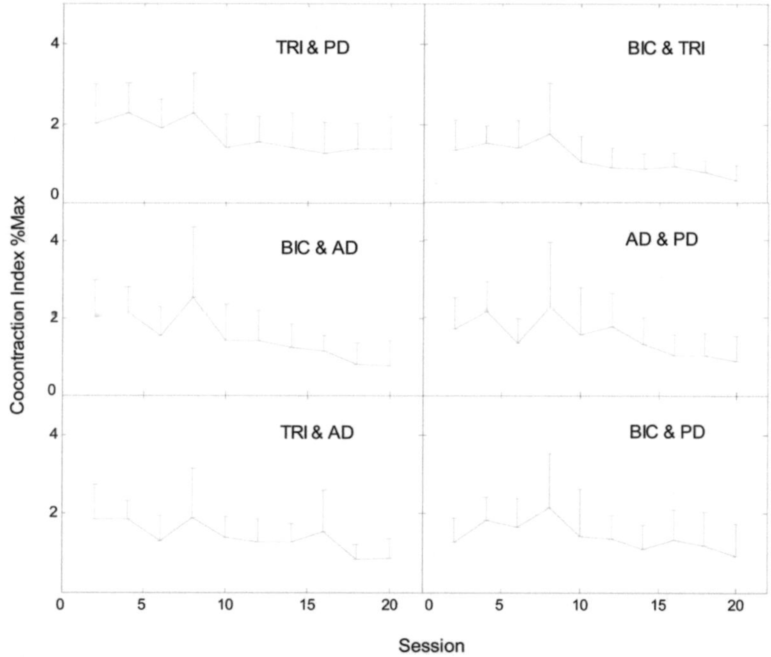

Fig 6. The variations in the overall cocontraction index of the muscle pairs, triceps brachii and posterior deltoid, biceps brachii and triceps brachii, biceps brachii and anterior deltoid, anterior deltoid and posterior deltoid, triceps brachii and anterior deltoid, and biceps brachii and posterior deltoid, represented by the values of mean and SD (error bar).

5. Acknowledgement

The authors would like to thank for the financial support from the Research Grants Council of the Hong Kong Special Administrative Region, China (grant no. PolyU 5271/05E) and the Hong Kong Polytechnic University (grant nos. G-U056, 1-BB50).

6. References

H. Nakayama, H. S. Jorgensen, H. O. Raaschou, and T. S. Olsen, "Recovery of upper extremity function in stroke patients: The Copenhagen stroke study," *Arch Phys Med Rehabil*, vol. 75, pp. 394-398, 1994.

V. M. Parker, D. T. Wade, and R. L. Hewer, "Loss of arm function after stroke: measurement, frequency, and recovery," *Int Rehabil Med*, vol. 8, pp. 69-73, 1986.

R. Colombo, F. Pisano, S. Micera, A. Mazzone, C. Delconte, M. C. Carrozza, P. Dario, and G. Minuco, "Robotic techniques for upper limb evaluation and rehabilitation of stroke patients," *IEEE T Neur Sys Reh*, vol. 13, pp. 311-323, 2005.

M. F. W. Ng, R. K. Y. Tong, and L. S. W. Li, "A Pilot Study of Randomized Clinical Controlled Trial of Gait Training in Subacute Stroke Patients With Partial Body-Weight Support Electromechanical Gait Trainer and Functional Electrical Stimulation: Six-Month Follow-Up," *Stroke*, vol. 39, pp. 154-160, 2008.

B. T. Volpe, M. Ferraro, D. Lynch, P. Christos, J. Krol, C. Trudell, H. I. Krebs, and N. Hogan, "Robotics and other devices in the treatment of patients recovering from stroke," *Current Atherosclerosis Reports*, vol. 6, pp. 314-319, 2004.

B. D. Schmit, J. P. A. Dewald, and W. Z. Rymer, "Stretch reflex adaptation in elbow flexors during repeated passive movements in unilateral brain-injured patients," *Arch Phys Med Rehabil*, vol. 81, pp. 269-278, 2000.

K. Y. Tong, M. F. Ng, and L. S. Li, "Effectiveness of gait training using an electromechanical gait trainer, with and without functional electric stimulation, in subacute stroke: A randomized controlled trial," *Arch Phys Med Rehabil*, vol. 87, pp. 1298-1304, 2006.

S. E. Fasoli, H. I. Krebs, J. Stein, W. R. Frontera, and N. Hogan, "Effects of robotic therapy on motor impairment and recovery in chronic stroke," *Arch Phys Med Rehabil*, vol. 84, pp. 477481, 2003.

C. Butefisch, H. Hummelsheim, P. Denzler, and K. H. Mauritz, "Repetitive training of isolated movements improves the outcome of motor rehabilitation of the centrally paretic hand," *J Neurol Sci*, vol. 130, pp. 59-68, 1995.

H. Hummelsheim, M. L. Maier-Loth, and C. Eickhof, "The functional value of electrical muscle stimulation for the rehabilitation of the hand in stroke patients," *Scand J Rehabil Med*, vol. 29, pp. 3-10, 1997.

N. Hogan, H. I. Krebs, J. Charnnarong, P. Srikrishna, and A. Sharon, "MIT-Manus: A workstation for manual therapy and training," presented at Proc. IEEE Workshop on Robot and Human Communication, 1992.

H. I. Krebs, N. Hogan, M. L. Aiesen, and B. T. Volpe, "Robot-aided neurorehabilitation," *IEEE Trans. Rehabil. Eng.*, vol. 6, pp. 75-87, 1998.

D. J. Reinkensmeyer, L. E. Kahn, M. Averbuch, A. N. McKenna_Cole, B. D. Schmit, and W. Z. Rymer, "Understanding and treating arm movement impairment after chronic brain injury: Progress with the ARM Guide," *J. Rehabil. Res. Dev.*, vol. 37, pp. 653-662, 2000.

D. J. Reinkensmeyer, J. P. A. Dewald, and W. Z. Rymer, "Guidance-based quantification of arm impairment following brain injury: A pilot study," *IEEE Trans Rehabil Eng*, vol. 7, pp. 1-11, 1999.

L. Dipietro, M. Ferraro, J. J. Palazzolo, H. I. Krebs, B. T. Volpe, and N. Hogan, "Customized interactive robotic treatment for stroke: EMG-triggered therapy," *IEEE Trans. Neural Syst Rehabil*, vol. 13, pp. 325-334, 2005.

B. T. Volpe, H. I. Krebs, N. Hogan, L. Edelstein, C. Diels, and M. Aisen, "A novel approach to stroke rehabilitation: robot-aided sensorimotor stimulation," *Neurology*, vol. 54, pp. 1938-1944, 2000.

J. Cauraugh, K. Light, S. Kim, M. Thigpen, and A. Behrman, "Chronic motor dysfunciton after stroke: recovering wrist and finger extension by electromyography-triggered neuromuscular stimulation," *Stroke*, vol. 31, pp. 1360-1364, 2000.

J. Rosen, M. Brand, M. B. Fuchs, and M. Arcan, "A myosignal-based powered exoskeleton system," *IEEE Trans Sys Man Cy B*, vol. 31, pp. 210-212, 2001.

J. Rosen, F. M.B., and M. Arcan, "Performances of hill-type and neural network muscle models: Towards a myosignal based exoskeleton," *Comput Biomed Res*, vol. 32, pp. 415-439, 1999.

H. S. Cheng, J. M.S., and C. C. K. Lin, "Improving elbow torque output of stroke patients with assistive torque controlled by EMG signals," *J Biomech Eng*, vol. 125, pp. 881-886, 2003.

S. Lee and Y. Sankai, "Power assist control for leg with HAL-3 based on virtual torque and impedance adjustment," presented at Proc. of IEEE SMC, 2002.

R. Song, K. Y. Tong, X. L. Hu, V. S. Tsang, and L. Li, "The therapeutic effects of myoelectrically controlled robotic system for persons after stroke-a pilot study," presented at IEEE EMBS 28th annual international conference, New York, USA, 2006.

X. L. Hu, K. Y. Tong, R. Song, X. J. Zheng, I. F. Lo, and K. H. Lui, "Muscle Synergies in Chronic Stroke during a Robot-Assisted Wrist Training," presented at IEEE 10th International Conference on Rehabilitation Robotics, Noordwijk, the Netherlands, 2007.

X. Hu, K. Y. Tong, R. Song, V. S. Tsang, P. O. Leung, and L. Li, "Variation of muscle coactivation patterns in chronic stroke during robot-assisted elbow training," *Arch Phys Med Rehabil*, vol. 88, pp. 1022-1029, 2007.

X. L. Hu, K. Y. Tong, V. S. Tsang, and R. Song, "Joint-angle-dependent neuromuscular dysfunctions at the wrist in persons after stroke," *Arch Phys Med Rehab*, vol. 87, pp. 671-679, 2006.

J. Powell, A. D. Pandyan, M. H. Granat, M. Cameron, and D. J. Stott, "Electrical stimulation of wrist extensors in poststroke hemiplegia," *Stroke*, vol. 30, pp. 1384-1389, 1999.

T. K. Koo, A. F. Mak, L. K. Hung, and J. P. Dewald, "Joint position dependence of weakness during maximum isometric voluntary contractions in subjects with hemiparesis," *Arch Phys Med Rehabil*, vol. 84, pp. 1380-1386, 2003.

J. P. A. Dewald, P. S. Pope, J. D. Given, T. S. Buchanan, and W. Z. Rymer, "Abnormal muscle coactivation patterns during isometric torque generation at the elbow and shoulder in hemiparetic subjects," *Brain*, vol. 118, 1995.

J. P. A. Dewald, V. Sheshadri, M. L. Dawson, and R. F. Beer, "Upper-limb discoordination in hemiparetic stroke: implications for neurorehabilitation," *Top Stroke Rehabil*, vol. 8, pp. 1-12, 2001.

A. R. Fugl-Meyer, L. Jaasko, I. Leyman, S. Olsson, and S. Steglind, "The post-stroke hemiplegic patient. I: A method for evaluation of physical performance," *Scand J Rehabil Med*, vol. 7, pp. 13-31, 1975.

R. A. keith, C. V. Granger, B. B. Hamilton, and F. S. Sherwin, "The functional independence measure: a new tool for rehabilitation," *Advances in Clinical Rehabilitation*, vol. 1, pp. 6-18, 1987.

M. Ferraro, J. H. Demaio, J. Krol, C. Trudell, K. Rannekleiv, L. Edelstein, P. Christos, M. Aisen, J. England, S. E. Fasoli, H. Krebs, N. Hogan, and B. T. Volpe, "Assessing the motor status score: a scale for the evaluation of upper limb motor outcomes in patients after stroke," *Neurorehabilitation & Neural Repair*, vol. 16, pp. 283-289, 2002.

B. Ashworth, "Preliminary trials of carisoprodol in multiple sclerosis," *Practitioner*, vol. 192, pp. 540-542, 1964.

G. Frost, J. Dowling, K. Dyson, and O. Bar-Or, "Cocontraction in three age groups of children during treadmill locomotion," *J Electromyogr Kinesiol*, vol. 7, pp. 179-186, 1997.

A. P. Georgopoulos, "On reaching," *Ann Rev Neurosci*, vol. 9, pp. 147-170, 1986.

T. E. Milner, C. Cloutier, A. B. Leger, and D. W. Franklin, "Inability to activate muscles maximally during cocontraction and the effect on joint stiffness," *Exp Brain Res*, vol. 107, pp. 293-305, 1995.

P. L. Gribble, L. I. Mullin, N. Cothros, and A. Mattar, "Role of cocontraction in arm movement accuracy," *J Neurophysiol*, vol. 89, pp. 2396-2405, 2003.

8

A One Hand Drive Wheelchair with New Manipulation Mechanism Constructed by a Triple Ring and with Assist Functions

Toshihiko Yasuda[1], Takeshi Nakanishi[2], Shougo Kuwabara[3], Daisuke Furikado[4], Naoyuki Kawakubo and Katsuyuki Tanaka[5]

[1]The University of Shiga Prefecture
[2]Daifuku Co., Ltd.
[3]Sanyo Electric Co., Ltd.
[4]Nissin Medical Industries Co., Ltd
[5]The University of Shiga Prefecture
Japan

1. Introduction

The wheelchair is a useful tool providing an aid to mobility for the elderly and handicapped. There exist various researches concerned with wheelchairs (Bourhis, G. et al., 2001; Veeger, H. E. J. et al., 2002; Boninger, M. L. et al., 2002; Adachi, Y. et al., 2003; Miura, H. et al., 2004; Urbano, J. et al., 2005). In order to improve the mobility of persons who can use only one hand for the manipulation of a wheelchair, for example, due to the hemiplegia, various one hand drive wheelchair, such as a lever type and a double ring type, are marketed. These equipments are extremely useful and are utilized in various welfare facilities or homes, because these equipments are effective in moving a short distance on the flat ground with comparatively little rolling resistance. However, because of a heavy physical load on the rough surface and on the slope, many users of these one-hand drive wheelchairs have to transfer to electric wheelchairs at the outdoors. Hence, users of one hand drive wheelchair often give up moving a long distance alone in the outdoors. In order to enjoy life and to maintain the remained ability by rehabilitation effect, it is effective to use the remained ability for moving oneself. Therefore, we consider that various assist functions, such as power assist function and so on, are required to the one-hand drive wheelchair in order to use it in the outdoors. Our research group have been produced a one-hand drive wheelchair with newly proposed manipulation mechanism and some assist functions (Yasuda, T. et al. 2006). Figure 1 shows the developed one-hand drive wheelchair. Dimensions of the wheelchair are presented in Table 1.

In this investigation, we demonstrate a one-hand drive wheelchair with the power assist function, the maximum velocity limiting function on the downward slope, the reverse prevention function on the upward slope and the function for turning on the spot. In order to realize these assist functions, we first developed new manipulation mechanism with a triple ring on a side of the wheelchair. Secondely, we made an assist system, which is constructed by sensing device for human's intention and wheelchair's running speed, micro compter and power supply device. These manipuation mechanism and assist system are able to realize various assist functions.

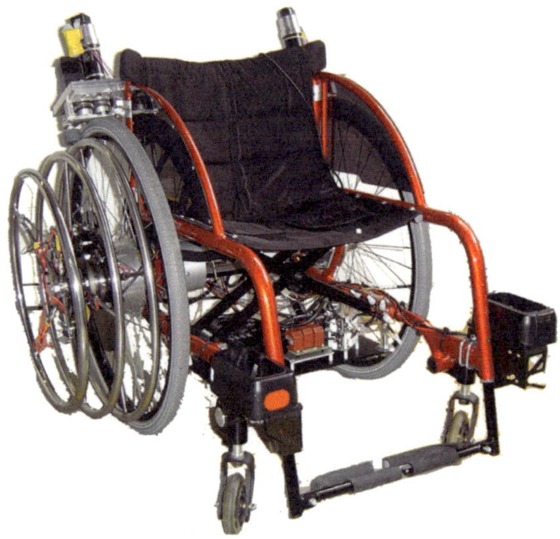

Fig.1. Developed one hand drive wheelchair

Dimension	
Length	850[mm]
Height	760[mm]
Width	690[mm]
Rear Wheel (Diameter)	560[mm]
Left turn hand rim diameter	540[mm]
Center hand rim diameter	500[mm]
Right turn hand rim diameter	455[mm]
Total mass	33[kg]

Table 1. Dimensions of the developed one hand drive wheelchair

2. Manipulation and assist mechanism of a triple ring type one hand drive wheelchair

2.1 Manipulation mechanism

An outline of manipulation mechanism of one-hand drive wheelchair developed in our research group is shown in Fig. 2. This wheelchair is manipulated by using right hand. The characteristic of this wheelchair is that all three hand rims are mounted on a right side of the wheelchair. A right turn hand rim, depicted in Fig. 2, is connected directly with the left wheel by an axis, so rotating this hand rim, the left wheel rotates and the wheelchair turns right (See Fig. 3(a)). A left turn hand rim is mounted on the right wheel, so rotating this hand rim, the right wheel rotates and the wheelchair turns left (See Fig. 3(b)). Rotating a middle hand rim, at first, a mechanical clutch is operated and right and left wheels unite like a spinning wheel. Furthermore, rotating a middle hand rim, the manipulation torque is transmitted to the both wheels at the same time and the wheelchair goes straight (See Fig. 3(c)). If the middle hand rim is released, then the mechanical clutch returns to the original state by restoring force of a spring automatically and left and right wheels separate. Therefore, all operations, such that "turn right", "turn left" and "go straight", are possible in this state. This middle hand rim is called the hand rim for going straight from now. This manipulation mechanism has the following two features:

1) The middle hand rim creates a state, in which right and left wheels are combined and "sure going straight" and "sudden stop" are realized by simple operation such as "rotating or fixing the hand rim for going straight".
2) the manipulator's intention, namely, "turn right", "turn left" and "go straight", is easily obtained by installing the sensor for the manipulation torque to three hand rims.

These features yeild the ability for long distance running in the outdoor and the safety. This manipulation mechanism is most important feature of the developed one hand drive wheelchair.

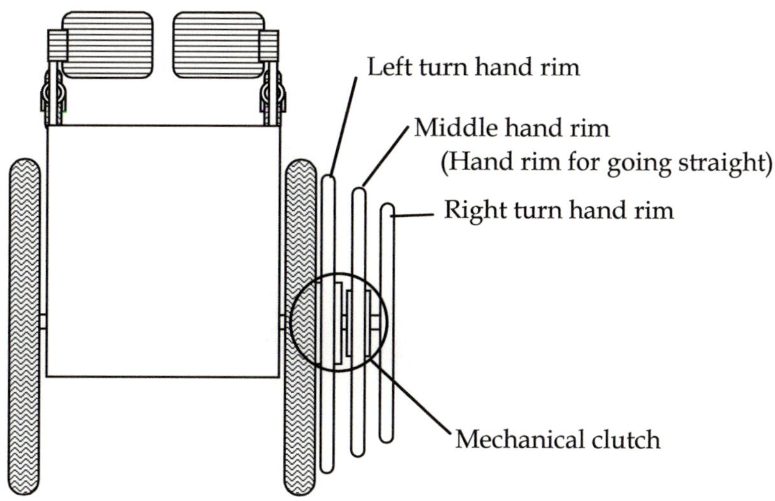

Fig.2. Manipulation mechanism (top view)

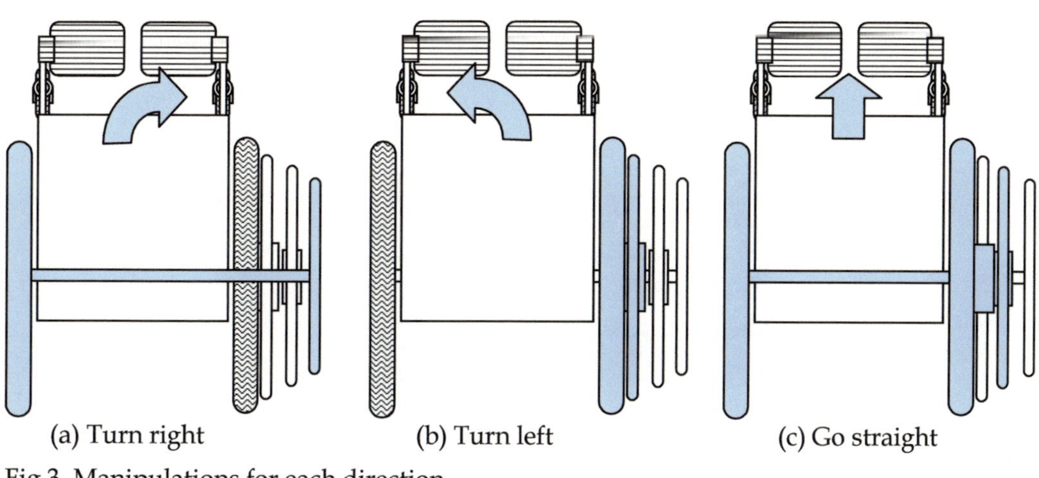

Fig.3. Manipulations for each direction

2.2. Assist unit

Figure 4 shows the assist unit, which realizes assist functions such as "power assist", "maximum velocity limiting on the downward slop" and so on. This assist unit is composed of a DC motor (150W), an electromagnetic brake, a stepping motor and so on. Two rubber rollers, gripping wheels, transmit the assist torque from the DC motor to wheels. The assist unit has a link mechanism which is shown in Fig. 5. By using the link mechanism, which is drived by the stepping motor, when the wheelchair dose not need the assist torque by the DC motor or the brake torque by the electromagnetic brake, rubber rollers separate from the wheel, as shown in Fig. 6(a). Hence, the DC motor and rubber rollers don't yield the rolling resistance. On the other hand, while the wheelchair needs the assist torque by the motor, rubber rollers grip the wheel, as shown in Fig. 6(b). Hence, the DC motor can give the assist torque to the wheelchair. When the wheel is fixed, rubber rollers grip the wheel and the electromagnetic brake stops the rotation of the wheel. By using these three modes, we realize "the power assist function", "the reverse prevention function" and so on. As shown in Fig. 7, this assist unit is independently equipped with both wheels.

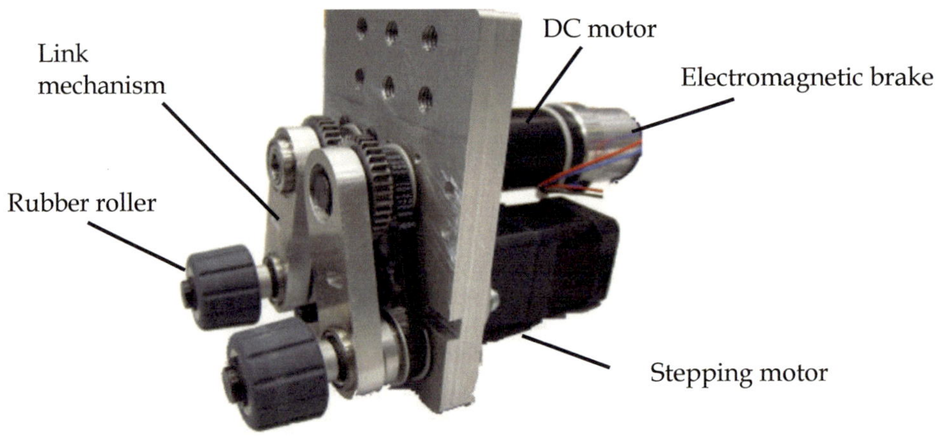

Fig.4. Assist unit

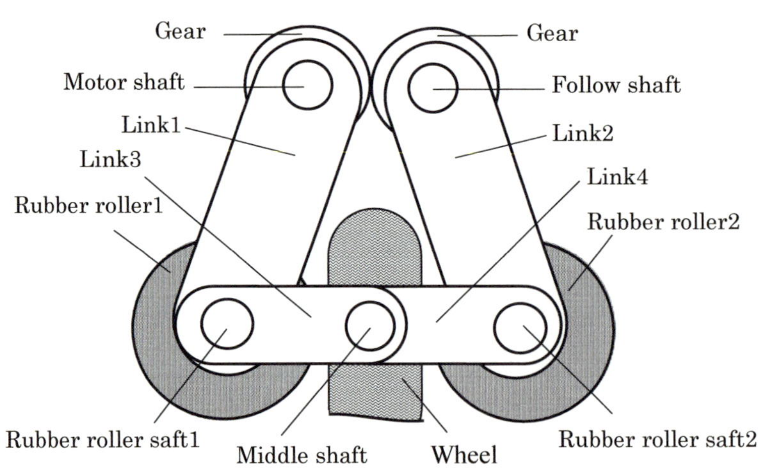

Fig.5. Link mechanism in assist unit

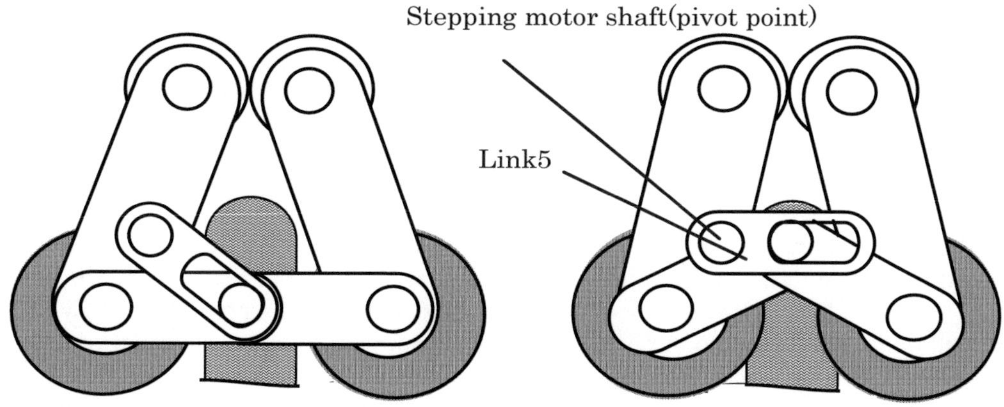

(a) Pull off state of rubber rollers (b) Put on state of rubber rollers

Fig.6. Manipulation of link mechanism

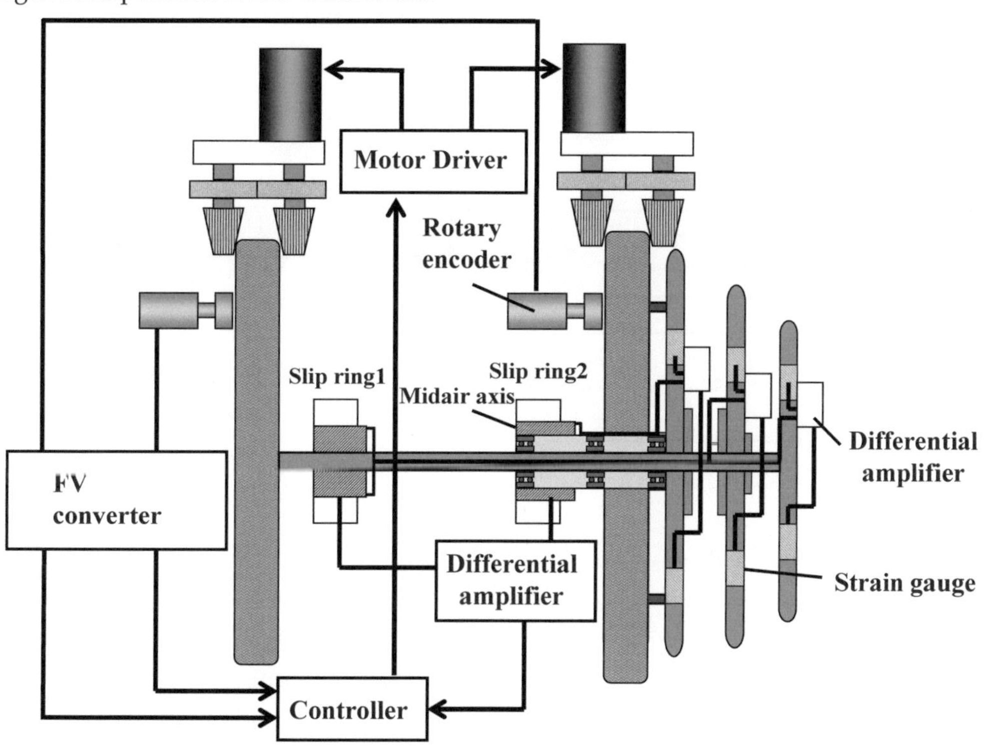

Fig.7. Structure of assist system

2.3. Assist system

Figure 7 shows the structure of the assist system, which includes a manipulation mechanism presented in section 2.1 and assist units presented in section 2.2 and, furthermore, includes some observation and control devices. "Manipulation torque inputted by using the hand rim" is acquired by the strain gauge mounted on hand rims. The detection range of the manipulation torque is from -60[Nm] (backward direction) to +60[Nm] (forward direction). This manipulation torque changes into the output voltage by using the bridge circuit of strain gauges, whose range is 0[V] (-60[Nm] backward direction) to 5[V] (60[Nm] forward direction). Slip rings are used to acquire the signal from the rotating part, namely, hand rims. The rotary encoder installed in right and left wheels acquires the rotation velocity of right and left wheels. The output of the rotary encoder is inputted to the FV converter. The output of FV converter for the forward velocity is 0[V] in the case where the forward velocity is 0[m/s] and the output of FV converter for forward velocity is 5[V] in the case where the forward velocity is 2.78[m/s]. The backward velocity is also observed by using the FV converter for backward velocity. The left and right wheels both have this velocity observation device. Based on these manipulation torques and rotation velocities of left and right wheels, various assist commands are automatically produced by the controller, built on a microcomputer H8/3048. The microcomputer controls the assist torque provided by the DC motor and ON-OFF of electromagnetic brake. Figure 8 presents the system flow of the assist system. The signal giving manipulation torques and rotation velocities are the analogue voltage, hence, these observation data are acquires through the A/D converter on the microcomputer. Motor drivers, electromagnetic brakes and stepping motors are controlled by commands, which are outputted from the D/A converter and the digital I/O ports.

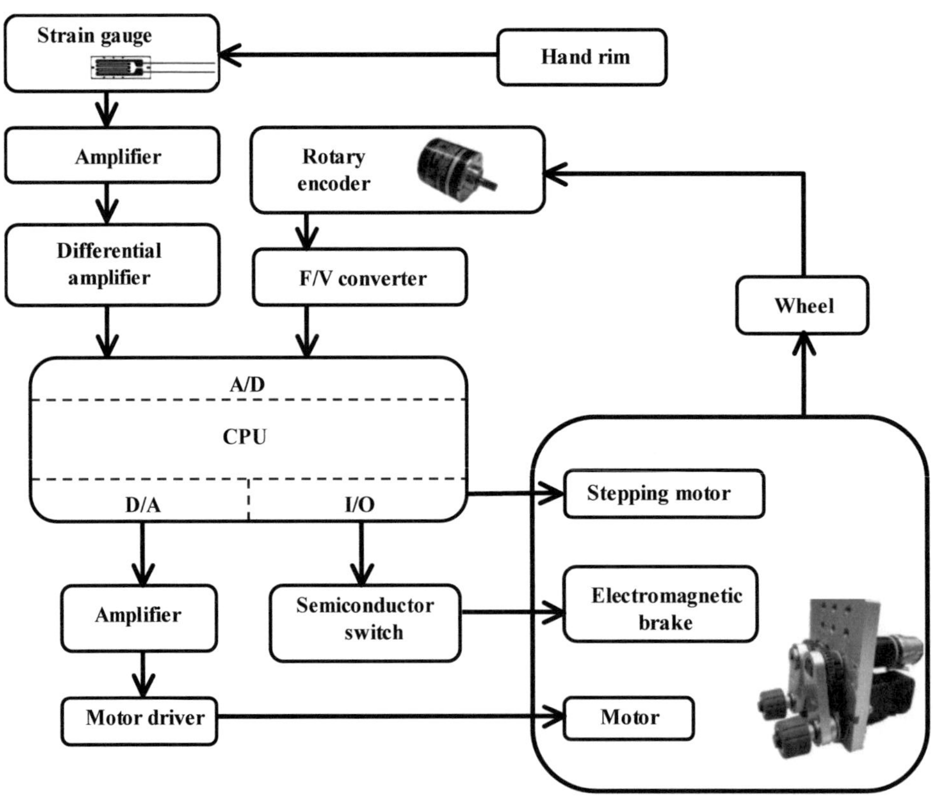

Fig.8. System flow of assist system

3. Assist functions and their evaluation

The manipulation mechanism and the assist unit, explained in Section 2, give the following four assist functions: the power assist function, the maximum velocity limiting function on the downward slope, the reverse prevention function on the upward slope and the function for turning on the spot. We consider that every four assist functions are required, in order to make the running ability of the one-hand drive wheelchair similar to the usual both hand drive wheelchair. The algorithm of each assist function is explained below and the effectiveness of assist functions is confirmed through experiments using a prototype, produced in our research group.

3.1 The power assist function

1) Aim of Assist

In order to realize the mobility similar to usual wheelchair manipulated by both hands, the manipulator's load of the one hand drive wheelchair for rotating handrim is heavy. For instance, the extremely heavy load is needed when the wheelchair goes straight on the upward slope. Therfore, in order to reduce the manipulator's load, the power assit function is indispensable to the one hand drive wheelchair.

2) Experimental Results of Long Distance Running

The purpose of the experiment is to evaluate the ability of the developed one hand drive wheelchair for the long distance running in the outdoor. The running route is on a flat ground in the outdoor and the distance of running is about 510[m] (from the entrance of school of engineering to the bus stop in our university). Table 2 shows experimental results, given by the one-hand drive wheelchair with the power assist function and a usual both hand drive wheelchair. Manipulators are five healthy men (from 22 to 24 years old). In this experiment, the assist torque is determined as

$$Assist\ Torque = Manipulation\ Torque.$$

Subject	Normal wheelchair [s] (a)	Developed one hand drive wheelchair [s] (b)	(b)/(a)
A	477	565	1.18
B	448	569	1.27
C	422	524	1.24
D	556	700	1.26
E	501	619	1.24
Average	480.8	595.4	1.238

Table 2. Experimental results of long distance running

The running time and the ratio are demonstrated in Table 2. Table 2 shows that, by using the normal wheelchair, subjects runs with the velocity about 4.04 [km/h] (average value) and, by using the developed one hand drive wheelchair, subjects runs with the velocity about 3.08 [km/h] (averaged value). Namely, about 75 percent of the velocity, that is obtained by using the normal wheechair, can be realized by the developed one hand drive wheelchair. Considering that human's average walking velocity is 4[km/h], this result shows the usability of the developed manipulation mechanism and power assist function. Furthermore, the running time of the one hand drive wheelchair doesn't take ten minutes. From these results, it is confirmed that we can use the developed one-hand drive wheelchair for a long distance running in the outdoor.

3.2 The maximum velocity limiting function on the downward slope

1) Aim and Assist Action

The running velocity of wheelchairs becomes excessive, when the wheelchair freely runs on the downward slope. Hence, manipulators keep on giving the manipulation torque in the opposite direction to the hand rim. Furthermore, the velocity and the direction of wheelchair must be adjusted with the suitable state by using both hands in usual wheelchairs. This manipulation increases the physical and the mental load of manipulators, especially, in the case where

- the distance of the downward slope is long.
- the downward slope is steep.

In the one hand drive wheelchair, the load becomes too heavy or it is impossibe to run the steep downward slope. Therefore, in order to reduce the manipulator's physical load and to achieve a safe running on the downward slope, "the maximum velocity limiting function on the downward slope" is added to the one hand drive wheelchair. Figure 9 shows the relation between "the manipulation torque and the velocity of the wheelchair" and "the maximum velocity limiting function". The limit of the wheelchair's velocity on the downward slope is set as 0.56[m/s] ≅ 2.0[km/s]. If no manipulation torque is given and the velocity of wheelchair increases and exceeds this limit of velocity 0.56[m/s], then the DC motors are automatically used as a deceleration brake to decelerate the velocity of the wheelchair.

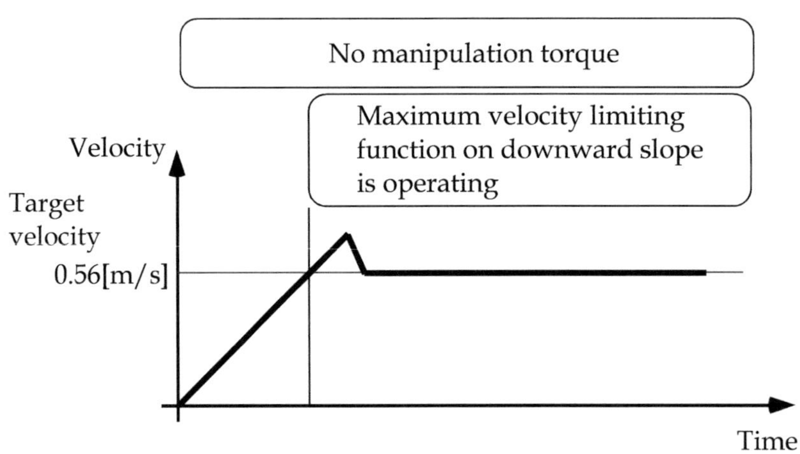

Fig.9. Assist action of the maximum velocity limiting function on the downward slope

In this investigation, we adopt PI controller for "the maximum velocity limiting function on the downward slope". The block diagram of the control system is shown in Fig.10. The propotional gain K_P and the integral constant K_I of the controller in Fig.10 are determined as

$$K_P = 3, \quad K_I = 3$$

by using simulation studies and experimental results.

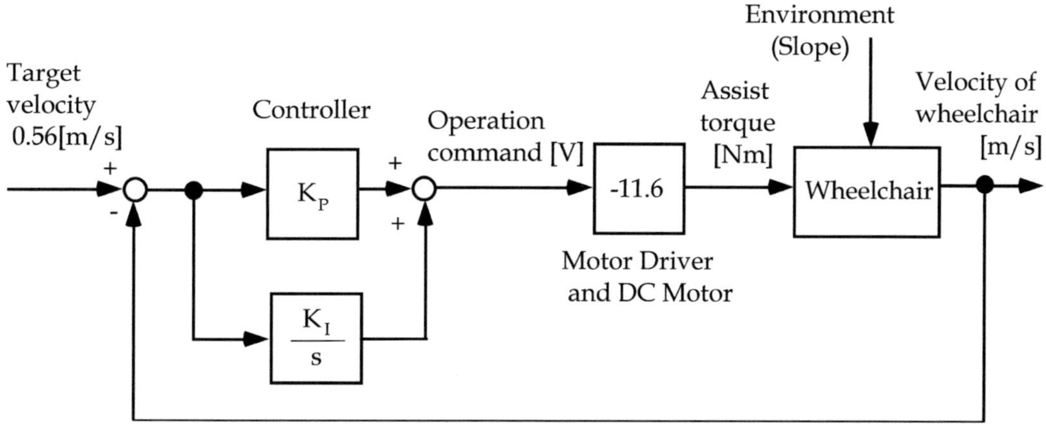

Fig.10. Block diagram of control system of the maximum velocity limiting function on the downward slope

2) Experimental Result of Maximum Velocity Limiting Function on Downward Slope

In order to confirm whether the running velocity in the downward slope can be maintained to 0.56[m/s] by using the maximum velocity limiting function, experiments using a prototype are carried out. The running velocity of wheelchair is measured, when the wheelchair runs from the top of the downward slope without and with "the maximum velocity limiting function". The length of the slope is about 5[m] and the average value of the slope angle is about 4.5[deg.] (See Fig.11 and Table 3). Subject is a healthy man (58[kg] in weight, 21 years old). Figure 12 shows the result of experiment.

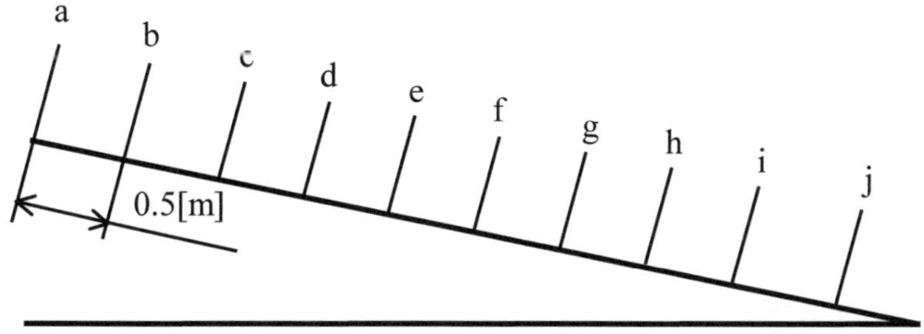

Fig.11. Experimental environment of the downward slope

	a	b	c	d	E	f	g	h	i	j	Average
Angle [deg.]	3.5	4	4.5	4.5	5	5.2	5	5	4.5	4	4.5

Table 3. Angle of each measuring point of the downward slope

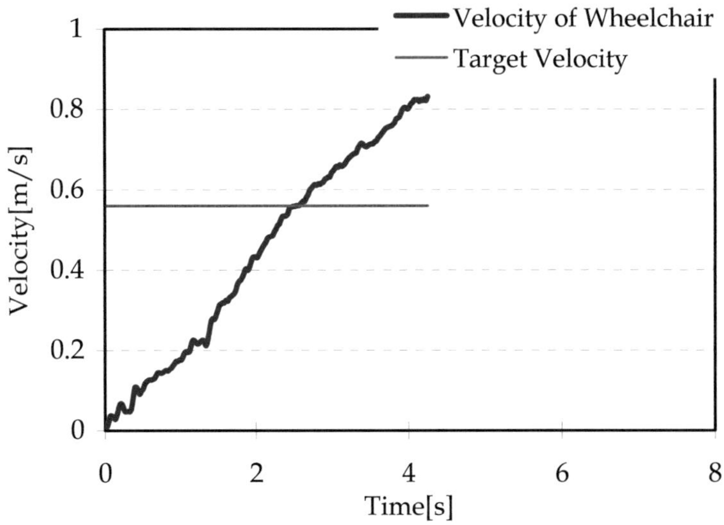

(a) Without maximum velocity limiting function

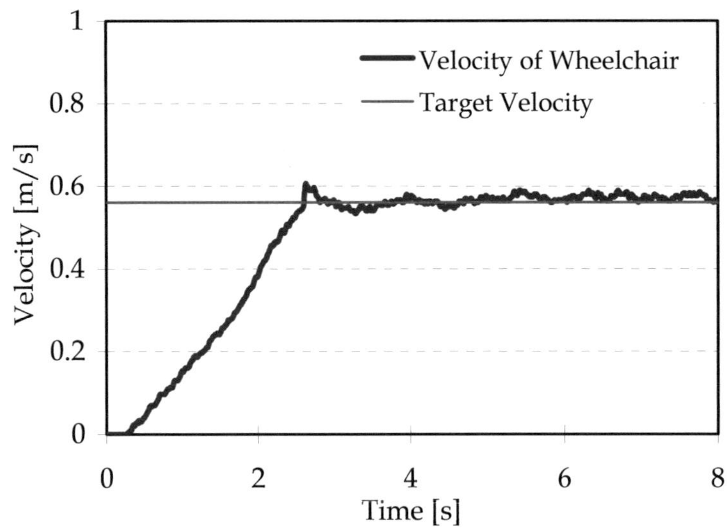

(b) With maximum velocity limiting function

Fig.12. Experimental result (average slope angle 4.5[deg.])

When the maximum velocity limiting function does not act, from the result of experiment shown in Fig.12(a), the running velocity exceeds the target velocity 0.56[m/s] and accelerates to about 0.82[m/s] at the end of the downward slope. On the other hand, adding the maximum velocity limiting function, as shown in Fig.12(b), the velocity of wheelchair can be controlled at constant velocity almost 0.56[m/s]. The velocity 0.56[m/s] is half of the average person's walking velocity, namely, the manipulator can run safely on the downward slope. Moreover, because this maximum velocity limiting function permits to decelerate the wheel speed under 0.56[m/s], the manipulator can freely changes the direction of the wheelchair by decelerating the wheel in one side. Thus, the maximum velocity limiting function achieves a safe running on the downward slope and the manipulator's physical and mental load are reduced.

3.3 The reverse prevention function on the upward slope.

1) Aim and Assist Action

In the case where the wheelchair runs on the upward slope, if the manipulator doesn't keep giving the operation torque to the hand rim toward the upward direction, the reverse running occurs. Because the reverse running is dangerous for manipulators, the safe running on the upward slope is realized by using "the reverse prevention function". The relation between "the manipulation torque and the velocity of the wheelchair" and "the power assist function and the reverse prevention function" is shown in Fig.13. Namely, the power assist function and the reverse prevention function act according to rules depicted in Fig.13.

After the reverse prevention, new rules concerned with the power assist function of turn left or turn right is needed, because, according to the rules on the flat ground concerned with the power assist function of turn left or turn right, the wheel, that manipulator doesn't give the torque, becomes free and the wheelchair turns on the upward slope. Table 4 summarizes rules of the power assist and the reverse prevention after the reverse prevention starts. For example, consider the case where the manipulator gives the manipulation torque to the right wheel for turning left after the reverse prevention. In this case, the reverse prevention function keeps on locking the left wheel, and the right wheel rotates by the manipulation torque and the assist torque. This situation is continued while the manipulation torque for the right wheel is given. If this torque disappears, then a right wheel is locked again by the reverse prevention function. The change of direction after the reverse prevention at the upward slope is realized by such a series of assist actions and the operability on the upward slope is improved. By the "go straight" operation, the reverse prevention mode is finished.

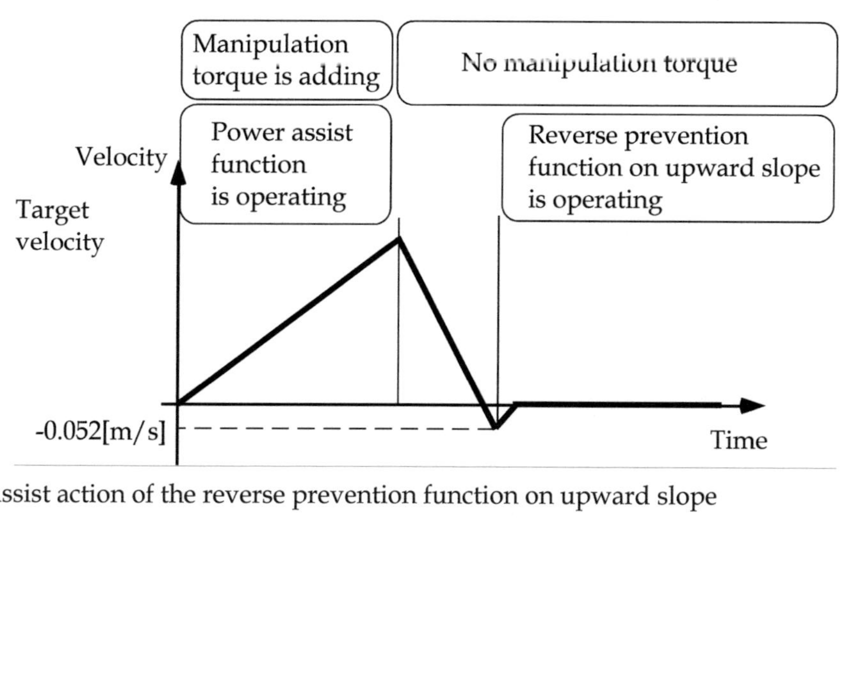

Fig.13. Assist action of the reverse prevention function on upward slope

Manipulation Torque	Left Wheel			Right Wheel		
	Brake	Roller	Motor	Brake	Roller	Motor
Nothing	ON	ON	Without Assist	ON	ON	Without Assist
Go Straight	OFF	ON	Assist	OFF	ON	Assist
Turn Right	OFF	ON	Assist	ON	ON	Without Assist
Turn Left	ON	ON	Without Assist	OFF	ON	Assist

Table 4. Assist action after the reverse prevention

2) Experimental Result of Reverse Prevention Function on Upward Slope

The effectiveness of the reverse prevention and the power assist function are confirmed by experiments. A manipulator gives the manipulation torque in forward direction to the hand rim for going straight and the wheelchair runs four upward slopes where the angle of the upward slope is from 3 to 6[deg.]. The running velocity and the manipulation torque are measured with and without "the power assist and the reverse prevention function". A manipulator is a healthy man (23 years old, 63[kg] in weight). Figure 14 shows the result of running in the case where the angle of upward slope is 5[deg.]. As shown in Fig.14, by adding the power assist function, the maximum manipulation torque, that manipulator gives to the hand rim for going straight, can be reduced to about 60 percent and the maximum running velocity increases about 1.2 times. The fact, that the maximum running velocity increases though the maximum manipulation torque decreases, indicates that manipulator's physical load on running upward slope is reduced.

In this investigation, a time from the beginning of the reverse running to the stopping of the wheelchair by the reverse prevention function is called "Stopping Time", and the maximum velocity in reverse running is called "Maximum Reverse Velocity". In Fig.15, the definition of "Stopping Time" and "Maximum Reverse Velocity" is illustrated. Experimantal results of "Stopping Time" and "Maximum Reverse Velocity" are demonstrated in Table 5, where subjects are three helthy men. In case of acting the reverse prevention function, the maximum value of the stopping time is 0.95[s] and the maximum value of reverse running velocity is 0.18[m/s] as shown in Table 5. The wheelchair stops surely on the upward slope by the aid of the reverse prevention function, as shown in Fig.14(a). On the other hand, without the reverse prevention function, even in the case where the slope angle is 5 [deg.], the maximun value of reverse running velocity is about 0.2[m/s] though the manipulation of the hand rim becomes extrimely quick, as shown in Fig.14 (b). From these facts, it is shown that the safe running on the upward slope is realized and the manipulator's mental load is reduced by the reverse prevention function. In addition, it is easy to freely change the direction of the wheelchair while the wheelchair stops.

As mentioned above, the manipulator's physical load can be reduced by the power assist function and the mental load can be reduced by the reverse prevention function. Thus, the comfortable drive on the upward slope is realized.

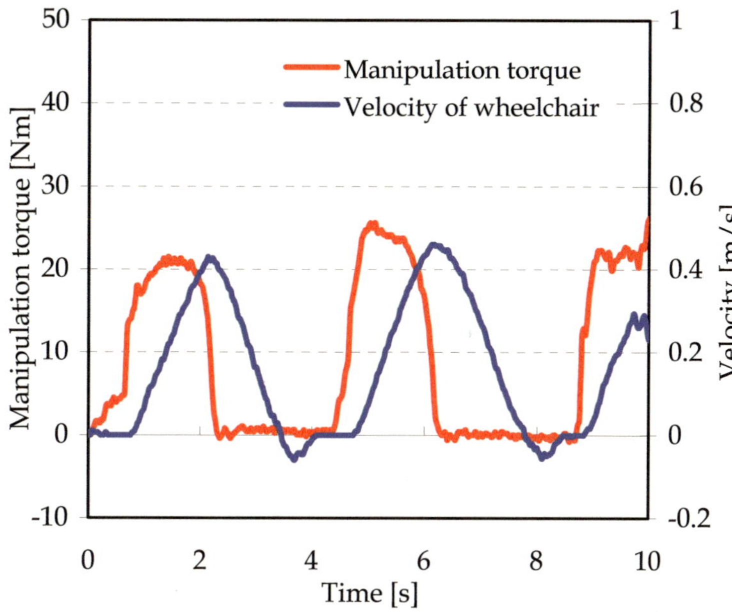

(a) With Assist Function

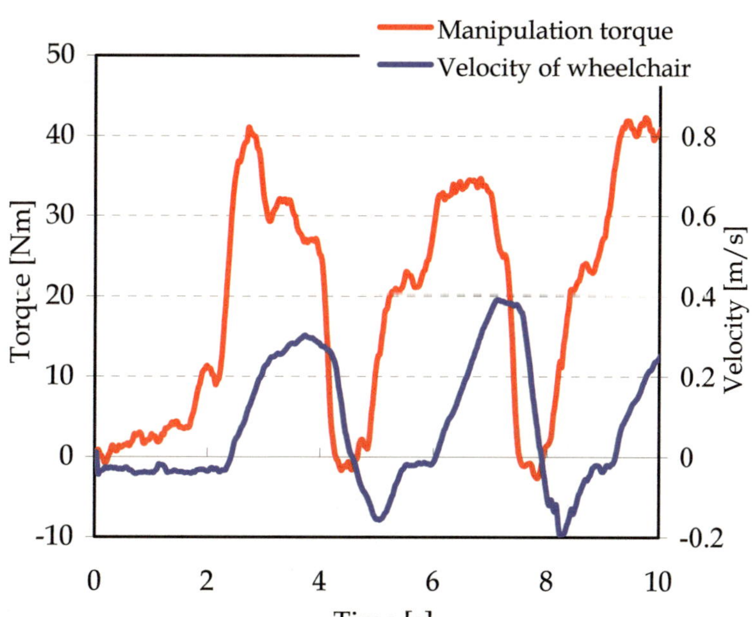

(b) Without Assist Function

Fig.14. Experimental Results of reverse prevention function (Slope Angle 5 [deg.])

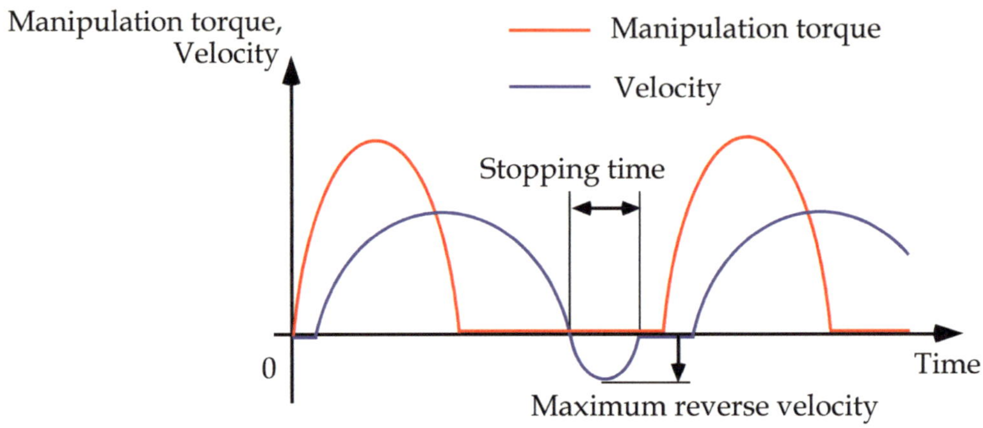

Fig.15. Definition of stopping time and maximum reverse velocity

		Slope angle [deg.]			
Subject Weight [kg]		3	4	5	6
A 61[kg]	Stopping time [s]	0.8	0.82	0.89	0.95
	Maximum reverse velocity [m/s]	0.06	0.09	0.11	0.18
B 61[kg]	Stopping time [s]	0.78	0.8	0.92	0.91
	Maximum reverse velocity [m/s]	0.07	0.09	0.10	0.15
C 85[kg]	Stopping time [s]	0.78	0.86	0.91	0.94
	Maximum reverse velocity [m/s]	0.06	0.09	0.14	0.15

Table 5. Stopping time and maximum reverse velocity until completing reverse prevention

3.4 The function for turning on the spot

1) Aim and Assist Action

In order to change the direction of wheelchair in narrow place and passage, the conventional both hand drive wheelchair usually turns on the spot. The manipulation for turning on the spot of normal wheelchairs is easy, because it is possible to rotate right and left wheels opposite direction at the same time by using both hands. However, for turning on the spot, the one hand drive wheelchair with a triple rings has to manipulate multi-step process like a car, which changes the direction in narrow place. Hence, we consider that it is effective to implement the assist function for turning on the spot to the one hand drive wheelchair.

A wheelchair turns on the spot, when right and left wheels rotate opposite direction at the same time, for example, the rotation direction of right wheel is forward and the rotation direction of left wheel is backward. Furthermore, it is desirable that right and left wheels rotate with same velocity. However, the motor driver in the assist unit of developed wheelchair does not act the velocity control mode, but acts the torque control mode. Namely, we cannot directly operate the velocity of wheels. Therefore, we obseve rotation angles of right and left wheels by using rotary encoders and we construct control system for the rotation angle of right or left wheel. The aim of the control system is to equalize the rotation angle of both wheels. The target value of the control system is the rotation angle of wheel, which rotates by manipulator and is called "input wheel". On the other hand, the opposite wheel is driven by the assist unit with reverse direction and is called "auto following wheel". Figure 16 shows the basic mechanism for turning on the spot. Figure 17 shows a block diagram of control system for turning on the spot. Suitable control scheme, implemented on controller, generates good performance of a function for turning on the spot.

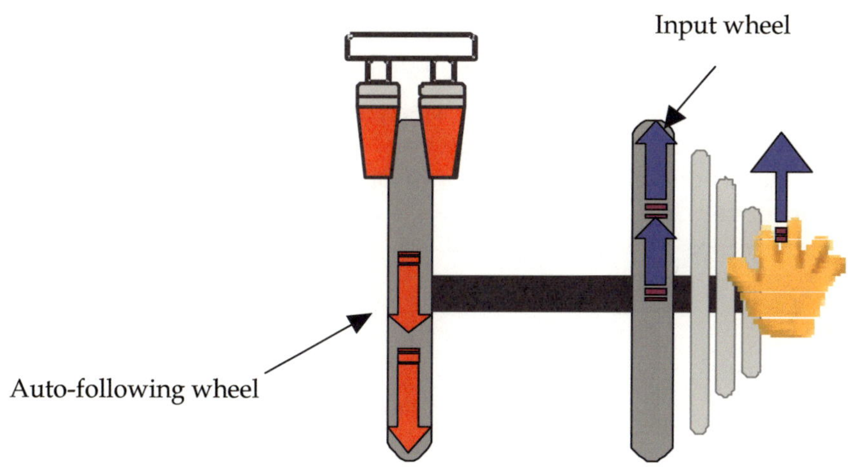

Fig.16. Method for turning on the spot

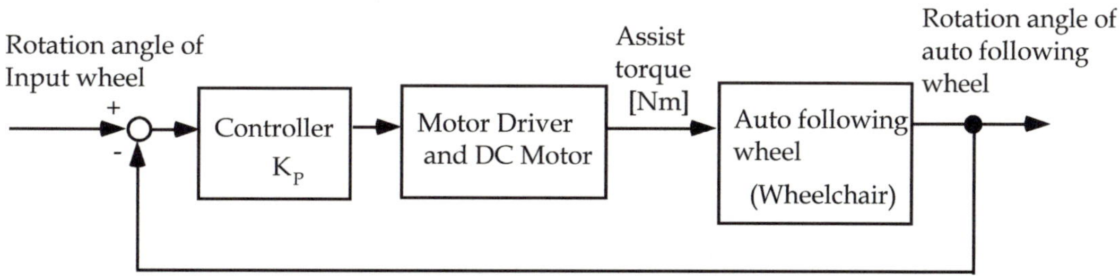

Fig.17. Block diagram of control system for turning on the spot

2) Experiments using Prototype

In order to realize the assist function for turning on the spot, we produced the simulator for one hand drive wheelchair with a triple ring, and simulated the motion of it in turning on the spot. The proportional plus integral action of the controller was tried, but we could not observe any useful effect of integral action. Hence, in this work, we decide to adopt proportion control for turning on the spot from simulation studies.

In this subsection, experimental results using the prototype are demonstrated. Preliminarily experiments clarify the following problems:

In the case where the manipulator finishes the manipulation of hand rim and some deviation of rotation angle of two wheels remains, if the auto-following wheel keeps rotating, then it is difficult for the manipulator to stop the wheelchair at the target direction. This phenomenon gives a manipulator unpleasantness. To solve the problems, we built the algorithm that the assist torque by the motor sets 0 when the manipulation of hand rim is over. As a result of this algorithm, manipulator can turn on the spot comfortably without feeling unpleasantness. This algorithm contradicts the purpose of controller, whose purpose is to decrease the deviation of rotation angle of two wheels. However, we think that comfortable feeling of the manipulator is more important.

From simulation results and experiments by using the prototype, we obtain that the suitable gain is $K_P = 5.0$ to 6.0. Subject is a healthy man (60.5[kg] in weight, 23 years old). Figure 18 shows the transition of the rotation angle of the input wheel and the auto-following wheel in the case where K_P =5.0. Figure 19 shows a movement of the wheelchair turning 180 degree. Rectangles with a line mark, which indicates the front of the wheelchair, show an outline of the wheelchair. Blue rectagle shows the start state and red rectagle shows the end state. The wheelchair turns inside area of a circle with radius 0.7[m]. The average displacement of middle point of both wheels is 135[mm]. For turning on the spot, we think that this performance is practical.

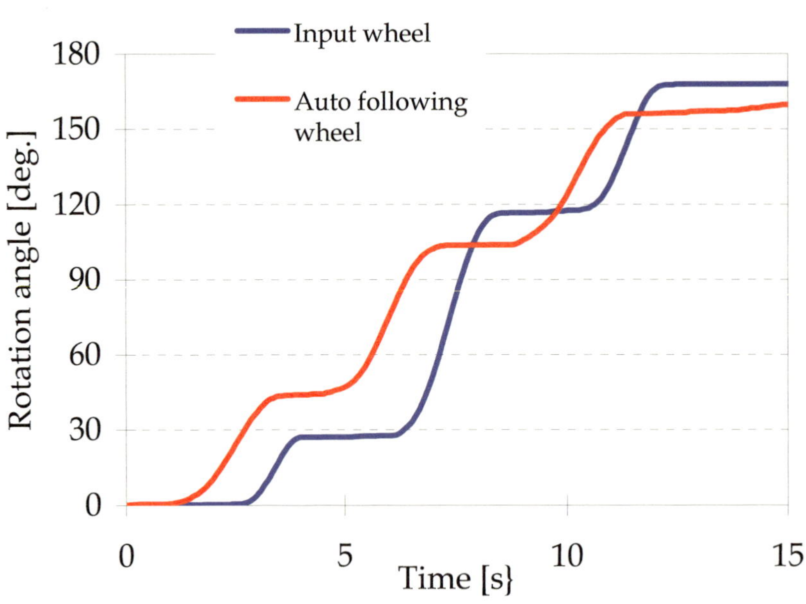

Fig.18. Rotation angle of input and auto following wheels

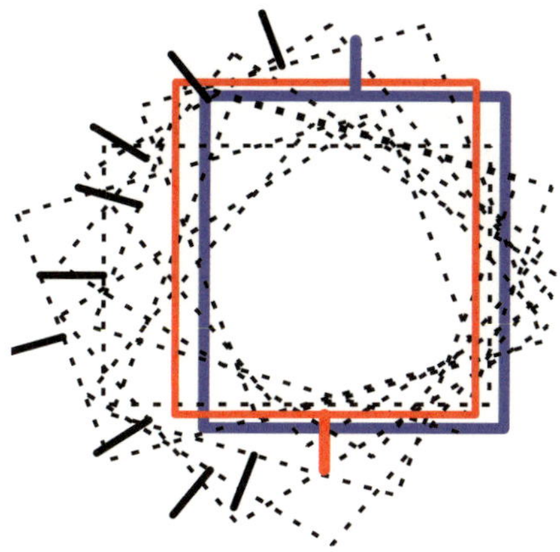

Fig.19. A movement of wheelchair in trun on the spot

4. Conclusion

In this investigation, a one-hand drive wheelchair has been proposed which has new manipulation mechanism with a triple ring and four assist functions such as "the power assist function", "the maximum velocity limiting function on the downward slope", "the reverse prevention function on the upward slope" and "the function for turning on the spot". It was shown that these assist functions reduce the manipulator's physical and mental load and realize the safe drive in the outdoor.

If the manipulator's remained ability is efficiently used, then the welfare robot may be utilized as the effective rehabilitation tool. From such a point of view, as future problems, we consider that it is important to construct the power assist function, which adapts the ability of manipulators and the change of environments. Furthermore, now, we are trying to equip more excellent control for the turning on the spot, which yields more useful motion for our wheelchair users.

5. References

Adachi, Y. et al. (2003), Estimation of user's attention based on Gaze and environment measurements for robotic wheelchair, *Proceedings of the 12th International IEEE Workshopon Robot and Human Interactive Communication*, pp.97-103

Boninger, M. L. et al. (2002). Propulsion patterns and pushrim biomechanics in manual wheelchair propulsion. *Arch. Phys. Med. Rehabil.*, Vol.83, pp.718-723

Bourhis, G. et al. (2001), An autonomous vehicle for people with motor disabilities, *IEEE Robotics & Automation Magazine*, pp. 20-28

Miura, H. et al. (2004). Task based approach on trajectory planning with redundant manipulators, and its application to wheelchair propulsion. *Proceedings of IEEE Conference on Robotics, Automation and Mechatronics*, pp.758-761

Urbano, J. et al. (2005), Velocity control of an Omni-directional Wheelchair Considering User's Comfort by Suppresing Vibration, *Proceedings of 2005 IEEE/RSJ International Conference on Intelligent Robots and Systems*, pp.2385 - 2390

Veeger, H. E. J. et al. (2002). Load on the shoulder in low intensity wheelchair propulsion. *Clinical Biomechanics,* Vol.17, pp.211-218

Yasuda, T. et al. (2006), One hand drive wheelchair with new manipulation mechanism and assist functions, *Proceedings of 2006 IEEE/RSJ International Conference on Intelligent Robots and Systems,* pp.2833 - 2838

9

Development of a Walking Assistive Service Robot for Rehabilitation of Elderly People

JaeHo Jang, SeungNam Yu, JungSoo Han and ChangSoo Han
Dept. of Mech. Eng., Hanyang University,
1271 Sa-1 dong, Sangnok-gu, Ansan, Kyeonggi-do, 426-791,
Dept. of Mech. Systems Eng., Hansung University,
389, Samseon-dong, Seongbuk-gu, Seoul,
Korea

1. Introduction

In order to realize a welfare community, where elderly and disabled people can comfortably live a normal life, it is essential to develop various devices and systems, using cutting-edge science and technology. In particular, there is an urgent need for the development of a human-friendly robotic system that is flexible and adaptable enough to meet the needs of our aging population, and the needs of disabled people living within the community. For walking exercises several items of auxiliary equipment are required to help the roles of the lower limbs. For traditional walking auxiliary equipment, these can be divided into the parallel bar (with the bottom part adjusted), which supports the walker at four points, crutches, which supports the walker at two points, and the cane which supports the walker at only one point. The parallel bar consists of two bars secured at an appropriate height in parallel: it is the most stable form of walking auxiliary equipment. The walker uses them to help him or her to stand straight, and balance the body while walking exercises are undertaken. With support at four points, parallel bars generally give stability, but since they are not externally secured, falls are possible. Its shortcomings include that the user cannot walk fast, and may have difficulty in walking up a slope. Crutches and canes are more commonly used. These buttress the weight of the user, thereby assisting weakened legs and lessening the weight carried, (which may also reduce pain) to enable more natural walking. However, these traditional rehabilitation devices depend upon use by and adjustment with the upper limbs and upper body, so that the degree of weight support is inconsistent and immeasurable. Furthermore, the energy consumption required in using these devices is considerable. Equipment such as the orthopedic scooter, the bath bike, etc., reduces the energy consumption and provides a larger degree of freedom of movement. However, they do not satisfy the basic needs in terms of adjustment of supporting strength.
BWS is an item of treatment equipment for rehabilitation exercises for patients with neurological damage. It was proposed in 1985 to assist the easy attainment of the natural walking pattern (L. Finch, 1991). In 1987, a cat that had its spine completely split was treated in a walking exercise program by using the weight support system repeatedly on a treadmill and it was then discovered that several aspects of walking in humans is similar to walking

in cats. However, before this method was used with human patients, a study was conducted with 10 ordinary people who had no mobility problems, walking on a treadmill with 70% or more of their weight supported. In this experiment there was no abnormal walking and up to 70% of BWS was found to be helpful. In rehabilitation BWS is now considered the most convenient method to overcome the difficulties in walking exercises (T. Pillar, 1991, A. Wernig, 1992, M. Visintin, 1994, S. Hesse, 1995). Patients with neurological problems have found exercises with BWS significantly better than exercising with no weight support. Walking in water would have a similar effect, but that, of course, would be very inconvenient. Barbeau(1987) designed and manufactured a broad range of rehabilitation devices. One involves the patient walk on the treadmill with an upper frame (that is adjusted by hand), supporting the weight of the patient. A robot assistant, AID-1 (T. I. Ide, 1986), uses compressed air to lift the patient's body and thereby lessen the weight. REHABOT, a device that lessens the patient's weight by the same method as AID-1, enables the patient to walk on a circular route. In the beginning, a parallel bar is used with this equipment, to allow patients who have orthopedic or central nervous system problems (which make the exercise more difficult) to use the equipment. The walking capability of healthy individuals is tested with 60% or more BWS. Norman (1995) redesigned the equipment, such that the treadmill and the BWS device used hydraulic power. When holding on the parallel bars to walk, the degree of support in any harness height could be adjusted, following the generated moment. In other systems, a spring is used to support the patient's weight with an almost consistent force. Many authors have studied the perpetuity of force that reduces weight bearing in lieu of the vertical movement of the harness. These efforts have been driven by the needs of orthopedic patients, where the force applied to a damaged skeleton must be controlled. Casalena (1995) proved that a mutual transformation is made between the location energy and the motion energy of gravity within each of the walking cycle. The percent of energy recovery under optimal walking speed reaches 65%. By using a spring, the external work of gravity on the muscle, which increases weight, is minimized.

In this paper, a new rehabilitation system is proposed, using a mobile robot. This system can provide rehabilitation exercises and a walking guidance system for patients suffering mobility problems or paralysis; it makes active exercise possible through the external force and independent strength of the patient. With this robot, the strengths and shortcomings of walking rehabilitation systems described above can be realized or resolved. The equipment is designed to be smaller and simpler than the existing systems, while securing the fullest degree of mobility for the user. Furthermore, this robotic system has the great advantage of making its use in remote treatment possible in the near future. Where a disabled or elderly person is currently treated at home or in a clinic with rehabilitation exercises, he or she may, in future, be able to do the exercise unassisted, and then transmit the result to the doctor, as necessary. Table.1 shows results of functional comparison of GRSR, developed in this study, and the other system.

2. Design architecture

2.1 Overview

The service robot consists of gait assistance and body weight support system. A user is attached from the wrist to the elbow, and moves with the machine. The mechanical design of the gait rehabilitation service robot was established through extensive analysis of: motion range requirements of the elderly; anthropomorphic data; kinematic and kinetic data from published clinical gait analyses; ergonomic constraints; analysis of the control law; extended periods of testing with former prototypes.

2.2 Design criteria

Before design the system, critical requirements and constraints had to be explored. Preliminary design work therefore involved planning how to increase the users' degree of freedom, as well as determining the necessary sensors for system control. The overall goal was to develop a system that would assist a user, weighing up to 75kg, in walking. First and foremost, the design has to be safe. Given the close proximity of man and machine, both the control and mechanical design must ensure the operator's safety at all times, even during failure. The second requirement is functionality.

Functions	PAM-AID	iRobot	ReMeRob	WARD	GRSR
Grip for the elderly	O	X	O	O	O
LRF Use	O	O	O	X	O
BWS mechanism	X	X	X	O	O
Obstacle avoidance	X	O	O	X	O
User command system	O	X	X	X	O
Touch screen control	X	O	O	X	O

Table.1 Results of functional comparison of GRSR and other system

Related to this is a further requirement for comfort in use: the device should not interfere with the behavior of the user. There are three possible prototypes in designing the overall architecture of an integrated of gait assisting and body weight support system. Table.2 shows the developmental history of prototype for gait rehabilitation and body weight support of elderly. Prototype#1 and #2 were developed by Park and Han (2004). The proposed system in this study is based on the Prototype #3.

The gait rehabilitation system design must allow enough freedom of movement for the machine to follow the body of the user during walking. While the user walk, the natural motion of the upper body swings from to top to bottom and right to left. The right and left motion is mainly important for gait stability; the system incorporates top and bottom motion for weight supporting. The revolving joint of the robot user is placed on an orthogonal (right angled) axis orthogonal to the user's sagittal plane (an imaginary plane that bisects the body from top to bottom). Essentially, the robot has 3 degrees of freedom; the gait assisting system has 2 degrees of freedom and body weight support system has 1 degree of freedom.

	GA System	BWS System
Proto-type #1	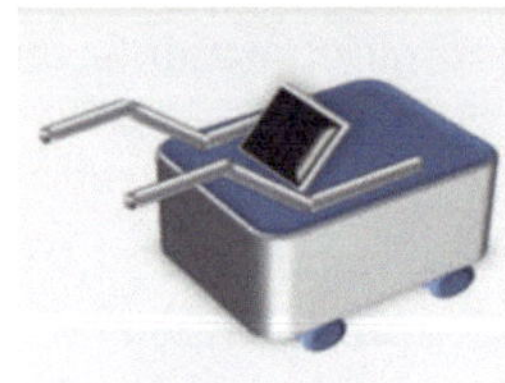	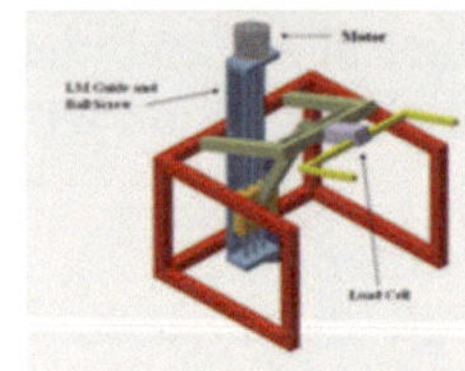
Proto-type #2	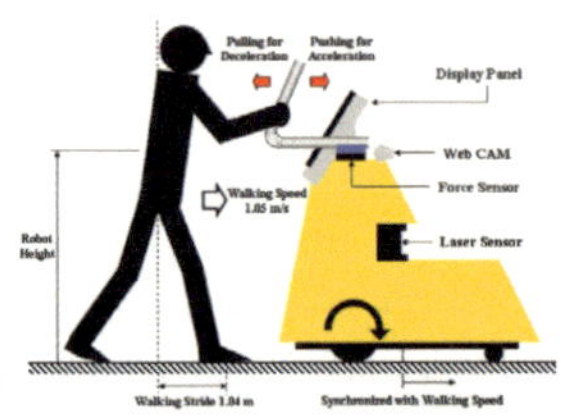	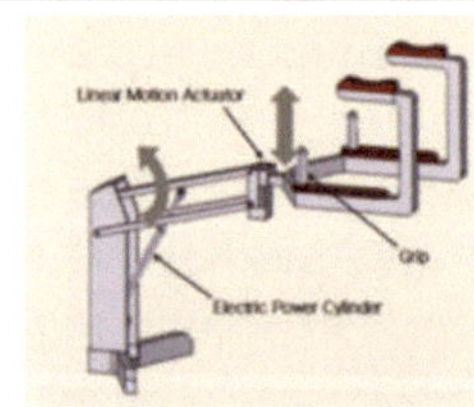
Proto-type #3	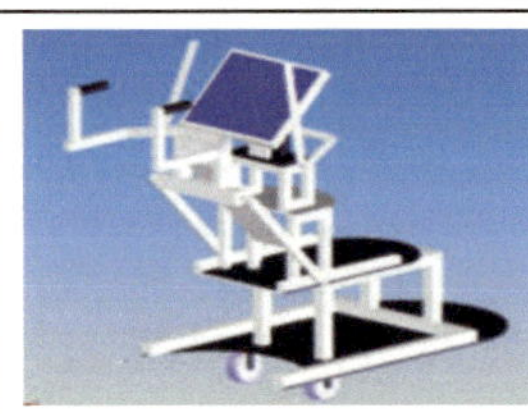	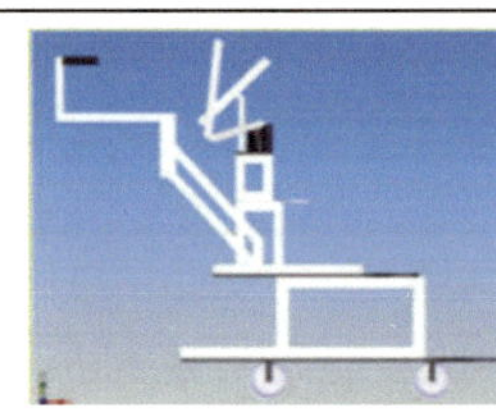

Table.2 Developmental history of prototype

2.3 Consideration of gait of elderly

Walking is a learned activity in which the moving body is supported successively by one leg and the other. Dynamic regulation of upright stance is essential to the safe and efficient performance of many activities of daily living. During the single limb support (approximately 40% of gait cycle for each limb) the body is in an inherent state of instability because the vertical projection of the center of mass passes along the medical border of the foot (D. A. Winter, 1991) and not within the base of support as suggested by Sudarsky (1990). For the swinging limb, the toe clears the ground by less than a centimeter as it travels at its highest forward velocity. The HAT(head, arm and trunk) segment represents about two thirds of the total body mass and has to be controlled by the hip muscle group to avoid tilting. The only period of stability is during double support phase and even during this period the two feet are not completely flat on the ground while the HAT segment travels at it is maximum velocity (V. T. Inman, 1981). Even if walking is performed almost unconsciously and largely automatically several sources of information help the subject to control walking. Dynamic walking balance is achieved by integrating sensory inputs from the visual, vestibular and proprioceptive systems (N. Teasdale, 1993, J. L. Poole, 1991, L. Wolfson, 1992) with adequate muscle strength, appropriate neuromuscular timing and free passive joint mobility (P. R. Trueblood, 1991). In normal aging, degeneration of one or more of these sensory systems occurs (R. W. Baloh, 1993, 2002) and may compromise balance during walking. Therefore, the elderly need some system that can help functions of these sensory systems and musculoskeletal systems for stable walking.

3. Control strategies

3.1 Basic approaches

In the walking-performance enhancing service robot, unlike traditional mobile robots, the human and the machine are integrated and in physical contact. This couples the dynamics of the hardware involved with control architecture. Several power assistance systems have been developed to amplify the strength of elderly people, but most successful research on gait assistance systems has been focused on rehabilitative devices for the physically disabled.

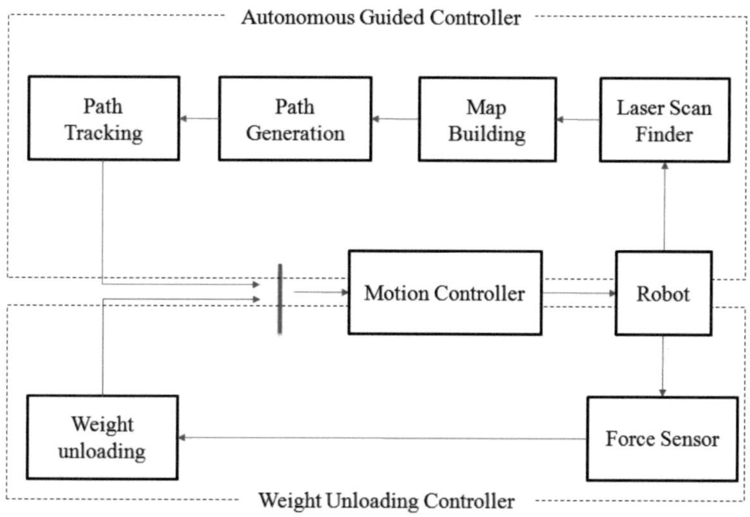

Figure.1 The controller of gait rehabilitation service robot

In these applications, the objective is usually to conduct a limited set of activities in a controlled environment, and many of the control schemes are used by machine. For integration of an appropriate control law, existing strategies are used. The controller of the gait rehabilitation service robot consists of an autonomous guided controller of the mobile robot for gait assistance system and a weight unloading controller of the body weight support system for gait rehabilitation as shown Fig.1.

First, in the case of the autonomous guided control, there are a lot of traditional control strategies for a mobile robot in terms of path generation and path tracking algorithms. The service robot has to achieve path generation and path tracking through the walking of the user, although it should incorporate obstacle avoidance in the environment. There could be a danger that if the user fell on or struck the service robot, it could turn with a short rotation radius or go backward. Therefore, for this service robot, a new control strategy has been designed using an integrating path tracking algorithm and a user-commend detecting algorithm. Next, in the case of weight unloading control, the force to reduce weight bearing following the walking cycle has to be consistent at all times with the force measured by the strength sensor, which is fed back and operates the motor to lift.

3.2 Autonomous guide control

The use of the autonomous guided control of the gait rehabilitation service robot consists of four stages; building a local map, calculating a goal position, calculating a user's speed command, and path tracking.

In stage 1, environmental information acquired from a Laser Range Finder (LRF) is passed to the digital filtering. Because the LRF may pick up errors resulting from a beam scan, this data should be filtered before use. In stage 2, a user's speed command is calculated using a distance between the human and the robot. Sensing of this distance is achieved using a linear potentiometer. Stage 3 involves turning the LRF data into a goal position in Cartesian coordinates, using the Vector Field Histogram (VFH) algorithm and fuzzy logic. The goal position is then placed at a set distance away from the robot, and updated each computational cycle. In stage 4, Modified goal position and speed command obtained in stage 2 and 3 are inputs of the path-tracking algorithm. Outputs of the path-tracking algorithm are the angular velocity of each wheel motor of mobile robot.

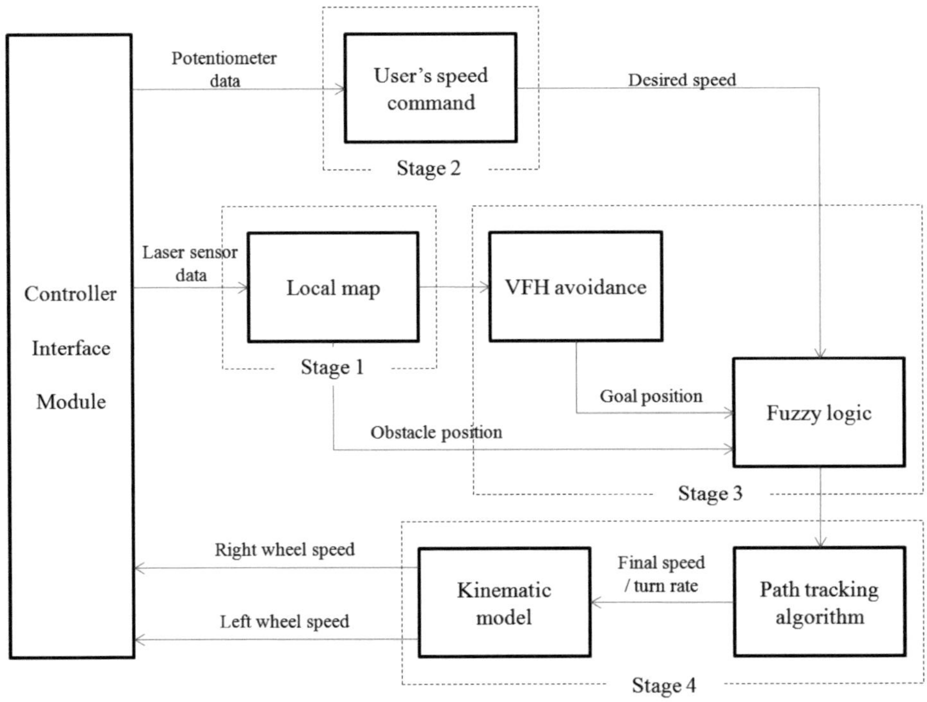

Figure.2 Data flow diagram and control stage block of autonomous guide control

3.3 Map building

Building an environment map is crucial subject for the mobile robot navigation and a number of researches on map building have been investigated so far (J. J. Leonard, 1991, D. Hahnel, 2003). There are two categories of a map building algorithm, one is a grid map and the other is a topological map. These methods can be implemented by a laser sensor as well as an ultra-sonic sensor, an infra red sensor and a vision sensor. Especially, 2-D LRF is good device to obtain quickly the accurate shape of an object. A higher level of control is needed or there would be increased a difficulty in driving at high speeds.

3.4 User-command system

User-command system is man machine interface that transfer to the elderly' decision about walking to service robot, for example walking is continue or not and walking speed is

increased or decreased. The Upper limbs of the elderly are hard to be used by communicator because upper the limbs are used to reduce body weight load. Therefore, user-command system without using upper limbs of the elderly is necessary.
Fig.3 shows user-command system that was designed considering describing contents in front.

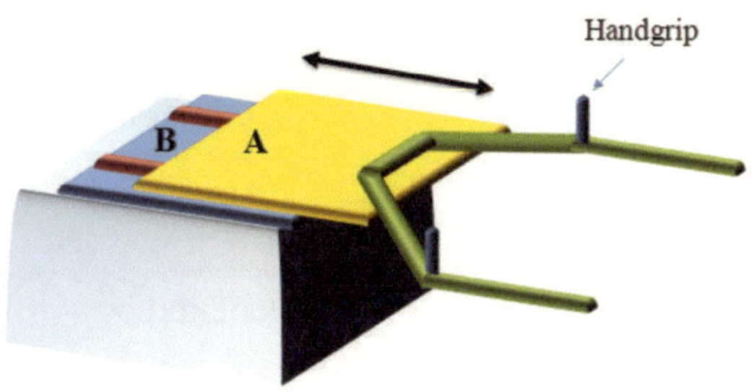

Figure.3 User-Command System of the GRSR

Using the mechanism such as Fig.3, it is kept the distance between service robot and human by speed control of service robot. But it does to prohibit that service robot travels backward for keeping the distance between service robot and human. Because the elderly can fall backward while use the service robot. The potentiometer is installed between the A plate and the B plate, then we measure degree of sliding between two plates.

3.5 VFH avoidance

We have used a method called the vector field histogram (VFH) (R. F. Vassallo, 2002). The VFH method employs a data-reduction technique. The data-reduction maps the active region C^* of the histogram grid C onto the polar histogram *H*, as follow: a window moves with the mobile robot, overlaying a square region of $w_s \times w_s$ cells in the histogram grid. The contents of each active cell in the histogram grid are now treated as an obstacle vector, the direction of which is determined by the direction β from the cell to the vehicle center point (VCP).

$$\beta_{i,j} = \tan^{-1}\frac{y_j - y_0}{x_i - x_0} \tag{1}$$

and the magnitude is given by

$$m_{i,j} = (C^*_{i,j})^2 (a - bd_{i,j}) \tag{2}$$

Where,
a, b are positive constants,
$C^*_{i,j}$ is the certainty value of active cell (i, j),
$d_{i,j}$ is the distant between active cell (i, j) and the VCP,

$m_{i,j}$ is the magnitude of the obstacle vector at cell (i, j),

x_0, y_0 are the present coordinates of the VCP,

x_i, y_j are the coordinates of active cell (i, j), and

$\beta_{i,j}$ is the direction from active cell (i, j) to the VCP.

H has a arbitrary angular resolution α such that $n = 180/\alpha$ is an integer. Each sector k corresponds to a discrete angle ρ quantized to multiples of α, such that $\rho = k\alpha$, where k=0,1,2, ⋯,n-1. Correspondence between $C^*_{i,j}$ and sector k is established through

$$k = INT(\beta_{i,j} / \alpha) \tag{3}$$

For each sector k, the polar obstacle density h_k is calculated by

$$h_k = \sum_{i,j} m_{i,j} \tag{4}$$

Each active cell is related to a certain sector by (1) and (3). A smoothing function is applied to H, which is defined by

$$h'_k = \frac{h_{k-l} + 2h_{k-l+1} + \cdots + 2h_{k+l-1} + h_{k+l}}{2l+1} \tag{5}$$

h'_k is the smoothed polar obstacle density.

3.6 Fuzzy logic

The mobile robot that is to be developed in this study has to perform the functions to adjust the speed depending on the route generation, obstacle avoidance and walking speed of user. For this purpose, the direction of the target point, the location of the obstacle and the output of the force sensor are accepted as the input variables. The fuzzy rule determines the speed of autonomic mobile robot and turning radius from it. The fuzzy value of the input and output variables are designed to have the same values for each Table.3 and Table.4.

Variable	Fuzzy Sets
Vel. of Robot Body	SM BG
Steering angle	NE ZE PO

Table.3 Output Variables and Fuzzy Sets in Them

Variable	Fuzzy Sets
Obstacle distance	NAR MED FAR
Obstacle orientation	NBG NMD NSM ZER PSM PMD PBG
Obstacle Size	SML BIG
Goal orientation	NEG ZRO POS

Table.4 Input Variables and Fuzzy Sets in Them

The input and the output values of the fuzzy are expressed through the fuzzy group. Another word, the value of each has the membership function.

Rule Base
The rule that determines the speed of the robot is as follows. →Move slowly if the obstacle is in front and near. →If not, move fast. The rule that determines the turning radius is as follows. →Avoid if an obstacle is near but is not located in the rear. →If an obstacle is in far distance, go toward the target point. →If an obstacle is in the rear, go toward the target point. →If an obstacle is in the mid-range and the size of the obstacle is small, go toward the target point.

The assumed rule used in this research is the max-min method of Mamdani.

$$z_0^i = \max(Z_1^i, Z_2^i, \cdots, Z_N^i)$$

$$z_k^i = \min(A_k, B_k, \cdots, C_k) \;\; (k = 1, 2, \cdots, N) \tag{6}$$

Here,

Z_0^i : Membership of first fuzzy group, z, of the output space

Z_k^i : Membership of i-th fuzzy group by the k-th rule

$A_k, B_k, \cdots, C$: Membership of each of the fuzzy groups of A, B, and C of the input space used for the k-th rule

The defuzzification method used in this research is the CG point method.

$$Z^* = \frac{\sum_i \int \mu_i(z) \cdot z dz}{\sum_i \int \mu_i(z) \cdot dz} \tag{7}$$

Here,

$\mu_i(z)$: λ of i-th fuzzy group of the output space – cut ($\lambda = Z_0^i$)

A computer simulation for obstacle avoidance and route tracing as described above is made for the assessment of the capability of the fuzzy controller. The obstacle is experimented for the static obstacle and the dynamic obstacle. Fig.4(a) and Fig.4(b) displays the avoidance on fixed obstacle and the simulation result of route tracing.

In this research, the presumption is made for the obstacle with the conic formation with the semi-diameter of r. Fig.4(a) and Fig.4(b) display the result in each case of r = 2 and r = 5, respectively. Fig.4(c) demonstrates the avoidance on dynamic obstacle and the result of route tracing. To experiment the dynamic obstacle, two robots with the same controller is placed and set the environment to cross each other to reach to the target point. As shown on Fig.4(c), the robot in horizontal progress waits for the robot in vertical progress to pass.

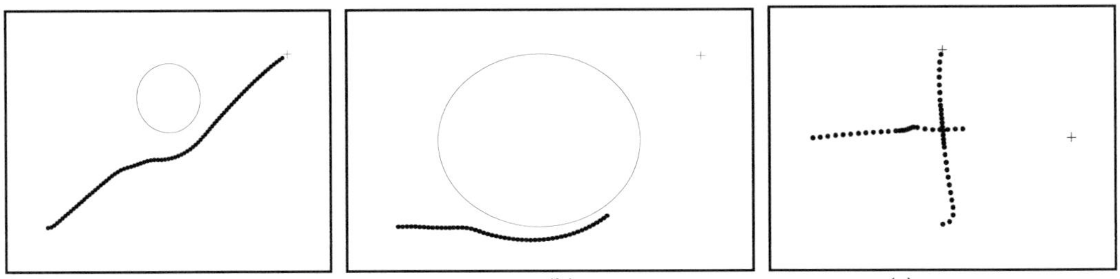

(a) (b) (c)

Figure.4 Simulation of dynamic obstacle avoidance

3.7 Path tracking

In order to obtain simple kinematics modeling, Tong-Jin et al. investigated using the assumption that no slip conditions exist between the wheels of a mobile robot and the ground (T. Fong, 2001, T. Fukao, 2000). A mobile robot kinematics is controlled through steering which is derived from the velocity difference between the two driving wheels. Based on the kinematics model, a controller is designed to track to the reference path through user command when a mobile robot is diverted by an obstacle. Let $q_r = (x_r, y_r, \psi_r)^T$ the configuration of a reference mobile robot and its control inputs which are the translational speed $\upsilon_r(t)$ and the curvature time-derivative $l_r(t)$ (T. Hamel, D, 2001, 1999).

The tracking error vector between the reference mobile robot, q_r and the tracking mobile robot, q_r expressed in the vehicle frame by

$$q_e = T_e(q_r - q_t) \tag{8}$$

Where, $T_e = \begin{bmatrix} \cos\psi_e & \sin\psi_e & 0 \\ -\sin\psi_e & \cos\psi_e & 0 \\ 0 & 0 & 1 \end{bmatrix}$

According to a mobile robot kinematic modeling and differentiating the equation (8), it yields

$$\begin{bmatrix} \dot{x}_e \\ \dot{y}_e \\ \dot{\psi}_e \\ \dot{l}_e \end{bmatrix} = \begin{bmatrix} \upsilon_r l_t^r y_e - \upsilon_r v_t + \upsilon_r \cos\psi_e \\ -\upsilon_r l_t^r x_e + \upsilon_r \sin\psi_e \\ \upsilon_r l_e \\ \dot{l}_r - u_t \end{bmatrix} \tag{9}$$

Where,

v_t : The relative velocity, $v_t = \upsilon_t / \upsilon_r$

l_r : The curvature of the reference trajectory, $l_r = \dot{\psi}_r / \upsilon_r$

l_t^r : The curvature of the real trajectory related to the reference trajectory,

$l_t^r = \dot{\theta}_t / \upsilon_t$

l_e : The error of curvature, $l_e = l_r - l_t^r$

In the nominal case (i.e., when the state variables are supposed to be exactly known), their proof of asymptotic stabilization of the proposed systems involves Lyapunov functions with negative semi-definite derivatives, and the convergence is proved by means of La Salle or Barbalat's theorem. Proving the stability of the origin $[x_e, y_e, \psi_e, l_e]$ can be easily done by considering the Lyapunov candidate function which is suggested as follows in this paper.

$$V(x_e, y_e, \psi_e, l_e) = x_e^2 + y_e^2 + l_e^2 + K_1(1-\cos\psi_e) + (K_2 l_2 + K_3 \sin(\upsilon_r)\cdot\psi_e)^2 \tag{10}$$

The time derivative of the equation (10) is as following.

$$V(x_e, y_e, \psi_e, l_e) = -|\upsilon_r|(2x_e^2 + K_1(K_2 l_e + K_3\cdot\sin(\upsilon_r)\cdot\psi_e)^2 + 2l_e^2 + K_1\cdot\psi_e\cdot\sin\psi_e) \tag{11}$$

K_1, K_2 and K_3 are control gains for handling the convergences of parameters. Equation (11) is negative definite for all ψ_e in $-\pi < \psi_e < \pi$. Using the Lyapunov theorem, the stationary point, $(x_e, y_e, \psi_e, l_e) = 0$ is the only stable equilibrium point inside the subset $\{\psi_e, \psi_e \in -\pi < \psi_e < \pi\}$ although $\dot{V}_e(x_e, x_e, \psi_e, l_e)$ is zero inside the set $\{x_e = 0,\ l_e = 0,\ y_e = -\eta,\ \sin(\upsilon_r)\cdot\psi_e,\ \psi_e \in -\pi < \psi_e \le \pi\}$.

It can be asserted in a large positively invariant domain. Equation (11) has three control gains, K_1, K_2 and K_3 that should be tuned under stability constraints $K_1 > 0$, $K_2 > 0$ and $0 < K_3 < 1$. In other words, these parameters are turning parameters for system's convergence. The good tuning - rule consists in specifying a well-damped behavior of the error system literalized around a situation where a reference mobile robot moves at a constant speed along a straight path.

A dynamic state feedback, stabilizing the dynamic error of equation (11) around zero, is as following:

$$v_t = -\frac{1}{2} l_c^r (K_2 \cdot l_e + K_3 \cdot \sin(\upsilon_r)\cdot\psi_e) + K_3 \cdot \sin(\upsilon_r)\cdot x_e + \cos\psi_e \tag{12}$$

$$u_t = \dot{l}_r + |\upsilon_r|(K_1(K_2\cdot l_e + K_3\cdot\sin(\upsilon_r)\cdot\psi_e) + 2l_e + K_3\cdot\sin(\upsilon_r)\cdot\sin\psi_e \tag{13}$$

The control inputs v_t, u_t are derived by satisfying robust condition, the Lyapunov stability. x_e, y_e are tracking minus l_e and ψ_e is tracking the orientation of a reference mobile robot applying a robust term, sinusoidal term. The control inputs equations (12) and (13) is applied to the parking problem, one of the tracking problems (M. L. Corradini, 2001). The inputs in a reference mobile robot are as following:

$$u_1 = \alpha\sin(wt) \tag{14}$$

$$u_2 = \beta\cos(wt) \tag{15}$$

The periods of x_1 and x_2 are $2\pi / w$ [sec] and the magnitude of x_3 is $\pi\alpha\beta / w^2$ using the inputs, equations (12) and (13). Then the kinematic equations of a reference mobile robot are as following:

$$\dot{x}_M = \cos\psi_M \cdot u_1 \tag{16}$$

$$\dot{y}_M = \tan\psi_M \cdot \cos\psi_M \cdot u_1 \tag{17}$$

$$\dot{\psi}_M = u_2 \tag{18}$$

If the desired coordinates x_{Md}, ψ_{Md} are selected by user, the magnitudes of equations (14) and (15) are as following:

$$\alpha = \frac{x_{Md} \cdot w}{1 - \cos wt} \tag{19}$$

$$\beta = \frac{\psi_{Md} \cdot w}{\sin wt} \tag{20}$$

3.8 Weight unloading control

In the body weight support system, the perpetuity that lessens the weight bearing than the vertical movement of the harness is more important. The aforementioned importance of force has the theoretical foundation in the event of the orthopedic patients who must control the strength applied to the damaged skeleton. Cavagna proved that the mutual transformation is made in between gravity location energy and exercise energy within each walking cycle (P. R. Cavanagh, 1995). Under the optimal walking speed, the percent of the energy recovery reaches to 65%. In this method, the external work of muscle on accelerating gravity of weight center is minimized. If the force (F_0) to lessen the applied weight bearing is consistent while walking, this mechanism is preserved. In such a case, there is no need of requiring this strength. If F_0 is expressed in a part of weight, S, the force for the earth gravity direction, R_s, that applies to a patient is

$$R_s = Mg(1 - S) \tag{21}$$

This makes the patient as if he feels like walking with less gravity. However, this is not the case if F_0 is inconsistent. The external undertaking would not be zero and the mechanism to save energy will be significantly in mess. Therefore, the force to lessen the weight bearing has to be consistent at all times. The motor is operated by feeding back the force measured by the strength sensor that the lifting is made with consistent force at all times. Furthermore, in the event that a patient lost the walking capability, it is recognized to prevent the falling by the loss of strength in the lower limbs to help the safety of the patient.

The user who is equipped with the body weight support system may feel as if walking in water or zero gravity condition while walking.

4. Experimental results

The gait rehabilitation service robot that uses the mobile robot and newly designed BWS system are manufactured in respective module to formulate the integrated walking rehabilitation system. (Fig.5)

In this section, two experiments for verifying the capability of the gait rehabilitation service robot are achieved about walking guide system and body weight support system.

Figure.5 Captured pictures of a robot actuation movie file

This experiment has numerous limitations to place on the patients as subjects that the ordinary persons are selected as the subject of experiment. An adult male of 30 years of age, 172cm of height, and 68kg of weight and an adult male of 27 years of age, 165cm of height, and 54kg of weight are selected as the subjects.

4.1 Walking guide system

The control input is the angular velocity of both side wheels. In a real system, the angular velocity is transformed to -5~5 voltage, and is inputted to the motor driver. The commands of velocity may increase the errors to the final arrival point; however, this error may be ignored because of its small value within large-scale navigation. With consideration for the limits of motor angular velocity, the maximum velocity is set to 5.23 [rad/sec].

Fig.6 shows that a walking guide system is tracking the reference path identical to the initial condition. In this experiment, control of the wheel was satisfactorily achieved under experimental conditions, but there were some errors. As a road surface is not perfectly flat, the error caused by the road surface can be regarded as unmodeled error between the road surface and a mobile robot's wheels. The error was measured as a distance between the final point after 40 sec and vertical distance from the service robot's center in navigation. A number of trials showed the average of the errors to be 0.52 m. The average error was 0.66 m in experiment. Considering unmodeled error, the value of error may be regarded as 0.14 m in the experiment. There are a number of causes for this error. The most significant cause is the friction and slip between the surface of the ground and a mobile robot wheels. A mobile robot with a 2WD type has a nonholonomic characteristic but this experiment shows that the controller is fairly good for path tracking.

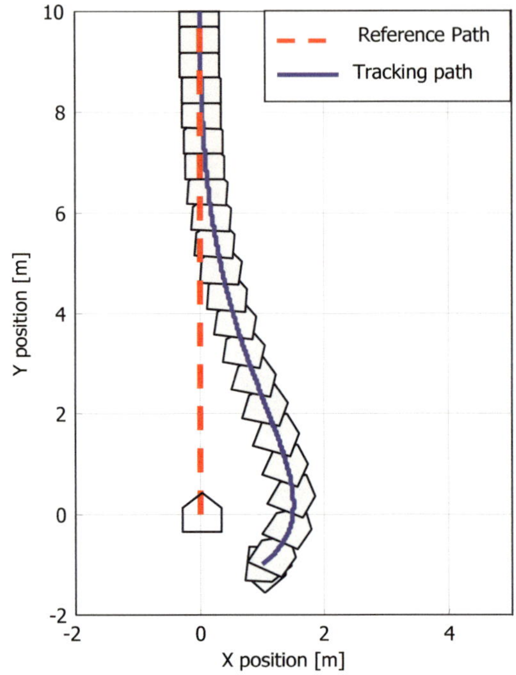

(a) Result of path navigation in an experiment for the path tracking algorithm

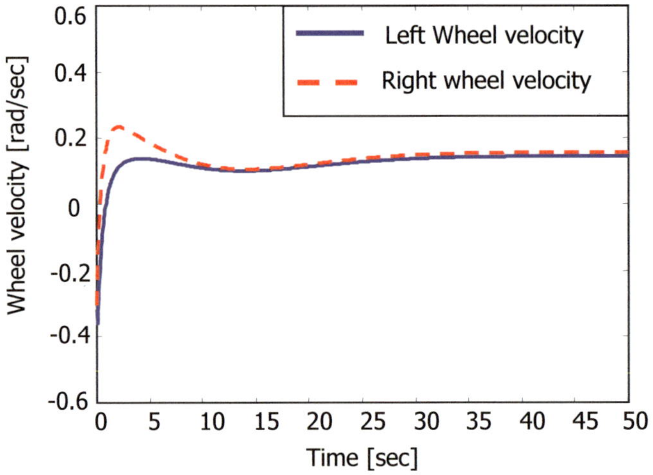

(b) Result of tracking wheel velocities in an experiment for the path tracking algorithm

Figure.6 Path tracking experiment

4.2 Body weight support system

For the details of the experiment, a comparison is made by the experiment on whether the force lifted from the body weight support system is consistently maintained and the EMG measurement following each body weight support system level. Table.5 is the experimental condition on this test.

	BWS Level	Velocity
Experiment condition	- Full weight - 20 % BWS - 40 % BWS	0.2 m/s

Table.5 Experiment on the EMG Signal

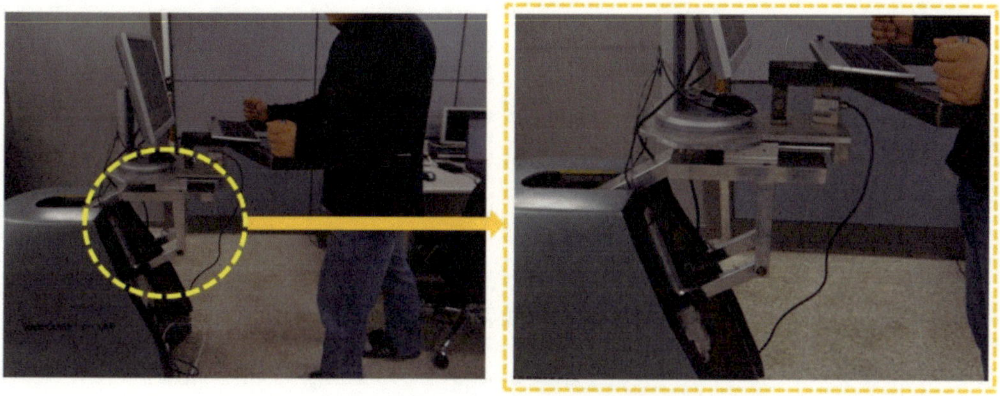

Figure.7 Captured pictures of a body weight support system actuation movie file

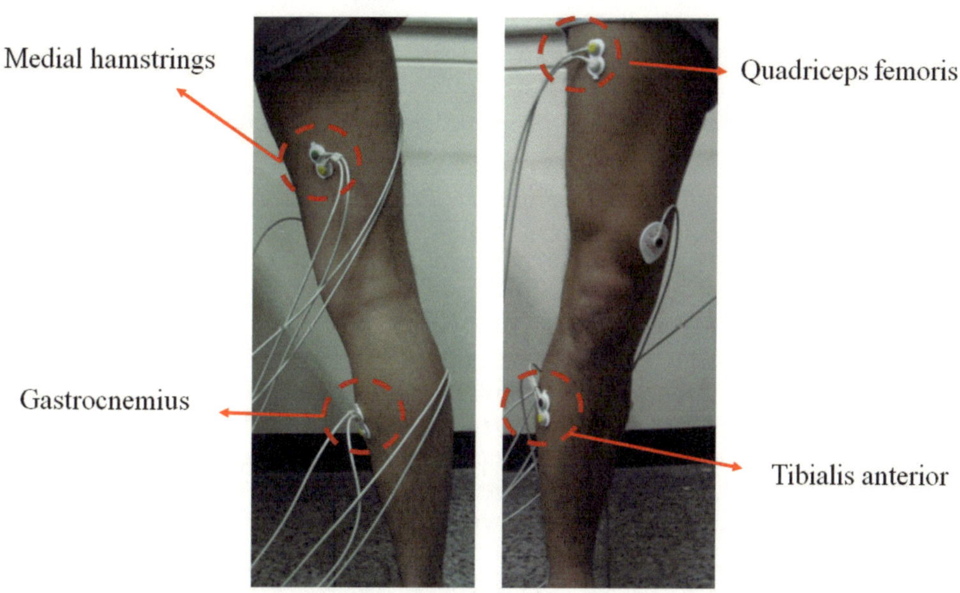

Figure.8 Picture of surface electrodes placed on lower extremity muscles

As shown in Fig.8, Surface electrodes were placed over the following muscles on the subject's dominant side: the quadriceps femoris(QF), medial hamstrings(MH), gastrocnemius(GA), and tibialis anterior(TA). Prior to the application of the surface electrodes, the subject's skin was shaved and cleaned with alcohol. Two electrodes were placed over each muscle with an interelectrode distance of approximately 1cm. During the testing session the subjects walked at 0.2m/s with 0% level during each of the following harness supported ambulation situations: full body weight (FWB), 20% body weight supported (BWS), and 40% body weight supported (BWS). Five seconds of EMG data (1024Hz) were recorded for each condition. The raw EMG was collected using Myosoft software (Noraxon USA®, Scottsdale, AZ®). An analog signal was recorded using the Myosoft software. (Fig.9)

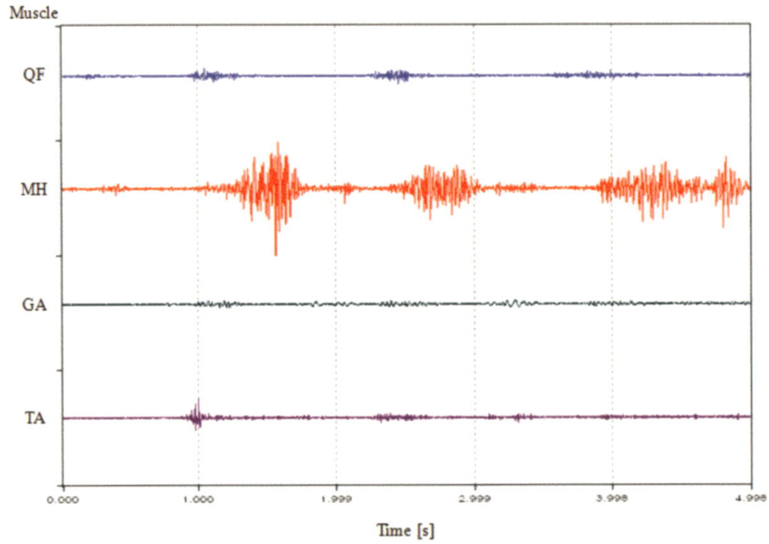

Figure.9 Raw data of EMG signals obtained on full body weight condition

Table.6 and table.7 list the average muscle activity for each muscle group expressed as a percentage of the FWB amplitude. Average EMG activity did not change significantly for any of the muscle groups when FWB was compared to 20% BWS ambulation. The average amplitude for quadriceps decreased to 91.6% of the FWB value, while average hamstring, gastrocnemius and tibialis anterior activity decreased to 96.9, 90 and 99.1% of the FWB value, respectively.

The average EMG activity decreased significantly from FWB to 40% BWS ambulation in the quadriceps muscles, but did not change significantly for the hamstring, gatrocnemius and tibialis anterior muscles. The average amplitude for quadriceps decreased to 72.8% of the FWB value, while the hamstring, gastrocnemius and tibialis anterior activity decreased to 85.9, 91.9 and 98.2% of the FWB value, respectively. The results suggest that muscle activation can be preserved while possibly decreasing load at the joint of lower extremity.

	Q*	H*	G*	T*
FWB	100	100	100	100
20%BWS	91.6	96.9	90	99.1
40%BWS	72.8	95	91.9	98.2

(*Q: Quadriceps, H: Hamstrings, G: Gastrocnemius, T: Tibialis anterior)

Table.6 Average muscle activity (EMG) expressed as a percentage of the FWB amplitude

	Q*	H*	G*	T*
Without robot	57.5	101	67.8	89.2
With ReMeRob	56.1	45.6	72.2	91.4
With RSR (20%BWS)	43	94.9	65	84.6

Table.7 %MVIC of muscle activity (EMG)

5. Conclusions

The gait rehabilitation service robot is developed for the elderly using mobile robot platform in this study. The action scope and the service mechanism of the robot are developed for consideration and analysis of the elderly gait. Also, the robot action is determined by environment information and the distance the elderly and robot. The environment information and distance the elderly and robot are acquired by measuring laser range finder and the linear potentiometer. Obstacle recognition in order to guide was presented and experimentally evaluated. The results of the experiments show that the mobile robot is tracking the reference path using a laser range finder. A robust path tracking algorithm for the path navigation of a 2WD mobile robot is proposed. Path tracking is performed by the robust controller based on a kinematic modeling. The reference path generation algorithm consisted of VFH avoidance and fuzzy logic for guide could be confirmed by the experiments. In gait assistance system's view, the conclusions of this study are as follows:
First, the performance of a local map building was enhanced using the Kalman filtering based on the dynamic modeling of a laser range finder. And VFH avoidance and fuzzy logic were suggested in order to apply the path tracking algorithm. Second, Robustness with a modifiable convergence was constructed by using a Liapunov function with control gains. The path algorithm was verified the availability by applying and implementing the path algorithm to the service robot. Third, User-command system was suggested by man machine interface that transfer to the elderly' decision about walking to service In body weight support system's view, the conclusions of this study are as follows:
First, the gait velocity was selected as an ergonomic design parameter of the service robot from investigating the elderly gait data. Second, the weight unloading force measured by the load cell was controlled in body weight support system and A 4-bar linkage was applied to mechanism of the body weight support system; that is, part of mechanism contacted user's upper arm keeps horizontality continuously.
GRSR is designed for elderly to communicate with remote site. The elderly can be treated by internet using web-cam image communication in remote site. Remote therapy or Doctor can be undertaken via tele-monitoring, thereby reducing length of hospital stay and avoiding unnecessary admission. Further, this system can be used as emergency calling and means of useful information acquisition for elderly. Fig.10 shows the wireless internet communication software and an image communication based on a web-cam through the internet. These are examined as ergonomic design parameters for the service robot.

(a) Realization of Basic Internet On-line System on GRSR

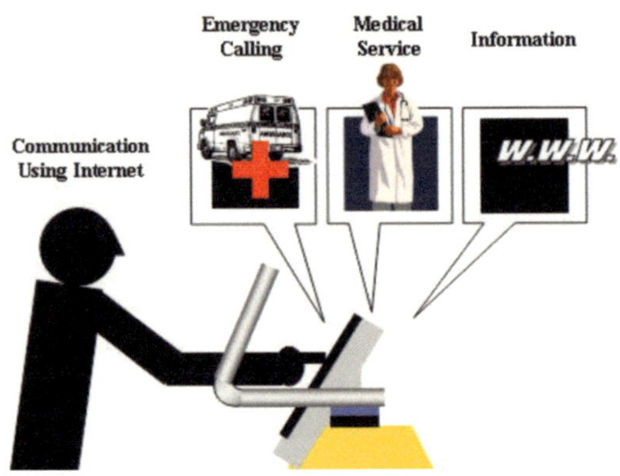

(b) Concept of Web Communication

Figure 10. Web Communication

6. Reference

A. Wernig and S. Muller, 1992, Laufband locomotion with body weight support improved walking in persons with severe spinal cord injuries, *Paraplegia*, Vol. 30, pp. 229-238

R. W. Baloh, K.M. Jacobson, T.M. Socotch, 1993, The effect of aging on visual-vestibular responses, *Exp Br Res*, Vol. 95, pp. 509-516.

R. W. Baloh, 2002, Disequilibrium and gait disorders in older people, *Rev Clin Gerontol*, Vol. 6, pp. 41-48

D. A. Winter, 1991, The Biomechanics and Motor Control of Human Gait: Normal, Elderly and Pathological, 2nd edition. Waterloo: University of Waterloo Press

D. Hahnel, W. Burgard, D. Fox, S. Thrun, 2003, An Efficient Fast SLAM Algorithm for Generating Maps of Large-Scale Cyclic Environments from Raw Laser Range Measurements, *Proceedings of the IEEE/RSJ International Conference on Intelligent Robotics and Systems*, Vol. 1, pp. 206-211

H. Barbeau, M. Wainberg, and L. Finch, 1987, Description of a system for locomotor rehabilitation, *Med. Biol. Eng. Comp.*, Vol. 25, pp. 341-344

J. J. Leonard and H. F. Durrant-Whyte, 1991, Simulatneous Map Building and Localization for an Autonomous Mobile Robot, *Proceedings of the IEEE/RSJ International Conference on Intelligent Robotics and Systems*, pp. 1442-1447

J. Borenstein, Y. Koren, 1991, The vector field histogram-fast obstacle avoidance for mobile robots, *IEEE Transactions and Automation*, Vol. 7, No. 3, pp. 278-288

J. L. Poole, 1991, Age relates changes in sensory system dynamics related to balance, *Phys Occup Ther Geriatr*, Vol. 10, pp. 55-66

K. E. Norman, A. Pepin, M. Ladouceur, H. Barbeau, 1995, A treadmill apparatus and harness support for evaluation and rehabilitation of gait, *Arch. Phys. Med. Rehab.*, Vol. 76, pp. 772-778

L. Finch, H. Barbeau, B. Arsenault, 1991, Influences of partial weight bearing on normal human gait: The development of a gait retraining strategy, *Can. J. Neurolog. Sci.*, Vol. 12, p. 183

L. Sudarsky, 1990, Geriatrics: Gait disorders in the elderly, *New Engl J Med*, Vol. 322, pp. 1441-1446

L. Wolfson, 1992, A dynamic posturography study of balance in healthy elderly, *Neurology*, Vol. 42, pp. 2069-2075

M. B. Smith, J. Casalena, D. A. Streit, and P. R. Cavanagh, 1995, Variable gravity rehabilitation system: Design and implementation of a full scale prototype, *J. Appl. Mechanisms Robot.*, Vol. 2, No. 3, pp. 7-12

M. L. Corradini, G. Orlando, 2001, Robust Tracking Control of Mobile Robots in the Presence of Uncertainties in the Dynamical Model, *Journal of Robotic Systems*, Vol. 18, No.6, pp. 317–323

M. Visintin and H. Barbeau, 1994, The effects of parallel bars, body weight support and speed on the modulation of the locomotor pattern of spastic paretic gait. A preliminary communication, *Paraplegia*, Vol. 32, pp. 540-553

N. Teasdale, C. Bard, J. Larue, M. Fleury, 1993, On the cognitive penetrability of the postural control, *Exp Aging Res*, Vol. 19, pp. 1-13

N. Teasdale, G. E. Stelmach, A. Breunig, H. J. Meeuwsen, 1991, Age differences in visual sensory integration, *Exp Brain Res*, Vol. 85, pp. 691-696

P. R. Trueblood, L.Z. Rubenstein, 1991, Assessment of instability and gait in elderly persons, *Comprehensive Ther*, Vol. 17, pp. 20-29

R. F. Vassallo, J. Santos-Victor, H. J. Schneebeli, 2002, Using motor representations for topological mapping and navigation, *IEEE/RSJ International Conference on Intelligent Robotics and Systems*, Vol. 1, pp. 478 – 483

S. Hesse, M. Malezic, A. Schaffrin, and K. H. Mauritz, 1995, Restoration of gait by combined treadmill training and multichannel electrical stimulation in nonambulatory hemiparetic patients, *Scand. J. Rehab. Med.*, Vol. 27, pp. 199-204.

S. McLachlan, J. Arblaster, D. K. Liu, J. V. Miro, L. Chenoweth, 2005, A multi-stage shared control method for an intelligent mobility assistant, *Proceedings of the 2005 IEEE International Conference on Rehabilitation Robotics, 2005*

S. J. Hall, 2003, Basic Biomechanics 4th edition, McGraw-Hill

T. Fukao, H. Nakagawa, N. Adachi, 2000, Adaptive Tracking Control of a Nonholonomic Mobile Robot, *IEEE Transactions On Robotics And Automation*, Vol. 16, No. 5, pp. 609-615

T Fong, C Thorpe, C Baur, 2001, Advanced Interfaces for Vehicle Teleoperation: Collaborative Control, Sensor Fusion Displays, and Remote Driving Tools, *Autonomous Robots*, Vol. 11, pp. 77–85

T. Hamel, D. Meizel, 2001, On robustness and precision of mobile robots missions, *Autometica*, Vol. 37, pp. 437-444

T. Hamel, D. Meizel, 1999, Robust control laws for wheeled mobile robots, *International Journal of Systems Science*, Vol. 27, No. 8, pp. 659-704

T. I. Ide, N. Akamatsu, and I. Nakajima, 1986, Development of a gait training robot (AID-1), *Orthop. Surg. Traumatol.*, Vol. 29, pp. 825-828

T. J. Park, J. S. Han, C. S. Han, 2004, Development of the gait assistant mobile robot using ergonomic design, *Proc. of the IEEE Int. Conf. on Robotics and Automation*, pp.2037-2042

T. Pillar, R. Dickstein, and Z. Smolinski, 1991, Walking reeducation with partial relief of body weight in rehabilitation of patients with locomotor disabilities, *J. Reh. Res. and Dev.*, Vol. 28, No. 4, pp. 47-52

V. T. Inman, H. J. Ralston, F. Todd, 1981, Human Walking. Baltimore: Williams-Wilkins

10

Experiences Developing Safe and Fault-Tolerant Tele-Operated Service Robots. A Case Study in Shipyards

Diego Alonso, Pedro Sánchez, Francisco J. Ortiz, Juan A. Pastor, Bárbara Álvarez and Andrés Iborra
Division of Electronics Engineering & Systems (DSIE)
Universidad Politécnica de Cartagena
Spain

1. Introduction

Human operators use tele-operated service robots for performing more or less hazardous operations (manipulation of heavy and/or dangerous products) in more or less hostile environments (nuclear reactors, space missions, warehouses, etc). Anyway, independently of the operation, the robot has to interact with both the environment it is working on and with human operators. Therefore, it is essential that the design (which include both software and hardware) of the robot involves no risk, or at least an acceptable level of risk, neither for the operators, nor for the environment nor for the robot itself.

Nevertheless, it is not always possible to make a system free of failures in its design or operation. Apart from the risk inherent to the use of the mechanisms themselves, these systems work in hazardous environments, where the probability of the risk is higher than normal. Should a failure happen, its consequences could even involve the loss of human lives. (Neumann, 1994) documents many cases of computer-related failures, such as the Therac-25 (a radiation-therapy device), the missiles shield in Saudi Arabia, etc.

Nevertheless, safety aspects are seldom included in the early phases of the system design process from the beginning, even though they are a critic aspect. Generally, safety has to conform and adapt to the already designed system and not vice versa, when it is widely known that safety involves not only the design of the software but also the hardware. Even more, a simple hardware solution can eliminate a hazard or simplify the software design in many situations.

However, the identification of safety requirements should not be different from the identification of the rest of requirements of the system. It only requires a more thorough study due to their importance (human lives and equipment integrity may depend on it). On the other hand, safety has a big repercussion in both the specification and design phases, especially when the time to define the architecture of the system arrives. Its impact is even bigger by the need to avoid common failure modes, which can propagate failures within different units of the system.

There are a number of standards and techniques for addressing safety requirements in electronic devices in general and for industrial arm robots, but none specifically designed for addressing the elicitation of safety requirements or for proposing software/hardware solutions in the domain of service robots.

With more than twelve years of experience in the design of software applications for tele-operated service robots (Iborra et al., 2003; Fernández et al, 2005), the DSIE research group at the Universidad Politécnica de Cartagena has developed several tele-operated service robots in the context of the **EFTCoR**[1] project (*Environmental Friendly and Cost-Effective Technology for Coating Removal*). This project was part of the European Industry effort to introduce environmental friendly ship maintenance. The EFTCoR project addressed the development of a solution to the problem of retrieval and confinement of the sub-products obtained from the ship maintenance operation (oxide, paint and adherences mainly). The variability of the domain (hull dimensions and shapes differ widely) and hard working conditions for robotic devices, imposed the design of different robotic systems, each adapted to the specific problem. In such a dangerous and hazardous environment, it is compulsory to address, from the beginning of the project, both the hardware and software safety requirements in order to design safe robots.

The main objectives of this chapter are (1) to stress the importance of capturing the safety requirements early in the design process, and (2) to present a systematic approach, based in both standards and some traditional safety techniques and solutions, that could guide other designers considering safety requirements from the beginning of their projects.

In order to illustrate such a systematic approach, we expose a thorough study of the safety requirements that a crane robot (a member of the EFTCoR project) had to conform to in order to work in such a hazardous environment as shipyards are. The design decisions adopted to conform such safety requirements will be also presented.

This chapter is structured in five sections. Section two details the state of the art in safety standards and techniques. The following two sections are devoted to the description of the safety requirements elicitation methodology and an example of its application in the context of the design and development of the EFTCoR crane robot. Finally, section five summarises the main results and conclusions extracted from our experiences and outlines future lines of work. The complete tables of the proposed elicitation process are presented in the appendix, at the end of this book chapter.

2. Survey of safety standards and techniques

Before going on, we introduce the meaning of some of the terms that appear in this chapter. According to (Douglass, 2003), a *risk* is an event or condition that can occur but is undesirable; *safety* is the characteristic of a system that does not incur too much risk to persons or equipment and an *accident* is damage to property o harm to persons, the happening of a risk. A *safety system* is, according to the definition of ANSI/RIA (Ansi/Ria, 1999), a system that has been tested, evaluated and proven to operate in a reliable and acceptable manner when applied in a function critical to health and welfare of personnel. According to (Leveson, 1995), a *hazard* is a state or set of conditions of a system (or object)

[1] The EFTCoR (http://www.eftcor.com) project was founded by the EU 5th Framework Programme (GROWTH G3RD-CT-00794) with 2M€. It included ten partners from six European countries.

that, together with other conditions in the environment of the system (or object) will inevitably lead to an accident (loss event).

There are several approaches to manage safety in the literature, but they are either too general or not specifically targeted at the tele-operated robot domain. Many deal with the problem of designing a standard that guides the whole process (from identification to solution) while others are simple tools or techniques. Among these standards, we want to stress the European Standard EN 61508:2001 (EN 61508, 2003) and the American ANSI/RIA R15.06-1999 (Ansi/Ria, 1999). Among the techniques for safety designs it is worth highlighting fault trees analysis (Hansen et al., 1998) and **ROPES** (Douglass, 1999) (*Rapid Object-oriented Process for Embedded Systems*).

- **EN 61508:2001**. This European standard sets up a generic approximation for dealing with all the activities related to the life cycle of the systems that use electric and/or electronic and/or programmable devices for safety functions. The other main purpose of this standard is to serve as a basis for the development of specific standards for each application sector, which would take into account techniques and solutions typical of the sector.
- **ANSI/RIA R15.06-1999**. The objective of this standard is to enhance the safety of personnel using industrial robot systems by establishing requirements for the manufacture (including remanufacture and overhaul), installation, safeguarding methods, maintenance and repair of manipulating industrial robots. It is the intention of this standard that the manufacturer (including re-manufacturer and re-builder), the installer and the end-user have specific responsibilities.
- **Fault trees**. It is one of the most popular approaches to identify, evaluate and manage safety requirements. These trees provide a graphical notation and a formal support that makes it easy to make the analysis from the perspective of the system failures and their origins. However, they do not offer a global framework for requirement specification as a discipline.
- **ROPES** is, in words of Douglass, *"a development process that emphasises rapid turnaround, early proofs of correctness and low risk"*. ROPES is an iterative process that organises the design process in small, incremental steps. Douglass proposes an eight-step methodology for dealing with the safety aspects of any system.

3. Description of the safety requirements elicitation process

As already outlined, there is no specific standard or methodology for addressing the elicitation of safety requirements for service robots, or for proposing software/hardware solutions that fulfil them. Instead, there are a number of standards and techniques for addressing safety requirements in electronic devices in general, and for industrial arm robots, but none specifically designed for service robots.

In this vein, this work presents a systematic approach, based in both standards and some traditional safety techniques and solutions, which could guide other designers considering safety requirements from the beginning of their projects.

As last section shown, until a new standard derived from EN 61508 and targeted at robotics appear, only the ANSI standard offers a specific guide that is close enough to the domain of tele-operated robots so that it can be adapted to this domain.

In order to complete those aspects not covered by the standard ANSI, the proposal *"eight steps to safety"* from Douglass (Douglass, 1999) has been adopted and adapted. In this work, Douglass proposes some design patterns oriented to the achievement of a particular safety objective, such as *multi channel voting pattern, watchdog pattern, safety executive pattern*, etc. By using these patterns, it is possible to design software solutions that conform to the needs imposed by the ANSI standard, according to the level of risk of a particular hazard.
Finally, *fault trees* could be used to obtain the possible causes of the failures that are analysed in second step of the methodology we propose. Fault trees analysis is a very used and mature technique, but it does not help measuring, classifying or solving failures, and thus, we have not applied this technique.
The four-step methodology presented in this chapter proposes the fusion of the standards and techniques presented in the previous section. It encourages the tracking of safety throughout the life cycle of the robot (as EN 61508 proposes) and uses the ANSI standard as a guide to classify hazards and to propose solutions. By completing ANSI with the contributions of ROPES, it is possible to deal with the design of software-based solutions that are more complex than a simple barrier. Figure 1 depicts a diagram showing the different steps that make the proposed methodology up.

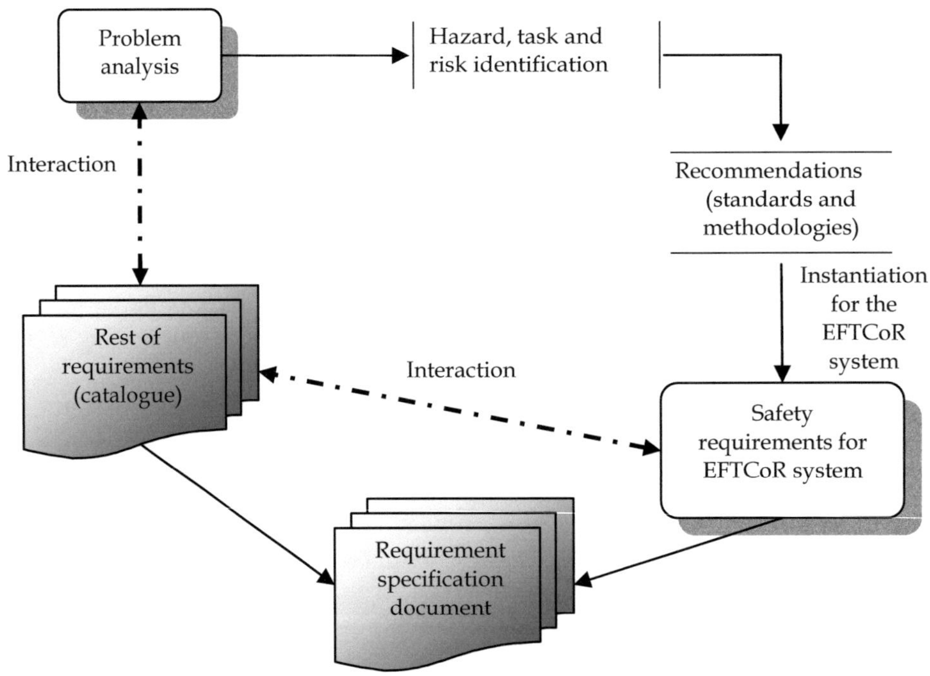

Fig. 1. Safety-driven requirements elicitation methodology.

STEP 1 ► Identify hazards

It is desirable that a system should normally work without imminent hazards. Therefore, the first step is to identify all the tasks that involve the use of the system and that have potential hazards. After that, each task is analysed for describing the hazards associated to it. Some possible sources for the identification of hazards, which can serve as a starting point in their identification, are the following ones (extracted from (Ansi/Ria, 1999)):

- The movement of mechanical components, especially those that can cause trapping or crushing.
- Stored energy in moving parts, electrical or fluid components.
- Power sources: electrical, hydraulic, pneumatic, etc.
- Hazardous atmospheres, material or conditions: explosive or combustible, radioactive, high temperature and/or pressure, etc.
- Acoustic noise, vibrations, EMI, etc.
- Human failures in design, construction, installation, and operation, whether deliberate or not.

This analysis of hazards also include the identification of the possible causes of the failure (hardware, software or human), the task in which it can happen, the reaction to the happening of the hazard, and some temporal data (adopted from ROPES (Douglass, 2002)). These time related parameters try to quantify (1) how long the hazard can be tolerated before it results in an accident (*tolerance time*), (2) the maximum amount of time to detect the happening (*detection time*), and (3) the maximum time to react to it (*reaction time*).

STEP 2 ► Identify risks

The objective of this second step is to (1) identify the possible risks of the system, (2) classify them according to the impact they have on the environment, and (3) link them to the hazards identified on the first step. The ANSI standard states that three characteristics have to be evaluated for each identified risk: its level of severity, its level of exposure and its level of avoidance. Each of these characteristics has two different values, resulting in eight possible combinations. Depending on the different values of these characteristics, a **RRC** *(Risk Reduction Category)* is obtained (see Table 1). Based on the RRC, ANSI requires a certain level of performance of the safeguard and circuits that are to be design to reduce the risk (simple, single channel, single channel with monitoring and control reliable). Moreover, ANSI also recommends the adoption of safety policies to help human operators avoid some risks (training, presence detectors, security barriers, etc).

After applying the safeguards designed for the specific RRC, a new analysis is performed to calculate the residual risk, just to be sure that the risk is kept at a tolerable level for both the system and the environment. This process does not end here, but has to be repeated during the life cycle of the robot to ensure that no new risk appears and that the risks already identified are kept under control.

STEP 3 ► Specify safety requirements

The purpose of this third step is to extract the system safety requirements from the results of the previous steps. This step is quite difficult to perform because neither the ANSI standard nor ROPES offer a methodology to deduce the requirements from the previous results, so this extraction has to be manually done. For addressing this third step, it is necessary to have or develop:

1. An appropriate process for requirement harvesting.
2. A way to catalogue them, in order for the requirements to be reused in other systems of the domain of application (tele-operated robots in this case).
3. Tools for tracing the use of the requirements throughout the development process and, in particular, until designing the architecture of the system.

RRC	Procedure	Control level
R4	Safeguarding, at a minimum, shall be by administrative means, awareness means including audio/visual warnings and training.	Simple Control (consult next table).
R3A, R3B	Non-interlocked barriers, clearance procedures and equipment.	Choose between Single channel and Simple Control (consult next table).
R1	Hazard elimination or hazard substitution	Control Reliable (consult next table).
R2A, R2B, R2C	Preventing access to the hazard or stopping the hazard (interlocked barrier guards, etc.)	Choose between Control Reliable, Single Channel or Single Channel with Monitoring (consult next table).

Description of the different control levels
1. Control reliable safety circuitry shall be designed, constructed and applied such that any single component failure shall not prevent the stopping action of the robot. These circuits shall be hardware based, and include automatic monitoring at the system level: a) The monitoring shall generate a stop signal if a fault is detected. A warning shall be provided if a hazard remains after cessation of motion; b) Following detection of a fault, a safe state shall be maintained until the fault is cleared. c) Common mode failures shall be taken into account when the probability of such a failure occurring is significant. d) The single fault should be detected at time of failure. If not practicable, the failure shall be detected at the next demand upon the safety function. 2. Single channel safety circuits shall be hardware based or safety related software and firmware based controllers, include components which should be safety rated, be used in compliance with manufacturers' recommendations and proven circuit designs. 3. Single Channel with Monitoring safety circuits shall include the requirements for single channel, shall be safety rated, and shall be checked (preferably automatically) at suitable intervals. a) The check of the safety function(s) shall be performed at machine start-up, and periodically during operation; b) The check shall either: allow operation if no faults have been detected, or generate a stop signal if a fault is detected. A warning shall be provided if a hazard remains after cessation of motion; c) The check itself shall not cause a hazardous situation; d) Following detection of a fault, a safe state shall be maintained until the fault is cleared. 4. Simple Control. Simple safety circuits shall be designed and constructed using accepted single channel circuitry, and may be programmable.

Table 1. Risk Reduction Category explanatory table (extracted and summarised from ANSI).

STEP 4 ► Make safe designs

The design of the software architecture of the system must consider the safety measures and avoid failures that spread through the whole system. A safe design must start with the previous security requirements (third step) to adopt a concrete architectural pattern that could be periodically reviewed when new hazards are identified. To be able to do it, to be able to be adaptable, a rigorous architectural approach that allows the evolution of the architectural model due to new requirements or by evolution of the conditions of work is necessary (which is also out of the scope of this book chapter).

4. Application of the elicitation process in the context of the EFTCoR project

This section describes the application of the safety requirements elicitation methodology in the context of the EFTCoR project in order to come up with a design that satisfies the safety requirements of the robot. The early analysis of the possible hazards and safety conditions allowed us to make the hardware and software design of the robot addressing the safety requirements from the beginning. Specifically, this section is divided into the following sub-sections:

- Section 4.1 briefly describes the objective of the EFTCoR project, the working environment, and the solution developed by the DSIE research group.
- Section 4.2 presents the EFTCoR commercial crane and the application of the proposed safety requirements elicitation methodology to refine its design. The exhaustive compilation and classification of the safety requirements presented in this sub-section can be consulted in the appendix.

4.1 Brief overview of the EFTCoR project

The EFTCoR family of robots offers a global environmentally friendly solution to the problems related to the most dangerous hull maintenance operations. These operations consist of periodic (every four to five years) removal of sea adherences and the hull coating followed by hull repainting. These operations preserve the hull integrity, guarantee safe sailing conditions, and maintain a smooth hull surface, which minimizes fuel consumption, reduces operating costs, and prevents excessive atmospheric contamination. Other maintenance operations are scheduled or even delayed to coincide with hull cleaning and repainting. The existing hull cleaning technology, grit blasting (see Figure 2-a) (IMO, 1999), is highly pollutant, environmentally unaffordable, dangerous for human operators, and it is progressively being banned in Europe. The solution developed in the context of the EFTCoR project comprises two families of robots:

- Tele-operated cranes with blasting tools for cleaning big areas (what is called "*full blasting*"), normally a ship vertical surface.
- Tele-operated climbing vehicles with a small blasting tool for cleaning small areas (what is called "*spotting*"), normally a ship bottom or bow.

All these robots consist of a primary positioning system, capable of covering large hull areas, and a secondary positioning system, mounted on the primary system, that can position a tool over a relatively small area (from 1 to 16 m^2). These robots have been developed to achieve the objective of performing the current hull cleaning operations in a way that avoids the emissions of residues to the environment and enhances the working conditions of the shipyard operators without worsening the current costs and operation times (see Figure 2-b).

a) Before EFTCoR

b) With EFTCoR crane

Fig. 2. Manual vs. automatic blasting operation.

The design of such a complex system as the EFTCoR involves the necessity of early detection and identification of failures so that correcting measures can be adopted early in the design of the robot. The fundamental characteristics of the EFTCoR that makes it compulsory to take into account the need of a safe approach when designing the robots are summarised by the following points:

- The operator uses a heavy mechatronic device whose range of movement can cause serious damage (see Figure 2-b).
- Some maintenance operations include the blasting of the hull with high-pressure abrasive particles. The energy of the jet makes it very dangerous for human operators and for the rest of the equipment, so it is necessary to train operators in the use of the tool, to maintain the equipment in perfect conditions and to install all the security components needed. In addition, the impact of the jet over the ship hull produces a lot of dust, worsening the condition of the working place.
- The system has to be designed for working outdoors, so it has to be able to deal with atmospheric agents that can alter its normal operation (rain, water on the ground, dust, noise, wind, etc.).
- The working environment of the robots (shipyards) is very dynamic: there are many cranes, load and unload of heavy equipments, many operators moving around (either working on the robot or conscious or not of its presence), etc.

4.2 Application of the proposed safety requirements elicitation methodology

This section describes in detail the application of the proposed methodology to the design of the EFTCoR crane robot. This robot is composed of a commercial crane as the primary positioning system and a cartesian robot (XYZ table) as the secondary positioning system (see Figure 3). The crane has its own control (provided by the manufacturer), a height of twelve meters and a weight of twenty tons, which make unavoidable the movement of the robot with the consideration of safety requirements. It also has, in its central zone, an articulated arm of two tons for holding the XYZ table (which includes a cleaning tool). The control system of the XYZ table has been designed to follow the cleaning instructions from a human operator or from a computer vision system, which locates the areas of the hull that have to be blasted and commands the crane robot.

Fig. 3. Crane robot for cleaning vertical surfaces in EFTCoR.

STEP 1 ► Identify hazards

Starting from the functional requirements of the EFTCoR system (which were part of the initial project proposal), thirty different tasks with a potential hazard were identified (see Table 2). As shown in the table, these tasks are performed not only by the operator of the robot but also by the maintenance and cleaning staff, they can have been planned or not, and their frequency can be daily, weekly, monthly, annually, etc.

Afterwards, the hazards related to the tasks to be performed by the robot (which are shown in Table 2) have to be identified. Table 3 shows (an excerpt of) the thirty one identified hazards and classifies them according to the tasks in which they can appear, the risk level associated to the hazard, the possible sources for the hazard, and their probability, as all these characteristics are needed to perform the following step. Table 3 also outlines some of the reaction measures that can be adopted to minimise the effects of the corresponding hazard. The *first step* of the proposed methodology ("identify hazards") is fulfilled after the elaboration of these two tables. Lastly, it is worth highlighting that:

- There is no direct relationship between the number of tasks a system must perform and the number of hazards that can be identified and characterised.
- There is no methodology that can help finding the corresponding hazards and their possible sources. This process depends on the domain and the requirements of the system, although a mature technique such as fault tree analysis could help finding them.

Task	Type	Description
T1	operator	Move the primary positioning system (rail)
T2	operator	Move the primary positioning system (vertical axis)
T3	operator	Move the primary positioning system (arm)
T4	operator	Coordinate the primary positioning systems for positioning the secondary positioning system (XZY table)
T5	operator	Move one of the axis of the secondary positioning system (XYZ table)
T6	operator	Coordinate the secondary positioning system axes (XYZ table)
T7	operator	Execute a sequence of movements of the secondary positioning system.
T8	operator	Execute a sequence of movements of the primary positioning system.
T9	operator	Activate the cleaning tool
T10	operator	Deactivate the cleaning tool
T11	operator	Execute an emergency stop of both positioning systems
T12	operator	Stop the primary positioning system (rail)
T13	operator	Stop the primary positioning system (arm and vertical axis)
T14	operator	Stop the primary positioning system (XYZ table)
T15	operator	Disable axis
T16	operator	Change movement/control parameters
T17	operator	Change working mode (manual/automatic)
T18	operator	Restock grit.
T19	operator	Test robot anchors
T20	maintenance	Calibrate primary positioning system
T21	maintenance	Calibrate secondary positioning system
T22	maintenance	Calibrate machine vision system
T23	maintenance	Repair one of the axis (primary or secondary positioning system)
T24	maintenance	Manually tool assembly
T25	maintenance	Manually tool disassembly
T26	maintenance	Substitute video wire
T27	maintenance	Substitute/repair the grit-blasting tool.
T28	maintenance	Substitute/repair the communication infrastructure
T29	maintenance	Substitute/repair the control infrastructure
T30	cleaning	Clean the machine vision system

Table 2. Task list of operations that can be performed when using the robot.

STEP 2 ► Identify risks

The *second step* starts from the results previously obtained in Table 3. In this step, different risks are evaluated and described for each hazard (see excerpt in Table 4). Each risk is characterised from different points of view (severity, exposure and avoidance, shown in the complete table in the appendix) and then associated a given RRC (consult Table 1), according to these characteristics.

Hazard	Task	Risk Level	Origins	Prob.	Reaction measures
H1. The cleaning tool hits the ship hull	T5-7, T21	Severe	Breakage or error in the axis controller. Comm. failure (logic or physic)	Low	Emergency alarm[2] Stop axis and move it away from the hull. Power axis off.
H2. Over exposure of the tool or delayed deactivation.	T10	Slight	Breakage of an axis of the secondary positioning system or the tool. Software scheduling error. Communication failure.	Low	Emergency alarm. Power tool off. Stop grit-blasting.
H3. Workers in the trajectory of the primary positioning system (rail).	T1	Very severe	There is a person in the trajectory of the movement of the primary positioning system.	Med.	Emergency alarm. Stop the robot. Emergency stop.
H4. Equipment in the trajectory of the primary positioning system (rail).	T1, T20	Severe	There is an obstacle in the trajectory of the primary positioning system.	Med.	Emergency alarm. Stop the robot. Emergency stop.
H5. Obstacle in the trajectory of the primary (vertical axis or arm).	T2-4, T20	Very severe	There is an obstacle in the trajectory of the primary positioning system.	High	Emergency alarm. Stop the robot. Emergency stop.
H6. Obstacle in the trajectory of the secondary positioning system.	T5-8, T21	Very severe	There is an obstacle in the trajectory of the secondary positioning system.	Low	Emergency alarm. Stop the robot. Emergency stop.

Table 3. Excerpt of the hazard identification list. Consult complete table in the appendix.

Afterwards, a series of solutions to leverage the risk are proposed, and the hazard is re-evaluated (severity, exposure and avoidance levels) to verify its new RRC. Depending on the final RRC value, additional measurements should be taken into account to leverage the remaining risk, re-evaluating the system again, until the needed residual RRC is finally achieved. For instance, the hazard H1 in Table 4 (consult appendix) was evaluated to have a severity level S1, an exposure level E2 and an avoidance level A1. According to the ANSI standard, these levels result in an RRC level R3A.

[2] Different alarm levels: visual, acoustic, etc, depending on the severity.

STEPS 3 and 4 ▶ Specify safety requirements and Make safe designs

The application of the steps 3 and 4 of the proposed methodology is out of the scope of this book chapter. Specifically, in step 3 a catalogue of safety requirements for the family of robots is extracted with the purpose of being able to reuse it when developing similar robots. We are currently developing a CASE tool for easing the reutilization of this catalogue.

Hazard	Risk	RRC	Solution	RRC
H1. The tool hits the ship hull.	R1. Tool breakage or damage to the secondary positioning system.	R3A	Add bumpers to the head of the tool. Limit Z-axis maximum torque.	R4
H2. Overexposure of the tool or delay in its deactivation.	R2. Damage to the ship hull surface.	R3A	Add a software timer (compliant with 6.4 ANSI) to control the tool working time. Add a motion sensor to the secondary positioning system and link it to the timer.	R4
H3. Workers in the trajectory of the primary positioning system (rail).	R3. Worker run over.	R2A	Add sensors to detect obstacle presence in the rail. These sensors are directly wired to the robot safety control infrastructure. Add a siren for signalling robot motion.	R3B
H4. Equipment in the trajectory of the primary positioning system (rail).	R4. Damage to the equipment and to the primary positioning system.	R2A	Same as H3	R3B
H5. Obstacle in the trajectory of the primary (vertical axis or arm).	R5. Damage to the primary and to the obstacle	R1	Add sensors for detecting obstacles in the trajectory of the arm and wire them to the robot safety control infrastructure.	R3B
H6. Obstacle in the trajectory of the secondary positioning system.	R6. Damage to the secondary or to the tool	R2B	Hw/Sw torque control for detecting overload.	R4

Table 4. Excerpt of the hazard identification list. Consult complete table in the annex.

Regarding step 4, the impact of considering the previously obtained safety solutions on the software architecture of the robot is detailed in Table 5, while the impact on the robot hardware design has already been summarised in Table 4.

4.3 Brief summary of the application of the methodology

This last sub section presents a brief summary of the conclusions that can be extracted from the whole study presented in this book chapter. To do so, the thirty one identified hazards (shown in Table 3 in the appendix) have been classified in six groups, depending on the type of safeguard adopted (consult Table 5). The percentage shown in the last column of Table 5 is relative to the total number of safety requirements that were present in the EFTCoR project initial proposal, sixty in total.

The following conclusions can be extracted from this study:

- **Forty five percent** of the safety requirements do not affect the architectural design neither its possible evolution.
- **Fifty five percent** of the safety requirements do affect software architecture, of which:
    - **Forty percent** imply the addition or extension of some components so that the state of the actuators can be double-checked.
    - **Six dot six percent** imply the design and adoption of redundant nodes.
    - **Eight dot six percent** imply the addition of new sensors to the robot to monitor the system (generally, safety-related sensors).

	Solution kind	Related hazard	Total safety req
Solution affects Sw architecture	Addition of safety Sw modules in the robot control unit.	H2, H14, H16, H17, H21	**8.33%**
	Addition of extra safety elements controlled by Sw modules (e.g. emergency stop, alarm bells and lights, etc).	H1, H2, H3, H3, H4, H4, H5, H10, H12, H12, H13, H13, H15, H15, H21, H22, H25, H25, H24, H26, H26, H28, H31, H27	**40%**
	Addition of redundant Sw safety modules and systems.	H10, H23, H5, H11	**6.66%**
			subtotal 55%
Solution does not affect Sw architecture	Addition of electrical/mechanical limiting elements.	H1, H6, H7, H8, H9, H18, H18, H19	**13.33%**
	Specification of usage and signalling procedures (e.g. cleaning, adding beacons, etc.).	H10, H24, H24, H28, H31, H25, H26, H27, H28, H28	**16.66%**
	Use of safety certified elements (e.g. connectors, wires, etc.).	H20, H20, H21, H21, H26, H29, H30, H30, H31	**15%**
			subtotal 45%

Table 5. Conclusions of the application of the methodology.

5. Conclusions and future research

When a system interacts with the environment and/or with humans, the analysis of possible hazards that could occur during its operation becomes indispensable. Nevertheless, this analysis is a complex process that needs the support of a methodology. The more domain-specific the methodology, the more accurate the results will be.

In the context of the EFTCoR project, the DSIE has developed a series of robots for ship hull cleaning in such a hazardous environment as shipyards are. In order to fulfil the safety requirements of the EFTCoR project, we have adopted a systematic approach for addressing them from the beginning of the design phase. In this book chapter, we have described a complete example of the application of the process to a real service robot that works in an aggressive industrial environment. This example considers not only the safety requirements elicitation process but also the classification of the different hazards and the proposed solutions to them.

The proposed safety requirements elicitation process is based on a mix of different techniques and standards, as no single standard or technique address all the different, hardware and software, aspects involved in the development of a service robot.

We have used the ANSI/RIA standard as the basis for the identification and classification of the hazards, risks and the safeguards to be adopted to reduce the risks to acceptable levels. This standard has being completed with the safety patterns extracted from Douglass when designing a more complex solution and the use of fault trees to identify the possible causes of failure. In this sense, we hope that a European standard, derived from EN 61508 and specifically targeted to robotics systems, will soon appear to fulfil the lack of a methodology for safety requirements specification and solutions in the EU.

Although it may seem that this work is the result of applying together ("glued" even) several standards, the contribution of this work goes further on because:

- It gathers the methodological experience of diverse authors, since this experience is usually absent in most of the standards.
- The range of application of the proposal is wider than that of one of a single standard or technique seen in section 3, because this work covers from requirements specification to the implementation patterns applied in architectural design.
- Lastly, a case study of a real application has been presented, where the safety requirements were naturally present from the beginning of the project, not added later.

Two important conclusions can be extracted from this work: (1) only half of the safety requirements really affect the software architecture of the system, and (2) only a few fractions of them require the use of external redundant control that must conform to the strictest level of safety. Nevertheless, since security requirements are, conceptually, independent of the functional ones, it would be more than desirable to have an architectural approach that allows designers to consider them in isolation.

We are currently working on the adoption of a *Model-Driven Engineering (*Stahl & Völter, 2006) approach to develop the software architecture of a robot starting from a model. In this vein, we plan to:

- Adapt and include the proposed safety requirements elicitation process as an orthogonal aspect, in order to map the solutions to these requirements to the software architecture of the robot.

- Develop a safety requirement catalogue and some heuristics to help designers take into account safety requirements from the beginning of the design.
- Keep a traceability record of the implementation of the safety solutions in the software architecture (and later in the generated code) of the robot.

Lastly, we want to conclude this book chapter stating that it contributes not only the experience of the DSIE research group in the application of different standards and methodologies, but also a complete example in the context of a real problem. We hope that the tables presented in the appendix could help designers and engineers as a starting point when developing similar systems. To our knowledge, there is no such a complete example in the literature as the one we have described in this book chapter.

6. Acknowledgements

This research has been funded by the Spanish CICYT project MEDWSA (TIN2006-15175-C05-02) and the Regional Government of Murcia Seneca Program (02998-PI-05).

7. References

ANSI/RIA R15.06 (1999), American National Standard for industrial robots and robot systems safety requirements. Robotic Industries Association.

Douglass, B. (1999), Doing hard time: developing real-time systems with UML, objects, frameworks and patterns. Addison-Wesley Longman. ISBN: 0-201-49837-5.

Douglass, B. (2002), Real-Time Design Patterns: Robust Scalable Architecture for Real-Time Systems. Addison-Wesley Professional. ISBN 0201699567.

EN 61508 (2003), Functional safety of electrical/electronic/programmable electronic safety-related systems. European Committee for Electro-technical Standardization.

Fernández, C.; Iborra, A.; Álvarez, B.; Pastor, J.; Sánchez, P.; Fernández, J. & Ortega, N. (2005). Ship shape in Europe: cooperative robots in the ship repair industry, IEEE Journal on Robotics and Automation, vol. 12, num. 3, pp. 65-77. ISSN 1070-9932.

Hansen, K.; Ravn, A. & Stavridou, V. (1998). From safety analysis to software requirements. IEEE Transactions on Software Engineering, vol. 24, num. 7, pp 573–584. ISSN: 098-5589.

Iborra, A.; Pastor, J.; Álvarez, B.; Fernández, C. & Fernández-Meroño, J. (2003). Robots in Radioactive Environments, IEEE Journal on Robotics and Automation, vol. 10, num. 4, pp. 12-22. ISSN 1070-9932.

International Maritime Organisation (IMO) (1999). Norm on Hull Integrity. IMO Publishing.

Leveson, N. (1995), Safeware: system safety and computers. ACM Press. ISBN: 0-201-11972-2.

Neumann, P. (1994). Computer-Related Risks. Addison-Wesley Professional. ISBN: 0-201-55805-X.

Stahl, T. & Völter, M. (2006). Model-Driven Software Development: Technology, Engineering, Management. Wiley. ISBN: 0470025700.

Appendix A. Complete Tables

Hazard	Task	Risk Level	Origins	Prob.	Reaction measures
H1. The cleaning tool hits the ship hull.	T5-7, T21	Severe	Breakage or error in the axis controller. Comm. failure (logic or physic)	Low	Emergency alarm[3] Stop axis and move it away from the hull. Power axis off.
H2. Over exposure of the tool or delayed deactivation.	T10	Slight	Breakage of an axis of the secondary positioning system or the tool. Software scheduling error. Communication failure.	Low	Emergency alarm. Power tool off. Stop grit-blasting.
H3. Workers in the trajectory of the primary positioning system (rail).	T1	Very severe	There is a person in the trajectory of the movement of the primary positioning system.	Med.	Emergency alarm. Stop the robot. Emergency stop.
H4. Equipment in the trajectory of the primary positioning system (rail).	T1, T20	Severe	There is an obstacle in the trajectory of the primary positioning system.	Med.	Emergency alarm. Stop the robot. Emergency stop.
H5. Obstacle in the trajectory of the primary (vertical axis or arm).	T2-4, T20	Very severe	There is an obstacle in the trajectory of the primary positioning system.	High	Emergency alarm. Stop the robot. Emergency stop.
H6. Obstacle in the trajectory of the secondary positioning system.	T5-8, T21	Very severe	There is an obstacle in the trajectory of the secondary positioning system.	Low	Emergency alarm. Stop the robot. Emergency stop.
H7. Primary pos. sys. limit sensor overcome (arm).	T3-4, T20	Very severe	Sensor breakage or software error. Comm. failure (either physical or logical).	Low	Emergency alarm. Emergency stop.

[3] Different alarm levels: visual, acoustic, etc, depending on the severity.

Hazard	Task	Risk Level	Origins	Prob.	Reaction measures
H8. Primary pos. sys limit sensor overcome (rail).	T1, T20	Very severe	Sensor breakage or software error. Comm. failure (either physical or logical).	Low	Emergency alarm. Stop the robot. Power axis off.
H9. Secondary pos. sys. limit sensor overcome (arm).	T5-7, T21	Slight	Sensor breakage or software error. Comm. failure (either physical or logical).	Low	Emergency alarm. Stop secondary pos. system. Power secondary off.
H10. The secondary pos. sys. hits the ground or the ship hull.	T1-3, T4,T8 T20	Severe	Sw error when calculating the trajectory of the primary. Human error when moving the primary. Error or breakage of the secondary proximity sensors. Sensor breakage or Sw error. Mechanical failure. Power failure. Comm. failure .	High	Emergency alarm. Emergency stop. Stop secondary and move the arm of the primary. Power robot off.
H11. The emergency stop does not work.	All	Very severe	Comm. failure . Emergency stop button breakage.	Low	Emergency alarm. Manually power robot off.
H12. The secondary pos. sys. does not stop.	T5-7, T14, T21	Severe	Error in the secondary positioning system controller. Comm. failure (either physical or logical). Power failure.	Low	Emergency alarm. Power robot off. Emergency stop.
H13. The primary pos. sys. does not stop.	T1-4, T8, T12, T13, T20	Very severe	Error in the primary positioning system controller. Comm. failure . Power failure.	Low	Emergency alarm. Power robot off. Emergency stop.
H14. The secondary pos. sys. movement sequence does not end.	T7, T21	Slight	Sw error when controlling the sequence of movements. Comm. failure (either physical or logical).	Low	Emergency alarm. Stop secondary and deactivate blasting tool.

Hazard	Task	Risk Level	Origins	Prob.	Reaction measures
H15. The primary pos. sys. movement sequence does not end.	T8, T12-T13, T20	Very severe	Sw error when controlling the sequence of movements. Comm. failure (either physical or logical).	Low	Emergency alarm. Stop primary axes.
H16. The axis does not get disabled.	T15	Slight	Axis Sw control error.	Low	Emergency alarm. Power robot off. Emergency stop.
H17. The robot does not behave has expected.	All	Very severe	Robot Sw control error. Hw failure. Power failure.	Low	Emergency alarm. Power robot off. Emergency stop.
H18. Free fall of the arm or the vertical axis of the primary.	All	Very severe	Axis controller Hw failure.	Low	Emergency alarm.
H19. Free fall of the secondary.	All	Severe	Axis controller Hw failure.	Low	Emergency alarm.
H20. Grit blasting hose released.	T5-7, T9, T10, T18	Very severe	Hose or connectors broken or hooked.	Med.	Emergency alarm. Emergency stop.
H21. Energy not released in maintenance mode (learning, calibration, etc)	T20-29	Very severe	Mechanical or electrical media necessary to release energy are not implemented correctly.	High	Emergency alarm. Emergency stop.
H22. The tool is activated outside the working area.	All	Severe	Communication failure. Human operator error. Calibration failure. Control Sw failure.	High	Emergency alarm. Emergency stop.
H23. The robot does no exit working mode after alarm.	All	Very severe	Control software failure. Communication failure.	Low	Emergency stop.
H24. Person in the robot working area.	All	Very severe	There is a person in the trajectory of the robot.	High	Stop the system. Emergency stop.

Hazard	Task	Risk Level	Origins	Prob.	Reaction measures
H25. The robot does not stop at operator command.	All	Very severe	Operator fails managing the robot. Control Sw or comm. failure. Hw or mechanical failure.	Low	Emergency alarm. Emergency stop.
H26. Higher parts of the crane oscillate dangerously	All	Very severe	Wind at higher speed that allowed for the crane (50 Km/h). Drastic changes in speed and direction in the movements of the primary. Rails in the primary with dust or strange elements.	Med.	Emergency alarm. Emergency stop.
H27. Irregular movement of the robot.	All	Severe	Dust, grit, humidity, lack of oil, waste of elements.	High	Emergency alarm. Emergency stop.
H28. Some part of the robot (hoses, cables) hooks or falls.	All	Very severe	Bad attachment of cables and hoses.	High	Emergency alarm. Emergency stop.
H29. Dust or grit in the environment.	All	Severe	Another dust generating task taking place near the robot. Bad aspiration of the blast tool. Escapes in connectors or inadequate hoses. Hw, Sw or comm. failure.	High	Take note of the incidence. Cleaning of the environment and the equipment.
H30. Water or humidity in the equipment.	All	Severe	Rain. Proximity of other operations using water.	High	Take note of the incidence. Secure stop of the equipment.
H31. Extreme temperatures over admissible ranges.	All	Severe	Extreme environmental conditions, cold or hot.	High	Take note of the incidence. Secure stop of the equipment.

Table 3. Complete hazard identification list.

Hazard	Risk	SEV. EXP. AV. RRC	Solution	SEV. EXP. AV. RRC
H1. The tool hits the ship hull.	R1. Tool breakage or damage to the secondary positioning system.	S1 E2 A1 R3A	Add bumpers to the head of the tool. Limit Z-axis maximum torque.	E1 A1 S1 R4
H2. Overexposure of the tool or delay in its deactivation.	R2. Damage to the ship hull surface.	S1 E2 A1 R3A	Add a software timer to control the tool working time. Add a motion sensor to the secondary positioning system and link it to the timer.	E1 A1 S1 R4
H3. Workers in the trajectory of the primary positioning system (rail).	R3. Worker runs over.	S2 E2 A1 R2A	Add sensors to detect obstacle presence in the rail. These sensors are directly wired to the robot safety control infrastructure. Add a siren for signalling robot motion.	E1 A1 S2 R3B
H4. Equipment in the trajectory of the primary positioning system (rail).	R4. Damage to the equipment and to the primary positioning system.	S2 E2 A1 R2A	Same as H3	E1 A1 S2 R3B
H5. Obstacle in the trajectory of the primary (vertical axis or arm)	R5. Damage to the primary and to the obstacle	S2 E2 A2 R1	Add sensors for detecting obstacles in the trajectory of the arm and wire them to the robot safety control infrastructure.	E1 A1 S2 R3B
H6. Obstacle in the trajectory of the secondary pos. sys.	R6. Damage to the secondary or to the tool	S2 E1 A1 R2B	Hw/Sw torque control for detecting overload.	E1 A1 S1 R4
H7. Primary pos. sys. limit sensor overcome (arm).	R7. Damage to the primary pos. sys. and/or workers.	S2 E1 A2 R2B	Add mechanical limits	E1 A1 S1 R4
H8. Primary pos. sys limit sensor overcome (rail).	R8. Damage to the primary positioning system and/or workers or equipment.	S2 E1 A2 R2B	Add mechanical limits	E1 A1 S1 R4

Hazard	Risk	SEV. EXP. AV. RRC	Solution	SEV. EXP. AV. RRC
H9. Secondary pos. sys. limit sensor overcome (arm).	R9. Damage to the secondary positioning system.	S1 E1 A2 R3B	Add mechanical limits	E1 A1 S1 R4
H10. The secondary pos. sys. hits the ground or the ship hull.	R10. Damage to the secondary positioning system and/or workers or equipment.	S2 E2 A2 R1	Define a precise procedure for positioning tool over hull surface. Add proximity sensors to the secondary (XYZ table), monitorised by control system.	E1 A1 S2 R3B
H11. The emergency stop does not work.	R8. Damage to the primary positioning system and/or workers or equipment.	S2 E1 A2 R2B	Physical redundancy of the emergency stop system.	E1 A1 S1 R4
H12. The secondary pos. sys. does not stop.	R11. Secondary positioning system breakage.	S1 E1 A2 R3B	Add additional emergency stop mechanisms. Add external motion sensors.	E1 A1 S1 R4
H13. The primary pos. sys. does not stop.	R8. Damage to the primary positioning system and/or workers or equipment.	S2 E1 A2 R2B	Add additional emergency stop mechanisms Add external motion sensors (6.4 ANSI).	E1 A1 S1 R4
H14. The secondary pos. sys. movement sequence does not end.	R12: The robot moves uncontrolled	S1 E1 A1 R4	Add a software module for monitoring the execution of the sequence.	E1 A1 S1 R4
H15. The primary pos. sys. movement sequence does not end.	R8: Robot knocks over	S2 E1 A1 R2B	Add emergency stop mechanisms Add motion sensors outside the control loop Add sensor to measure the crane slope. Stop system is slope surpasses a threshold.	E1 A1 S1 R4
H16. The joint does not get disabled.	R13: Damage to the joint.	S1 E1 A1 R4	Add current sensor to detect joint enable state. Add alarm to inform operator.	E1 A1 S1 R4

Hazard	Risk	SEV. EXP. AV. RRC	Solution	SEV. EXP. AV. RRC
H17. The robot does not behave has expected.	R14. Damage to environment, equipment or workers.	S2 E1 A2 R2B	Add mode constraints verification measures (commands allowed, max speed and acceleration, ...) in control software.	E1 A1 S1 R4
H18. Free fall of the arm or the vertical joint of the primary.	R14. Damage to environment, equipment or workers.	S2 E1 A2 R2B	Add safety brake (stops joint in case of power failure). Add anti-fall mechanism (block stop) to vertical joint.	E1 A1 S2 R3B
H19. Free fall of the secondary.	R14. Damage to environment, equipment or workers.	S2 E1 A2 R2B	Add safety brake (stops joint in case of power failure).	E1 A1 S1 R4
H20. Grit blasting hose released.	R14. Damage to environment, equipment or workers.	S2 E2 A2 R1	Add sensors to detect hose release. Identify hoses and nozzles by means of codes of colors.	E1 A1 S2 R3B
H21. Energy not released in maintenance mode (learning, calibration, etc)	R14. Damage to environment, equipment or workers.	S2 E1 A2 R2B	Add means to discharge stored static electricity. Use proper earth connections and test them before operation. Use adequate protections against electrical risk. Limit robot speed in maintenance or programming modes.	E1 A1 S2 R3B
H22. The tool is activated outside the working area.	R14. Damage to environment, equipment or workers.	S2 E1 A2 R2B	Add mode constraints verification measures (tool activation...) in control software.	E1 A1 S1 R4
H23. The robot does no exit working-mode after alarm.	R14. Damage to environment, equipment or workers.	S2 E1 A2 R2B	Add redundancy: independent module to process alarms and start safety actions. Add means to coordinate control system and alarm processing module. Control system and alarm processing module should be aware of their respective behaviors and stop the system if detect a failure in the other.	E1 A1 S1 R4

Hazard	Risk	SEV. EXP. AV. RRC	Solution	SEV. EXP. AV. RRC
H24. Person in the robot working area.	R15: Damage to people due to mechanical parts or grit impact.	S2 E1 A1 R2B	Mark working area. Identify hoses and nozzles by means of codes of colors. Add sensors to detect obstacle presence in the rail. Add a siren for signalling robot motion.	E1 A1 S1 R3B
H25. The robot does not stop at operator command.	R14. Damage to environment, equipment or workers.	S2 E1 A2 R2B	Add emergency stop mechanisms. Add redundancy: independent module to process alarms and start safety actions.	E1 A1 S1 R4
H26. Higher parts of the crane oscillate dangerously	R8: Robot knocks over R14. Damage to environment, equipment or workers.	S2 E2 A2 R1	Add motion constraint: horizontal movements over rails should be done with secondary in lower position. Add sensor to measure wind speed. Add sensor to measure the crane slope. Stop system is slope surpasses a threshold.	E1 A1 S1 R4
H27. Irregular movement of the robot.	R15: Breakage of joints mechanical parts. R16: Vibrations cause Hw breakage or malfunctioning.	S1 E2 A1 R3A	Periodical maintenance. Cleaning of equipment after working. Add sensor to measure the crane slope. Stop system is slope surpasses a threshold.	E1 A1 S1 R4
H28. Some part of the robot (hoses, cables) hooks or falls.	R8: Robot knocks over. R17: Hoses and wire breakage. R15: Damage to people due to mechanical parts or grit impact.	S2 E1 A2 R2B	Add emergency stop mechanisms. Plug and fix properly hoses and wires. Inspect working area.	E1 A1 S1 R3B
H29. Dust or grit in the environment.	R15: Damage to people due to mechanical parts or grit impact. R18: Damage to spotlights. R19: Vision system malfunctioning.	S2 E2 A2 R1	Plug and fix properly hoses and wires. Use dust-resistant spotlights and protect them. Protect video cameras.	E1 A1 S1 R3B

Hazard	Risk	SEV. EXP. AV. RRC	Solution	SEV. EXP. AV. RRC
H30. Water or humidity in the equipment.	R19: Vision system malfunctioning. R20: Electrical shortcut. R21: spotlights explosion.	S2 E2 A2 R1	Use water-resistant connectors. Use dust-resistant spotlights and protect them. Use water resistant video cameras.	E1 A1 S1 R3B
H31. Extreme temperatures over admissible ranges.	R22: Hoses freezing. R23: Hw breakage or malfunctioning.	S1 E1 A2 R3B	Add temperature sensor. Add ventilation to drivers and hardware devices.	E1 A1 S1 R4

Table 4. Complete hazard identification list.

11

Service Robot Operated by CDMA Networks for Security Guard at Home

JeGoon Ryu, ByeongChoel Yoo and Toshihiro Nishimura
The Graduate School of Information, Production, and Systems, Waseda University
Japan

1. Introduction

Mobile robotic technology and its application in various field is currently an area of high interest. Especially, the progression of a mobile robot and communication techniques enables intelligent robots to be developed with more versatile applications. Telerobotic techniques combine robotic systems with remote control technology to enable the performance of task which is impossible or hazardous for human to perform directly. Through these ubiquitous robot controls, the paradigm of robot which performs specific tasks remotely has changed into intelligent robots to perform public and individual tasks.

The purpose of the SG-Robot(*Security Guard Robot*) project is in-line with on-going activities towards the development of autonomous robots for security guard design at home. In security industry, ubiquitous robots have provided a variety of information to the user and achieve its duty through the Web. The mentioned project aims at the development of interactive security guard robots, which are able to be operated with a mobile phone.

Last year, the interface for telepresence robot has developed over the Web or RF between robots, but these systems have the demerits of limited distance or geographical limit to be established to an internet link. In this paper, we propose the SG-Robot that can be operated and conduct surveillance of the environment around itself anytime/anywhere using wireless communication networks such as CDMA (Code Division Multiple Access) Networks.

This robot has the functionalities of detecting and avoiding obstacles, recognizing face, navigating using a map, capturing and compressing image at high speed, transferring image at real time to CDMA phone, and operating remotely by a mobile phone. The goal of these functionalities is to tele-operate robot using a mobile phone anytime/anywhere and to construct the systems that has the home network function, ubiquitous computing, and home security.

2. Related works

A variety of service robots were developed that are designed to remotely operate in the various environments over the last decade. Moreover, Web-based tele-operation interfaces for robots have been developed and have gained serious interest over the last few years.

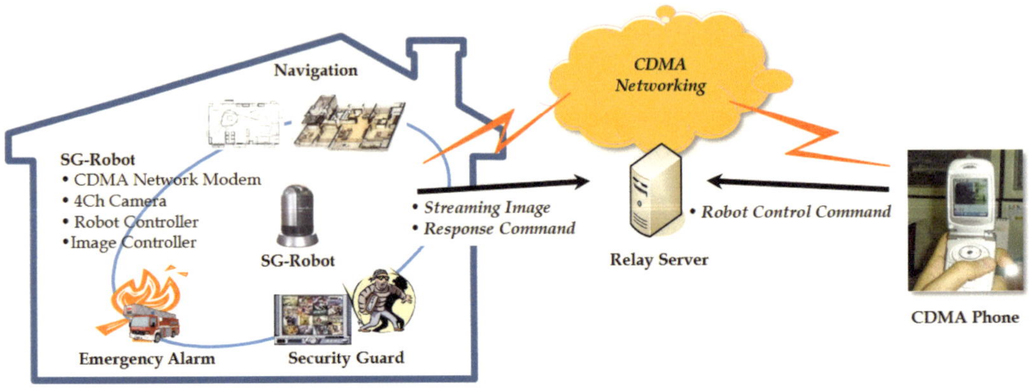

Fig. 1. Configuration of the SG-Robot using CDMA networks.

The Web interfaces for robots allow users to issue single commands which are executed immediately. Three of the earlier systems are the Mercury Project installed in 1994 (Goldberg et al., 2002), Austria's Tele-robot on the Web (Taylor & Trevelyan, 1995), which came on-line nearly at the same time, and the Tele-Garden (Goldberg et al., 1995), which replaced the Mercury robot in 1995. While the Mercury robot and the Tele-Garden allow Web user to perform different types of digging tasks, such as excavation of artifacts as well as watering and seeding flowers, the tele-robot on the Web gives Web users the opportunity to build complex structures from toy blocks. These systems also suffice to provide still images from a camera mounted on the robot arm after a requested movement task has been completed.

Mobile robots which can be controlled over the Web provide exclusive remote control to a single person or provide queues to schedules user requests. KhepOnTheWeb (Saucy & Mondada, 1998) is a typical representative of mobile robots with a Web interface. Users can give elementary movement actions to the robots and observe them using several Web-cameras. Xavier (Simmons et al., 1997) was probably the first mobile robot which operates in a populated office building controlled through the Web. Xavier can be advised by Web users to move to an office and to tell a knock-knock joke after arrival. The mobile robotic platforms Rhino (Buhmann et al., 1995 ; Burgard et al., 1998) and Minerva (Thrun et al., 1999 ; Thrun et al., 2000) could also be operated over the Web for museum tour guide. Their interfaces relied on client-pull and server-push techniques to provide visual feedback of the robot's movements. Cyberguard (Massios & Voorbraak, 1999) is a mobile security robot marketed by Cybermotion. The aim of Cyberguard is to patrol indoor building areas for intruders, fires and chemical threats. The robot would collect the images with inexpensive camera that could be sent over very low speed radio links. The CAPM (Suthakorn et al., 2001) project of the Johns Hopkins University have built a robot working for a library. The robot system is designed and implemented in a friendly library environment. But, it needs another robot system for browsing the selected book.

The SG-Robot system uses video streams to convey observed information to the user. Additionally, Multi SG-Robots can provide their information to single user or multiple users. Therefore, the users can control, with selecting the SG-Robots in ubiquitous computing environment. The SG-Robot was able to conduct the surveillance task and communicate between the SG-Robot and users over the CDMA2000-1x communication network efficiently.

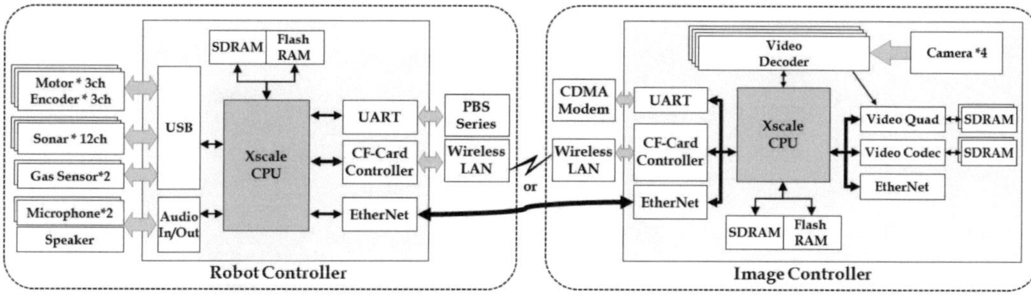

Fig. 2. Architecutre of SG-Robot.

3. System configuration

The SG-Robot project, a research and development activity entitled "Tele-operated robot using CDMA networks", introduces a new communication interface approach to telepresence and security surveillance of robot at home or at outdoor. The user is offered the capability to remotely issue commands to navigate specific place. In order to execute such commands, the SG-Robot moves autonomously or manually in the home site, reaching thus the desired navigation targets. At the same time, imaged scenes by cameras are transmitted to the user as a video stream, enabling tele-observation of the real home site and the public institution. Fig. 1 represents the concept of communication link of the developed SG-Robot.
The SG-Robot entirely consists of the image controller and the robot controller. This structure is easy to establish the image controller for other robot systems. Any robot controller is able to simply communicate with this image controller by specific data protocol.

3.1 Architecture of SG-Robot

SG-Robot is characterized by differential drive, turret drive for turning body without driving the wheels, communication interface and various sensors. The sensory equipment includes auditory, vision, ultrasonic, and IR scanner sensors, which allow the robot to behave autonomously and to observe environments around it.
Fig. 2 represents the architecture of the developed SG-Robot. The controller of SG-Robot entirely consists of the image controller and the robot controller. Main CPU of each controller is PXA255, ARM architecture, which has the characteristics suitable for the embedded robot system, such as low power and high performance. Each controller exchanges command data through the wireless adhoc LAN. The reason of that is to easily establish the image controller for other robot systems.
The robot controller is divided into three branches by function: communication, sensory, and drive function. And the image controller is divided into two parts by function: internal/external communication and image processing function. The image controller directly communicates with mobile phone through CDMA modem placed in robot system. Received/transmitted packets are streaming images and robot control commands. Input images through a camera placed in front of the SG-Robot are compressed with the QCIF size and transmitted to mobile phone by CDMA modem using CDMA2000-1x network. Image controller allows data to be sent to mobile phone, after CDMA modem connects to the relay server.

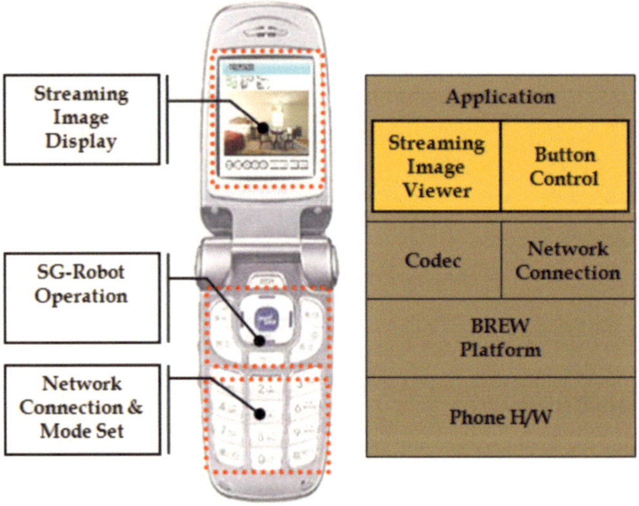

Fig. 3. Mobile phone for controlling SG-Robot.

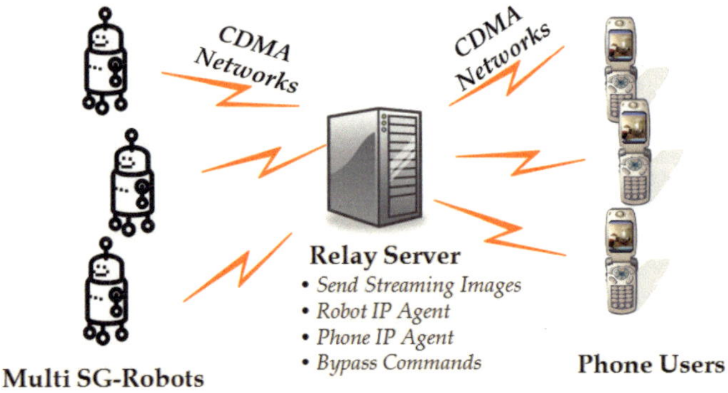

Fig. 4. Relay server for connection management.

3.2 CDMA mobile phone for robot control

For decades the telephone has always been within easy reach to provide us with real-time voice communication (Fischer, 1992). More recently, CDMA communication has proved itself as a remarkable communication network, allowing us to exchange message and images with others and to download VOD from multimedia server. Therefore, a user can be offered various services and contents anytime/anywhere.

Fig. 3 shows the mobile phone for controlling SG-Robot. CDMA mobile phone has two functions for controlling SG-Robot. First, it can display streaming images. When a camera is used for SG-Robot, motion picture data need to be transferred to mobile phone through the CDMA networks. CDMA2000-1x network has very low bit rate of the bandwidth.

Due to the mismatch between the data rates of the CDMA networks and the motion picture, data compression should be performed before the transmission from the mobile phone. Image controller of the SG-Robot provides MPEG-4 and JPEG compression of the motion picture. While MPEG-4 algorithms require large intermediate memory space, the JPEG algorithm does not need to store an entire frame for the compression. As mobile phone has low capacity memory, the baseline JPEG compressing methods are used in image coding over the CDMA networks. Next, mobile phone allows user to control SG-Robot. SG-Robot is operated by pushing the buttons. Therefore, a user can monitor the environment around SG-Robot and control to navigate it through the LCD screen of mobile phone.

3.3 Relay server for connection management

The CDMA modem in SG-Robot uses PPP Protocol based on Mobile IP for wireless network access. For continuing connection between SG-Robot and mobile phone, relay server need to be established. Multi SG-Robots and users can be also constructed in this configuration. The relay server allows multiple users to connect with SG-Robot in accordance with their permission degree. In the whole SG-Robot configuration, a relay strategy is designed to increase the system efficiency and to satisfy the isochronous requirements of showing motion pictures and providing command control of SG-Robot to multiple users. Fig. 4 shows the relay server for connection management. Table 1- 4 also represent protocols between the SG-Robot and relay server and between the relay server and mobile phone.

Byte1	Byte 2	Byte 3
0x33	COMMAND_TYPE	COMMAND_DATA

Table 1. Packet for data tranmission to robot (ro relay server)

COMMAND_TYPE	COMMAND_DATA	Description
0x00	0x01 0x03	ACK for phone connection Close the connection
0x10	0x08 0x02 0x04 0x06 0x05	Forward go Backward go Turn left Turn right Stop
0x20	0x01 0x02 0x03	Initial mode Manual mode Free navigation mode
0x30	0x32 0x33	Request to start image transmission Request to stop image transmission
0x40	0x01	Forced termination

Table 2. Command from mobile phone to robot (or relay server)

Byte1	Byte 2	Byte 3	Byte 4
0x44	COMMAND	DATA 1	DATA 2

Table 3. Packet for data transmission to mobile phone (or delay server)

COMMAND	DATA 1	DATA 2	Description
0x50	High byte of image	High byte of image	If COMMAND equals 0x50, transmit key frame (I-picture)
0x51	High byte of Image	High byte of image	If COMMAND equals 0x51, transmit P-picture
0x00	0x01	0x01	ACK for robot connection
	0x03	0x03	Close the connection
0x40	0x01	0x01	State that SG-Robot is performing the command after receiving it
		0x02	State that SG-Robot can't perform the command
	0x02	0x01	State that SG-Robot don't connect to the Relay server

Table 4. Response from robot to mobile phone (or relay server)

4. Implementation of SG-robot

The objectives of SG-Robot are to guard home or public places, to inform user of the state of emergency, and to be controlled for specific mission. To accomplish these objectives, robot controller uses the RTAI (Real Time Application Interface) to provide real-time performance with embedded Linux. It provides a high-performance real-time kernel which supports both deadline and highest-priority-first scheduling.

4.1 Framework of SG-robot

Sensor-based robotic systems contain both general and special purpose hardware, and thus the development of applications tends to be a very tedious and time consuming task. In the SG-Robot system, obstacle avoiding behavior, various robot motion, efficient user interface, communication interface, and system fail check are easily implemented by using multitasking function, intertask communication mechanism, and real-time runtime libraries of RTAI. Fig. 5 and Fig. 6 represent the implemented framework in this system.

The framework for the SG-Robot controller consists of 5 layers, such as hardware layer, firmware layer, task layer, thread layer, and GUI layer. Especially, in the thread layer and task layer, data share uses the shared memory and data transfer uses the FIFO. Each thread communicates with each other using the message event. Data interface of this robot framework was designed with similarly a nervous system and memory system of human. As SG-Robot simultaneously processes the command of user and the navigation procedure, it can guarantee a high efficiency and stability the designed system. PBS series and sonar array is updated each 200ms, and odometry is only updated each 100ms. After gained information is stored in shared memory, free navigation thread and find obstacle thread reprocess the information to avoid obstacle and to navigate to specific goal.

Multi threads in framework for image controller also have the soft real-time performance. Captured image information is updated each 0.1ms and stored in Queue. At the same time, face recognition thread conducts the process to find who was known through comparing with sample face in DB, and CDMA communication thread send the captured images in queue to mobile phone.

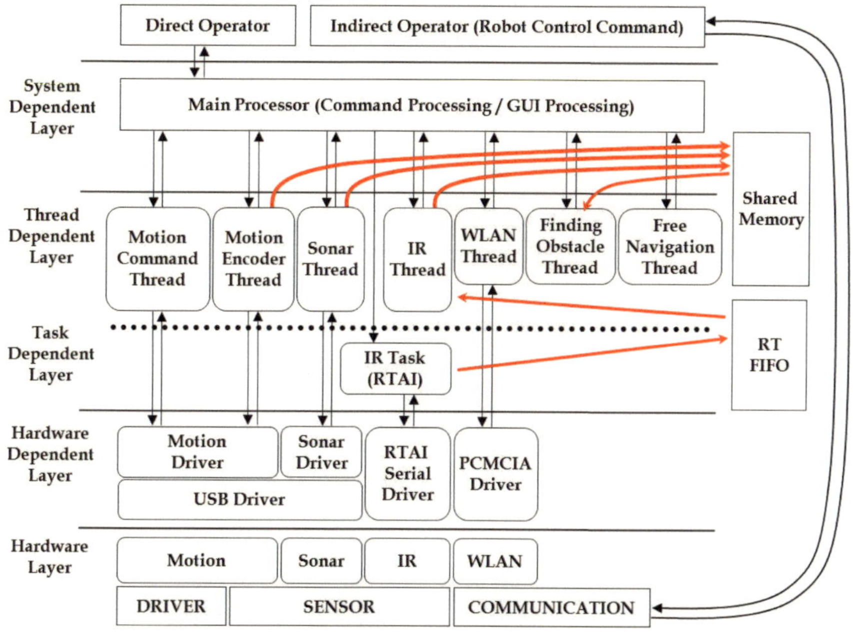

Fig. 5. Framework for robot controller.

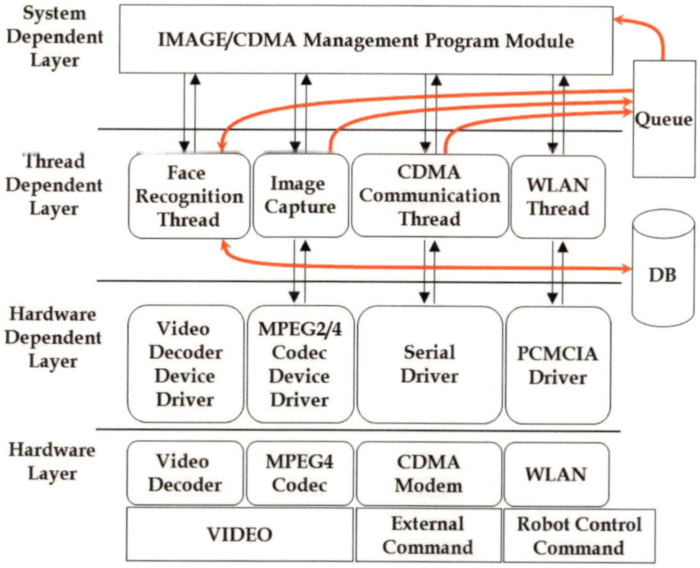

Fig. 6. Framework for image controller.

4.2 Locomotion of SG-robot

SG-Robot has the differential drive and the turret drive. As depicted in Fig. 7, differential drive is simple drive mechanism for a ground-contact mobile robot. Under differential drive, for each of the two drive wheels to exhibit rolling motion, the robot must rotate about a point that lines on the command axis of the two drive wheels. By varying the relative velocity of the two wheels, the point of this rotation can be varied, and different trajectories chosen.

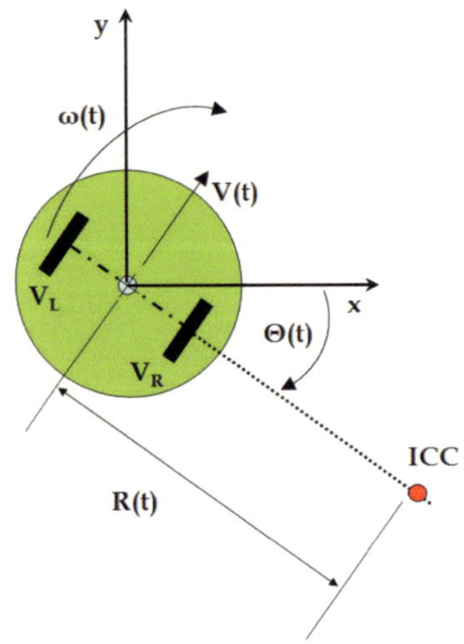

Fig. 7. Kinematics of SG-Robot using differential drives.

Differential drive has a nonholonomic constraint and is very difficult to solve in general. If it is assumed that $v_l(t) = v_l$, $v_r(t) = v_r$, and $v_r = v_l$, then equation (1) yields.

$$x(t) = \frac{l}{2}\frac{v_r + v_l}{v_r - v_l}\sin[\frac{t}{l}(v_r - v_l)]$$

$$y(t) = -\frac{l}{2}\frac{v_r + v_l}{v_r - v_l}\cos[\frac{t}{l}(v_r - v_l)] + \frac{l}{2}\frac{v_r + v_l}{v_r - v_l} \tag{1}$$

$$\theta(t) = \frac{t}{l}(v_r - v_l)$$

where l is the distance along the axle between the centers of the two wheels, the left wheel moves with velocity v_l along the ground and the right with velocity v_r, and $(x, y, \theta)_{t=0} = (0,0,0)$. Given a goal time t and goal position (x, y). SG-Robot considers two special cases of the motion of the differntial drive vehicle. If $v_l = v_r = v$, then the robot's motion simplifies as (2).

$$\begin{pmatrix} x' \\ y' \\ \theta' \end{pmatrix} = \begin{pmatrix} x + v\cos(\theta)\delta t \\ y + v\sin(\theta)\delta t \\ \theta \end{pmatrix} \tag{2}$$

Fig. 8. Operation of SG-Robot using mobile phone.

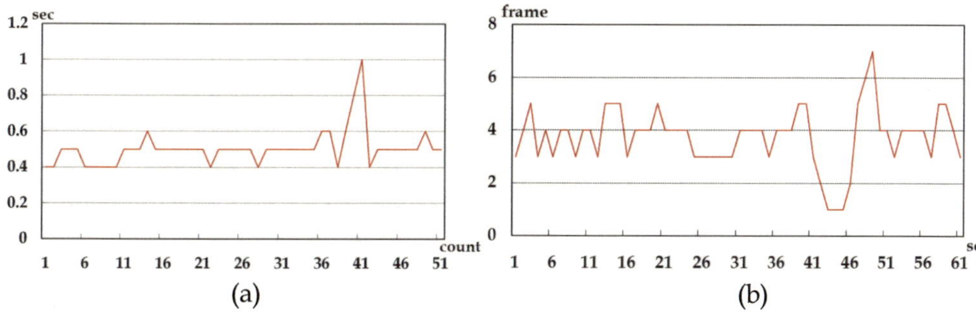

Fig. 9. Response time of robot control command and tranmission rate of streaming image to the mobile phone.

And if SG-Robot choose $-v_l = v_r = v$, then equation (3) is simplified as follow.

$$\begin{pmatrix} x' \\ y' \\ \theta' \end{pmatrix} = \begin{pmatrix} x \\ y \\ \theta + \frac{2v\delta t}{l} \end{pmatrix} \qquad (3)$$

To drive SG-Robot to some goal pose (x, y, θ), the robot can be spun in place until it is aimed at (x, y), then driven forward until it is at (x, y), and then spun in place until the required goal orientation θ is met.

5. Experimental result

To evaluate the designed system, we measure the transmission rate of motion pictures and command data in the home environment. Fig. 8 shows the degined SR-Robot and CDMA mobile phone. Fig. 9 represents the response time of the robot after being issued the command by mobile phone and the transmission rate of streaming images from a camera in the robot to mobile phone.

In general, the average response time was 0.3~0.4ms. But, the response time takes about 1sec from time to time, due to the irregular wave interference or the spatial characteristics. The image transmission rate is 4~5frame/sec. Sometimes, it was also delayed about 2~3frames by the same reason. Fig. 10 shows the scene which streaming images are transferred from camera in front of SG-Robot to mobile phone, when user operates the SG-Robot by mobile phone. Through the LCD screen displaying the images, user could recognize the scene and operate the robot to other place. When SG-Robot meet the obstacles while user issues the command of "Forward go", it don't go any more and send the message, "Can't accomplish the command because of obstacles", to the mobile phone.

Fig. 10. Images transmitted to mobile robot as SG-Robot moves.

Fig. 11 represents the navigated path when the user at outdoors operates the SG-Robot located in modeling house in an indoor environment. We obtained a rate of 95% of successive control during conducted experiments. The successive rate is calculated using an experimental time and the time that is subtracted by delay time due to irregular wave interference. In this experiment, the operator could control the SG-Robot to navigate to a specific place through the LCD screen of mobile phone.

6. Conclusion

We have created a surveillance robot that allows SG-Robot to be operated by mobile phone through the CDMA networks. The experiments have shown that CDMA networks communication link can be used for exploration, navigation and real-time tele-operation task.

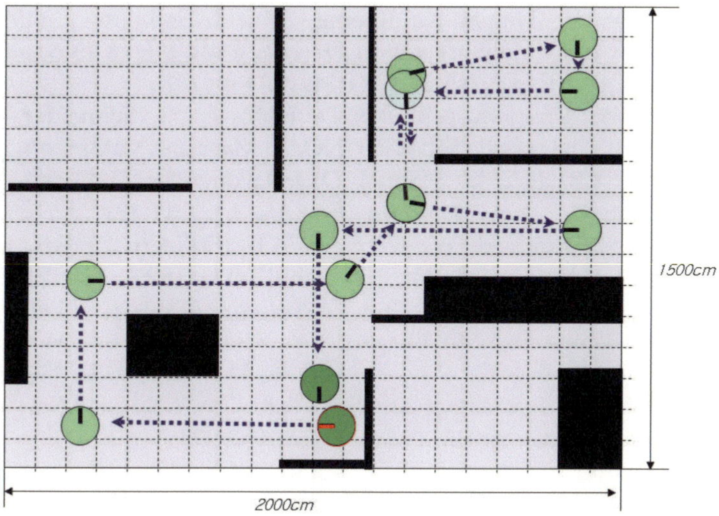

Fig. 11. SG-Robot navigation operated by the mobile phone of the user.

In most tele-robotic applications, communication line delays are limiting factor. Our experiments have shown that even in extreme delay situations streaming images over CDMA networks can successfully be used for robot control. That was possible because SG-Robot also has the decision to choice whether a "move and wait" strategy has to be adopted, or whether it is safe to move continuously in spite of line delays. Important another point addressed in this system lies in the fact that it can be used and monitored for multiple users and it is possible to operate multiple robots by one user. For these functions, relay server is constructed. The relay server authorizes multiple users to connect with SG-Robot in accordance with their permission degree, and enables multi SG-Robots to transmit the environment information to users in accordance with their registered level.

The outcomes of this study can be utilized for intelligent service robot and URC (Ubiquitous Robotic Companion) industries and will be considered to be a valuable technology to apply robot technology to mobile communication.

An ongoing and further area of investigation is to add the visual localization and map building to the SG-Robot. Using simultaneously localization and map building techniques with a single camera or camera set, the operator will be able to control the robot over visual map and robot will be able to possess powerful visual intelligence.

7. References

Goldberg, K.; Gentner, S.; Sutter, C.;Wiegley, J. & Farzin, B. (2002). The mercury project : A feasibility study for online robots, *Beyond Webcams :An Introduction to Online Robots*, MIT Press.

Goldberg, K.; Santarromana, G.; Bekey, G.; Gentner, S.; Moris, R. & Wiegley, J. (1995). The telegarden, *Proceeding of ACM SIGGRAPH*, pp. 135-140.

Taylor, K. & Trevelyan, J. (1995). A telerobot on the world wide web, *Proceeding of 1995 National Conference of the Australian Robot Association*, pp. 108-120.

Saucy, P. And Mondada, F. (2000). KhepOnTheWeb : open access to a mobile robot on the internet, *IEEE Robotics and Automation Magazine*, pp.41-47.

Simmons, R.; Goodwin, R.; Haigh, K.; Koenig, S. & O'Sullivan, J. (1997). A layered architecture for office delivery robot. *Proceeding of the First International Conference on Autonomous Agents*, pp. 245-252, Marina del Rey, CA.

Buhmann, J.; Burgard, W.; Cremers, A.B.; Fox, D.; Hofmann, T.; Schneider, F.; Strikos, J. & Thrun, S. (1995). The mobile robot RHINO, *AI Magazine*, vol. 16, no. 2, pp. 31-38.

Burgard, W.; Cremers, A.B.; Fox, D.; Hahnel, D. & Lakemeyer, G. (1998). The interactive museum tour-guide robot, *Proceeding of AAAI-98*, pp. 11-18.

Thrun, S.; Bennewits, M.; Burgard, W.; Cremers, A.B.; Dellaert, F.; Fox, D.; Hahnel, D.; Rosenberg, C.; Roy, N.; Schulte, J. & Schulz, D. (1999). MINERVA : A second-generation museum tour-guide robot. *Proceeding of ICRA-99*, vol. 3, pp. 1999-2005.

Thrun, S.; Beetz, M.; Bennewitz, M.; Burgard, W.; Cremers, A.B.; Dellaert, F.; Fox, D.; Hahnel, D. ; Rosenberg, C.; Roy, N.; Schulte J. & Schulz, D. (2000). Probabilistic Algorithms and the Interactive Museum Tour-Guide Robot Minerva, *International Journal of Robotics Research*, vol. 19, no. 11, pp. 972-999.

Suthakorn, J.; Lee, S.; Zhou, Y.; Thomas, T. & Choudhury, S. (2001). A robotic library system for an off-site shelving facility, *Proceeding of ICRA'01*, pp. 3589-3594.

Massios, N. A. & Voorbraak, F. (1999). Hierarchical decision-theoretic robotic surveillance, *Technical Report*, University of Amsterdam, pp-1999-21,ILLC.

Fischer, C.S. (1992). *America Calling : A social history of the telephone to 1940*, Berkeley : Univeristy of California Press.

Burgard, W. & Schulz, D. (2001). Robust visualization for web-based control of mobile robot, *Beyond Webcams :An Introduction to Online Robots*, MIT-Press.

Dellaert, F.; Burgard, W.; Fox, D. & Thrun, S. (1999). Using the condensation algorithm for robust, vision-based mobile robot localization, *Proceeding of the IEEE Conference on Computer Vision and Pattern Recognition*, vol. 2, pp. 588-594.

Trahanias, P.; Burgard, W.; Argyros, A.; Hahnel, D.; Baltzakis, H.; Pfaff, P. & Stachniss, C. (2005). TOURBOT and WebFAIR : Web-operated mobile robots for tele-presence in populated exhibitions, *IEEE Robotics & Automation Magazine*, vol. 12, no. 2, pp.77-89.

Tomizawa, T.; Ohya, A. & Yuta, S. (2002). Book browsing system using an autonomous mobile robot teleoperated via the internet, *proceeding of the IEEE/RSJ Conference on IROS'2*, pp.1284-1289.

Burgard, W.; Trahanias, P.; Hahnel, D.; Moors, M.; Schulz, D.; Baltzakis, H. & Argyros, A. (2003). Tele-presence in populated exhibitions through web-operated mobile robots, *Autonomous Robots*, Kluwer academic publisher, vol. 15-3, pp. 299-316.

Paulos, E. & Canny, J. (1997). Ubiquitous tele-embodiment : Application and Implications, *International Journal of Human Computer Studies*, Academic Press, vol. 46, no. 6, pp. 861-877.

12

Safety Intelligence and Legal Machine Language: Do We Need the Three Laws of Robotics?

Yueh-Hsuan Weng, Chien-Hsun Chen and Chuen-Tsai Sun
Conscription Agency, Ministry of the Interior
National Nano Device Laboratories (NDL)
Dept. of Computer Science, National Chiao Tung University
Taiwan

1. Introduction

In this chapter we will describe a legal framework for Next Generation Robots (NGRs) that has safety as its central focus. The framework is offered in response to the current lack of clarity regarding robot safety guidelines, despite the development and impending release of tens of thousands of robots into workplaces and homes around the world. We also describe our proposal for a *safety intelligence* (SI) concept that addresses issues associated with open texture risk for robots that will have a relatively high level of autonomy in their interactions with humans. Whereas Isaac Asimov's Three Laws of Robotics are frequently held up as a suitable foundation for creating an artificial moral agency for ensuring robot safety, here we will explain our skepticism that a model based on those laws is sufficient for that purpose. In its place we will recommend an alternative *legal machine language* (LML) model that uses non-verbal information from robot sensors and actuators to protect both humans and robots. To implement a LML model, robotists must design a biomorphic nerve reflex system, and legal scholars must define safety content for robots that have limited "self-awareness."

2. Service robots

Since the Japanese already show signs of a special obsession with robots, it is no surprise that many new ideas on robot regulation are also emerging from Japan. Regarded as a "robot kingdom," it will most likely be the first country to produce and sell large numbers of NGRs for private use (Ministry of Economy, Trade and Industry [METI], 2004). That day is expected to emerge within the next two decades, raising both expectations and concerns among safety-conscious Japanese (Cabinet Office, Government of Japan [COGJ], 2007). Issued in February 2004, the Fukuoka World Robot Declaration contains details on Japanese expectations for emerging NGRs that will co-exist with and assist human beings, physically and psychologically. The guiding principle behind the document is to contribute to the realization of a safe and peaceful society (European Robotics Research Network [EURON], 2006); however, it fails to describe what NGRs should be.

In a report predicting the near future (2020-2025) in robot development, the Japanese Robot Policy Committee (RPC, established by METI) discusses two NGR categories: (a) next generation industrial robots capable of manufacturing a wide range of products in variable batch sizes, performing multiple tasks, and (unlike their general industrial predecessors) working with or near human employees; and (b) service robots capable of performing such tasks as house cleaning, security, nursing, life support, and entertainment—all functions that will be performed in co-existence with humans in businesses and homes (METI, 2004). The report authors predict that humans will gradually give NGRs a growing number of repetitive and dangerous service tasks, resulting in increased potential for unpredictable and dangerous actions. In a separate report published the following year, the RPC distinguished between the two NGR categories by listing three unique characteristics of service robots: strong mobility, close physical proximity to humans, and fewer repetitive operations (METI, 2005).

The Japanese Robot Association (JARA; http://www.jara.jp) predicts that next generation robots (NGRs) will generate up to 7.2 trillion yen (approximately $64.8 billion US) of economic activity by 2025, with 4.8 trillion yen going to production and sales and 2.4 trillion yen to applications and support (JARA, 2001). The Association divides production and sales into three domains: personal living (3.3 trillion yen), medical and social welfare (900 billion yen), and public service (500 billion yen). In summary, the NGR industry is currently emerging, and we can expect NGRs to enter human living spaces and start serving human needs within twenty years, making it imperative that we address safety and control issues now.

3. Robot safety

3.1 Robot sociability problems

Since 2000, Japanese and South Korean technocrats have been discussing and preparing for a human-robot co-existence society that they believe will emerge by 2030 (COGJ, 2007; Lovgren, 2006). Based on the content of policy papers and analyses published by both governments, researchers are currently studying potential robot sociability problems (RSP) that—unlike technical problems associated with design and manufacturing—entail robot-related impacts on human interactions in terms of regulations, ethics, and environments. Regulators are assuming that within the next two decades, robots will be capable of adapting to complex and unstructured environments and interacting with humans to assist with daily life tasks. Unlike heavily regulated industrial robots that toil in isolated settings, NGRs will have relative autonomy, thus allowing for sophisticated interactions with humans. That autonomy raises a number of safety issues that are the focus of this article.

3.2 NGR safety overview

In 1981, a 37-year-old factory worker named Kenji Urada entered a restricted safety zone at a Kawasaki manufacturing plant to perform some maintenance on a robot. In his haste, he failed to completely shut down the unit. The robot's powerful hydraulic arm pushed the engineer into some adjacent machinery, thus making Urada the first recorded victim to die at the hands of a robot ("Trust Me," 2006). This example clearly supports Morita et al.'s (1998) observation that when task-performing robots and humans share the same physical space, the overriding goal must be to ensure human safety. They note that several safety principles have already been adopted for industrial robots—for example, separation of operation space, fail-safe design, and emergency stop buttons. However, when NGRs and humans share the same physical space, it will be necessary to give robots the capabilities to protect biological beings and intelligence to use them.

Toward that end, METI invited experts from Japan's service and industrial robotics industries, lawyers and legal scholars, and insurance company representatives to participate in discussions of NGR safety issues. In July 2007 they published *Guidelines to Secure the Safe Performance of Next Generation Robots* (METI, 2007). According to Yamada (2007), the new guidelines are similar to those previously established for industrial robots in terms of risk assessment and risk reduction, but also contain five additional principles that make them unique to NGR safety requirements:

- In addition to manufacturers and users, parties who need to be involved in making decisions about NGR-related issues include company managers and sellers. In certain service domains, pedestrians and patients (in other words, those who will have the most contact with NGRs) should be included under the umbrella of NGR risk assessment.
- Accomplishing the goals of risk reduction requires multiple processing and testing procedures.
- Manufacturers need to cooperate with users, sellers, and managers when designing and making robots, especially during the risk assessment and reduction stages.
- Risk assessment procedures, safety strategies, and safety measures need to be clearly and thoroughly documented.
- Managers and sellers are obligated to notify manufacturers about accidents and to provide complete records related to those accidents for use in improving robot safety.

According to the METI guidelines, NGR manufacturers are obligated to enforce risk assessment rules and procedures during production. However, Kimura (2007) notes two major challenges to service robot risk assessment: their *openness* makes it difficult to clearly define users and task environments, and their *novelty* makes it hard to predict and calculate risk. Here we will add a third challenge: industrial robot safety involves machine standards, while autonomous NGR safety involves a mix of machine standards and *open texture risk* resulting from unpredictable interactions in unstructured environments (Weng et al., 2007).

In a May 2006 paper on legislative issues pertaining to NGR safety, Japanese METI committee members described the difference between industrial robots and NGRs in terms of *pre-* and *post-human-robot interaction responsibilities* (METI, 2006). In the following discussion we will refer to them as *pre-* and *post-safety regulations*.

For industrial robots, safety and reliability engineering decisions are guided by a combination of pre-safety (with a heavy emphasis on risk assessment) and post-safety regulations (focused on responsibility distribution). Pre-safety rules include safeguards regarding the use and maintenance of robot systems from the design stage (e.g., hazard identification, risk assessment) to the training of robot controllers. As an example of pre-safety rules, the United Kingdom Health and Safety Executive Office (2000) has published a set of guidelines for industrial robot safety during the installation, commissioning, testing, and programming stages. Another example is International Standardization Organization (ISO, 2006a) rules—especially ISO 10218-1:2006, which covers safety-associated design, protective measures, and industrial robot applications. In addition to describing basic hazards associated with robots, ISO rules are aimed at eliminating or adequately reducing risks associated with identified hazards. ISO 10218-1:2006 spells out safety design guidelines (e.g., clearance requirements) that extend ISO rules covering general machine safety to industrial robot environments (ISO, 2006b). Those rules thoroughly address safety issues related to control systems and software design, but since their primary focus is on robot arms and manipulators (ISO, 1994), they have limited application to NGRs.

3.3 Human based intelligence and open-texture risk

Neurologists view the human brain as having three layers (primitive, paleopallium, and neopallium) that operate like "three interconnected biological computers, [each] with its own special intelligence, its own subjectivity, its own sense of time and space, and its own memory" (MacLean, 1973). From an AI viewpoint, the biomorphic equivalents of the three layers are *action intelligence, autonomous intelligence,* and *human-based intelligence* (Fig. 1).

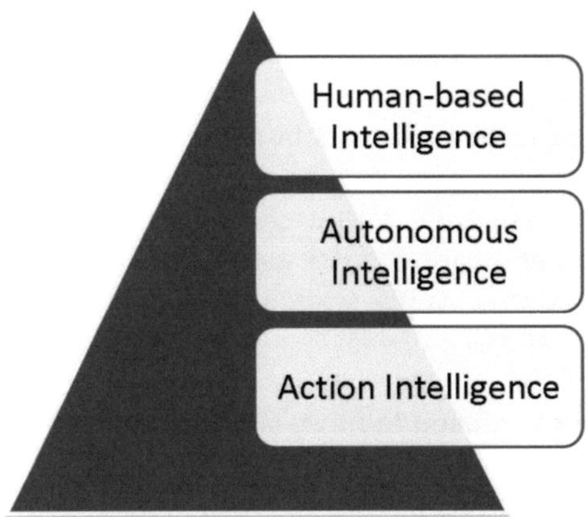

Figure 1. Robot Intelligence Layers.

Action intelligence functions are analogous to nervous system responses that coordinate sensory and behavioral information, thereby giving robots the ability to control head and eye movement (Hager et al., 1995), move spatially (Lewis, 1992), operate machine arms to manipulate objects (Chapin et al., 1999), and visually inspect their immediate environments (Dickmanns, 1988). Autonomous intelligence refers to capabilities for solving problems involving pattern recognition, automated scheduling, and planning based on prior experience (Koditschek, 1989). These behaviors are logical and programmable, but not conscious.

The field of robotics is currently in a developmental period that bridges action and autonomous intelligence, with robots such as ABIO (http://support.sony-europe.com/abio/), QRIO (Yeomans, 2005), and Roomba (http://www.irobot.com/sp.cfm?pageid=122/) on the verge of being lab tested, manufactured, and sold. These simple, small-scale robots are strong indicators of NGR potential and the coming human-robot co-existence age. Even as "pioneer" robots, they have remarkable abilities to perform specific tasks according to their built-in autonomous intelligence. For example, ABIO and QRIO robots have been programmed to serve as companions for the elderly, and Roomba robots are capable of performing housecleaning chores. However, none of them can make decisions concerning self-beneficial actions or distinguish between right and wrong based on a sense of their own value.

At the third level is human-based intelligence (HBI)—higher cognitive abilities that allow for new ways of looking at one's environment and for abstract thought; HBI is also referred to as "mind" and "real intelligence" (Neisser et al., 1996). Since a universally accepted definition of human intelligence has yet to emerge, there is little agreement on a definition for robot HBI. Many suggestions and predictions appear to borrow liberally from science fiction—for instance, Asimov's (1976) description of HBI robots forming a new species with the long-term potential of gaining power over humans. In real-world contexts, researchers are experimenting with ways of combining action, autonomous, and human-based intelligence to create robots that "comprehend complex ideas, learn quickly, and learn from experience" (Gottfredson, 1997).

HBI research started in the 1950s—roughly the same time as research on artificial intelligence (AI), with which HBI is closely associated. One of the earliest and most famous efforts at examining HBI potential entailed what is now known as the "Turing Test" (Turing, 1950). Taking a behaviorist approach, Turing defined human intelligence as the ability "to respond like a human being," especially regarding the use of natural language to communicate. There have been many efforts to create programs that allow robots to respond to words and other stimuli in the same manner as humans ("Emotion Robots," 2007), but no AI program has ever passed the Turing test and been accepted as a true example of HBI (Saygin et al., 2000). The legal and RSP issues that will arise over the next few decades are intricately linked with AI, which was originally conceived as "the science and engineering of making intelligent machines, especially intelligent computer programs" (McCarthy, 2007).

	Safety Design by Risk Assessment	**Safety Intelligence**
Risk	Machine Risk	Autonomous Behavior Risk (Open Texture Risk)
Limit	Machine Standards	Robot's Intelligence Architecture
Effect	Decrease Risk	Prevent dangerous behaviors

Table 1. A Comparison of Safety Regulation Methods.

Designed and constructed according to very specific standards, industrial robots are limited to performing tasks that can be reduced to their corresponding mechanisms—in other words, they cannot alter their mechanisms to meet the needs of changing environments. Therefore, the primary purpose for performing industrial robot risk assessments is to design mechanisms that match pre-approved safety levels (Table 1).

Complex NGR motions, multi-object interactions, and responses to shifts in environments resulting from complex interactions with humans cannot be reduced to simple performance parameters. Furthermore, residual risk will increase as robot intelligence evolves from action to autonomous to human-based, implying that current assessment methods will eventually lose their value. Instead, NGR designers/manufacturers and HBI developers will have to deal with unpredictable hazards associated with the legal concepts of *core meaning* and *open texture risk*. While all terms in a natural language have core meanings, the open texture characteristic of human language allows for interpretations that vary according to specific domains, points of view, time periods, and other factors—all of which can trigger uncertainty and vagueness in legal interpretations (Lyons, 1999). In addition to clearly defining core meanings, autonomous NGR designers and programmers must predict their acceptable and useful ranges.

In a policy paper published in May of 2007, the Japanese METI predicted that the addition of NGRs into human society will occur in three stages (METI, 2007). The focus of the first (current) stage is on "the diffusion of specific working object robots" to perform tasks such as cleaning, security, carrying documents or objects, and performing desk clerk duties. The second (expected to begin between now and 2010) will consist of "specific user and autonomously working object robots"—for example, nurse robots that can perform tasks to support the elderly. The final stage (beginning between 2010 and 2015) will primarily consist of multi-task autonomous ("universal") robots. METI describes the first stage as a "point" diffusion (meaning a low degree of human contact), the second as a "line" diffusion (medium degree), and the third as a "facet" diffusion (high degree). The extent to which each stage affects the daily lives of humans depends on the speed and content of AI development. Regardless of the specific characteristics of human-NGR interaction within each stage, open-texture risk will expand significantly with each transition. The inherent unpredictability of unstructured environments makes it virtually impossible that humans will ever create a fail-safe mechanism that allows autonomous robots to solve all open-texture problems.

4. Safety intelligence

4.1 Safety intelligence (SI) and Asimov's Three Laws.

NGR safety regulations will require a mix of pre- and post-safety mechanisms, the first entailing programming and using a robot's AI content to eliminate risk, and the second a product liability system to deal with accidents that do occur. A clear security issue will be limiting NGR "self-control" while still allowing them to perform designated tasks. As one *Roboethics Roadmap* author succinctly states, "Operators should be able to limit robot autonomy when the correct robot behavior is not guaranteed" (EURON, 2006).

Giving operators this capability requires what we will call *safety intelligence* (SI), a system of artificial intelligence restrictions whose sole purpose is to provide safety parameters when semi-autonomous robots perform tasks. Researchers have yet to agree on a foundation for a SI system, but the most frequently mentioned is Isaac Asimov's (1950) "Three Laws of Robotics," created for his science fiction novel, *I, Robot*:

- First Law: A robot may not injure a human being or, through inaction, allow a human being to come to harm.
- Second Law: A robot must obey orders given it by human beings, except when such orders conflict with the First Law.
- Third Law: A robot must protect its own existence as long as such protection does not conflict with the First or Second Law.

The first two laws represent a human-centered approach to SI that agrees with the current consensus of NGR designers and producers. As robots gradually take on greater numbers of labor-intensive and repetitive jobs outside of factories and workplaces, it will become increasingly important for laws and regulations to support SI as a "mechanism of human superiority" (Fodor, 1987). The third law straddles the line between human- and machine-centered approaches. Since the purpose of their functionality is to satisfy human needs, robots must be designed and built so as to protect themselves as human property, in contrast to biological organisms that protect themselves for their own existence. As one magazine columnist has jokingly suggested, "A robot will guard its own existence ... because a robot is bloody expensive" (Langford, 2006).

In his introduction to *The Rest of the Robots* (1964), Asimov wrote, "There was just enough ambiguity in the Three Laws to provide the conflicts and uncertainties required for new stories, and, to my great relief, it seemed always to be possible to think up a new angle out of the 61 words of the Three Laws." While those ambiguities may be wonderful for writing fiction, they stand as significant roadblocks to establishing workable safety standards for complex NGRs. *Roboethics Roadmap* authors note that the Three Laws raise many questions about NGR programming: Which kinds of ethics are correct and who decides? Will roboethics really represent the characteristics of robots or the values of robot scientists? How far can and should we go when we program ethics into robots? (EURON, 2006)

Other robot researchers argue that Asimov's laws still belong to the realm of science fiction because they are not yet applicable. Hiroshi Ishiguru of Osaka University, the co-creator of two female androids named Repliee Q1 and Repliee Q2 (Whitehouse, 2005), believes it would be a mistake to accept Asimov's laws as the primary guiding principle for establishing robot ethics: "If we have a more intelligent vehicle [i.e., automobile], who takes responsibility when it has an accident? We can ask the same question of a robot. Robots do not have human-level intelligence" (Lovgren, 2007). Mark Tilden, the designer of a toy-like robot named RoboSapien, says "the problem is that giving robots morals is like teaching an ant to yodel. We're not there yet, and as many of Asimov's stories show, the conundrums robots and humans would face would result in more tragedy than utility" (ibid.). Ian Kerr (2007), law professor at the University of Ottawa, concurs that a code of ethics for robots is unnecessary:

> Leaving aside the thorny philosophical question of whether an AI could ever become a moral agent, it should be relatively obvious from their articulation that Asimov's laws are not ethical or legal guidelines for robots but rather about them. The laws are meant to constrain the people who build robots from exponentially increasing intelligence so that the machines remain destined to lives of friendly servitude. The pecking order is clear: robots serve people.

4.2 Three problems tied to Asimov's laws

Currently there are two competing perspectives on dealing with the mix of AI and safety: creating artificial agents with safety-oriented reasoning capabilities, or programming robots with as many rules as required for ensuring the highest level of safe behavior. Which perspective wins out will depend on how policy makers, designers, and manufacturers address three issues that we will address in the following subsections.

4.2.1 Legalities: safety intelligence

Future robot-related planning and design decisions will involve human values and social control, and addressing them will require input from legal scholars, social scientists, and public policy makers. A great amount of the data they will use to support their decisions must come from researchers familiar with robot legal studies. Levy (2006) argues convincingly that a new legal branch of robot law is required to deal with a technology that by the end of this century will be found in the majority of the world's households. Robot safety will be a major component of robot law. From an engineering viewpoint, SI entails self-control; from a legal viewpoint, SI entails automatic compliance and obedience to human rules. The need to add individual restrictions to autonomous robot behavior will be limited to those problems that pre-safety risk assessments cannot successfully address in order to meet two major goals of robot law: restricting human behaviors to prevent the misuse of robots, and restricting autonomous robot behaviors to mitigate open texture risk. Most likely a combination of these restrictions and carefully crafted legislation to let robots enforce the legal norm will be required to achieve a maximum state of robot safety.

As noted earlier, it is unrealistic to assume that SI will ever be able to eliminate all risk; instead, it is better to view SI as one part of a regulatory system that also contains a strong technological component. We believe the respective roles of designers, users, creators, and robots must be spelled out before the human-robot co-existence society emerges, while acknowledging that liability questions will require many court decisions to refine. For example, if Asimov's laws are used as a guideline, how will we enforce the idea that robots must not allow human beings to come to harm *through inaction*? Further, what about situations in which a NGR follows an incorrect human command?

4.2.2 Decision-making: do NGRs need doctrinal reasoning?

Giving autonomous robots the ability to define their own safety concepts means giving them the power to decide both when and how to react to stimuli. At some point such decisions will require artificial ethical and morality reasoning—the ability to distinguish between right and wrong. Robotists such as Shigeo Hirose (1989) argue that in conflicts involving doctrinal reasoning and morality, Asimov's Three Laws may become contradictory or be set aside in favor of poorly chosen human priorities. Using an extreme example, robots could be programmed to commit homicide under specific circumstances based on the wishes of a human majority. This example touches on two fears that many people have when they consider autonomous robots: they are troubled by the idea of letting robots obey rules that are difficult or impossible to express legislatively, and fearful of letting them defend laws established by imperfect humans.

The robot decision-making problem requires debate on both morality reasoning and legal content. Asimov himself acknowledged the potential for multiple contradictions between his Three Laws and doctrinal reasoning—many of them based on natural language, the medium used by humans to access legal content. Despite the many problems that human legal language present regarding vagueness and abstraction, laws and rules have the quality of being understood and followed via doctrinal reasoning. To argue that the same process can be used to accomplish SI means finding a universally accepted answer to the question, "Do NGRs need doctrinal reasoning to accomplish SI, and is it possible for them to acquire it?"

The focus of the new field of human-robot interaction (HRI) is communication and collaboration between people and robots. HRI researchers view human-robot communication as having verbal (using natural language) and non-verbal components (e.g., actions, gestures, non-linguistic sounds, and environment). Humans communicate using a mix of verbal and non-verbal information, with non-verbal information representing up to

60-70% of total communication (Kurogawa, 1994). We believe that non-verbal communication is a more reliable means for implementing SI and avoiding contradictions tied to doctrinal reasoning. The issue is necessarily complex because it entails AI, control engineering, human values, social control, and the processes by which laws are formed. In addition to being beyond the scope of this chapter, SI decisions are beyond the range of robotists' training and talent, meaning that they must work hand-in-hand with legal scholars.

4.2.3 Abstract thinking and the meaning of "safety"

The ability to think abstractly is uniquely human, and there is no way of being absolutely certain of how robots will interpret and react to abstract meanings and vague terms common to human communication. For example, humans know how to distinguish between blood resulting from a surgical operation and blood resulting from acts of violence. Making that distinction requires the ability to converse, to understand abstract expressions (especially metaphors), and to use domain knowledge to correctly interpret the meaning of a sentence. There are many examples that illustrate just how difficult this task is: Chomsky's (1957) famous sentence showing the inadequacy of logical grammar ("Colorless green ideas sleep furiously"), and Groucho Marx's line, "Time flies like an arrow, fruit flies like a banana" (http://www.quotationspage.com/quote/26.html). Such examples may explain why Asimov (1957) described robots as "logical but not reasonable."

If one day robots are given the power to think abstractly, then humans will have to deal with yet another legal problem: robo-rights. In 2006, the Horizon Scanning Centre (part of the United Kingdom's Office of Science and Innovation) published a white paper with predictions for scientific, technological, and health trends for the middle of this century ("Robots Could," 2006). The authors of the section entitled "Utopian Dream, or Rise of the Machines" raise the possibility of robots evolving to the degree that they eventually *ask* for special "robo-rights" (see also "Robot Rights?" 2006). In this paper we will limit our discussion to how current human legal systems (in which rights are closely tied to responsibilities) will be used when dealing with questions tied to early generations of non-industrial robots.

Whenever an accident involving humans occurs, the person or organization responsible for paying damages can range from individuals (for reasons of user error) to product manufacturers (for reasons of poor product design or quality). Rights and responsibilities will need to be spelled out for two types of NGRs. The system for the first type—NGRs lacking AI-based "self-awareness"—will be straightforward: 100% human-centered, in the same manner that dog owners must take responsibility for the actions of their pets. In other words, robots in this category will never be given human-like rights or rights as legal entities.

The second type consists of NGRs programmed with some degree of "self-awareness," and therefore capable of making autonomous decisions that can result in damage to persons or property. Nagenborg et al. (2007) argue that all robot responsibilities are actually human responsibilities, and that today's product developers and sellers must acknowledge this principle when designing first-generation robots for public consumption. They use two codes of ethics (one from the Institute of Electrical and Electronics Engineers and the other from the Association of Computing Machinery) to support their view that for complex machines such as robots, any attempt to remove product responsibility from developers, manufacturers, and users represents a serious break from human legal systems. We may see a day when certain classes of robots will be manufactured with built-in and retrievable "black boxes" to assist with the task of attributing fault when accidents occur, since in practice it will be difficult to attribute responsibility for damages caused by robots—especially those resulting from owner misuse. For this reason, Nagenborg et al. have proposed the following meta-regulation:

> If anybody or anything should suffer from damage that is caused by a robot that is capable of learning, there must be a demand that the burden of adducing evidence must be with the robot's keeper, who must prove her or his innocence; for example, somebody may be considered innocent who acted according to the producer's operation instructions. In this case it is the producer who needs to be held responsible for the damage.

If responsibility for robot actions ever reaches the point of being denied by humans, a major issue for legal systems will be determining punishment. In wondering whether human punishment can ever be applied to robots, Peter Asaro (2007) observes that :

> They do have bodies to kick, though it is not clear that kicking them would achieve the traditional goals of punishment. The various forms of corporal punishment presuppose additional desires and fears of being human that may not readily apply to robots—pain, freedom of movement, morality, etc. Thus, torture, imprisonment and destruction are not likely to be effective in achieving justice, reform and deterrence in robots. There may be a policy to destroy any robots that do harm, but as is the case with animals that harm people, it would be a preventative measure to avoid future harms rather than a true punishment. ... [American law] offers several ways of thinking about the distribution of responsibility in complex cases. Responsibility for a single event can be divided among several parties, with each party assigned a percentage of the total.

If we go this route, we may need to spell out robo-rights and responsibilities in the same manner that we do for such non-human entities as corporations; this is a core value in human legal systems. The question is whether or not we will be able to apply human-centered values to robots in the same manner. To practice "robot justice," those systems will be required to have separate sets of laws reflecting dual human/robot-centered values, and robot responsibilities would need to be clearly spelled out.

4.3 Three criteria and two critical issues for SI

As we have shown so far, Asimov's Three Laws of Robotics face several significant legal and engineering challenges, especially the need for collaboration among designers, AI programmers, and legal scholars to address robot sociability problems with safety as the guiding principle. Based on our belief that the importance of SI will increase as we rapidly approach a period in which NGRs are developed, tested, built, and sold on a large scale, we are currently working on a safety regulation model that emphasizes the role of SI during the pre-safety stage. In its current form, our proposal rests on three criteria for safe human-NGR interaction: (a) the ongoing capability to assess changing situations accurately and to correctly respond to complex real-world conditions; (b) immediate protective reactions in human-predictable ways so as to mitigate risks tied to language-based misunderstandings or unstable autonomous behaviors; and (c) an explicit interaction rule set and legal architecture that can be applied to all NGRs, one that accommodates the needs of a human-robot co-existence society in terms of simplicity and accountability. In order to apply the three criteria, there is a need to break down the safety intelligence concept into the two dimensions that are the focus of the next two sections. The first, which is ethics-centered, involves a special "third existence" status for robots or a complete ban on equipping NGRs with human based intelligence. The second involves the integration of third existence designation with a *legal machine language* designed to resolve issues associated with open texture risk.

5. Future considerations

In "ROBOT: Mere Machine to Transcendent Mind" (1999), Carnegie Mellon professor Hans Moravec predicts that robot intelligence will "evolve" from lizard-level in 2010, to mouse-level in 2020, to monkey level in 2030, and finally to human level in 2040—in other words, some robots will strongly resemble *first-existence* (biological) entities by mid-century. If his prediction is correct, determining the form and content of these emerging robots will require broad consensus on ethical issues in the same manner as nuclear physics, nanotechnology, and bioengineering. Creating consensus on these issues may require a model similar to that of the Human Genome Project for the study of Ethical, Legal and Social Issues (ELSI) sponsored by the US Department of Energy and National Institutes of Health (http://www.ornl.gov/sci/techresources/Human_Genome/research/elsi.shtml). Each agency has earmarked 3-5 percent of all financial support for genome research for addressing ethical issues. ELSI's counterpart across the Atlantic is the European Robotics Research Network (EURON), a private organization devoted to creating resources for and exchanging knowledge about robotics research (http://www.euron.org/). As part of its effort to create a systematic assessment procedure for ethical issues involving robotics research and development, EURON has published *Roboethics Roadmap* (2006), a collection of articles outlining potential research pathways and speculating on how each one might develop.

According to the *Roadmap* authors, most members of the robotics community express one of three attitudes toward the issue of roboethics:

- Not interested. They regard robotics as a technical field and don't believe they have a social or moral responsibility to monitor their work.
- Interested in short-term ethical questions. They acknowledge the possibility of "good" or "bad" robotics and respect the thinking behind implementing laws and considering the needs of special populations such as the elderly.
- Interested in long-term ethical concerns. They express concern for such issues as "digital divides" between world regions or age groups. These individuals are aware of the technology gap between industrialized and poor countries and the utility of developing robots for both types.

We (the authors of this paper) belong to the third category, believing that social and/or moral questions are bound to accompany the emergence of a human-robot co-existence society, and that such a society will emerge sooner than most people realize. Furthermore, we agree with the suggestions of several *Roboethics Roadmap* authors that resolving these ethical issues will require agreement in six areas:

- Are Asimov's Three Laws of Robotics usable as a guideline for establishing a code of roboethics?

- Should roboethics represent the ethics of robots or of robot scientists?
- How far can we go in terms of embodying ethics in robots?
- How contradictory are the goals of implementing roboethics and developing highly autonomous robots?
- Should we allow robots to exhibit "personalities"?
- Should we allow robots to express "emotions"?

This list does not include the obvious issue of what kinds of ethics are correct for robots. Regarding "artificial" (i.e., programmable) ethics, some *Roadmap* authors briefly touch on needs and possibilities associated with robot moral values and decisions, but generally shy away from major ethical questions. We consider this unfortunate, since the connection between artificial and human-centered ethics is so close as to make them very difficult to separate. The ambiguity of the term *artificial ethics* as used in the EURON report ignores two major concerns: how to program robots to obey a set of legal and ethical norms while retaining a high degree of autonomy (Type 1), and how to control robot-generated value systems or morality (Type 2). Since both will be created and installed by humans, boundaries between them will be exceptionally fluid.

5.1 "Learning" ethics

Two primary research categories in the field of artificial intelligence are *conventional* (or symbolic) and *computational*. Conventional AI, which entails rational logical reasoning based on a system of symbols representing human knowledge in a declarative form (Newell & Simon, 1995), has been used for such applications as chess games employing reasoning powers (Hsu et al., 1995), conversation programs using text mining procedures (http://www.alicebot.org), and expert systems that are used to organize domain-specific knowledge (Lederberg, 1987). While conventional AI is capable of limited reasoning, planning, and abstract thinking powers, researchers generally agree that the use of symbols does not represent "mindful" comprehension and is therefore extremely limited in terms of learning from experience (Dreyfus & Dreyfus, 1980).

Computational (non-symbol) AI (Engelbrecht, 2002) mimics natural learning methods such as genetic (Mitchell, 1996) or neural (Abdi, 1994). It supports learning and adaptation based on environmental information in the absence of explicit rules—an important first existence capability. Some advantages of computational AI are its capacities for overcoming noise problems, working with systems that are difficult to reduce to logical rules, and performing such tasks as robot arm control, walking on non-smooth surfaces, and pattern recognition. However, as proven by chess programs, computational AI is significantly weaker than conventional AI in abstract thinking and rule compliance. Still, many robotics and AI researchers believe that HBI in robots is inevitable following breakthroughs in computational AI (Warwick, 2004), while others argue that computational and conventional AI are both examples of behaviorism, and therefore will never capture the essence of HBI (Dreyfus & Dreyfus, 1980; Penrose, 1989). Those in the latter group claim that reaching such a goal will require a completely new framework for understanding intelligence (Hawkins & Blakeslee, 2006).

We believe the odds favor the eventual emergence of HBI robots, and therefore researchers must take them into consideration when predicting future robot safety and legal issues. They may agree with Shigeo Hirose of the Tokyo Institute of Technology that a prohibition on HBI is necessary (quoted in Tajika, 2001). Hirose is one of a growing number of researchers and robot designers resisting what is known as the "humanoid complex" trend, based on his strict adherence to the original goal of robotics: inventing useful tools for human use (quoted in Kamoshita, 2005). For Alan Mackworth, past president of the Association for the Advancement of Artificial Intelligence, the HBI issue is one of "should or shouldn't we" as opposed to "can or can't we" ("What's a Robot?" 2007). Arguing that goal-oriented robots do not require what humans refer to as "awareness," Mackworth challenges the idea that we need to create HBI for machines.

5.2 Third existence

In an earlier section we discussed the idea that robot responsibility should be regarded as human-owner responsibility. If we allow HBI robots to be manufactured and sold, the potential for any degree of robot self-awareness means dealing with issues such as punishment and a shift from human-centered to human-robot dual values. This is one of the most important reasons why we support a ban on installing HBI software in robots—perhaps permanently, but certainly until policy makers and robotists arrive at a generally accepted agreement on these issues. We also believe that creating Type 1 robots—in other words, programming robots to obey a set of legal and ethical norms while retaining a high degree of autonomy—requires agreement on human-centered ethics based on human values. The challenge is determining how to integrate human legal norms into robots so that they become central to robot behavior. The most worrisome issue is the potential of late-generation HBI robots with greater degrees of self-awareness to generate their own values and ethics—what we call Type 2 artificial ethics. Implementing Type 2 robot safety standards means addressing the uncertainties of machines capable of acting outside of human norms. We are nowhere near discussing—let alone implementing—policies for controlling HBI robot behavior, since we are very far from having HBI robots as part of our daily lives.

However, if the AI-HBI optimists are correct, the risks associated with HBI robots will necessitate very specific guidelines. A guiding principle for those guidelines may be the categorization of robots as "third existence," a concept created by Waseda University's Shuji Hashimoto (2003). Instead of living/biological (first existence) or non-living/non-biological (second existence), he describes third existence entities as machines that *resemble* living beings in appearance and behavior. We think this definition overlooks an important human-robot co-existence premise: most NGRs will be restricted to levels of autonomous intelligence that fall far short of HBI, therefore their similarities with humans will be minor. In addition, the current legal system emphasizes the status of robots as second-existence human property, which may be inadequate for those semi-autonomous NGRs that are about to make their appearance in people's homes and businesses, especially in terms of responsibility distribution in the case of accidents. In their analyses, Asaro (2007) suggests the creation of a new legal status for robots as "quasi persons" or "corporations," while Nugenborg (2007) emphasizes the point we made earlier about how robot owners must be held responsible for their robots' actions in the same way as pet owners.

6. Legal machine language

We believe there are two plausible strategies for integrating NGR safety intelligence with legal systems, one that uses natural languages and one that uses artificial machine languages. We will respectively refer to these as *legal natural language* (LNL) and *legal machine language* (LML) (Weng, et al., unpublished manuscript). The main goal of an LNL approach is to create machines capable of understanding ranges of commands and making intelligent decisions according to laws and rules written in a natural language. Since LNL is the primary medium for accessing legal content in human society, using it to program robots eliminates the need for an alternative medium, which is considered a difficult task requiring a combination of HBI and solutions to the structured-vs.-open texture issues discussed earlier. For the foreseeable future, NGRs will not be capable of making autonomous decisions based on a combination of legal natural language and an underlying principle such as Asimov's Three Laws.

Constructed languages (e.g., programming languages and code) can be used as bases for legal machine languages (LMLs) that express legal content. Two common examples already in use are *bots* (also called *agent software*) and service programs employed in agency network environments for providing content, managing resources, and probing for information. It is possible to use LML to control robot behavior—in Lawrence Lessig's (1999) words, "We can build, or architect, or code cyberspace to protect values that we believe are fundamental." We will give three examples of how LMLs are currently being used. First, the Internet Content Rating Association (ICRA) (http://www.fosi.org/icra) has created a standard content description system that combines features of a Resource Description Framework (RDF) and a Platform for Internet Content Selection (PICS). The system permits the marking of website content according to categories such as nudity, sex, or violence, and then allows parents and teachers to control online access. RDF makes it possible for machines to understand and react to ICRA labels embedded in websites as meta-tags. Whenever a browser with a built-in content filter (e.g., Microsoft Explorer's "Content Advisor") reads ICRA label data, it has the power to disobey user instructions based on "legal standards" established by parents or other authority. The second example is Robot Exclusion Standards (RES) (http://www.robotstxt.org/orig.html), also known as Robots Exclusion Protocol or robots.txt protocol. RES is currently being used to prevent web bots ("spiders") from accessing web pages that owners want to keep private, with the privacy policy written in a script or programming language. The third is Creative Commons (CC) (http://creativecommons.org/licenses/), used by copyright holders who want to control access in terms of sharing written information that they place on web sites. In addition to open content licensing, CC offers a RDF/XML metadata scheme that is readable by web bots. These examples show how humans can use constructed languages to communicate with robots in highly autonomous environments and to control their behavior according to agreed-upon legal standards.

Constructed languages are being examined as a means of overcoming the problem of ambiguity and the need to understand a range of commands associated with natural languages. Perhaps the best-known example is Loglan (http://www.loglan.org), identified as potentially suitable for human-computer communication due to its use of predicate logic, avoidance of syntactical ambiguity, and conciseness. The constructed language approach to programming commands in robots can be viewed as an intermediate step between human commands and autonomous robot behavior. As we stated above, NGRs in the foreseeable future will not be capable of making autonomous decisions based on a combination of legal natural language and an underlying principle such as Asimov's Three Laws. Since legal machine language does not give robots direct access to legal content, machines can be programmed to behave according to legal constraints without the addition of artificial ethics. Most likely a mix of LNL and LML will be required as NGRs interact more with humans, with human-machine communication based on simple rule sets designed to enhance safety when robots work around humans.

In an earlier section we discussed the neo-mammalian brain (HBI equivalent) functions of processing ethics, performing morality reasoning, and making right/wrong decisions. However, such a high level of reasoning is not required to recognize situations, avoid misunderstandings, and prevent accidents; a combination of action intelligence (unconscious) and autonomous intelligence (subconscious) is sufficient for these purposes. Nature gives us many examples of animals interacting safely without complex moral judgments or natural languages. When birds of the same species migrate, their shared genetic background allows them to fly in close or v-shaped formations without colliding—an important feature for collective and individual survival. This kind of safe interaction requires adherence to a set of simple non-verbal rules: avoid crowding neighboring birds, fly along the same heading, or at least along the same average heading as neighboring birds (Klein et al., 2003). The same flocking concept observed in birds, schools of fish, and swarms of insects is used to control unmanned aircraft and other machines (Gabbai, 2005; Reynolds, 1987). HBI-level moral and ethical decisions still require autonomous and action intelligence to correctly recognize environmental and situational factors prior to reacting in a safe manner.

6.1 Reflex control

Currently the focus of some exciting research, *reflex control* holds major potential as a tool allowing for the use of non-verbal communication and LML to command and control NGRs. Reflex control is a biomorphic concept based on the behaviorist principles of stimulus and response and the evolutionary principles of cross-generation inheritance. The basic function of reflexes, which are genetically built-in and learned by humans and animals, is to trigger protective behaviors within and across species, as well as between a species and its environment. Today we have a much clearer understanding of reflex action (and how it might be used to control robot behavior) than we do of complex intelligence. Some researchers are experimenting with sensors linked to reflexive controls—a combination known as "experiential expert systems" (Bekey & Tomovic, 1986)—for motion control (Zhang et al., 2002), automatic obstacle avoidance (Newman, 1989), collision avoidance (Wikman et al., 1993), and other simple behaviors associated with autonomous and action intelligence. Instead of requiring complex reasoning or learning capacities, robots equipped with reflex action can navigate fairly complex environments according to simple sets of common rules. Other robotists are working on a combination of reflex control and the ability to learn new protective behaviors for biomorphic robots, perhaps with a mechanism for transferring new knowledge to other machines. From a SI standpoint, reflex control allows for a high degree of safety in robot behavior because reactions can be limited to direct and immediate danger and programmed according to explicitly mapped responses. This meets our criteria for SI as a fast, clear, non-verbal, and predictable approach. Different sets of simple rules may be developed for NGRs operating in different environments (or countries) while still adhering to a "safety gene" standard (an explicit legal architecture applicable to all kinds of NGR robots), thereby satisfying another SI criteria.

An important distinction between human-designed robot reflex control and a NGR safety intelligence unit is that the SI unit would include content based on societal requirements to address open texture risk. Again, such content will require input from specialists to ensure compliance with accepted legal principles. If robotists ever reach a level of technological expertise that allows for HBI-programmed machines, a decision will have to be made about whether or not robots should be given human capabilities, or if reflex actions are sufficient for the needs and purposes of a human-robot co-existence society.

7. Conclusion

By the middle of this century, artificial intelligence and robot technology will no longer be considered science fiction fantasy. At the same time that engineers will be addressing all kinds of technical issues, a combination of engineers, social scientists, legal scholars, and policy makers will be making important decisions regarding robot sociability. In all cases, the topmost concern must be robot safety, since the emphasis for the future will be on human-robot *co-existence*.

In this paper we described a *safety intelligence* concept that meets three criteria: understanding situations, making decisions, and taking responsibility. The SI concept can also be broken down into two dimensions, the first involving special "third existence" ethical guidelines for robots plus a ban on equipping NGRs with human-based intelligence, and the second involving a mix of third existence designation and legal machine language designed to resolve issues associated with open texture risk. If current reflex control research proves successful, robots will someday be equipped with a safety reflex system to ensure that they avoid dangerous situations, especially those involving humans.

8. Acknowledgements

The authors would like to thank Mr. Jon Lindemann for his comments on the original manuscript.

9. References

Abdi, H. (1994). A neural network primer. *Journal of Biological Systems,* Vol. 2, 247-281.

Asaro, P. (2007). Robots and responsibility from a legal perspective. Paper presented at the *IEEE ICRA '07 Workshop on Roboethics,* Rome. Available online at http://www.roboethics.org/icra07/contributions/ASARO%20Legal%20Perspective.pdf.

Asimov, I. (1950). *I, Robot.* Doubleday: New York.

Asimov, I. (1957). *The Naked Sun.* Doubleday: New York.

Asimov, I. (1964). *The Rest of the Robots.* Doubleday: New York.

Asimov, I. (1976). *Bicentennial Man.* Ballantine Books: New York.

Robots Could Demand Legal Rights. (2006). *BBC News,* December 21. Available online at http://news.bbc.co.uk/2/hi/technology/6200005.stm.

Bekey, G. A. & Tomovic, R. (1986). Robot control by reflex actions. *Proceedings of the 1986 IEEE International Conference, Part 3: Robotics and Automation*, San Francisco, pp. 240-247.

Cabinet Office Government of Japan (COGJ). (2007). *Long-term strategic guidelines "Innovation 25"* (unofficial translation). Available online at http://www.cao.go.jp/innovation/index.html.

Chapin, J. K., Moxon, K. A., Markowitz, R. S. & Nicolelis, M. A. L. (1999). Real-time control of a robot arm using simultaneously recorded neurons in the motor cortex. *Nature Neuroscience*, Vol. 2, 664-670.

Chomsky, N. (1957). *Syntactic Structures*. Mouton: The Hague/Paris.

Dickmanns, E. D. (1998). Dynamic computer vision for mobile robot control. *Proceedings of the 19th* International *Symposium Expos. Robots*, Sydney, Australia, pp. 314-327.

Dreyfus, H. L. & Dreyfus, S. E. (1980). *From Socrates to expert systems: The limits and dangers of calculative rationality*. Available online at http://Socrates.berkeley.edu/%7Ehdreyfus/html/paper_socrates.html.

Emotion Robots Learn from People. (2007). BBC News, February 23, available online at http://news.bbc.co.uk/2/hi/technology/6389105.stm.

Engelbrecht, A. P. (2002). *Computational intelligence: An introduction*. John Wiley & Sons: Hoboken, New Jersey.

European Robotics Research Network (EURON). (2006). *Roboethics Roadmap, Release 1.1*, Author: Genova, Italy.

Fodor, J. A. (1987). Modules, frames, fridgeons, sleeping dogs and the music of the spheres. In Z. Pylyshyn (ed.), *The robot's dilemma: The frame problem in artificial intelligence*. Ablex Publishers: Norwood, New Jersey.

Gabbai, J. M. E. (2005). *Complexity and the aerospace industry: Understanding emergence by relating structure to performance using multi-agent systems*. Unpublished doctoral dissertation, University of Manchester.

Gottfredson, L. S. (1997). Mainstream science on intelligence. *Intelligence*, Vol. 24, Issue 1, 13–23.

Hager, G. D., Chang, W.-C. & Morse, A. S. (1995). Robot hand-eye coordination based on stereo vision. *IEEE Control Systems*, Vol.15, Issue 1, 30-39.

Hashimoto, S. & Yabuno, K. (2003). *The book of Wabot 2*. Chuko: Tokyo (in Japanese).

Hawkins, J. & Blakeslee, S. (2006). *On intelligence*. Yuan-Liou: Taipei (Chinese translation).

Hirose, S. (1989). A robot dialog. *Journal of Robotics Society of Japan*, Vol. 7, Issue 4, 121-126 (in Japanese).

Hsu, F. H., Campbell, M. & Hoane A. J. Jr. (1995). Deep Blue system overview, *Proceedings of the Ninth International Conference on Supercomputing*, Barcelona, pp. 240-244.

Huxley, A. L. (1932). *Brave New World*. Harper & Brothers: New York.

International Organization for Standardization (ISO). (1994). ISO 8373. Manipulating industrial robots – Vocabulary.

International Organization for Standardization (ISO). (2006a). ISO 10218. Robots for industrial environments - Safety requirements - Part 1: Robot.

International Organization for Standardization (ISO). (2006b). ISO 13849-1. Safety of machinery - Safety-related parts of control systems - Part 1: General principles for design.

Kamoshita, H. (2005). *The present and future of robots*. X-media: Tokyo (in Japanese).
Kerr, I. (2007). Minding for machines. *Ottawa Citizens News*. Available online at http://www.canada.com/ottawacitizen/news/opinion/story.html?id=e58202bb-f737-4ba7a0as-79e8071a1534.
Kimura, T. (2007). Risk assessment of service robots and related issues. *Journal of Robotics Society of Japan*, Vol. 25, Issue 8, 1151-1154 (in Japanese).
Koditschek, D. E. (1989). Robot planning and control via potential functions. *The Robotics Review*, Vol. 1, 349-367.
Kurogawa, T. (1994). *Nonverbal interface*. Ohmsha: Tokyo (in Japanese).
Langford, D. (2006). "It's the law" (magazine column). *SFX*, Issue 146. Available online at http://www.ansible.co.uk/sfx/sfx146.html
Lederberg, J. D. (1987). How Dendral was conceived and born. *Proceedings of the 1987 Association for Computing Machinery Conference on History of Medical Informatics*, Bethesda, Maryland, pp. 5-19.
Lessig, L. (1999). *Code and other laws of cyberspace*. Basic Books Press: New York.
Levy, D. (2006). *Robots unlimited: Life in a virtual age*. A. K. Peters: Wellesley, Massachusetts.
Lewis, M. A. (1992). Genetic programming approach to the construction of a neural network for control of a walking robot, *Proceedings of the IEEE International Conference on Robotics and Automation*, Nice, France, pp. 2118-2623.
Lovgren, S. (2006). A Robot in Every Home by 2020, South Korea Says, *National Geographic News* Sep. 6, 2006, available online at http://news.nationalgeographic.com/news/2006/09/060906-robots.html.
Lovgren, S. (2007). Robot Codes of Ethics to Prevent Android Abuse, Protect Humans, *National Geographic News*, March 16, 2007, available online at http://news.nationalgeographic.com/news/2007/03/070316-robot-ethics.html.
Lyons, D. (1999). Open texture and the possibility of legal interpretation. *Law and Philosophy*, Vol. 18, Issue 3, 297-309.
MacLean, P. D. (1973). A triune concept of the brain and behavior. In T. Boag & D. Campbell (eds.), *Hincks Memorial Lectures*, University of Toronto Press.
McCarthy, J. (2007). *What is Artificial Intelligence?* Available online at http://www-formal.stanford.edu/jmc/whatisai.html.
METI Robot Policy Council. (2005). *Robot policy midterm report*. Available online at http://www.meti.go.jp/policy/robotto/chukanhoukoku.pdf (in Japanese).
METI Robot Policy Council. (2006). *Robot policy report*. Available online at http://www.meti.go.jp/press/20060516002/robot-houkokusho-set.pdf (in Japanese).
Ministry of Economy, Trade and Industry (METI). (2004). *Toward 2025 and the human-robot co-existence society: The next generation robot vision seminar report*. Available online at http://www.meti.go.jp/kohosys/press/0005113/0/040402robot.pdf (in Japanese).
Ministry of Economy, Trade and Industry (METI). (2007). *Overview of METI robot policy*. Available online at http://www.meti.go.jp/policy/robotto/0705gaiyo.pdf (in Japanese).
Ministry of Economy, Trade and Industry (METI). (2007). *Guidelines to secure the safe performance of Next Generation Robots*. Available online at http://www.meti.go.jp/press/20070709003/02_guideline.pdf (in Japanese).
Mitchell, M. (1996). *Introduction to genetic algorithms*. MIT Press: Cambridge, Massachusetts.

Moravec, H. (1999). *ROBOT: Mere machine to transcendent mind.* Oxford University Press: New York.

Morita, T., Suzuki, Y., Kawasaki, T. & Sugano, S. (1998). Anticollision safety design and control *methodology* for human-symbiotic robot manipulator. *Journal of Robotics Society of Japan*, Vol. 16, Issue 1, 102-109 (in Japanese).

Nagenborg, M., Capurro, R., Weber, J. & Pingel, C. (2007). Ethical regulations on robotics in Europe. *AI & Society* (online edition), available online at http://www. springerlink. com/content/3t2m260588756t7v/ .

Neisser, U., Boodoo, G., Bouchard, T. J., Boykin, A. W., Brody, N., Ceci, S. J., et al. (1996). Intelligence: Knowns and unknowns. *American Psychologist*, Vol. 51, 77–101.

Newell, A. & Simon, H. A. (1995). GPS: A program that simulates human thought. In Feigenbaum, E. A. & Feldman, J. (eds.) *Computers and Thought.* MIT Press: Cambridge, Massachusetts.

Newman, W. S. (1989). Automatic obstacle avoidance at high speeds via reflex control. *Proceedings of the 1989 IEEE International Conference, Part 2: Robotics and Automation,* Scottsdale, Arizona, pp. 1104-1109.

Penrose, R. (1989). *The emperor's new mind.* Oxford University Press: New York.

Saygin, A. P., Cicekli, Y. & Akman, V. (2000). Turing Test: 50 years later. *Minds and Machines,* Vol. 10, Issue 4, 463-518.

Searle, J. R. (1980). Minds, brains and programs. *Behavioral and Brain Sciences*, Vol. 3, Issue 3, 417-457.

Spector, L., Klein, J., Perry, C. & Feinstein, M. (2005). Emergence of collective behavior in evolving populations of flying agents. *Genetic Programming and Evolveable Machines,* Vol. 6, Issue 1, 111-125.

Tajika, N. (2001). *The future Astro Boy.* ASCOM: Tokyo (in Japanese).

Robot Rights? It Could Happen, U.K. Government Told. (2006). *CBC News*, December 21, available online at http://www.cbc.ca/technology/story/2006/12/21/tech-freedom.html.

Trust me, I'm a robot. (2006). *The Economist*, June 8, available online at http://www. economist.com/displaystory.cfm?story_id=7001829.

Turing, A. M. (1950). Computing machinery and intelligence. *Mind*, Vol. LIX, Issue 236, 433-460.

United Kingdom Health and Safety Executive Office. (2000). HSG 43:2000. Industrial robot safety.

Warwick, K. (2004). *March of the machines.* University of Illinois Press: Chicago.

Weng, Y. H., Chen, C. H. & Sun, C.T. (2007). The legal crisis of next generation robots: On safety intelligence. *Proceedings of the 11th International Conference on Artificial Intelligence and Law (ICAIL'07)*, Palo Alto, California, pp. 205-210.

Weng, Y. H., Chen, C. H. & Sun, C.T. (n. d.). Toward The Human-Robot Co-Existence Society: On Safety Intelligence for Next Generation Robots, *ExpressO (Unpublished Manuscript),* Available online at http://works.bepress.com/weng_yueh_hsuan/1.

What's a Robot? (2007). *CBC News*, July 16, http://www.cbc.ca/news/background/tech /robotics/definition.html.

Whitehouse, D. (2005). Japanese Develop "Female" Android. *BBC News*, July 27. Available online at http://news.bbc.co.uk/1/hi/sci/tech/4714135.stm.

Wikman, T. S., Branicky, M. S. & Newman, W. S. (1993). Reflexive collision avoidance: A generalized approach. *Proceedings of the 1993 IEEE International Conference, Part 3: Robotics and Automation,* Kobe, Japan, pp. 31-36.

Yamada, Y. (2007). Currently existing international/domestic safety standards associated with service robots and ongoing tasks. *Journal of Robotics Society of Japan,* Vol. 25, Issue 8, 1176-1179 (in Japanese).

Yeomans, M. (2005). Qrio dances into spotlight at Carnegie Mellon Center. *Pittsburgh Tribune-Review,* January 29. Available online at http://www.pittsburghlive.com/x/pittsburghtrib/s_298114.html.

Zhang, X., Zheng, H., Duan, G. & Zhao, L. (2002). Bio-reflex-based robot adaptive motion controlling theory. *Proceedings of the 4th World Congress on Intelligent Control and Automation,* Vol. 3, pp. 2496-2499.

13

A Novel Modular Mobile Robot Prototype for Urban Search and Rescue

Houxiang Zhang[1], Wei Wang[2], Guanghua Zong[2] and Jianwei Zhang[1]
[1]*University of Hamburg*
[2]*Beijing University of Aeronautics and Astronautics*
[1]*Germany*
[2]*China*

1. Introduction

The last few years have witnessed an increasing interest in reconfigurable mobile robotic technologies (Yim, M.; et al., 2000). The applications include the following areas: industrial inspection (Granosik, G.; Hansen, M. G.; Borenstein, J. 2005), conducting surveillance, urban search and rescue, military reconnaissance and space exploration (Yim, M.; et al., 2003). Reconfigurable robots consist of several modules which are able to change the way they are connected. The modular approach enables mobile robots to reconfigure, which is essential for tasks that are difficult for a fixed-shape robot (Kamimura, A.; et al.; 2001). During movement the robot confronts an unstructured environment and handles the uncertainties by reconfiguring its structure (Castano, A.; et al., 2000). The basic requirement for this kind of robotic system moving on rough terrains is extraordinary motion capabilities.
The emphasis for discussion of this chapter is on the field of urban search and rescue (Miller, P.G. 2002) (Takayama, T.; Hirose, S. 2000). The history of human development has always been a struggle with natural disasters such as earthquakes, storms and floods. Recently the number of disasters by accidents or terrorism has evidently been increasing too (Casper, J.; 2002). Urban search and rescue is a domain that involves a great amount of manpower; and it is quite dangerous and laborious in a hostile environment (Matsuno, F.; Tadokoro, S. 2004). The development of mobile robots offers an intelligent alternative solution to the above-mentioned problems. The application of the robotic system can relieve people of this challenging work and provide an opportunity for robots to play a pivotal support role, realize automatic manipulation in a complex environment and improve the level of civil and military technology.
In this chapter, a novel modular reconfigurable mobile robot named JL-I is presented, which to date consists of three identical modules. After a short survey, the basic functions of a reconfigurable mobile robot are summarized systematically based on research into work targets. JL-I features active joints formed by serial and parallel mechanisms which endow the robot with the ability of changing shapes in three dimensions. With the docking mechanisms, the modules can connect or disconnect flexibly and automatically. After that the discussion focuses on the various locomotion capabilities, such as crossing high vertical

obstacles and getting self-recovery when the robot is upside-down. The results of a series of successful on-site tests are given to confirm the principles described above and the robot's ability.

2. Reconfigurable robots in literature

In recent years considerable progress has been made in the field of reconfigurable modular robotic systems, which usually comprise three or more rigid segments that are connected by special joints (Rus, D.; Vona, M.; 2000) (Suzuki, Y.; et al. 2006). The common kinematics modes include multiple legs, wheeled and chain-track vehicles. However, the robots with multiple-legs kinematics are too complex due to a structure with many degrees of freedom (Zhang, H.; et al., 2004). These robots do not meet the requirements of miniaturization, flexible and quick movement, so that multi-legs kinematics is rarely adapted.

The common characteristics of reconfigurable robotic systems lie in two points. Firstly, the robotics system comprises several similar modules which are independent units with full locomotion functions (Shen, W.; et al. 2006). Secondly, specially designed joints connect individual units to form a more flexible prototype. The first prototype (Hirose et al. 1990) with powered wheels was designed by Hirose and Morishima in 1990, which consists of several vertical cylindrical segments. Another robot with wheels on each module to provide the driving force was developed by Klaassen for the inspection of sewage pipes (Klaassen et al. 1999). A serpentine robot from Takayama and Hirose consists of three segments. Each segment is driven by a pair of tracks, but all tracks are powered simultaneously by a single motor located in the centre segment (Takayama, T.; Hirose, S.; 2000). The special ability of adapting to irregular terrain is passive and provided by springs. KOHGA (Kamegawa, T.; et al. 2004) has recently been developed by IRS in Japan. It consists of eight serially interconnected individual units with two tracks except the first and last modules.

Another group of reconfigurable robots features passive modules (Yim, M.; et al. 2001) (Murata, S.; et al. 2002) (Shen. W.; et al. 2002) (Kurokawa, H.; et al. 2003). It can only move after the modules are assembled (Moechel, R.; et al, 2005) (Gonzalez-Gomez, J.; et al. 2005) (Gonzalez-Gomez, J.; et al. 2004). In (Gonzalez-Gomez, J.; et al. 2007), 1D, 2D and 3D chain robots are classified according to their topology. As an example, PolyBot is able to optimize its parts to fit the specific task. Generally, this kind of reconfigurable robots is relatively simple so that the locomotion capability is not as efficient as the above-mentioned kind with powered tracks.

However, for urban rescue and search, the fact that the known reconfigurable robots can only assume few configurations due to relatively simple connecting and pose-adjusting mechanisms is a ubiquitous deficiency. For example, the Millibot Train robot from Carnegie Mellon University consists of seven compact segments, which can connect by couplers with one DOF (Brown, H.B.; et al., 2002). A reconfigurable mobile robot designed by M. Park is not able to change its configuration actively (Park, M.; et al., 2004). Another recent module robot (Wang, T.M.; et al., 2005) has only one DOF in its pose-adjusting mechanism. The robot from Université Libre de Bruxelles (Sahin, E.; et al., 2002) has a one-DOF pose-adjusting mechanism and one coupler to change the configuration between the neighboring modules as well.

Since 1999 our group has been focusing on the design and development of mobile robots for urban search and rescue purposes. A smart mobile robot was proposed as a flexible mobile platform carrying a CCD camera and other sensors. A more flexible structure with two linked-track vehicles was proposed (Wang, W.; et al. 1999). The structure can be reconfigured so that the robot can move between surfaces standing at an angle of 0 - 90 degrees due to the pitching DOF actuated by the joint to increase the flexibility. The project presented here has the aim of developing an automatic field robot to meet the requirements of high flexibility, robustness.

3. System design and realization

3.1 Design considerations

Urban search and rescue is one important application for mobile robots in a complex environment. The basic functions of urban search and rescue robots include five aspects. (Zhang, H.; et al. 2006a)

1) Locomotion capability is the lowest basic functionality of the robotic system, which includes the following details.

The robot works not only indoors but outdoors as well. It should have a flexible mobility in rugged terrain to get to every point in the work space. In order to finish a task in an unstructured environment, the ability to cross high obstacles and span large gaps is indispensable. Sometimes the working environment is very complicated, including not only high steps and deep ditches but also narrow fences and floors cluttered with debris. As a result, the robot should have the capability of adopting different configurations to match various tasks and suit complex environments.

2) Enough intelligence for the discrimination of a variety of obstacle situations: Multiple sensing and control systems are incorporated to handle the uncertainties in a complex environment. Software should be dexterous enough to identify the various geometries and intelligent enough to autonomously reconstruct the environment. Sensor fusion is an important capability since no single sensor can identify all aspects of the environment.

3) Working autonomously with the corresponding effective treatment: As a rescue robot, it should move as fast as possible in order to get real-time information. Once the global task commands are entered by the user, the robot should move while accomplishing the rescue task. Path planning and behavior organization is built on global prior knowledge and local sensory information.

4) No connection with the environment: In order to move freely, it is important for the mobile field robot not to be wired or otherwise connected to the environment. The robot should carry all it needs: onboard power, the controller, and wireless communication.

5) Cooperation ability: The rescue robot is designed to operate with humans. The level of interaction may vary significantly, depending on the robot's design and on the circumstances. The controlling and monitoring of the robot is achieved through a GUI to allow an effective and user-friendly operation.

Usually urban searching and rescuing is based on the cooperation in a team. The given targets will be assigned separately. Every robot should communicate with the others and perform distributed activities.

3.2 Prototype design and realization

The proposed mobile system should have various moving modes. The JL-I system consists of three connected, identical modules for crossing grooves, steps, obstacles and traveling in complex environments. Design of the robot includes five parts:

1) Design of the independent mobile module which includes movement mechanisms and driving systems;
2) Development of the docking system and the reconfigurable driving mechanisms;
3) Development of the control system;
4) Kinematics analysis;
5) Experimental testing.

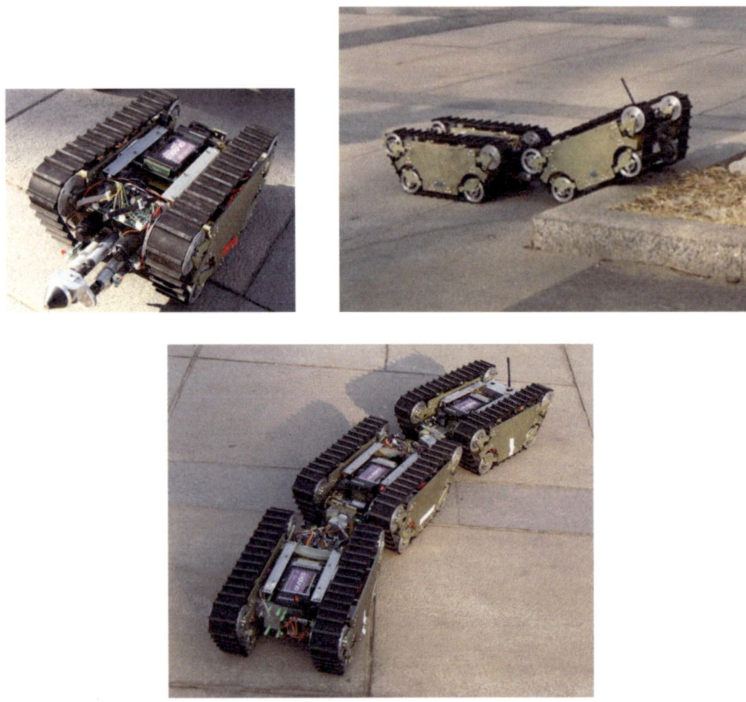

Fig. 1. A photo of JL-I

The mechanical structure of JL-I is flexible due to its identical modules and special connection joints (Fig. 1). Actually, each module is an entire robot system that can perform distributed activities (Fig. 2). Three DOF active spherical joints between two modules and the docking mechanism enable the adjacent modules to adopt optimized configurations to negotiate difficult terrain or to split into three small units to perform tasks simultaneously.

Fig. 2. Performing distributed activities

By combining such locomotion capabilities, JL-I will move in almost all kinds of rough environments. The principle of terrain adaptability is shown in Fig. 3. The robot can change its posture by pitching around the Y axis, yawing around the X axis and rotating around the Z axis. The yawing and pitching movements are achieved by the parallel mechanism. The third rotation DOF around the joint's Z axis is achieved by the serial mechanism.

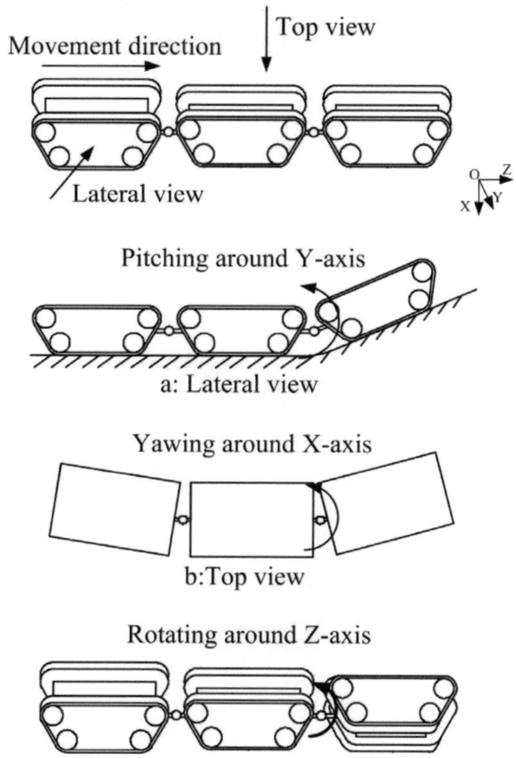

Fig. 3. Adapting to terrains

In contrast to the results of previous research, this robot includes the following innovative aspects (Zhang, H.; et al. 2006b):

1. *Smooth movement:* It is known that a common mobile robot will lose its moving ability if it is not able to keep its balance. However, the JL-I with its many active spherical joints can smooth the undulating motion of the gravity centre of the whole system.

2. *High adaptability:* Identical modules in the JL-I robot have a large variety of configurations owing to the pose-adjusting joints and the docking mechanisms. Therefore, the robot can adopt many shapes which make the JL-I system able to carry out various different tasks and move in diverse environments.

3. *Self recovering ability:* Furthermore, JL-I has the ability to implement self-recovery. By disconnecting the malfunctioning module, the system can repair itself if one module does not work normally. In this way, working safety and efficiency are increased when the robot moves in a complex environment.

The single module is about 35 centimetres long, 25 centimetres wide and 15 centimetres high. Fig. 4 shows the mechanical structure of the module which comprises two powered tracks, a serial mechanism, a parallel mechanism, and a docking mechanism. Two DC motors drive the tracks providing skid-steering ability in order to realize flexible movement.

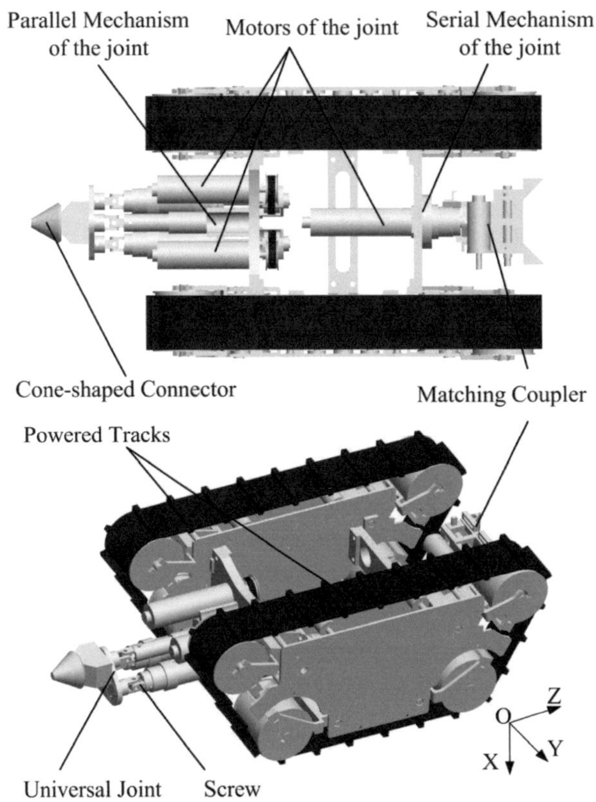

Fig. 4. The mechanical structure of the module

The docking mechanism consists of two parts: a cone-shaped connector at the front and a matching coupler at the back of the module, as shown in Fig. 5. The coupler is composed of two sliders propelled by a motor-driven screw. The sliders form a matching funnel which guides the connector to mate with the cavity and enables the modules to self-align with certain lateral offsets and directional offsets. After that, two mating planes between the sliders and the cone-shaped connector constrain the movement, thus locking the two modules. This mechanism enables any two adjacent modules to link, forming a train-like configuration. Therefore the independent module has to be rather long in order to realize the necessary docking function. In designing this mechanism and its controls, an equilibrium between flexibility and size has to be reached (Wang, W.; et al. 2006) (Zong, G.; et al. 2006).

The robot features the serial and parallel mechanisms which form a three-DOF active spherical joint. There are two reasons for using serial and parallel mechanisms for our mobile robot. Firstly, the JL-I robot can be made lightweight and dexterous while allowing for a larger payload. Secondly, the advantages of the high rigidity of a parallel mechanism, as shown in Fig. 6, and the extended workspace of a serial mechanism can be combined, thus improving the flexibility of the system.

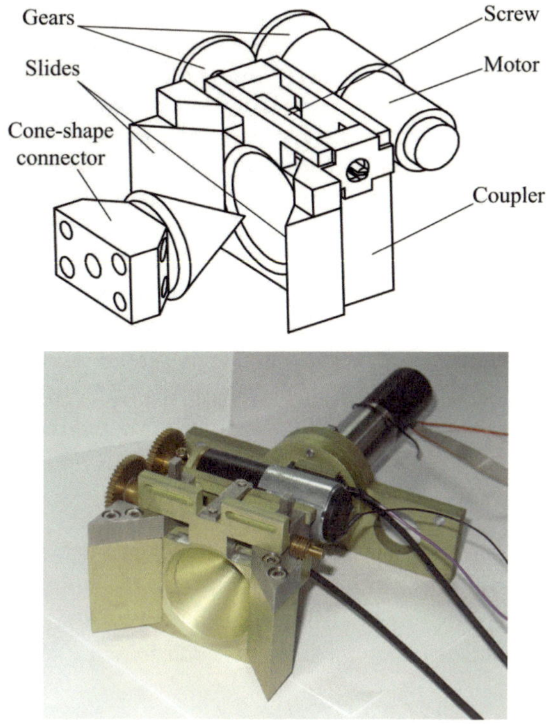

Fig. 5. The docking mechanism

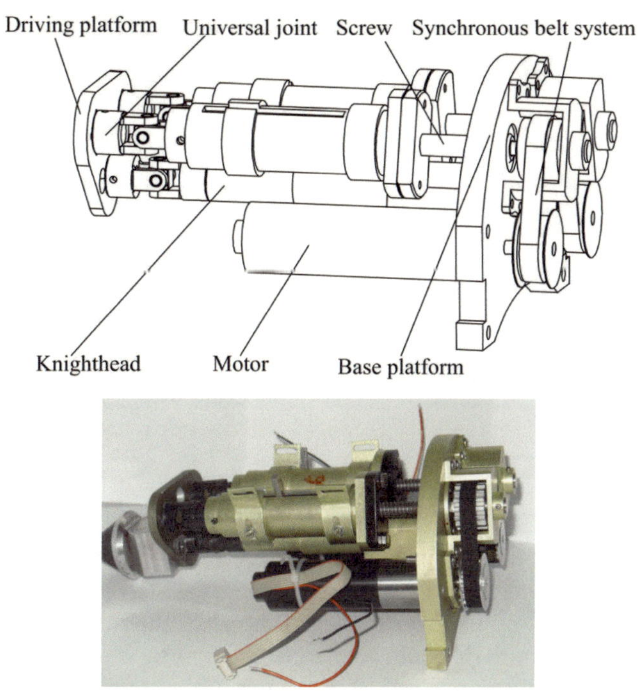

Fig. 6. The parallel mechanisms

The serial mechanism can rotate 360° around the Z axis. This joint is actuated by a geared minimotor which provides a continuous torque of 3.5 Nm at a speed of 30 rpm. The parallel mechanism can pitch around the Y axis and yaw according to the X axis. Each leg of this parallel joint consists of a driving platform, a universal joint, a screw, a synchronous belt system, a DC motor and a base platform. The universal joint connects the driving platform and the knighthead. The other end of the knighthead is fixed to the base platform. By revolving the screw, the driving platform can be manipulated relatively to the base platform. By controlling the active joints and the docking mechanisms, the JL-I can change its shape in three dimensions.

To ensure its ability of performing tasks individually, there is enough space in each module for sensors, the onboard controller, and batteries. Considerable stress is laid on weight reduction as well as on construction stiffness to achieve a dexterous movement mechanism. Most of the mechanical parts are designed specifically and mainly manufactured from aluminium. A module weighs approximately 7 kg including the batteries.

4. Required locomotion capability

Due to the uncertainty of the practical environment, it is important for a robot to be able to carry out various complicated locomotion processes for performing urban search and rescue tasks. The JL-I is capable of almost all necessary actions that can be required in real situations, e.g. crossing obstacles such as steps and roadblocks, self-recovery (Zhang, H.; et al. 2007).

4.1 Crossing a step

Fig. 7 shows the process of crossing a step from a view in the Y direction. Here the step is almost twice as high as the robot.

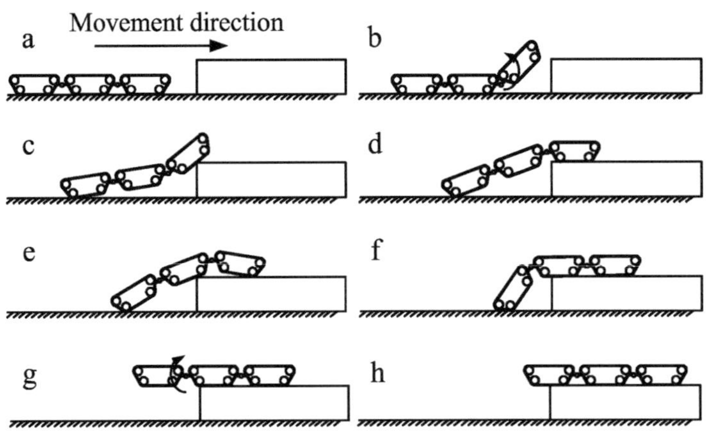

Fig. 7. The sequence of crossing a step

(a) The robot is in its home state, and the sensor detects the step in the movement direction.
(b) The first module is pitching up around the Y axis while the robot is moving forward.
(c) The approaching movement does not stop until the first module touches the step.
(d) The first module pitches down to attach to the top of the step.
(e) The robot is moving forward continuously.

(f) The robot is moving until the first two modules are attached to the step.
(g) The last module is pitching up around the Y axis while the robot is moving forward.
(h) The robot is now in its home state again, and the process of crossing the step is over.

4.2 90° self-recovery

It is possible for the robot to implement a 90° recovering movement by adopting the proper configuration sequence as shown in Fig. 8.

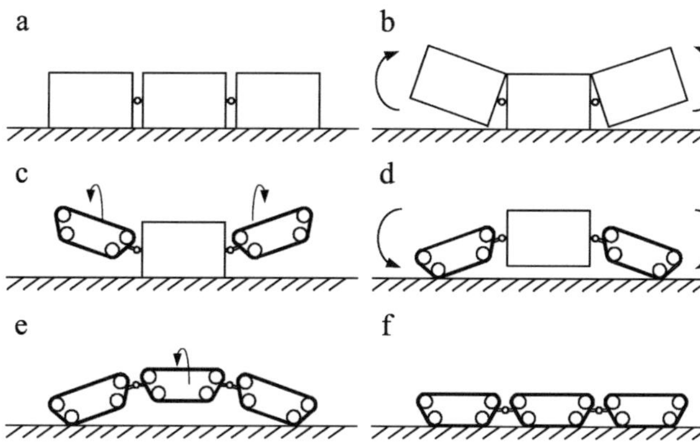

Fig. 8. The 90° recovering movement

(a) The robot is lying on its side.
(b) The first module and the last module are yawing up around the X axes of the active joints.
(c) Then the first module and the last module rotate 90° around the Z axes.
(d) After that, they are pitching down around the Y axes of the active joints until they attach to the ground in order to raise the middle module up.
(e) The middle module rotates around the Z axis until it is parallel to the ground.
(f) In the end, the module is pitching down around the Y axes of the active joints until all three modules attach to the ground together. The robot is now in its home state again, and the process of 90° self-recovery is over.

4.3 180° self-recovery

It is also possible for the robot to tip over and realize the 180° recovery movement as shown in Fig. 9.
(a) The robot is in its home state.
(b) The first and the last modules are pitching down around the Y axes of the active joints until the middle module is in the air.
(c) The middle module rotates 180° according to the Z axis.
(d) The first module and the last module are pitching down around the Y axes of the active joints until the middle module attaches to the ground.
(e) The first module and the last module are pitching up around the Y axes of the active joints again.
(f) The first module and the last module are rotating 180° around the Z axes of the active joints.

(g) Then the first module and the last module are pitching down around the Y axes of the active joints again until all three modules attach to the ground.
(h) The process of 180° self-recovery is over.

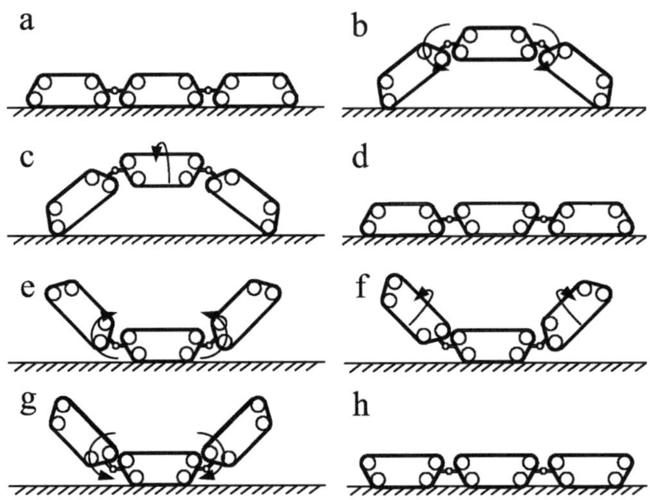

Fig. 9. The 180° recovering movement

4.4 Crossing a narrow fence

As shown in Fig. 10, the train configuration robot is able to cross a fence narrower than the width of its modules.

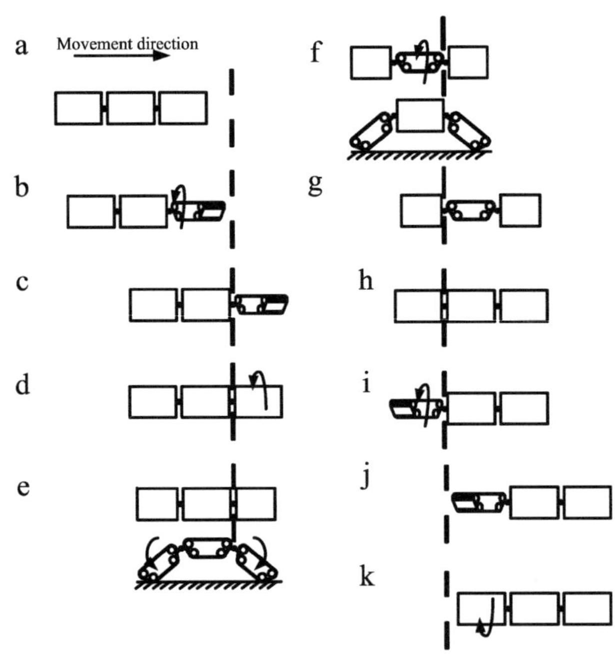

Fig. 10. The sequence of crossing a fence

(a) The robot is in its home state, and the sensor detects the fence in the moving direction.
(b) The robot stops before the fence, and then the first module pitches up around the Y axis and then rotates 90° according to the Z axis.
(c) The crossing movement does not stop until the first module passes through the fence.
(d) The first module rotates and pitches to get back into the home state, and then the three modules attach to the ground together again.
The following steps (e) to (k) of the second and third modules are similar to those of the first one. The process will be achieved until the robot crosses the fence entirely. In order to show the principle clearly, the lateral views of steps (e) and (f) are also given.

5. Kinematics analysis

5.1 The DOF of the active joint

To demonstrate the reconfiguring capability, the kinematics analysis of two connected modules should be studied. Fig. 11 shows the kinematics model of the joint between two modules. Where *OXYZ* is the world coordinate fixed at the plane QEF which represents the front unmovable module during the reconfiguration. The origin is located at the universal joint *O*, the *Z*-axis coincides with the axis of the serial mechanism and the X-axis points to the middle point of line *AB*. Another reference coordinate *O'X'Y'Z'* is fixed at triangular prism *OABPCD* which represents the back moveable module. The *O'X'Y'Z'* is coincident with the *OXYZ* when the spherical joint is in its home state.
The required orientation for the reference frame *O'X'Y'Z'* on the back module is achieved by a rotation of θ_z, a pitching angle θ_y and a yawing angle θ_x according to the relative axes. From the mechanical point of view, actually the pitching and yawing motions are implemented by the outstretching and returning movement of the L_1, L_2 of the parallel mechanism, and the rotation of θ_z is actuated by the serial mechanism. The freedom of the reconfiguring movement is three and can be described with the generalized coordinate θ (1). The joint variants of the movement are named q, described as (2).

$$\theta = [\theta_x, \theta_y, \theta_z]^T \quad (1)$$

$$q = [L_1, L_2, \theta_z]^T \quad (2)$$

According to the mechanical principle, the DOF can be concluded first, which will lay the foundation for later discussions. As shown in Fig. 11, there are altogether 8 joints, out of which 3 joints are active and actuated by respective DC motors. In this figure, there are three Hooker joints at points *O*, *A*, and *B*; two linear movement joints at links AC and BD; one rotating joint along the axis Z_1Z_2 and two spherical joints at C and *D*. According to (3), the DOF can be concluded. Where m means the DOF; n means the movable links of the active joint, there are seven links totally; g means the total number of the joints
Where f_i is the DOF of the relative joints. It is noted that the DOF of a rotating joint and a linear movement joint is one; the DOF of the Hooker joint is two while the spherical joint has three DOF. All these results are inserted into (3) to get the DOF.

$$m = 6(n - g - 1) + \sum_{i=1}^{g} f_i = 3 \quad (3)$$

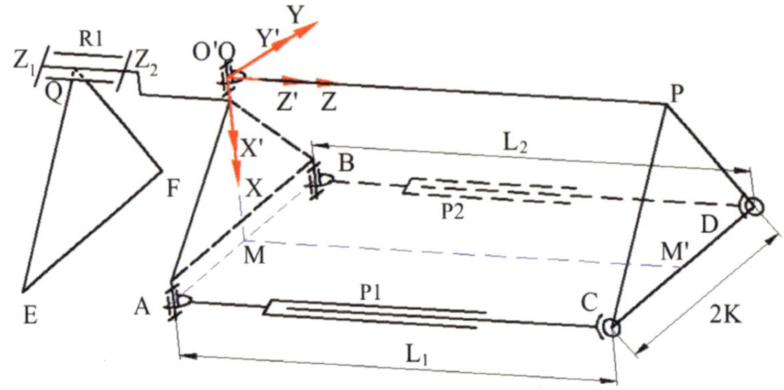

Fig. 11 The kinematics mode of the active spherical joint.

5.2 Preliminary analysis of working space

The next question is the working space analysis in the world coordinate in order to implement all locomotion capabilities since there are three general DOFs. As discussed in the former section, the angle θ_z is required to have 360° rotation around the Z axis. It is noted that θ_z is an independent DOF actuated by the serial mechanism and it normally occurs after the pitching movement or yawing movement. So we only need to focus on the working spaces of θ_x and θ_y since both pitching and yawing movements are dependent on the extending or contacting cooperation of L_1 and L_2 in the parallel mechanism. All these are important for designing the system structure and the parallel mechanism.

For a general movement, the JL-I robot only needs to pitch up at a tiny angle in order to cross a low step or some small obstacles; it should also only yaw at a small angle to turn left or right during the movement. In practice the maximum positions should be calculated considering the mechanical constraints and collision avoidance.

Fig. 12 shows the analytic draft of the working space for the pitching movement, in which the collision and structure constants are taken into account. Here three rectangles represent the modules to simplify the discussion in the lateral view. The left module is fixed and the right one is moveable. Situation a_1 in black is the same home state as that shown in Fig. 9(h); b_1 in blue is the maximum pitching position taking the mechanical constraints into account in order to avoid collision, similarly as in Fig. 9(g).

From Fig. 12, the maximum pitching angle θ_{ymax} can be found as in (4).

$$\theta_{y\max} = 180 - 2arctg\frac{y}{l} \tag{4}$$

Where y is the half vertical height when the module stands on flat ground; l is the length of the equivalent connecting joint of JL-I; x is the half width of the largest rotating section. Taking the example parameters of the robot implemented in practice, i.e. y is 75 mm and l is 35 mm, finally the general working space of the pitching movement can be concluded to be

$$\theta_{y\max} = 50.0° \tag{5}$$

This result comes from the designing point of view. However, in order to implement a 90-180° recovering movement, the JL-I robot has to adopt a proper configuration sequence as shown in the previous section. In Fig.12, c_1 in red is the minimum pitching position in order to implement the DOF of θ_z. This minimum angle limitation is nevertheless very important for the practical implementation of the robot's reconfiguration.

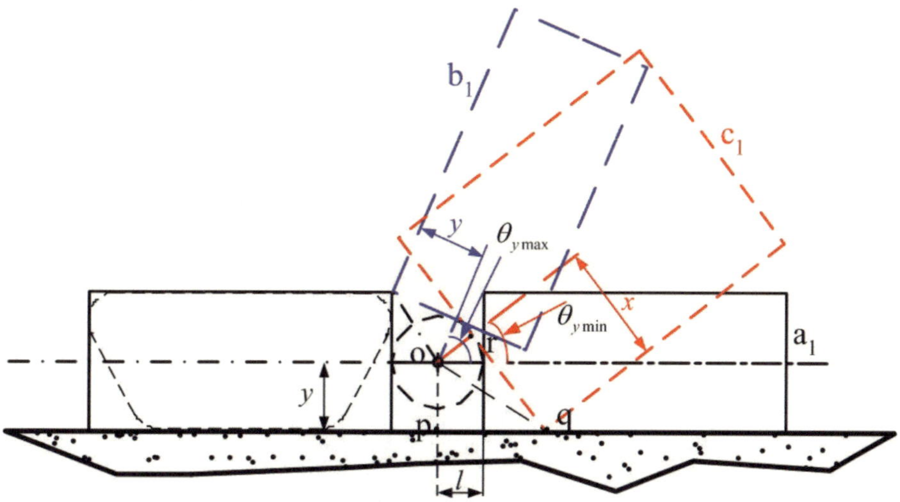

Fig. 12. The working space of the pitching movement.

Furthermore, as shown in Fig.9 (f), in order to actuate one module to rotate 90-180° around the Z axes of the active joints, it can only operate without any collision. The largest rotating section is illustrated in Fig. 13 in red, which is the same red rectangle in Fig. 12.

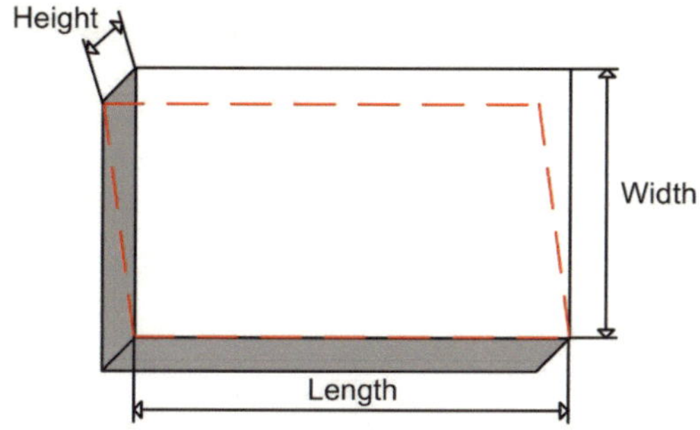

Fig. 13. The maximum rotating section during pitching movement.

Fig. 14 illustrates the analytic quadrangle of the robot working space of Fig. 12 at a large scale. According to the geometric analysis, we have

$$\begin{cases} \theta_1 = \arccos \dfrac{y}{\sqrt{l^2 + x^2}} \\ \theta_2 = \theta_3 = 90 - \theta_1 \\ \theta_4 = \arccos \dfrac{x}{\sqrt{l^2 + x^2}} \end{cases} \tag{6}$$

$$\theta_{min} = 90 - \theta_4 - \theta_3 = 90 - \theta_4 - (90 - \theta_1)$$

$$= \theta_1 - \theta_4 = \arccos\frac{y}{\sqrt{l^2 + x^2}} - \arccos\frac{x}{\sqrt{l^2 + x^2}} \tag{7}$$

$$\begin{cases} x = \sqrt{Height^2 + Width^2} \\ y = Height / 2 \end{cases} \tag{8}$$

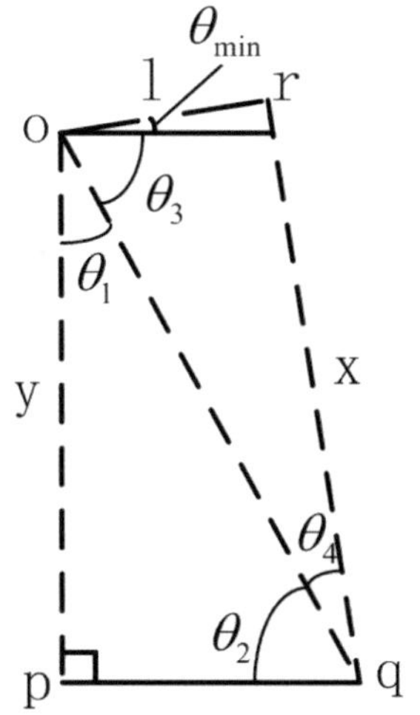

Fig. 14. The analytic quadrangle of the working space.

Given the height and width of the module, putting (8) into (7) we can get the minimum working space of the pitching DOF for robotic runtime reconfiguration in order to avoid collisions. Finally the working space for θ_y is as in (9).

$$\theta_y \in [-(180 - 2arctg\frac{y}{l}), -(\arccos\frac{y}{\sqrt{l^2 + x^2}} - \arccos\frac{x}{\sqrt{l^2 + x^2}})\ ] \cup$$
$$+(\arccos\frac{y}{\sqrt{l^2 + x^2}} - \arccos\frac{x}{\sqrt{l^2 + x^2}}),\ +(180 - 2arctg\frac{y}{l})] \tag{9}$$

Similarly for the working space of the yawing movement, θ_x can be also described like θ_y (9) while y is the half width of the module at the moment. According to the prototype structure, the pitching and yawing working spaces are obtained. As in our implementation, when the module width is 250 mm, the height is 150 mm and l is 35 mm, the general working space is described in (10); while the restricted working space for avoiding collision is (11)

$$\begin{cases} \theta_x \in [-32.0°,\ +32°] \\ \theta_y \in [-50.0°,\ +50.0°] \end{cases} \tag{10}$$

$$\begin{cases} \theta_x \in [-32.0°, \ -8.0°] \cup [+8.0°, \ +32.0°] \\ \theta_y \in [-50.0°, \ -24.0°] \cup [+24.0°, \ +50.0°] \end{cases} \tag{11}$$

In order to simplify the mechanical structure, we can design the working space of θ_x, θ_y to be the same, e.g. both within -50 to 50 degrees, which not only reduces the implementation cost but also slightly increases redundancy for practical operation (Zhang, H.; et al. 2007).

6. On-site experiments

Relevant successful on-site experiments with the JL-I were carried out recently, confirming the principles described above and the robot's ability.

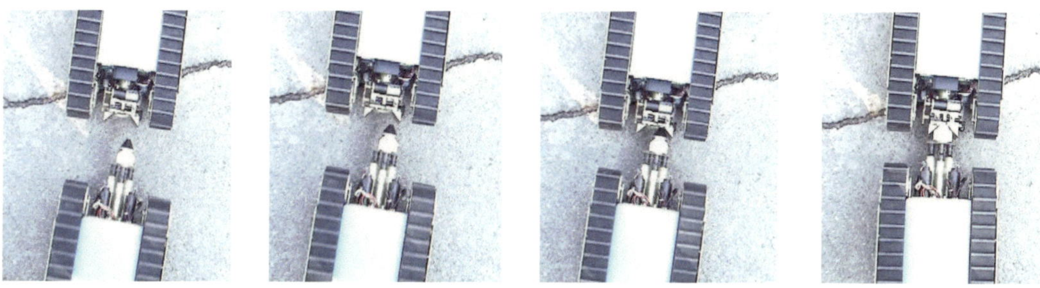

(a) (b) (c) (d)

Fig. 15. The docking process

Fig. 15 shows the docking process of the connection mechanism whose most distinctive features are its ability of self aligning and its great driving force. With the help of the powered tracks, the cone-shaped connector and the matching coupler can match well within ±30mm lateral offsets and ±45°directional offsets.

Compared with many configurable mobile robots, the JL-I improves its flexibility and adaptability by using novel active spherical joints between modules. The following figures 16, 17, 18, 19 show the typical motion functionalities of the robot one by one, whose principles are discussed above.

(a) (b) (c)

Fig. 16. Climbing stairs

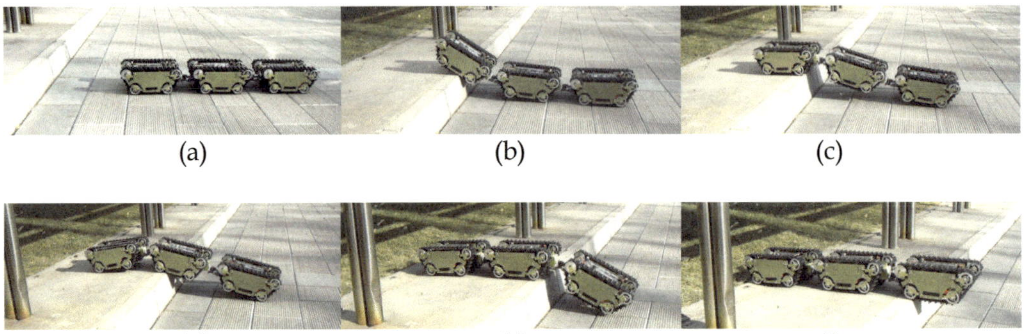

(a) (b) (c)

(d) (e) (f)

Fig. 17. Snapshots of crossing a step

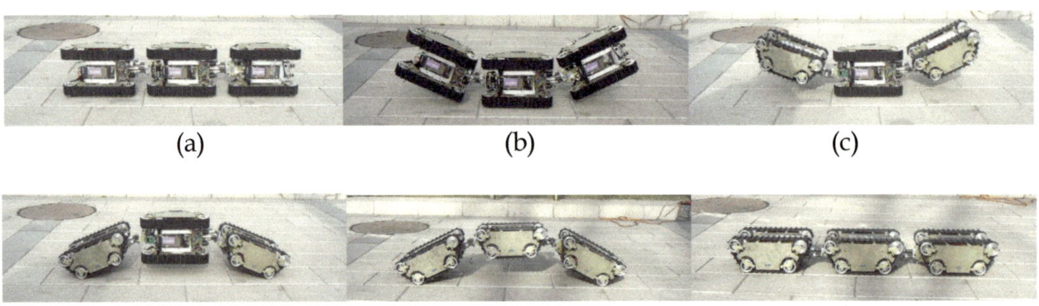

(a) (b) (c)

(d) (e) (f)

Fig. 18. Snapshots of the 90° self-recovery

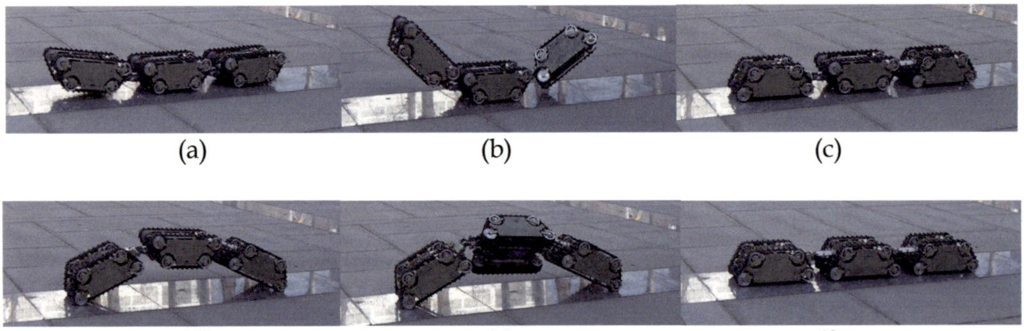

(a) (b) (c)

(d) (e) (f)

Fig. 19. Snapshots of the 180° self-recovery

The experimental results show that the 3 DOF active joints with serial and parallel mechanisms have the ability to achieve all the desired configurations. The performance specifications of JL-I are given in Table 1.

Parameters	Values
Posture adjustment angle around X-axis	±45°
Posture adjustment angle around Y-axis	±45°
Posture adjustment angle around Z-axis	0~360°
Maximum lateral docking offset	±30 mm
Maximum directional docking offset	±45°
Maximum height of steps	280 mm
Maximum length of ditches	500 mm
Minimum width of the fence	200 mm
Maximum slope angle	40°
Self-recovering ability	0~180°
Maximum climbing velocity	180 mm/s
Maximum unchangeable working time	4 hours

Table 1. Performance specifications

8. Conclusions

In contrast to conventional theoretical research, the project introduced in this project successfully completes the following innovative work:
1. It proposes a robot named JL-I which is based on a module reconfiguration concept. The robot features a docking mechanism with which the modules can connect or disconnect flexibly. The active spherical joints formed by serial and parallel mechanisms endow the robot with the ability of changing shapes in three dimensions.
2. The DOF and working space of reconfiguration between two modules is given. Related experimental tests have shown that the JL-I can implement a series of various reconfigurations. This implies the mechanical feasibility, the rationality of the analysis and the outstanding movement adaptability of the robot.
Actually all of the locomotion capabilities are pre-programmed at the moment. In future, our research will focus on the realization of real autonomy. At the same time, the dynamic analysis of the movement functions is another important issue in this project.

9. Acknowledgement

The work in this chapter is proposed by National High-tech R&D Program (863 Program) of China (No. 2006AA04Z241). The authors would like to thank Dr. Z.C. Deng and Mr. W.P. Yu at Robotics Institute at Beijing University of Aeronautics and Astronautics for a great amount of work and on-site testing.

10. Reference

Yim, M.; David, G. (2000). PolyBot: a module reconfigurable robot. *Proceedings of the 2000 IEEE International Conference on Robotics and Automation*, pp.514-520, ISBN: 0-7803-5886-4, San Francisco, CA, USA, April 2000, IEEE Service Center, Piscataway

Granosik, G.; Hansen, M. G. & Borenstein, J. (2005). The omnitread serpentine robot for industrial inspection and surveillance. *Industrial Robot: An International Journal*, Vol.32, No.2, Feb. 2005, pp.139-148, ISSN: 0143-991X

Yim, M.; Roufas, K.; Duff, D.; Zhang, Y.; Eldershaw, C. & Homans, S. (2003). Module Reconfigurable Robot in Space Application, *Autonomous Robots*, Vol. 14, Issue 2-3, 2003, pp.225-237, ISSN: 0929-5593.

Kamimura, A.; Kurokawa, H.; Yoshida, E.; Murata, S.; Tomita, K. & Kokaji, S. (2005). Automatic locomotion design and experiments for a modularrobotic system, *IEEE/ASME Transactions on Mechatronics*, Vol. 10, No. 3, March 2005, pp.314-325, ISSN: 1083-4435

Castano, A.; Shen, W.M. & Will, P. (2000). CONRO: towards miniature self-sufficient metamorphic robots. *Autonomous Robots*, Vol.13, No.4, April 2000, pp.309-324, ISSN: 0929-5593

Miller, P.G. (2002). Snake robots for search and rescue, In: *Neurotechnology for Biomimetic Robots*. Ayers, J.; Davis, L.; Rudolph, A. (Ed.), pp. 271-284, 2002, MIT Press, ISBN-13:978-0-262-01193-8.

Takayama, T. & Hirose, S. (2000). Development of Souryu-I connected crawler vehicle for inspection of narrow and winding space, *Proceedings of the 26th Annual Conference of the IEEE Industrial Electronics Society*, pp. 143-148 ISBN: 0-7803-6456-2, Nagoya, Aichi, Japan, Oct. 2000, IEEE Service Center, Piscataway

Matsuno, F.; Tadokoro, S. (2004). Rescue Robots and Systems in Japan. Proceedings of Robio2004, pp.12-20, ISBN: 0-7803-8614-8, Shenyang, China, 22-26 Aug. 2004, IEEE Service Center, Piscataway.

Casper, J. (2002). Human-Robot Interactions during the Robot-assisted Urban Search and Rescue Response at the World Trade Center, MS Thesis, Computer Science and Engineering, USF, April, 2002.

Rus, D.; Vona, M. (2000). A basis for self reconfigurable robots using crystal modules, *Proceedings of the 2000 IEEE Conference on Intelligent Robots and Systems*, pp. 2194-2202, ISBN: 0-7803-6348-5 Takamatsu, Japan, October 2000, IEEE Service Center, Piscataway

Suzuki, Y.; Inou, N.; Kimura, H. & Koseki, M. (2006). Reconfigurable group robots adaptively transforming a mechanical structure – Crawl motion and adaptive transformation with new algorithms. *Proceedings of IEEE International Conference on Intelligent Robots and Systems*, pp. 2200-2205, ISBN: 1-4244-0259-X, Beijing, China, October 2006, IEEE Service Center, Piscataway

Zhang, H.; Zhang, J.; Zong, G. (2004). Requirements of glass cleaning and development of climbing robot systems, Proceedings of the 2004 International Conference on Intelligent Mechatronics and Automation, pp. 101-106, ISBN: 0-7803-8748-1, Chengdu, China, August 26-31, 2004, IEEE Service Center, Piscataway.

Shen, W.; Krivokon, M.; Chiu, H.; Everist, J. & Rubenstein, M. (2006). Multimode Locomotion via SuperBot Reconfigurable Robots. *Autonomous Robots*, Vol.20, 2006, pp165-177, ISSN: 0929-5593

Hirose, S. & Morishima, A. (1990). Design and control of a mobile robot with an articulated body. *The International Journal of Robotics Research*, Vol. 9 No. 2, Feb. 1990. pp.99-113, ISSN: 0278-3649

Klaassen, B. & Paap, K.L. (1999). GMD-SNAKE2: a snake-like robot driven by wheels and a method for motion control. *Proceedings of the 1999 IEEE International Conference on Robotics and Automation*, pp. 3014 - 3019, ISBN: 0792356012, Detroit, MI, USA, May 1999, IEEE Service Center, Piscataway

Takayama, T.; Hirose, S. (2000). Development of Souryu-I connected crawler vehicle for inspection of narrow and winding space, Proceeding of the 26th Annual Conference of the IEEE Industrial Electronics Society, Vol.1, pp.143-148, ISBN: 0-7803-6456-2, Nagoya, Japan, Oct, 2000, IEEE Service Center, Piscataway.

Kamegawa, T.; Yamasaki, T.; Igarashi, H.; and Matsuno, F.(2004). Development of the Snake-like Rescue Robot "KOHGA", *Proceedings of the 2004 IEEE International Conference on Robotics and Automation*, pp. 5081-5086, ISSN: 1050-4729,ISBN: 0-7803-8232-3, New Orleans, LA, USA, April-May, 2004, IEEE Service Center, Piscataway.

Yim, M.; Zhang, Y.; Mao, E. (2001). Distributed control for 3D shape Metamorphosis, *Autonomous Robots*, Vol.10, No.1, pp.41-56, 2001, ISSN: 0929-5593.

Murata, S.; Yoshida, E.; Kamimura, A.; Kurokawa, H.; Tomita, K.; Kokaji, S. (2002). M-TRAN:Self-Reconfigurable Module Robotic System, *IEEE/ASME Transactions on Mechatronics*, Vol.7, No.4, pp.431-441, 2002, ISSN: 1083-4435.

Shen, W.; Salemi B. & Will, P. (2002). Hormone-Inspired adaptive communication and distributed control for CONRO self-reconfigurable robots. *IEEE Transactions on Robotics and Automation*, Vol. 18, No. 4, Oct. 2002, pp.700-712, ISSN: 0882-4967

Kurokawa, H.; Kamimura, A.; Yoshida, E.; Tomita, K. & Kokaji, S. (2003). M-TRAN II: Metamorphosis from a Four-Legged Walker to a Caterpillar. *Proceedings of the 2003 IEEE/RSJ Intl. Conference on Intelligent Robots and Systems*, pp. 2454-2459, ISBN: 0-7803-7860-1, October 2003, Las Vegas, Nevada, IEEE Service Center, Piscataway.

Moechel, R.; Jaquier, C.; Drapel K., Dittrich E. & Upegui A. (2005). Yamor and Bluemove-an Autonomous Modular Robot with Bluetooth Interface for Exploring Adaptive Locomotion, *Proceeding of the 8th International Conference on Climbing and Walking Robots, CLAWAR 2005*, pp. 685-692, ISBN: 978-3-540-26413-2,September, 2005, London, U.K. Springer Berlin Heidelberg.

Gonzalez-Gomez J.; Aguayo, E; & Boemo, E. (2004). Locomotion of a Modular Worm-like Robot using a FPGA-based embedded MicroBlaze Soft-processor. Proceeding of the 7th International Conference on Climbing and Walking Robots, CLAWAR 2004, pp. 869-878, ISBN: 3540229922, September, 2004, Madrid, Spain, Springer.

Gonzalez-Gomez, J. & Boemo E., (2005). Motion of Minimal Configurations of a Modular Robot: Sinusoidal, Lateral Rolling and Lateral Shift, *Proceeding of the 8th International Conference on Climbing and Walking Robots, CLAWAR 2005*, pp. 667-674, ISBN: 978-3-540-26413-2,September, 2005, London, U.K. Springer Berlin Heidelberg.

Gonzalez-Gomez, J.; H. Zhang, Boemo, E. (2007). Locomotion Principles of 1D Topology Pitch and Pitch-Yaw-Connecting Modular Robots, In: *Bioinspiration and Robotics: Walking and Climbing Robots*, Habib, Maki K (Ed.), pp.403-428. Advanced Robotic System and I-Tech, ISBN:978-3-902613-15-8.

Brown, H. B. & et al. (2002). Millibot trains for enhanced mobility. *IEEE/ASME Transactions on Mechantronics*, Vol.7, No.4, March 2002, pp.452-461, ISSN: 1083-4435

Park, M.; Chung W. & Yim M. (2004). Control of a mobile robot with passive multiple trailers. *Proceedings of the 2004 IEEE International Conference on Robotics and Automation*, pp. 4369-4374, ISBN: 0-7803-8232-3, New Orleans, LA, United States, April-May 2004, IEEE Service Center, Piscataway

Wang, T.; Zou, D.; Chen, D. (2005). Mechanism design and control method of reconfigurable tracked robot, Journal of Beijing University of Aeronautics and Astronautics, Vol.32, No.7, pp. 705-708, ISSN: 1001-5965.

Sahin, E.; Labella, T.H. & et al. (2002). SWARM-BOT: Pattern formation in a swarm of self-assembling mobile robots. *Proceedings of the 2002 IEEE International Conference on Systems, Man and Cybernetics*, pp. 145-150, ISBN: 0-7803-7437-1, Yasmine Hammamet, Tunisia, October 2002, IEEE Service Center, Piscataway

Wang, W.; Zong, G. (1999) Analysis on the Mechanics Stability for a New Kind of Robot, *Journal of Robot*, Vol.21, No.7, pp. 642-648, ISSN: 1002-0446.

Zhang, H.; Wang, W.; Deng, Z.C. & Zong, G.H. (2006a). A novel reconfigurable robot for urban search and rescue. *International Journal of Advanced Robotic Systems*, Vol.3 No.4, 2006, PP.259-366, ISSN: 1729-8806

Zhang, H.; Deng, Z.; Wang, W.; Zhang, J. & Zong, G. (2006b). Locomotion Capabilities of a Novel Reconfigurable Robot with 3 DOF Active Joints for Rugged Terrain, *Proceedings of the 2006 IEEE/RSJ International Conference on Intelligent Robots and Systems, IROS 2006*, pp.5588-5593, Beijing, China, October, 2006

Wang, W.; Zhang H.X; Zong, G.H. & Zhang, J.W. (2006). Design and realization of a novel reconfigurable robot with serial and parallel mechanisms. *Proceedings of 2006 IEEE International Conference on Robotics and Biomimetics*, pp. 697-702, ISBN: 1-4244-0571-8, Kunming, China, Dec. 2006, IEEE Service Center, Piscataway

Zong, G.; Deng, Z.; & Wang, W. (2006) Realization of a Modular Reconfigurable Robot for Rough Terrain, *Proceedings of the 2006 IEEE International Conference on Mechatronics and Automation*, pp. 289-294, ISBN: 1-4244-0466-5 Luoyang, Henan, China, June,2006, , IEEE Service Center, Piscataway.

Zhang, H.; Chen, S.; Wang, W.; Zhang, J. & Zong, G. (2007). Runtime Reconfiguration of a Modular Mobile Robot with Serial and Parallel Mechanisms, *2007 IEEE/RSJ International Conference on Intelligent Robots and Systems*, pp.2999-3004, ISBN: 1-4244-0912-8, San Diego, CA, USA, Oct. 29 - Nov. 2, 2007, IEEE Service Center, Piscataway

14

Imitation-Based Task Programming on a Low-Cost Humanoid Robot

Jacopo Aleotti and Stefano Caselli
University of Parma,
Italy

1. Introduction

Humanoid robots are complex service platforms with anthropomorphic features, specifically designed for close interaction with humans. Conventional programming strategies are hardly applicable to humanoids due to the high number of degrees of freedom that must be coordinated concurrently. Therefore, exploiting humanoids' potential in service tasks remains an elusive goal. One of the most promising techniques for dealing with humanoid robots is programming by demonstration, which allows even unexperienced users to easily interact with the robot based on the teaching by showing or imitation paradigm. In particular, the ability to imitate human gestures and follow task-relevant paths are essential skills for legged humanoid robots, as they provide the fundamental techniques for physical human-robot interaction. This chapter investigates the potential of imitation in programming humanoid motor skills. As target platform, we have adapted a Robosapien V2 (RSV2), a low-cost small humanoid available in the toy market. The chapter focuses on the teaching of basic, humanoid-relevant skills such as body postures and walking paths. We have explored and combined multiple sensing sources to capture human motion for imitation purposes, namely a dataglove, an electromagnetic motion tracker, and a monocular vision system for landmark recognition. The imitation approach illustrated in this chapter is rather general, even though its implementation is constrained by limitations of RSV2 and by sensor inaccuracies. In particular, the chapter reports successful experiments on gesture imitation, including arms motion as well as upper body and head movements. The gesture repertoire learned by the robot can serve both as a body language for understanding human requests in human-robot interaction and as a set of primitives which can be combined for programming more complex tasks. We believe that a deep assessment of a low-cost humanoid robot is extremely important for the robotic research community since the technological requirements and the costs to develop more advanced humanoid robots still prevent them to become broadly available. Currently, most high-end humanoids are developed as prototypes platforms under the supervision of important private companies. Therefore, low-cost humanoid platforms such as RSV2 provide an exciting and affordable opportunity for research in humanoid integration in service tasks.

2. Background and related work

The development of humanoid systems is a tremendous challenge for robotics research. Many academic laboratories as well as industrial companies are devoting substantial efforts and resources in building advanced humanoid robots. Several authors (Schaal, 1999; Mataric, 2000; Breazeal & Scassellati, 2001) have pointed out that the principal challenges in building effective humanoid robots involve different aspects such as motor actuation, perception, cognition and their integration. Typically, complex mechanical systems must be regulated and a high number of degrees of freedom must be controlled. Study of appropriate materials, motors, power supplies and sensors is also required. Moreover, humanoid systems require advanced algorithms and software solutions to achieve learning skills for autonomous operations. The development of natural user interfaces for human robot interaction is also required since one of the main goals is to overcome the traditional approaches of robot design by creating robots able to interact with humans in everyday life. However, the technological requirements and the costs to develop effective humanoid robots still prevent such advanced systems to become available to the public. Few low-cost humanoids have been designed for research purposes. One of them is Robota (Billard, 2003), a small humanoid doll used as educational toy for robotic classes and playing vision-based imitation games. Recently, several low-cost humanoid platforms have appeared in the consumer market. Besides being considered as high quality toys for personal entertainment, these devices can also be exploited for research projects. One of such robots is Robosapien V2 (RSV2) developed by WowWee and released at the end of 2005. RSV2 has been chosen for the experimental evaluation reported in this chapter as it is one of the most promising low-cost robots available in the market. Our research on imitation is motivated by the fact that robot programming by demonstration is an effective way to speed up the process of robot learning and automatic transfer of knowledge from a human to a robot. As it is well known that robot programming using traditional techniques is often difficult for untrained users, especially in the context of service robotics, programming by demonstration provides an intuitive solution by letting the user act as a teacher and the robot act as a learner (Ikeuchi & Suehiro, 1994; Zöllner et al., 2002). The need for a high-level approach in robot programming is even more apparent with humanoid robots, where conventional programming of a high number of degrees of freedom would be clearly unacceptable.

In this work we present a system for robot programming by demonstration where a Robosapien V2 is programmed to imitate human gestures and walking paths. In particular, we have explored the idea of augmenting an entertainment humanoid with multiple sensors to accomplish rather complex imitative tasks. The system combines a dataglove, a motion tracker and a monocular vision system. The first contribution of the chapter is the experimental evaluation of various imitation games where the robot follows the gestures of a human teacher. A second contribution is the proposal of an algorithm for walking imitation. The chapter shows how humanoid walking paths can be synthesized by fitting a parametric curve with data observed in human-demonstrated trajectories, thereby providing an effective technique for path imitation. Walking paths can be further adapted by the robot based on online visual guidance.

Hereafter, prior work on humanoid imitation and motion learning and is discussed. Many authors have investigated the problems of human body tracking and mapping of body postures to humanoid robot. The paper by (Bandera et al., 2004) proposes a method for real-time estimation of upper body postures based on computer vision and an inverse kinematic model of the robot. In (Riley et al., 2003) a similar approach was used incorporating 3D vision and a kinematic model of the human teacher. A single camera system for data acquisition with a particle filtering technique for parameter estimation was adopted in (Menezes et al., 2005). Other works were mainly focused on the problem of adaptation of human motion to humanoid motion with kinematics constraints such as joint and velocity limits (Pollard et al., 2002) or ground contact conditions for humanoid locomotion (Matsui et al., 2005). In (Matsui et al., 2005) a method was proposed for the generation of natural motions in an android by measuring the robot posture at its visible surfaces and comparing it to the posture of a human teacher. The work in (Inamura et al., 2001) focused on mimesis learning using primitive symbol observation with Hidden Markov Models to abstract the dynamics of human motion and to generate natural motion patterns.

In (Inamura et al., 2005) an imitation strategy was adopted based on attention points and intent imitation. A neural learning system was proposed in (Kuniyoshi et al., 2003) for learning motion patterns from the observation of self robot movement. A mirror neuron model was implemented in (Ito & Tani, 2004) for imitative interactions. In (Shon et al., 2006) a nonlinear regression algorithm was used for mapping motion capture data from a human actor to a humanoid robot. Calinon et al. (Calinon et al., 2006) presented a method for extracting the goals of a demonstrated task and determining the best imitation strategy that satisfies the goals. As proposed in (Nakazawa et al., 2002; Nakaoka et al., 2005), studies were also conducted for learning and reproducing human dances through human observation.

Our work is among the first attempts to investigate the capabilities of RSV2 for humanoid research. The proposed imitation mechanism is strongly constrained by the limitations of the toy robot. The work in (Behnke et al., 2006) is the only one that has considered the use of Robosapien for research purposes. The authors have developed an augmented version of Robosapien V1. A Pocket PC and a color camera were added to the robot to make it autonomous and its capabilities were tested for basic soccer skills.

The problem of tracking and following demonstrated navigational routes has been investigated mainly in the context of wheeled mobile robots. Dixon and Khosla (Dixon & Khosla, 2004) presented a system for learning motor skill tasks by observing user demonstrations with a laser range finder. The system is able to extract subgoals and associate them with objects in the environment generalizing across multiple demonstrations. The work in (Hwang et al., 2003) proposed a touch interface method to control a mobile robot in a supervised manner. The algorithm extracts a set of significant points from the user-specified trajectory, produces a smooth trajectory using Bezier curves and allows on-line modification of the planned trajectory in a dynamic environment. In (Hon Nin Chow et al., 2002) a reactive sensor-motor mapping system is described where a mobile robot learns navigational routes and obstacle avoidance. In (Wanitchaikit et al., 2006) a self-organizing approach for robot behaviour imitation is presented. A demonstrator mobile robot collects visual information about the environment; then the movement features are presented to the imitation engine. Tang et al. (Tang et al., 2001) proposed a vision based autonomous navigation system for mobile robots in an indoor environment by teaching and playing-back scheme. The system memorizes a sequence of omnidirectional images using them to compute the trajectory to track the taught route. Arakawa et al. (Arakawa et al., 1995) developed a trajectory generation method for wheeled vehicles. The algorithm uses Bezier curves for smoothly reducing the deviation from the guideline. Morioka et al. (Morioka et al., 2004) focused on the problem of human-following for a mobile robot in an intelligent environment with distributed sensors. The control law is based on a virtual spring model.

Several interesting research projects have considered the use of humanoid robots for navigation and autonomous mapping of indoor environments. The work in (Michel et al.,

2005) presented an approach to autonomous humanoid walking for the Honda ASIMO in the presence of dynamically moving obstacles. The system combines vision processing for real-time environment mapping and footsteps planning for obstacle avoidance. A similar approach was presented in (Kagami et al., 2003). In (Michel et al., 2006) an online environment reconstruction system is presented. The system utilizes both external sensors for global localization, and on-body sensors for detailed local mapping for the HRP-2 humanoid robot.
The chapter is organized as follows. Section 3 describes the experimental set-up including a description of the augmented RSV2 and the input devices. Sections 4 and 5 describe the adopted gesture and path imitation techniques. Section 6 discusses the trend toward low-cost humanoid and imitation systems. The chapter closes in section 7 discussing the obtained results.

3. Experimental set-up

3.1 Robosapien V2

The humanoid robot used in this work is a Robosapien V2 developed by WowWee and shown in figure 1 (left image). It is a low cost toy robot expressly designed for the consumer market. RSV2 is 60cm tall and is driven by 12 DC motors. It has realistic joint movements and onboard sensors. Some RSV2 technical specifications are reported in Table 1.

Length	43cm
Width	32cm
Height	60cm
Weight	7Kg
Body batteries	6 D
Brain batteries	4 AAA

Table 1. RSV2 specifications.

RSV2 is fully controllable by a remote infrared controller and has capabilities which make it suitable for entertainment but also for research purposes. The main functionalities of the robot include true bipedal walking with multiple gaits, turning, bending, sitting and getting up. RSV2 can also pick up, drop and throw small objects with articulated fingers. The robot is equipped with an infrared vision system, a color camera, stereo sonic sensors and touch-sensitive sensors in its hands and feet. These sensors can be used to trigger simple reactive behaviors to environmental stimuli through preprogrammed motion sequences, which can be stored in the onboard memory. For example RSV2 can recognize and react to primary colors and skin tones. Moreover, it is able to react to sounds, to track close moving objects with its head and to avoid obstacles while walking. The locomotion of RSV2 is achieved by alternatively tilting the upper body and moving the leg motors in opposite directions. The lateral swinging movement of the upper body generates a periodic displacement of the center of mass between the two feet. To achieve fully autonomous capabilities RSV2 must be augmented with external computing power. There are two possible ways for augmenting RSV2. The first method is hacking the on board electronics to get access to the motor and sensor signals. The second method is to bypass the remote controller with an infrared transmitter connected to a processing unit. This strategy is less intrusive but it only allows transmission of commands to the robot, therefore it also requires external sensors like a camera, as proposed in [3]. In this work we chose a similar approach. We adopted a USB-UIRT (Universal Infrared Receiver Transmitter) device, which is connected to a host PC. The device is able to transmit and receive infrared signals. Reception has been used to learn the IR codes of the RSV2 remote controller. As a result, the IR codes of the motor commands for the RSV2 have been decoded and stored in a database.

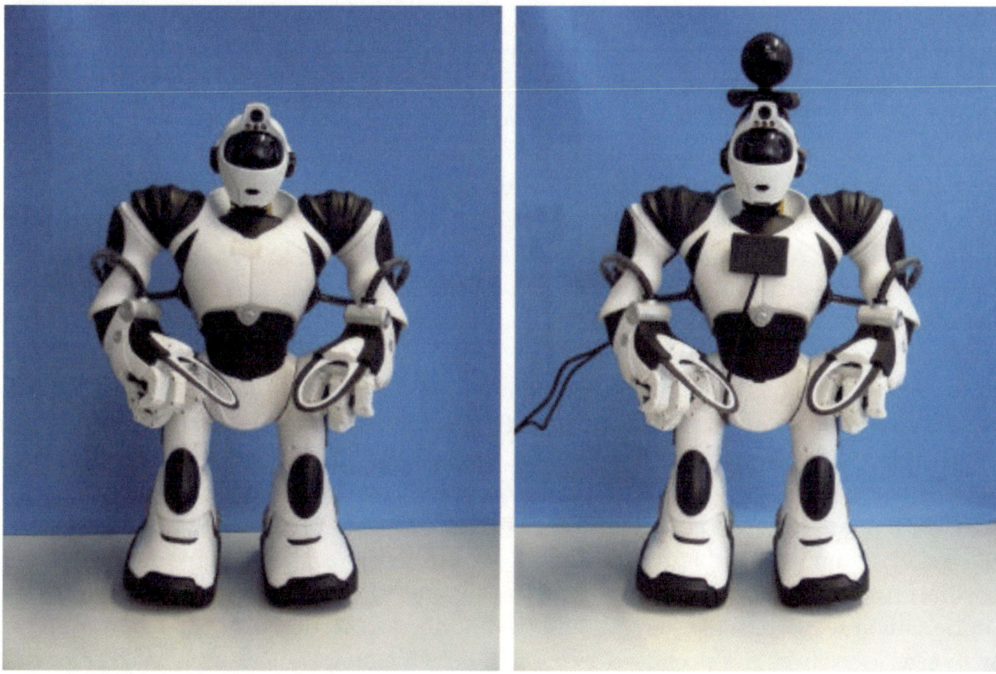

Fig. 1. Original RSV2 (left image) and augmented RSV2 (right image).

3.2 Sensor devices

The experimental set-up of the proposed system comprises a CyberTouch glove (by Immersion Corporation), a FasTrack 3D motion tracking device (by Polhemus, Inc.) and a monocular vision system. The CyberTouch glove used in the experiments has 18 sensors for bend and abduction measurements. The FasTrack is a six degrees of freedom electro-magnetic sensor that tracks the position and orientation of a small receiver relative to a fixed transmitter. The vision system exploits a standard webcam with a CMOS image sensor with 640x480 resolution and USB 2.0 compatible connection. The camera has 30fps nominal refresh rate. In the current hardware set-up the actual measured frame rate is 12fps. For evaluation of gesture imitation, sensor devices have been configured to collect information about joint angles of the human teacher as described in section 4. In particular, the camera has been located in a fixed position to observe the human upper body. Figure 1 (right image) shows the RSV2 in its augmented configuration used for walking imitation. The webcam has been mounted on the head of the robot to minimize negative effects on its balance. This solution provides an additional advantage as the pan-tilt head motors can be exploited to rotate the camera. Camera rotation along two axes is crucial for enhancing the robot marker recognition capabilities, as it will be pointed out in section 5. The Fastrak receiver has been mounted on the back of RSV2 and is used for global localization. The USB-UIRT device has been mounted on RSV2 as well. The device is located on the torso and, being very close to the RSV2 infrared receiver, it helps in avoiding packet loss in transmission of motor commands.

3.3 Software

The application which runs the imitation system has a multithreaded architecture and has been built upon two software libraries, namely the Virtual Hand Toolkit (VHT), and the ARToolKit (ART). The main thread is devoted to sensor data acquisition from the devices, visual processing, and computation of the motor commands for the Robosapien. A second thread is in charge of sending the motor commands to the humanoid robot through the IR transmitter. The Virtual Hand Toolkit is the application development component of the VirtualHand Suite 2000 by Immersion Corporation. VHT includes a Device Configuration Utility used to initialize and calibrate the devices and a Device Manager for data acquisition. The toolkit offers high level functionalities and low level translators for specific I/O devices such as the Cybertouch and the Fastrack. ARToolKit [16] is an open-source multiplatform software library designed for augmented reality applications. The library exploits a pattern recognition algorithm and allows overlaying of computer graphics images on the video stream captured by a camera in real-time. In our experiments we used the toolkit to recognize markers on the ground acting as via points, markers attached to static obstacles and markers attached to the body of the human demonstrator for gesture imitation.

The overlay option was enabled for testing the effectiveness of the recognition algorithm. In particular, ARToolKit uses computer vision techniques to compute the real camera position and orientation relative to markers, as will be described in section 4. Markers are squares of known size (8×8cm in the proposed experiments). One snapshot of each marker must be provided to the system in advance as training pattern. The recognition algorithm is a loop which consists of different steps. First the live video image is binarized using a lighting threshold. Then the algorithm searches for square regions. The squares containing real markers are then identified and the corresponding transformation matrix is computed. OpenGL is used for setting the virtual camera coordinates and drawing the virtual images.

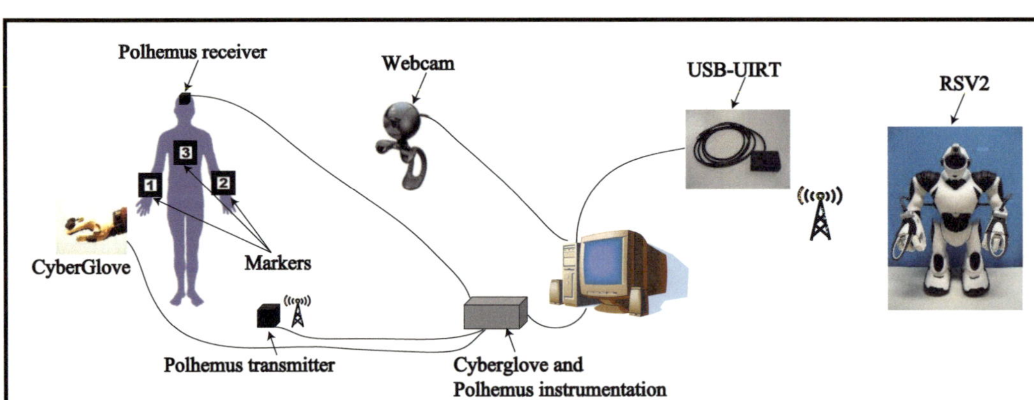

Fig. 2. System architecture for gesture imitation.

4. Imitation of human gestures

In this section we present the proposed imitation strategy for mapping the movements of the human demonstrator to RSV2. In the current setup six degrees of freedom have been considered, namely one degree of freedom for the two arms plus wrist rotation of the right arm, one degree of freedom for torso yaw and two degrees of freedom for head pitch and yaw. The associated joints can be controlled independently. The imitation strategy consists in a proportional real-time mapping between each computed sensor data and the corresponding joint. The system requires the demonstrator to perform an initial calibration routine to identify the range of motion for each movement. Each range is discretized into as many intervals as the number of feasible discrete configurations of the corresponding joint. Due to the kinematics limitations of RSV2, the resolution of the joints is quite limited and never goes beyond 8 intervals as remarked in section 4.4. After the calibration phase RSV2 starts imitating the gestures of the human demonstrator. The user can decide to play with all the six degrees of freedom concurrently or with a restricted subset by deactivating some of the sensor inputs.

Figure 2 shows the structure of the system used for gesture imitation and highlights the hardware components along with the sensors and their interconnections. Markers are attached to the body of the human demonstrator and the camera tracks the motion of the tags. The motion tracker is used to detect the motion of the head.

4.1 Arms and upper-body visual tracking

The vision system was used to track the arms motion of the human demonstrator along with the upper body. One marker is attached to each arm and a third marker is attached to the human torso. The vision algorithm is able to identify and track the geometrical configuration of multiple markers concurrently. Hereafter the algorithm used by ARToolkit for estimating the transformation matrix of markers is described. The algorithm finds for each marker the transformation T_{cm} which relates the coordinates (X_m, Y_m, Z_m) in the marker reference frame to the coordinates (X_c, Y_c, Z_c) in the camera reference frame. The rotation component is estimated by projecting the two pairs of parallel lines contouring the marker. These lines in the camera screen coordinates (x, y) can be expressed as

$$\begin{aligned} a_1x + b_1y + c_1 &= 0 \\ a_2x + b_2y + c_2 &= 0 \end{aligned} \tag{1}$$

From camera calibration, the projection matrix which relates the camera coordinates to the screen coordinates can be computed as follows

$$\begin{bmatrix} hx \\ hy \\ h \\ 1 \end{bmatrix} = \begin{bmatrix} P_{11} & P_{12} & P_{13} & 0 \\ 0 & P_{22} & P_{23} & 0 \\ 0 & 0 & 1 & 0 \\ 0 & 0 & 0 & 1 \end{bmatrix} \begin{bmatrix} X_c \\ Y_c \\ Z_c \\ 1 \end{bmatrix} \tag{2}$$

where h is a constant. By substituting (x, y) of equation 2 into equation 1 the equations of two planes in space can be found as shown next:

$$a_1P_{11}X_c + (a_1P_{12} + b_1P_{22})Y_c + (a_1P_{13} + b_1P_{23} + c_1)Z_c = 0$$
$$a_2P_{11}X_c + (a_2P_{12} + b_2P_{22})Y_c + (a_2P_{13} + b_2P_{23} + c_2)Z_c = 0 \quad (3)$$

The outer product of the normal vectors of these planes $n_1 \times n_2$ gives the direction vector v_1 of the two parallel sides of the marker. The two direction vectors of the two pairs of parallel lines v_1 and v_2 are then compensated for errors. The outer product $v_1 \times v_2$ provides the rotation component of the transformation matrix. The translation component of T_{cm} is obtained by the correspondence between the coordinates of the four vertices of the marker in the marker coordinate frame and the coordinates of the vertices in the camera coordinate frame. The z component of the translation vector is used for tracking the movement of the arms of the human demonstrator. The rotation component R of the transformation matrix is used to compute the orientation of the marker attached to the torso. The Euler angles are extracted from R as shown in equation 4 (giving two possible solutions). The yaw angle of the marker is set equal to θ_1. The yaw angle is then mapped to the joint controlling the yaw rotation of the upper body of RSV2.

$$\begin{aligned} \theta_1 &= -\mathrm{asin}(R_{31}) \\ \theta_2 &= \pi - \theta_1 \\ \varphi_1 &= \mathrm{atan2}\left(\frac{R_{32}}{\cos\theta_1}, \frac{R_{33}}{\cos\theta_1}\right) \\ \varphi_2 &= \mathrm{atan2}\left(\frac{R_{32}}{\cos\theta_2}, \frac{R_{33}}{\cos\theta_2}\right) \end{aligned} \quad (4)$$

Fig. 3. Human gestures (first row), results after video processing (second row) and corresponding RSV2 configurations (third row).

4.2 Head motion tracking

The motion of the head of the human demonstrator is tracked using the FasTrack. The tracker receiver is attached to the nape of the human. The transformation matrix from the receiver frame (rx) to the world reference frame (W) is given by

$$T_{W \leftarrow rx}(t) = T_{W \leftarrow tx} T_{tx \leftarrow rx}(t) \quad (5)$$

where $T_{W \leftarrow tx}$ is the transformation between the transmitter frame to the world frame. The pitch and yaw angles of the marker are extracted from the rotation component R of $T_{W \leftarrow rx}$ and set equal to θ_2 and φ_2 respectively.

4.3 Head motion tracking

The rotational motion of the right hand wrist of the human operator is linearly mapped to the wrist of the right hand of RSV2. The motion is measured by the wrist sensor of the CyberGlove. The wrist angle γ of the CyberGlove is computed using the relation

$$\gamma = G * (Digital\_Value - O) \quad (6)$$

where G and O are the gain and offset calibration values.

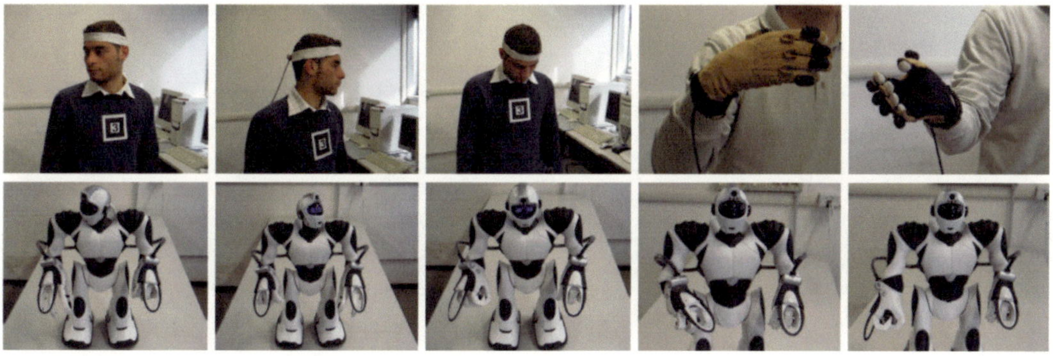

Fig. 4. Imitation of head and right wrist movements.

4.4 Results

Experiments involving gesture imitation were carried out to test the effectiveness of the system. Figure 3 shows RSV2 imitating arms and upper-body movements by visual tracking of trained patterns attached to the user's body. The same figure shows also the results of the video processing algorithm. Three squares with different colors (or gray shadings) are superimposed to the recognized markers. Figure 4 shows an experiment involving the imitation of head and wrist movements.

The evolution of the command signals sent to the humanoid robot compared to the sensed data for each degree of freedom has been reported in figure 5. The curves representing the command signals are step functions obtained by sampling the input data signals. The cardinality of the set of steps of each curve equals the resolution of the corresponding degree of freedom of the robot. For example the arms and the head yaw movements have the highest resolution with 8 and 12 steps respectively. It can be seen from figure 5 that the robot degrees of freedom track with acceptable fidelity commanded values, up to the limited resolution available.

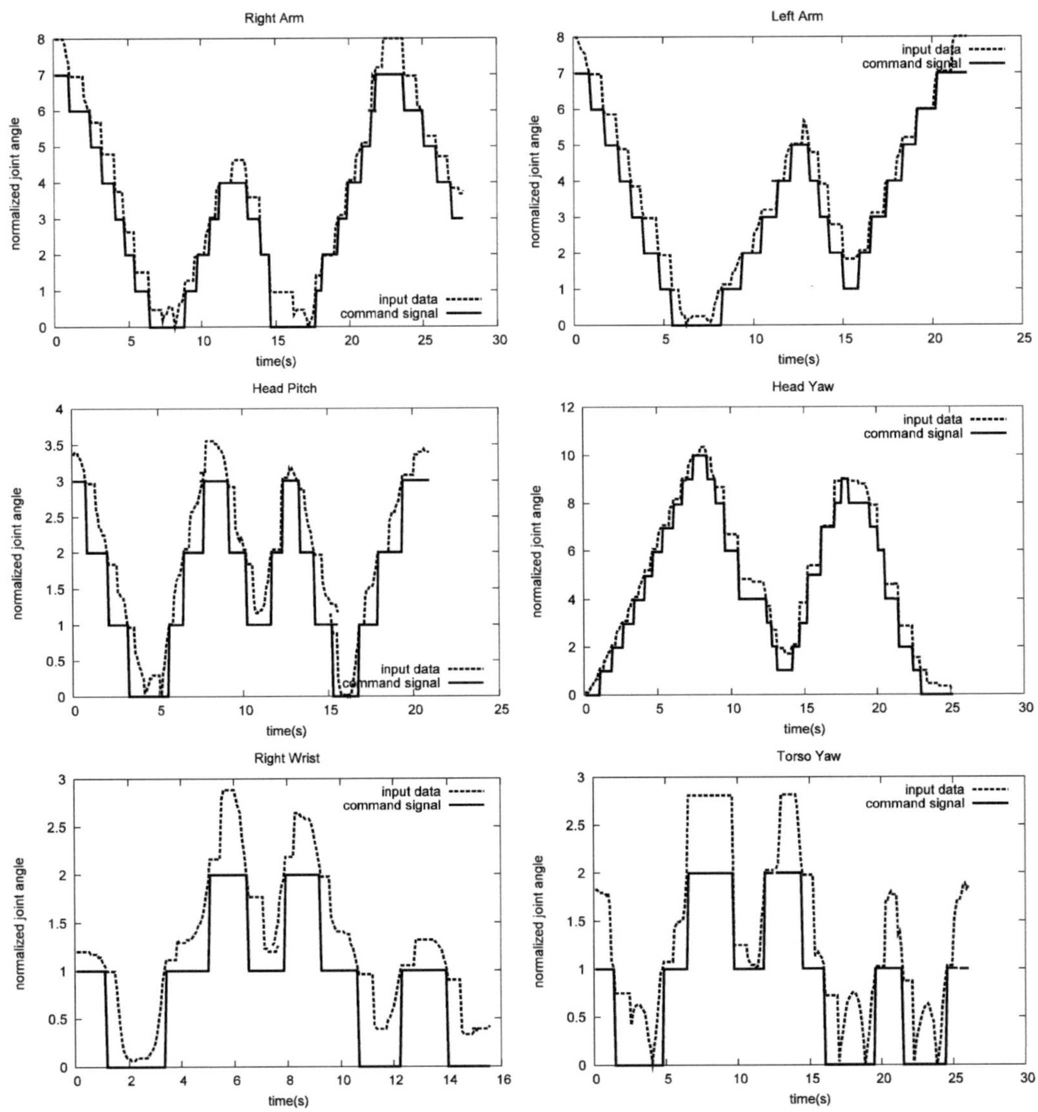

Fig. 5. Evolution of the command signal compared to the sensed data for each degree of freedom.

5. Walking imitation

Figure 6 shows the experimental setup used for walking imitation tasks. The system comprises a demonstration phase, where the user demonstrates a walking path, and an imitation phase, where the robot follows the provided trajectory. The setup also includes a fixed ceiling camera for initial recognition of the topology of the markers on the ground. In the first experiment a simple walking imitation task has been tested where the robot has to follow a taught route. In the demonstration phase a human performs a walking path while his movements are tracked by the Fastrak receiver, which is attached to one of the legs of the demonstrator. After the demonstration phase RSV2 imitates the demonstrated path relying only on the tracker sensor without visual guidance. The transformation matrix from the Fastrak receiver frame (rx) to the world reference frame (W) is given by equation 5. Figure 7 shows a sequence of images taken from both the demonstration phase and the imitation phase. The demonstrated path is a smooth U-turn around a static obstacle. Figure 8 reports the evolution of RSV2 walking trajectory compared to the demonstrated trajectory. The demonstrated trajectory is approximated as a NURBS (Non Uniform Rational B-Spline) curve (Piegl, 1991). The systems computes a piecewise linear approximation of the NURBS generating a set of viapoints and RSV2 follows the approximated path through a sequence of rotations and translations. The computed mean error of the path following task is about 15cm. The demonstrated path, shown in figure 8, appears deformed on the right side due to sensor inaccuracies of the tracker, which increase as the distance between the transmitter and the receiver increases. The same inaccuracies are the main cause of the worsening of path following performance in the last stage of the task after the U-turn.

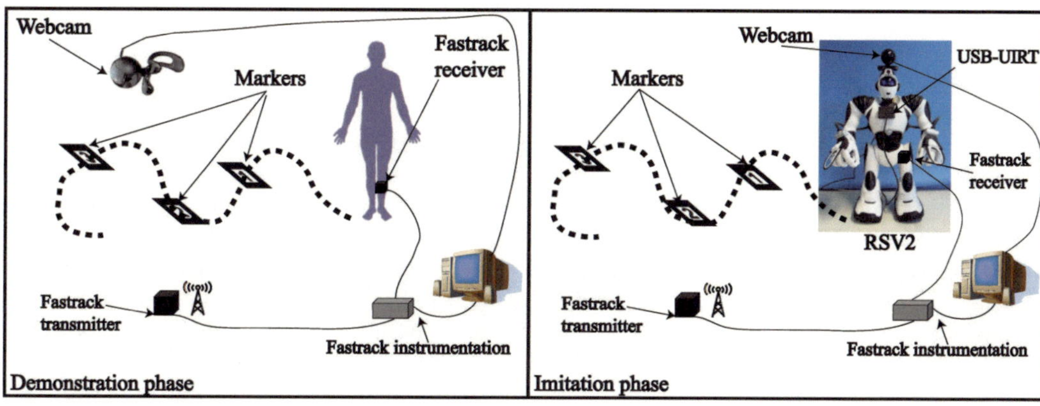

Fig. 6. System architecture for walking imitation.

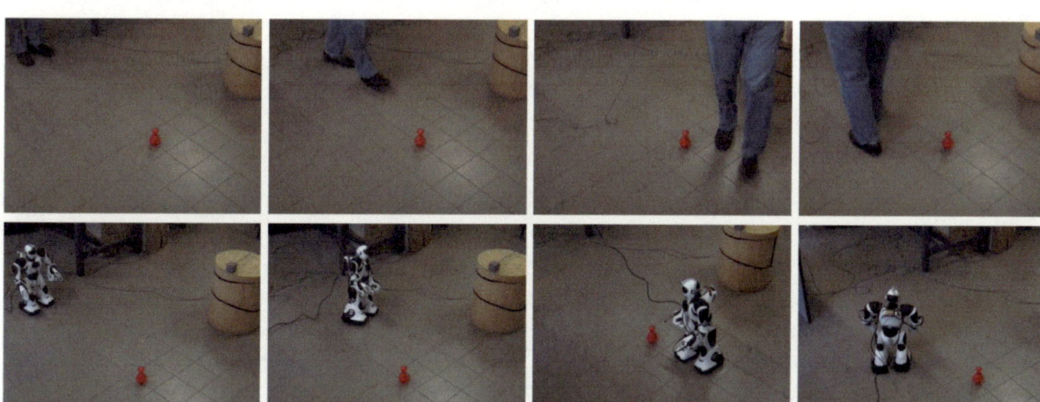

Fig. 7: Walking experiment 1: demonstration phase (top row) and imitation phase (bottom row).

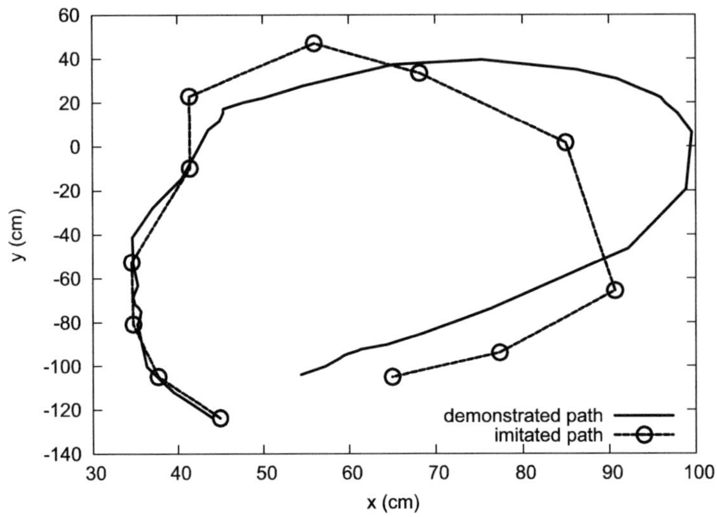

Figure 8: Walking experiment 1: demonstrated path and imitated path.

5.1 Dynamic obstacle avoidance

In a second experiment, a vision-guided imitation task was performed which requires dynamic obstacle avoidance. After the demonstration phase an obstacle is positioned on the original path. RSV2 is programmed to follow the path and to scan the environment for possible obstacles by turning its head whenever it reaches a via-point. A marker is attached to the obstacle for detection. Figure 9 shows a sequence of images taken from the imitation phase of the task. Figure 10 reports the demonstrated path together with the replanned path and the actual imitated path (the obstacle is detected when the robot reaches the second via-point). Figure 11 shows the output of the onboard camera when the marker is recognized, with a colored square superimposed to the recognized marker. The algorithm used for replanning the trajectory performs a local deformation of the original NURBS around the obstacle. Initially the NURBS is resampled and a deformation is applied to each sample point $(x(u_i), y(u_i))$ which falls within a certain distance from the obstacle according to the following equation

$$\begin{cases} \tilde{x} = x(u_i) + \alpha e^{-\frac{d^2}{\beta}} (x(u_i) - x_m) \\ \tilde{y} = y(u_i) + \alpha e^{-\frac{d^2}{\beta}} (y(u_i) - y_m) \end{cases} \quad (7)$$

where α and β are constants. The deformation stretches the samples with a gaussian modulation factor which depends on the distance d between the robot and the obstacle. The value of d is approximated by the measured distance d_{cm} between the camera and the marker which are approximately orthogonal. The obstacle coordinates (x_m, y_m) are set equal to the coordinates of the marker in the world reference frame and are given by

$$\begin{cases} x_m = x_r + d\cos\theta_m \\ y_m = y_r + d\sin\theta_m \end{cases} \quad (8)$$

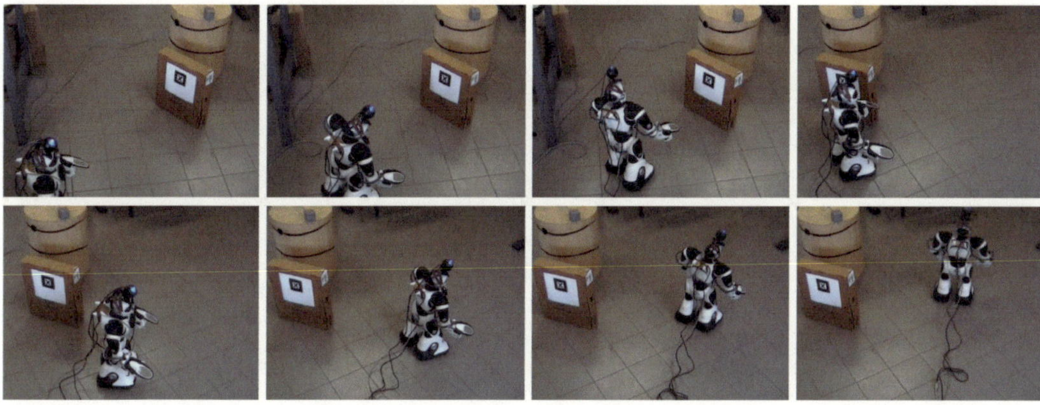

Figure 9: Walking experiment 2: imitation phase with dynamic vision-guided obstacle avoidance.

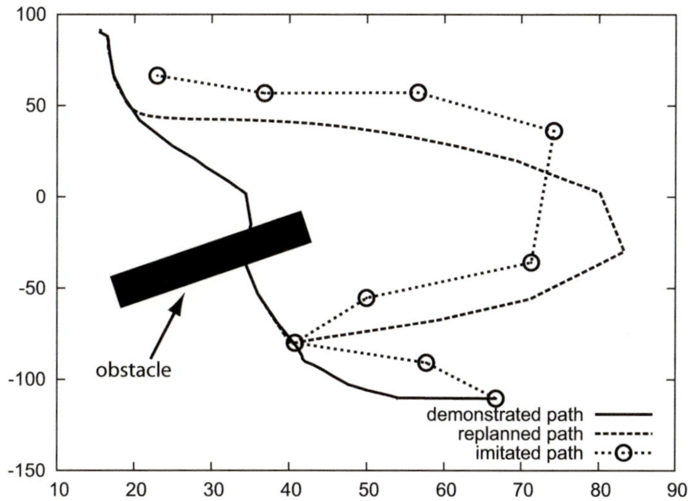

Figure 10: Walking experiment 2: demonstrated path, replanned path and imitated path.

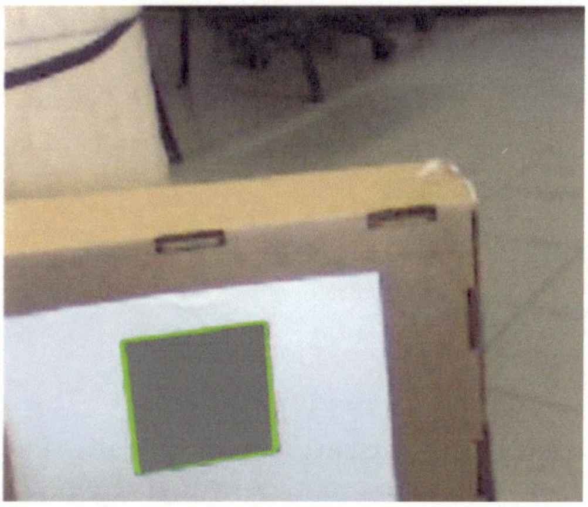

Figure 11: Walking experiment 2: obstacle detection.

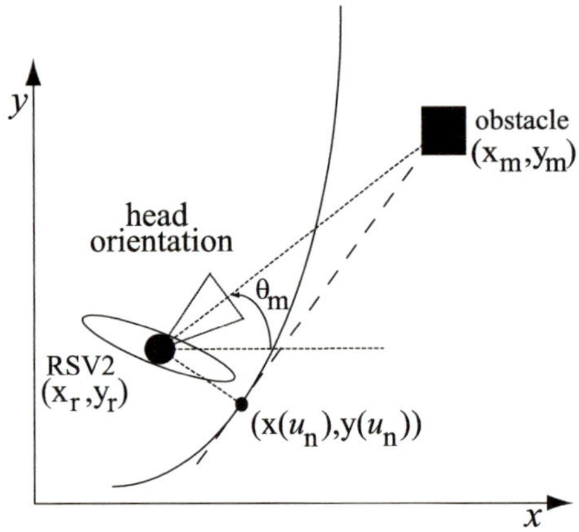

Figure 12: Walking experiment 2: approximation of obstacle coordinates.

where (x_r, y_r) is the current position of the robot and θ_m is the angle between the x-axis and the obstacle, which depends on both RSV2 body and head rotation. The angle θ_m can not be reliably computed, as the actual head rotation is hard to estimate due to the non ideal alignment between the head and the body when the robot stops. Therefore, the obstacle coordinates are approximated as follows

$$\begin{cases} x_m \approx x(u_n) + d_{cm} \dfrac{\dot{x}(u_n)}{\sqrt{\dot{x}^2(u_n) + \dot{y}^2(u_n)}} \\ y_m \approx y(u_n) + d_{cm} \dfrac{\dot{y}(u_n)}{\sqrt{\dot{x}^2(u_n) + \dot{y}^2(u_n)}} \end{cases} \quad (9)$$

where $(x(u_n), y(u_n))$ is the point on the original NURBS nearest to the current position of the robot. This means that the obstacle is assumed to lie on the tangent to the curve at the closest point to the robot.

Figure 13: Walking experiment 3: imitation phase with vision guided marker following.

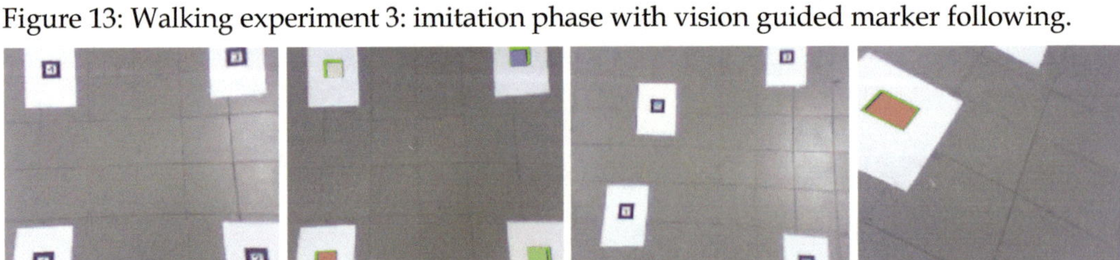

Figure 14: Walking experiment 3 form left to right: initial marker configuration, marker detection from ceiling camera, modified marker configuration and marker detection from onboard camera.

5.2 Vision-guided landmark following

In a third experiment another vision-guided imitation task has been investigated. A set of markers is positioned on the ground of the workspace and a fixed ceiling camera detects the initial configuration of the markers (the estimation error is approximately 7cm). The demonstration consists of a walking task where a human reaches a sequence of markers. The correct sequence of reached markers is computed from the measured trajectory and from the topology of the markers. After the demonstration phase, the markers are slightly moved into a new configuration and RSV2 is programmed to reach the demonstrated sequence of markers. In the proposed experiment the task is to reach marker 1 and then marker 3. Figure 13 reports images of the imitation phase. Figure 14 shows the initial marker configuration, the modified marker configuration, and the marker detection from onboard camera. Figure 15 shows the demonstrated path, the configuration of the markers and the imitated path. RSV2 imitates the reaching task by following the demonstrated path. Once the robot detects a marker on the floor it tries to approach the marker through a sequence of small movements by estimating the position and the distance of the target with the onboard camera with equation 10. The estimation error is less critical than in experiment 2 since the robot performs small movements while approaching the marker as shown in figure 15.

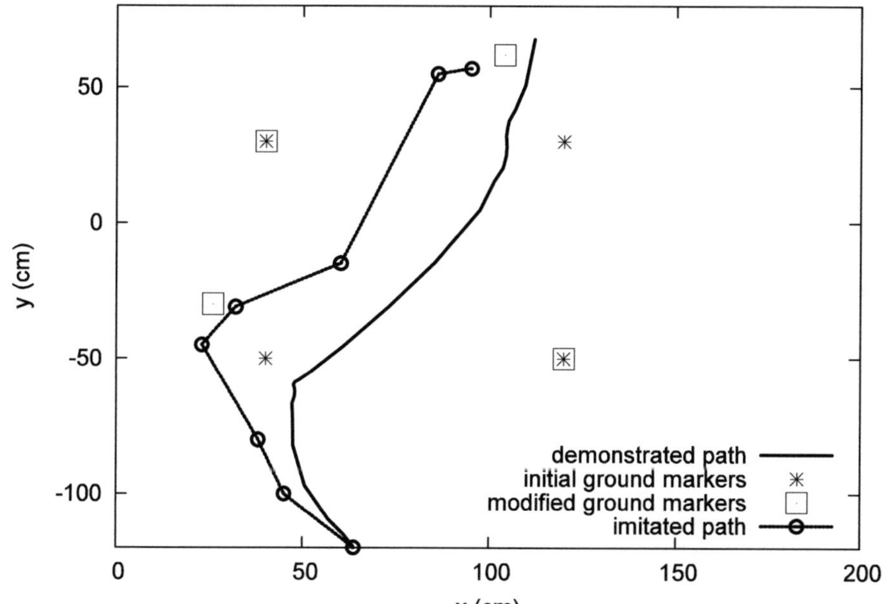

Figure 15: Walking experiment 3: demonstrated path, marker positions and imitated path.

5.3 Vision-guided grasping

RSV2 autonomous grasping skills have been evaluated in a further experiment where the task is to reach a landmark on the floor indicating the position of a set of graspable objects. The robot starts approximately 80cm away from the landmark and in its initial configuration the head and the body are properly aligned. Hence RSV2 is able to detect the landmark using the onboard camera with a tolerable error and moves towards the objects. Once the robot gets closer to the landmark it tries to reach a proper alignment with the target with small rotations and translations. Figure 16 shows the execution of the grasping task while figure 17 shows the marker detection phase at two different configurations. Due to the limitations of RSV2 the grasping procedure is not controllable and allows only to lower the upper body and one arm of the robot trying to pick up objects which are located near the foot of the robot. To overcome these kinematics limitations of the toy and the sensor errors, multiple objects have been placed near the landmark for redundancy. However, the rate of successful grasping operations is still low (approximately 40%). This result suggests that improved grasping capabilities might be required to exploit a low-cost humanoid for experimental research involving grasping.

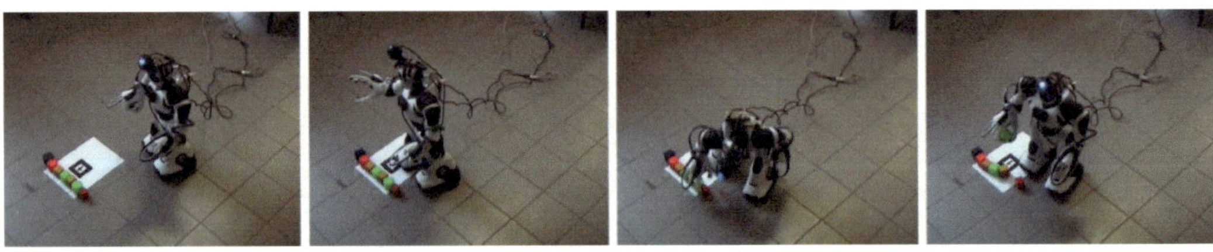

Figure 16: Walking experiment 4: vision guided grasping task.

Figure 17: Walking experiment 4: marker detection.

6. Towards a complete low-cost platform

In the previous sections we have presented a system for humanoid task programming based on imitation which is focused on the use of a low-cost robot. The cost of the robot is only few hundreds of dollars, which is far less than the cost of more advanced humanoids. However, it is still necessary to discuss whether the proposed platform can be considered as a fully low-cost platform as a whole. The software developed is mostly open-source, except for low-level drivers connecting with glove and tracker devices which could be replaced by functionally equivalent open-source drivers. Additional costs arise from the sensor devices used for imitating human gestures and walking path. The USB camera and the USB-UIRT device are also very cheap (about 50$ each). By contrast, the Polhemus FasTrack and Immersion CyberTouch are high quality measurement devices which come at high price. Therefore, the developed platform cannot be considered as a fully low-cost system.

We are therefore investigating alternative tracking devices that can be exploited for motion capture. In this section we report our initial investigation on the exploitation of a Nintendo Wii remote controller for gesture recognition. This device costs less than 100$ and hence would keep the entire system low-cost.

A peculiar feature of the Wii console is its main wireless controller, the Wiimote, and the secondary controller, the Nunchuk, which can be used in conjuction to detect acceleration and orientation in three dimensions (figure 18, left image). The Wii controllers use a combination of accelerometers for tracking human motion (the Wiimote also supports infra-red sensors to detect pointing tasks). The controllers support Bluetooth connection which can be used to connect the device to a normal PC. Figure 18 (right image) shows RSV2 imitating upper arm gestures in a similar experiment to the one presented in figure 3. The user is handling the Wii controllers in her hands and the system detects the posture of the arms (one degree of freedom for each arm) by sensing accelerations. Although the accuracy of the Nintendo controller is not comparable to the high accuracy of the FasTrack, we believe that the use of such a simple wearable device can lead to interesting results, especially if employed for entertainment or education.

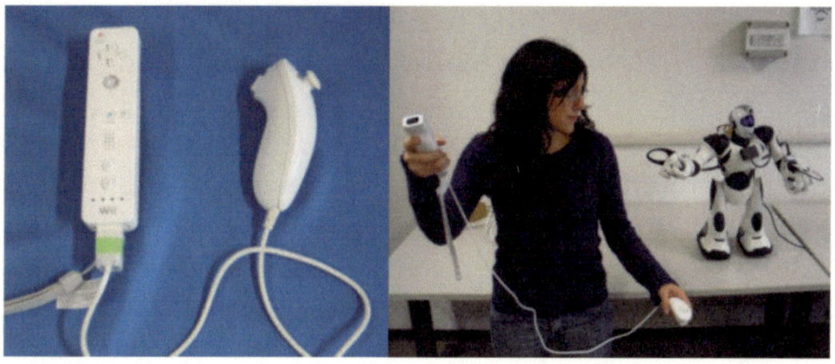

Figure 18: Nintendo Wii controllers (left image) and imitation experiment (right image).

7. Conclusion

In this chapter, a robot programming by demonstration system oriented to imitation of human gestures for a humanoid robot has been presented, along with a path specification system based on imitation of human walking paths. The experimental system investigated comprises a Robosapien V2 humanoid and includes multiple sensor devices such as an electromagnetic tracker and a monocular vision system. The novelty of the approach is the investigation of a high level programming paradigm such as imitation with a low cost humanoid toy robot. Indeed, the ability to teach elementary motions is a prerequisite step toward programming more structured tasks. Vision guided trajectory imitation and dynamic obstacle avoidance have also been successfully experimented, along with autonomous grasping skills. The experiments show that the ability to teach motion paths enables a toy robot to achieve rather complex navigation tasks. The proposed imitation approach is quite general, even though its implementation is constrained by some limitations of RSV2 and by sensor inaccuracies. The usable workspace is currently restricted to a square of approximately $2m^2$ due to the range resolution of the tracking device. Moreover, the tilting motion of the upper body while the robot is walking prevents the use of a continuous visual feedback. Therefore the vision system can be exploited only when the robot is not moving. Finally, the non ideal alignment between the head and the body of the robot when it stops strongly affects the estimation of landmarks pose used for visual guidance. The kinematics limitations of RSV2 also affect the imitation accuracy. Such limitations include the difficulty of performing sharp turns, the absence of side steps commands and the simplified grasping abilities. Nonetheless, we believe that low-cost humanoid platforms such as RSV2 provide an exciting and affordable opportunity for research in humanoid programming based on imitation.

8. Acknowledgment

This research is partially supported by Laboratory LARER of Regione Emilia-Romagna, Italy.

9. References

A. Arakawa, M. Hiyama, T. Emura, and Y Kagami. Trajectory Generation for Wheeled Mobile Robot Based on Landmarks. In Proceedings of the International Conference on Systems, Man and Cybernetics, Vancouver, Canada, Oct. 1995.

J.P. Bandera, L. Molina-Tanco, R. Marfil, and F. Sandoval. A model-based humanoid perception system for real-time human motion imitation. In Proceedings of the IEEE Conference on Robotics, Automation and Mechatronics, volume 1, pages 324–329, 1-3 Dec. 2004.

S. Behnke, J. Müller, and M. Schreiber. Playing soccer with robosapien. In Proceedings of The 9th RoboCup International Symposium, Osaka, Japan, Jul. 2005.

A. Billard. Robota: Clever toy and educational tool. Robotics and Autonomous Systems, 42(3-4):259–269, 2003.

C. Breazeal and B. Scassellati. Challenges in building robots that imitate people. In K. Dautenhahn and C. Nehaniv, editors, Imitation in Animals and Artifacts. MIT Press, 2001.

S. Calinon, F. Guenter, and A. Billard. Goal-directed imitation in a humanoid robot. In Proceedings of the IEEE Int'l Conference on Robotics and Automation (ICRA), Barcelona, Spain, Apr. 18-22, 2005.

Hon Nin Chow, Yangsheng Xu, and S.K. Tso. Learning human navigational skill for smart wheelchair. In Proceedings of the IEEE/RSJ Int'l Conference on Intelligent Robots and Systems, (IROS), Sept. 2002.

K. R. Dixon and P. K. Khosla. Learning by Observation with Mobile Robots: A Computational Approach. In Proceedings of the IEEE Int'l Conference on Robotics and Automation (ICRA), New Orleans, LA, Apr., 2004.

J. H. Hwang, R. C. Arkin, and D. S. Kwon. Mobile robots at your fingertip: Bezier curve on-line trajectory generation for supervisory control. In Proceedings of the IEEE/RSJ Int'l Conference on Intelligent Robots and Systems, (IROS), Las Vegas, Nevada, Oct., 2003.

K. Ikeuchi and T. Suehiro. Toward an assembly plan from observation, Part I: Task recognition with polyhedral objects. 10(3):368–385, 1994.

T. Inamura, N. Kojo, T. Sonoda, K. Sakamoto, K. Okada, and M. Inaba. Intent imitation using wearable motion capturing system with on-line teaching of task attention. In Proceedings of the IEEE-RAS Int'l Conference on Humanoid Robots, Tsukuba, Japan, Dec., 2005.

T. Inamura, Y. Nakamura, H. Ezaki, and I. Toshima. Imitation and primitive symbol acquisition of humanoids by the integrated mimesis loop. In Proceedings of the IEEE Int'l Conference on Robotics and Automation (ICRA), Seoul, Korea, May 2001.

M. Ito and J. Tani. On-line imitative interaction with a humanoid robot using a mirror neuron model. In Proceedings of the IEEE Int'l Conference on Robotics and Automation (ICRA), New Orleans, LA, Apr., 2004.

M. J. Mataric. Getting humanoids to move and imitate. IEEE Intelligent Systems, 15(4):18–24, 2000.

S. Kagami, K. Nishiwaki, J.J. Kuffner, K. Okada, M. Inaba, and H. Inoue. Vision-Based 2.5D Terrain Modeling for Humanoid Locomotion. In Proceedings of the IEEE Int'l Conference on Robotics and Automation (ICRA), Taipei, Taiwan, Sept., 2003.

H. Kato and M. Billinghurst. Marker Tracking and HMD Calibration for a video-based Augmented Reality Conferencing System. In Proceedings of the 2nd International Workshop on Augmented Reality (IWAR), Oct., 1999.

Y. Kuniyoshi, Y. Yorozu, M. Inaba, and H. Inoue. From visuo-motor self learning to early imitation a neural architecture for humanoid learning. In Proceedings of the IEEE Int'l Conference on Robotics and Automation (ICRA), Taipei, Taiwan, Sept. 2003.

D. Matsui, T. Minato, K. MacDorman, and H. Ishiguro. Generating natural motion in a android by mapping human motion. In IEEE/RSJ Int'l Conference on Intelligent Robots and Systems (IROS), Edmonton, Canada, Aug., 2005.

P. Menezes, F. Lerasle, J. Dias, and R. Chatila. A single camera motion capture system dedicated to gestures imitation. In Proceedings of the IEEE-RAS Int'l Conference on Humanoid Robots, Tsukuba, Japan, Dec., 2005.

P. Michel, J. Chestnutt, S. Kagami, K. Nishiwaki, J. Kuffner, and T. Kanade. Online Environment Reconstruction for Biped Navigation. In Proceedings of the IEEE Int'l Conference on Robotics and Automation (ICRA), Orlando, Florida, May, 2006.

P. Michel, J. Chestnutt, J. Kuffner, and T. Kanade. Vision-Guided Humanoid Footstep Planning for Dynamic Environments. In Proceedings of 5th IEEE-RAS International Conference on Humanoid Robots, 2005.

K. Morioka, Joo-Ho Lee, and H. Hashimoto. Human-Following Mobile Robot in a Distributed Intelligent Sensor Network. 51(1):229–237, 2004.

S. Nakaoka, A. Nakazawa, F. Kanehiro, K. Kaneko, M. Morisawa, and K. Ikeuchi. Task model of lower body motion for a biped humanoid robot to imitate human dances. In Proceedings of the IEEE/RSJ Int'l Conference on Intelligent Robots and Systems (IROS), Edmonton, Canada, Aug., 2005.

A. Nakazawa, S. Nakaoka, K. Ikeuchi, and K. Yokoi. Imitating human dance motions through motion structure analysis. In Proceedings of the IEEE/RSJ Int'l Conference on Intelligent Robots and Systems (IROS), Lausanne, Switzerland, Oct., 2002.

L. Piegl. On NURBS: A Survey. IEEE Computer Graphics and Applications, 11(1):55–71, Jan., 1991.

N.S. Pollard, J.K. Hodgins, M.J. Riley, and C.G. Atkeson. Adapting human motion for the control of a humanoid robot. In Proceedings of the IEEE Int'l Conference on Robotics and Automation (ICRA), Washington, DC, May, 2002.

M. Riley, A. Ude, K. Wade, and C.G. Atkeson. Enabling real-time full-body imitation: a natural way of transferring human movement to humanoids. In Proceedings of the IEEE Int'l Conference on Robotics and Automation (ICRA), Taipei, Taiwan, Sept., 2003.

S. Schaal. Is imitation learning the route to humanoid robots? Trends in Cognitive Sciences, 3(6):233–242, Jun., 1999.

A.P. Shon, K. Grochow, and R.P.N. Rao. Robotic imitation from human motion capture using gaussian processes. In Proceedings of the IEEE-RAS Int'l Conference on Humanoid Robots, Tsukuba, Japan, Dec., 2005.

L. Tang and S. Yuta. Vision Based Navigation for Mobile Robots in Indoor Environment. In Proceedings of the IEEE Int'l Conference on Robotics and Automation (ICRA), Seoul, Korea, May, 2001.

S. Wanitchaikit, P. Tangamchit, and T. Maneewarn. Self-Organizing Approach for Robots Behavior Imitation. In Proceedings of the IEEE Int'l Conference on Robotics and Automation (ICRA), Orlando, Florida, May, 2006.

R. Zöllner, O. Rogalla, R. Dillmann, and M. Zöllner. Understanding Users Intention: Programming Fine Manipulation Tasks by Demonstration. In Proceedings of the IEEE/RSJ Int'l Conference on Intelligent Robots and Systems, (IROS), Sept., 2002.

15

Psychological Evaluation for Rough Shape and Biped Walking of Humanoid Robots Using Virtual Reality

Kenji Inoue[1] and Tatsuo Arai[2]
[1]*Yamagata University,*
[2]*Osaka University,*
Japan

1. Introduction

Japan is becoming an aging society composed largely of elderly people, and the proportion of aged people in the population is increasing year by year. There is an increasing need for robots that can coexist with people and help them in their daily lives: housekeeping robots, nursing care robots, etc. Such robots are required to have "physical safety" and "mental safety". Physical safety means that robots do not injure humans. Mental safety means that humans do not feel fear or anxiety toward robots. In addition, it is important that humans do not have feelings of unpleasantness or aversion toward robots. Accordingly, when designing robots coexisting with people and planning their motions, it is necessary to consider the influences of the robots on human impressions and psychology. Mental safety has not yet been fully discussed. This is because the parameters of robots (shape, size, color, motion, speed, etc.) that may affect human psychology have not been clarified, and the method of measuring and evaluating human psychology for the robots has not been established.

There have been some researches on the evaluation of human psychology about robots and the interaction between robots and humans. Goetz et al. conducted questionnaire about matching between tasks and appearance of humanoid robots (Goetz et al., 2003). Robins et al. investigated the attitudes of autistic children for robots (Robins et al., 2004), and Woods et al. discussed the design of robots from children's viewpoint (Woods et al., 2004). Some psychological experiments for wheeled humanoid robot "Robovie" were conducted (Kubota et al., 2003; Kanda & Ishiguro, 2004; Sakamoto et al., 2004). Seal robot "Paro" was also psychologically evaluated worldwide (Shibata et al., 2003; Shibata et al., 2004). Kanda et al. investigated the impressions on real robots with different appearance by the semantic differential method when they did the same tasks (Kanda et al., 2005). These researches are significant because they evaluate the psychological suitability of the existent robots as partner robots or human-friendly robots. But they are not enough to analyze which parameters of robots bring desirable psychological effects.

For the purpose of clarifying the relationship between the parameters and their psychological effects, we have proposed the evaluation of human impressions and

psychology for robots coexisting with people using virtual reality (Nonaka et al., 2004; Inoue et al., 2005a; Inoue et al., 2005b; Ujiie et al., 2006; Inoue et al., 2007). CG robots are presented to human subjects using head-mounted displays or projection-based immersive 3D visualization system "CAVE" (Cruz-Neira et al., 1993), and the subjects and the robots coexist in the virtual world. The subjects answer questionnaire about their impressions or psychology for the robots and their motions; the questionnaire answers lead to the psychological evaluation. By using virtual reality, the parameters of robots can be easily changed and tested. It is also possible to experiment in various situations and environments.
In the present article, we describe the recent two subjective evaluations about humanoid robots using CAVE: human impressions for rough shape of humanoid robots (Inoue et al., 2007) and human impressions for the way of biped walking (Ujiie et al., 2006) of one type of humanoid robot. Based on these results, we comprehensively discuss the rough shape and the way of biped walking of humanoid robots which bring desirable psychological effects as service robots.

2. Psychological evaluation of robots using virtual reality

When designing robots coexisting with people and planning their motions, it is necessary to consider the influences of the robots on human impressions and psychology. The purpose of the psychological experiments is to analyze the relationship between robot parameters (shape, size, color, motion, speed, etc.) and human psychology: what kind of psychological effects the parameters have. If we can obtain such knowledge, we can determine the parameters so that they may bring desirable psychological effects. For this purpose, it is required to investigate and compare human psychological reactions to many kinds of robots and their various patterns of motion. But, because we aim at designing new robots, real robots do not exist. Hence, if we perform psychological experiments using real robots, we must make many new robots with different parameters and control them only for the experiments, not for practical use. Making and controlling real robots, however, requires much cost and time. In addition, human psychology for the robots may depend on the situation. But preparing various real environments for the tests is difficult, and some situations are dangerous (e.g., on a street).
For these reasons, we have proposed to evaluate human impressions and psychology for robots using virtual reality. Fig. 1 shows the experimental system using CAVE. CAVE is one of immersive visualization systems. It consists of four screens and a projector for each. The screens are placed on the front, left, right and floor, thus surrounding a subject. The size of this space is about 3[m]x3[m]x3[m]. The projectors project 3D computer graphics on the screens. The subject wears stereoscopic glasses and stands inside the CAVE to see a stereoscopic view. By measuring the pose of the subject's head, the graphics seen from the point of view of the subject is always projected. As a result, the subject feels like he or she is existing inside the virtual world. CAVE can give a higher realistic sensation to humans than head-mounted displays. As shown in Fig. 1, a 3D CG robot in real scale is presented to a human subject using CAVE. The subject answers the questionnaire about his or her impressions or psychology about the robot; that leads to the psychological evaluation.
Because we do not have to make real robots, the proposed method allows us to change the parameters of robots easily and to evaluate many kinds of robots and their different motions

in various situations. We do not have to measure the locations of the robots using some sensors or measurement systems. We can keep the same experimental conditions and display the same motions repeatedly; this is suitable for the experiments which require many subjects. Accordingly, this method is suitable to evaluate new robots and to analyze the psychological effects of the robot parameters. On the other hand, this system cannot deal with situations where the subject has physical contact with the robot: for example, a nursing-care robot. The movable area of the subject is limited inside CAVE.

As we mentioned, this method applies virtual reality to the problem where experiments using real objects are difficult to perform. But, in order to obtain similar results to human psychology in real world, experimental settings using virtual reality should resemble real case as much as possible. Especially when we evaluate robots which work around people, the robots should be displayed to subjects in correct size and 3-dimensionally, and the distance between the robots and subjects should be the same as real case. Accordingly this method uses CAVE as the virtual reality environment.

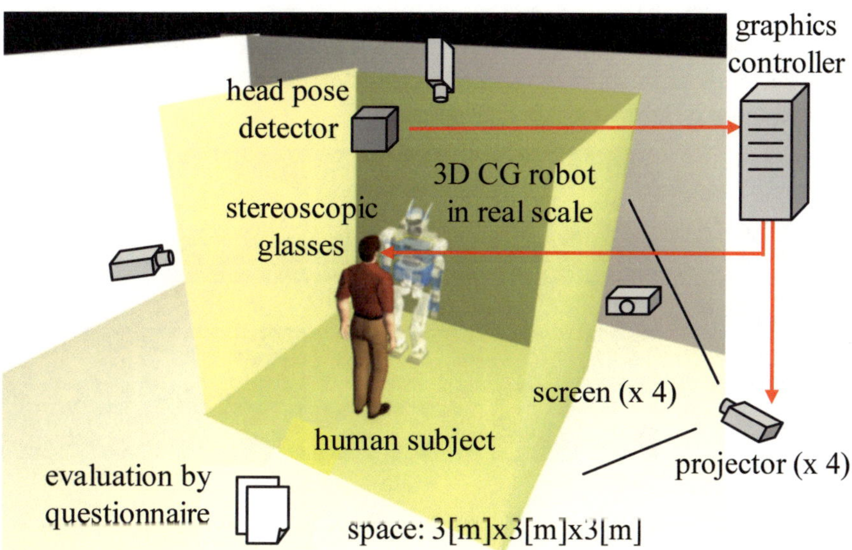

Fig. 1. Psychological evaluation of robots using CAVE

In the previous work (Nonaka et al., 2004), we evaluated holding-out motion of humanoid robots: a robot reaches out for the cup placed on a side table, and holds out it for a subject sitting in front of the robot (this experiment used a head-mounted display). Then we showed the effects of the rotation of the robot's head and body on the human psychology. In the previous work (Inoue et al., 2005a), we evaluated passing-by motion of humanoid robots: a robot approaches a subject from the front and passes near him or her in a corridor. Then we showed the effects of the head motion and walking speed on the human psychology; these could be a sign of indicating that the robot is aware of the subject. The work (Inoue et al., 2005b) compared virtual and real mobile manipulators.

Humanoid robots are candidates for robots coexisting with people because they have human-like structures. We describe the recent two experiments about humanoid robots.

3. Psychological evaluation for rough shape of humanoid robots

3.1 Robot model

Appearance is an important factor for impressions. Here we investigate human impressions for rough shape of humanoid robots.

The appearance of a humanoid robot is determined by a large number of parameters: height, width, ratio of links, ratio of link thickness, shape of links, color, material, and so on. But it is impossible to change all parameters. In this study we select a) thickness of head, b) thickness of body, and c) thickness of legs. The thickness of arms is changed together with the thickness of body. Thus the parameter b) includes the thickness of arms. We call these parameters a), b) and c) "rough shape parameters".

When we investigate psychological effects of rough shape parameters, it is better to reduce the influences of the other parameters on human psychology. From this reason, we make 3D CG robot models in real scale by combining simple solid elements: gray spheres and cylinders. We change the thickness of head, body and legs by changing the diameters of the spheres and cylinders. The purpose of this experiment is to find how human impressions change when the rough shape parameters change. It is not absolute impressions on the simplified robot models. Accordingly we define a standard robot and make other robots with different thickness of head, body and legs. Then we evaluate the impressions for these robots relative to the standard robot.

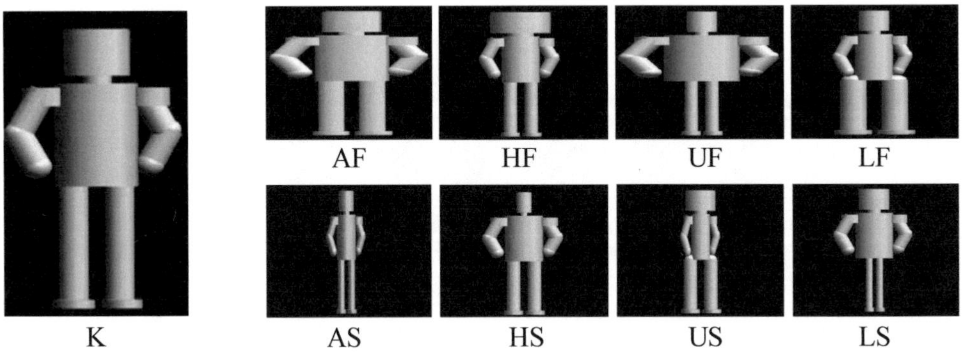

Fig. 2. Standard robot "K" and eight robots to be evaluated

robot	head	body	legs
K	standard	standard	standard
AF	**fat**	**fat**	**fat**
AS	*slim*	*slim*	*slim*
HF	**fat**	standard	standard
HS	*slim*	standard	standard
UF	standard	**fat**	standard
US	standard	*slim*	standard
LF	standard	standard	**fat**
LS	standard	standard	*slim*

Table 1. Rough shape parameters of eight robots

Fig. 2 shows the standard robot "K". Its height (1.54[mm]) and the ratio of the links are the same as those of existing humanoid robot "HRP-2"(Kaneko, 2004). Notice that the standard robot does not mean a typical humanoid robot. It defines standard for relative evaluation of impressions. We generate eight robots shown in Fig. 2 by doubling the thicknesses of head, body and legs of the standard robot or reducing them to half. All robots have the same height of the standard robot. In this figure, the first symbol represents the changed parameter: "A" = all parameters, "H" = head, "U" = body, "L" = legs. The second symbol represents thickness: "F" = fat (double), "S" = slim (half). Table 1 summarizes the rough shape parameters of these robots. Changing three parameters (head/body/legs) in three levels (standard/flat/slim) generates 27 robots in total. But evaluating all of them by psychological experiment requires too much time per subject; it is impossible because of subject's fatigue. Thus we select 8 robots to see the effects of each parameter and overall thickness.

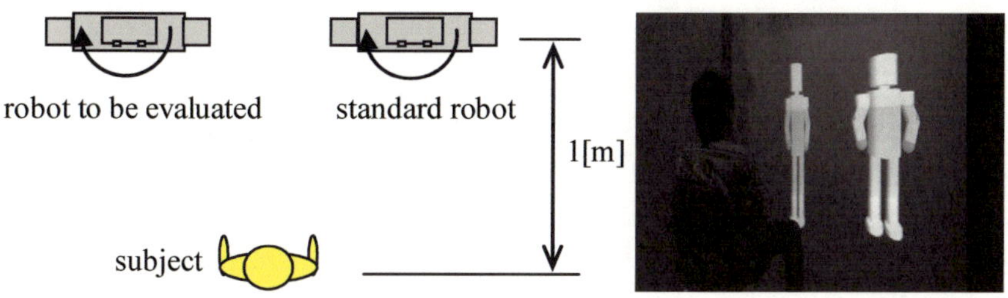

Fig. 3. Setting of experiment for evaluating rough shape

3.2 Experimental method

Fig. 3 shows a bird's-eye view of the setting of the virtual environment and a scene of the experiment using CAVE. Each of the eight robots and the standard robot stand upright and side by side. Each subject sits on a chair in front of the robots; the distance between the robots and the subject is 1[m]. The relationship between the height of the robots and the eye level of the subject may affect his or her impression. Sitting on the chair reduces the differences of the eye level among the subjects and makes all subjects look up at the robots. The robots are rotating so that the subject can see them from all angles. The standard robot is always displayed, and the eight robots are displayed one by one. All subjects see the eight robots. In order to cancel out the order factor, the eight robots are presented in randomized order for each subject. The subjects are 8 men and 17 women (total 25 subjects) between the ages of 19 and 66.

3.3 Psychological evaluation

After seeing each of the eight robots, the subject answers the questionnaire about his or her impression of the presented robot relative to the standard robot. The questionnaire by the semantic differential method consists of 38 adjective pairs, summarized in Table 2. We pick up some of these adjective pairs from those already used in the similar psychological experiments, and add the others which seem suitable to this experiment. The subject evaluates each adjective pair according to seven rating grades. In this table, positive adjectives are arranged on the left side. Some adjective pairs in the questionnaire sheet are swapped so that the pairs may be balanced. The sheet is written in Japanese.

Adjective pairs in Japanese		Factor 1	Factor 2	Factor 3
yasashii	kowai	-0.391	**0.678**	0.263
ningentekina	kikaitekina	-0.152	0.412	0.276
kouitekina	tekitaitekina	-0.354	**0.705**	0.240
karui	omoi	-0.526	0.148	**0.732**
chiseitekina	chiseinokaita	-0.119	0.382	**0.661**
shigekitekina	taikutsuna	0.598	0.061	0.102
yoi	warui	0.111	**0.665**	0.353
atarashii	furui	0.022	0.229	0.589
chiisai	ookii	-0.567	0.113	**0.683**
anshin-na	fuan-na	0.390	0.469	-0.234
hakkirishita	bonyarishita	0.531	-0.079	-0.096
yuukan-na	okubyouna	**0.774**	-0.111	-0.500
sensaina	soyana	-0.512	0.351	**0.653**
shitashimiyasui	shitashiminikui	-0.021	**0.754**	0.078
kimochinoyoi	kimocihnowarui	-0.058	**0.756**	0.226
youkina	inkina	0.517	0.329	-0.236
uchitoketa	katakurushii	0.144	**0.683**	0.195
chikazukiyasui	chikazukigatai	-0.076	**0.778**	0.211
yukaina	fuyukaina	0.149	**0.654**	0.029
sukina	kiraina	0.150	**0.824**	0.211
hayai	osoi	-0.081	0.109	**0.785**
subayai	noroi	-0.117	0.150	**0.759**
hadena	jimina	**0.608**	-0.141	0.020
binkan-na	donkan-na	-0.272	0.063	**0.787**
sekkyokutekina	shoukyokutekina	**0.801**	-0.061	-0.030
surudoi	nibui	-0.134	0.198	**0.776**
anzen-na	kiken-na	-0.150	**0.660**	-0.050
utsukushii	minikui	-0.020	0.567	0.462
omoiyarinoaru	wagamamana	-0.118	**0.604**	0.048
nigiyakana	sabishii	**0.664**	0.095	-0.219
tsuyoi	yowai	**0.718**	-0.129	-0.559
katudoutekina	fukappatuna	**0.711**	0.116	0.059
rippana	hinjakuna	**0.747**	0.014	-0.441
danseiteki	joseiteki	0.546	-0.274	-0.552
tanomoshii	tayorinai	**0.702**	0.008	-0.465
odayakana	hageshii	-0.541	0.517	0.228
shihaitekina	fukujuutekina	**0.678**	-0.403	-0.300
namerakana	gikochinai	-0.016	0.497	0.481

Table 2. Adjective pairs and factor loadings for rough shape experiment

3.4 Factor analysis

We quantify the results of the questionnaire in the range of 1 to 7 so that the positive adjective has higher score. Then we apply factor analysis. Factor analysis is a statistical data reduction technique used to explain variability among observed variables in terms of fewer unobserved variables called factors. The observed variables are modeled as linear combinations of the factors:

$$\boldsymbol{x} = \boldsymbol{A}\boldsymbol{f} + \boldsymbol{e} \quad (1)$$

$$\boldsymbol{x} = [x_1, x_2, \cdots, x_n]^T \quad (2)$$

$$\boldsymbol{f} = [f_1, f_2, \cdots, f_m]^T \quad (3)$$

$$\boldsymbol{e} = [e_1, e_2, \cdots, e_n]^T \quad (4)$$

$$\boldsymbol{A} = \begin{bmatrix} a_{11} & a_{12} & \cdots & a_{1m} \\ a_{21} & a_{22} & \cdots & a_{2m} \\ \vdots & \vdots & \ddots & \vdots \\ a_{n1} & a_{n2} & \cdots & a_{nm} \end{bmatrix} \quad (5)$$

where $x_i(i = 1,\cdots,n)$: observed variable, $f_j(j = 1,\cdots,m)$: common factor, $e_i(i = 1,\cdots,n)$: specific factor, $a_{ij}(i = 1,\cdots,n; j = 1,\cdots,m)$: factor loading. In this case, the observed variable is score of adjective pair, and n=38.

After extracting factors by the repetitive principal factor method, we determine that a three-factor solution, m=3, is suitable based on the following factors and the difference in eigenvalues. Table 2 summarizes the factor loadings of factor 1, 2, and 3 after normalized Varimax method. Focusing on the adjective pairs whose absolute value of factor loading is 0.60 or more, we interpret the meanings of the three factors.

(a) The adjective pairs which have higher factor loadings of factor 1 and lower factor loadings of the other factors are divided into two groups. The first group contains "rippana (grand)", "tanomoshii (dependable)", etc. This group is related to the impression of leaders. The second group contains "sekkyokutekina (aggressive)", "katsudoutekina (active)", etc. This group is related to the impression of active person. Hence we call factor 1 "leadership-activity" factor.

(b) Factor 2 has higher factor loadings for "sukina (favorite)", "chikazukiyasui (accessible)", "kimochinoyoi (amiable)", etc. Accordingly, we interpret factor 2 as a familiarity with or good feelings about the robot, and call factor 2 "friendliness" factor.

(c) The adjective pairs which have higher factor loadings of factor 3 and lower factor loadings of the other factors are "hayai (fast)", "subayai (quick)", "binkan-na (sensitive)", etc. These are related to the speed of the motion; we call factor 3 "quickness" factor.

3.5 Discussions about effects of rough shape on human impressions

Fig. 4 shows the mean value of the standard factor score for each of the eight robots with respect to each factor. The meanings of the symbols in the graphs are described in 3.1 and Table 1. We apply the T-test (5% level) to all combinations of these results.

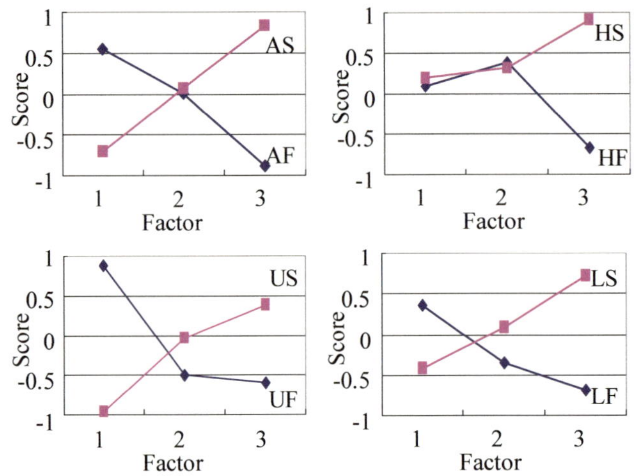

Fig. 4. Standard factor scores of eight robots

(a) With respect to the quickness factor (factor 3), the fat one (F) and the slim one (S) differ significantly for all parameters (A, H, U, L). Regardless of the parameters, the slim one (S) gives the impression of higher quickness than the standard robot "K", and the fat one (F) gives the impression of lower quickness than the standard robot "K". When the overall thickness of robot (A) is changed, the difference between the fat robot "AF" and the slim robot "AS" is largest.

(b) With respect to the leadership-activity factor (factor 1), changing the thickness of head (H) makes little difference. For other parameters (A, U, L), the fat one (F) and the slim one (S) differ significantly. The fat one (F) gives the impression of higher leadership-activity than the standard robot "K", and the slim one (S) gives the impression of lower leadership-activity than the standard robot "K". When the thickness of body (U) is changed, the difference between the fat body "UF" and the slim body "US" is largest. That is more effective than changing the overall thickness of robot (A).

(c) With respect to the friendliness factor (factor 2), the difference between the fat one (F) and the slim one (S) is smaller than the cases of the quickness and leadership-activity factors. Only for the legs (L), the fat legs "LF" and the slim legs "LS" differ significantly.

(d) When the overall thickness of robot (A) is changed, the factor scores of the friendliness nearly equal zero. It means that the impressions for the fat robot "AF" and the slim robot "AS" are almost same as the standard robot "K". This result suggests that keeping the ratio of the overall thickness gives little influences on the friendliness factor.

(e) The slim body "US" and the slim legs "LS" give almost the same impression of the friendliness as the standard robot "K". But the fat body "UF" and the fat legs "LF" give the impression of lower friendliness than the standard robot "K".

(f) Both the fat head "HF" and the slim head "HS" have greater factor scores of the friendliness than zero. They give the impression of higher friendliness than the standard robot "K". Fig. 5 shows the average rating grades of each robot on the adjective pairs related to the friendliness factor. The slim head "HS" gets high grade with respect to "yoi (good)", and the fat head "HF" gets high grade with respect to "yukaina (pleasant)". Accordingly the reason why the friendliness increases is different when the head is made fat (F) and slim (S).

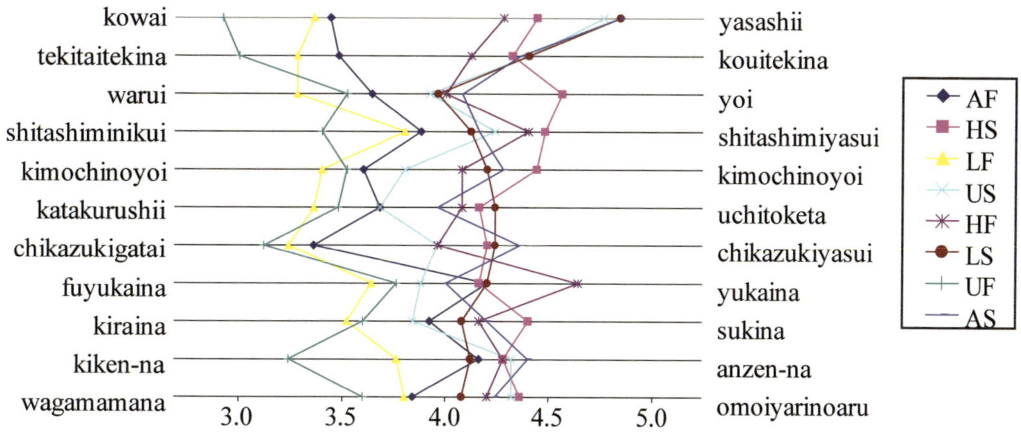

Fig. 5. Rating grades of eight robots on adjective pairs related to friendliness

4. Psychological evaluation for biped walking of humanoid robots

4.1 Purpose of experiment

When robots are introduced into human society in the future, the robots and humans will pass each other frequently. Hence, it is important that the humans do not feel uncomfortable or insecure around the moving robots. While the work (Inoue, 2005a) investigates the head motion and walking speed as a sign of awareness, this experiment evaluates the way of biped walking for its own sake. The walking motion of current humanoid robots is mainly intended for stable walking. But various walking motions will be possible for humanoid robots: walking with the knees bent, walking with the body swaying, walking with long strides, and so on. How do the differences in walking motions influence human impressions? If we can obtain some knowledge on this relationship, this knowledge will be useful for designing and planning suitable walking motions for the humanoid robots which will move around in our society.

It is not easy to develop a new method of biped walking and to make real humanoid robots walk stably. This is because the performances of current humanoid robots are limited and the effects of a disturbance (e.g., irregularity of the ground) must be compensated. But psychological experiments aim at investigating visual effects of walking motions on human psychology. For these reasons, we evaluate human impressions for biped walking of humanoid robots using virtual reality.

4.2 Experimental method

Fig. 6 shows the model of humanoid robot HRP-2 used in the experiment and a bird's-eye view of the settings. The robot is 1.54 [m] in height and 0.62 [m] in width. A 3D CG model of this robot in real scale is presented to subjects through CAVE. The width of the virtual corridors is 2.3 [m]. Each subject sits on a chair placed at the intersection of two corridors. The robot starts from the position of 1 [m] in front and 3 [m] on the left of the subject, and it cuts in front of him or her along the corridor. The distance between the subject and the robot's path is 1 [m]. The subject sees the walking robot from its side (not from the front) so that he or she can see the whole body motion of the robot. The walking speed is constant at 0.17 [m/s]. The sound of the robot's walking, which is a recording of a real HRP-2, is also presented.

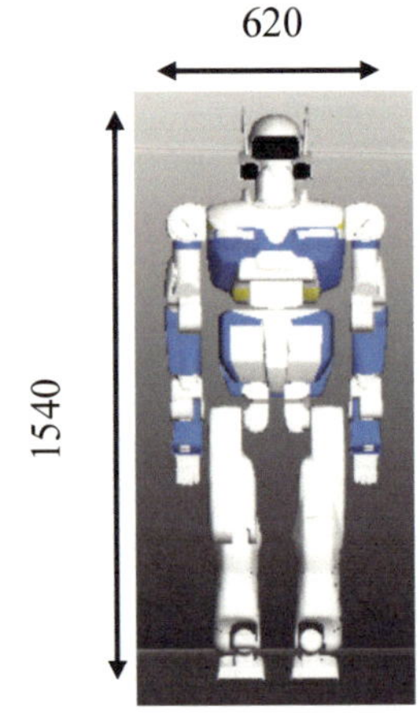

arm DOF:	6
hand DOF:	1
leg DOF:	6
body DOF:	2
neck DOF:	2
total DOF:	30

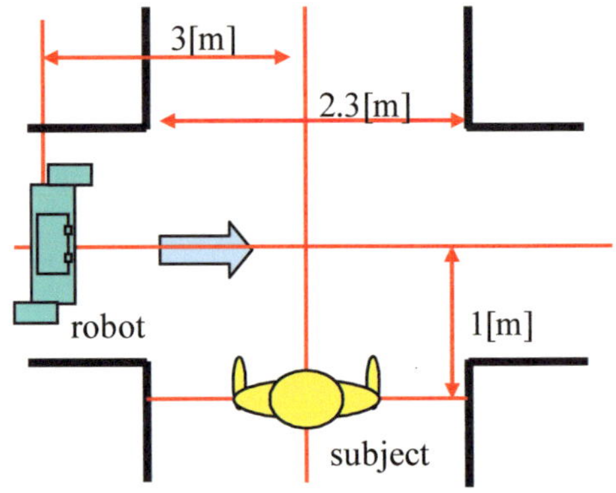

Fig. 6. Humanoid robot HRP-2 and setting of experiment for evaluating biped walking

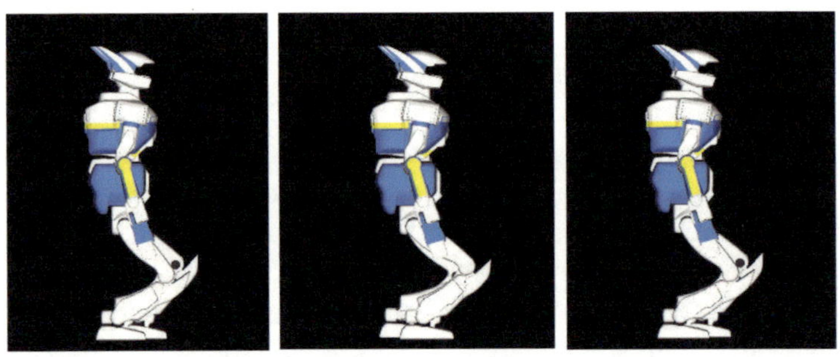

(a) Knees bent

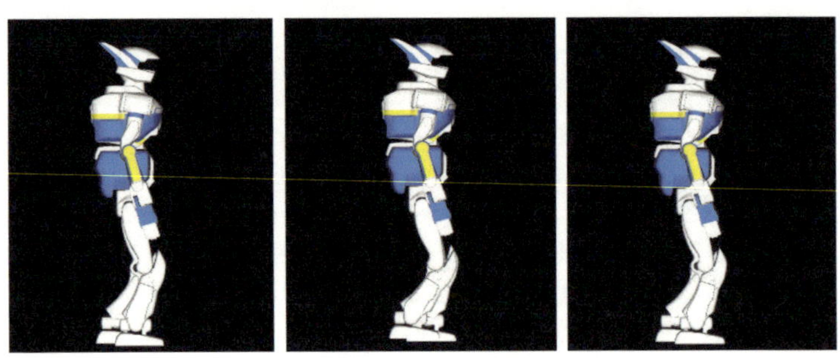

(b) Knee stretched

Fig. 7. Biped walking motions with knees bent or stretched

There are many parameters that define biped walking motion. In this experiment, the following two parameters, which are commonly seen in the walking of current humanoid robots, are selected: 1) the knees are bent or stretched, and 2) the body is side-swaying or in an upright posture. Fig. 7 illustrates the difference in (a) knees bent and (b) knees stretched. Fig. 8 illustrates the difference in (a) body side-swaying and (b) body in an upright posture. Combining them, we generate four walking motions. In all motions, the head and body are facing in the walking direction. All subjects see these four walking motions. In order to cancel out the order factor, these are presented to subjects in randomized order.

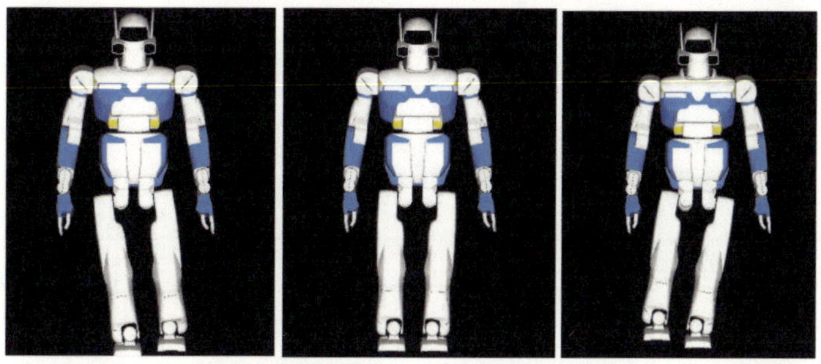

(a) Body side-swaying

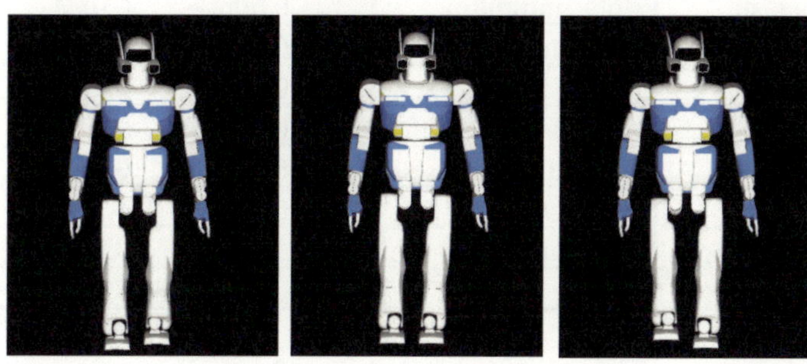

(b) Body in upright posture

Fig. 8. Biped walking motions with or without body side-swaying

4.3 Psychological evaluation

After seeing each motion, the subject answers the questionnaire about his or her impression on the robot's motion. The questionnaire by the semantic differential method consists of 29 adjective pairs, summarized in Table 3. Each adjective pair is evaluated according to five rating grades. In this table, positive adjectives are arranged on the left side. Some adjective pairs of the questionnaire sheet are swapped so that the pairs may be balanced. The sheet is written in Japanese.

The subjects are 34 men and 13 women between the ages of 14 and 32; they do not get a chance to see real humanoid robots.

4.4 Factor analysis

We quantify the results of the questionnaire in the range of 1 to 5: the positive adjective is 5, and the negative adjective is 1. Then we apply the factor analysis. After extracting factors by the repetitive principal factor method, we determine that a three-factor solution is suitable based on the following factors and the difference in eigenvalues. Table 3 summarizes the factor loadings of factor 1, 2, and 3 after Varimax normalized. Focusing on the adjective pairs whose absolute value of factor loading is 0.60 or more, we interpret the meanings of the three factors.

(a) Factor 1 has high factor loadings for many adjective pairs. Among them, the factor loadings of other factors for "shitashimiyasui (friendly)," "kawairashi-i (lovable)," and "yukaina (pleasant)" are small. Accordingly, we call factor 1 the "friendliness" factor.

(b) Factor 2 has high factor loadings for "hayai (fast)," "subayai (quick)," and "binkan-na (sensitive)," which concern the speed of the robot motion; we call factor 2 the "quickness" factor.

(c) Factor 3 has high factor loadings for "hageshi-i (intense)" and "hadena (showy)," which concern the activity of the robot motion; we call factor 3 the "activity" factor.

Adjective pairs (in Japanese)		Factor 1	Factor 2	Factor 3
Yasashi-i	Kowai	0.699	0.039	-0.299
Kanji no yoi	Kanji no warui	0.641	0.188	-0.165
Shitashimiyasui	Shitashiminikui	0.600	0.085	0.004
Anzen-na	Kiken-na	0.410	0.177	-0.447
Atatakai	Tsumetai	0.673	0.021	0.106
Kawairashi-i	Nikurashi-i	0.608	-0.014	-0.017
Uchitoketa	Katakurushi-i	0.648	0.052	0.173
Anshin-na	Fuan-na	0.436	0.279	-0.408
Chikazukiyasui	Chikazukigatai	0.714	0.215	-0.158
Akarui	Kurai	0.630	0.188	0.172
Omoiyari no aru	Wagamamana	0.455	0.025	-0.338
Ningentekina	Kikaitekina	0.436	0.348	0.046
Shizukana	Souzoushi-i	0.126	-0.145	0.724
Chu-uibukai	Fuchu-uina	0.589	0.119	0.277
Yukaina	Fuyukaina	0.716	0.065	-0.016
Sukina	Kiraina	0.675	0.090	-0.135
Kyoumibukai	Taikutsuna	0.380	0.094	0.151
Yoi	Warui	0.588	0.180	-0.147
Kouritsutekina	Muda no o-oi	0.026	0.490	-0.363
Hayai	Osoi	-0.066	0.735	0.183
Subayai	Noroi	-0.003	0.804	0.184
Hageshi-i	Odayakana	-0.292	0.065	0.697
Sekkyokutekina	Shoukyokutekina	0.264	0.374	0.542
Noudoutekina	Jyudoutekina	0.134	0.294	0.480
Hadena	Jimina	0.042	0.112	0.691
Karoyakana	Omo-omoshi-i	0.183	0.499	0.070
Binkan-na	Donkan-na	0.124	0.677	0.137
Kashikoi	Orokana	0.177	0.551	-0.216
Shizen-na	Fushizen-na	0.312	0.563	-0.122

Table 3. Adjective pairs and factor loadings for biped walking experiment

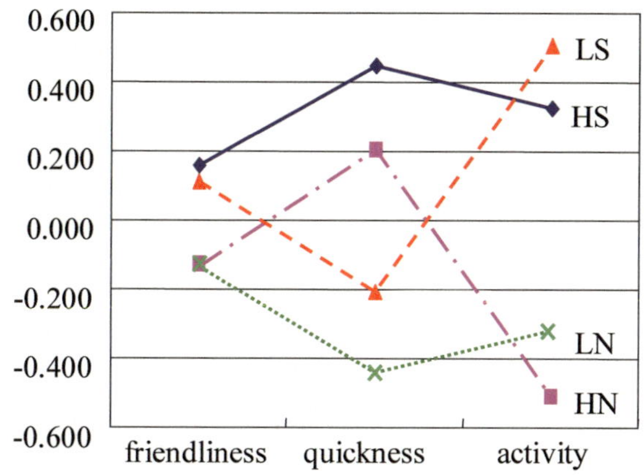

Fig. 9. Standard factor scores of four biped waking patterns

4.5 Discussions about effects of biped walking on human impressions

Fig. 9 shows the mean value of the standard factor score for each walking motion with respect to each factor. Here "L" means knees bent, "H" means knees stretched, "S" means body side-swaying, and "N" means body in an upright posture. Thus, the four walking motion patterns are expressed as "LS," "LN," "HS," and "HN."

It seems that the walking motions are divided into two groups with respect to each factor. We apply the T-test (5% level) to these results.

(a) With respect to the friendliness factor, the motions seem divided into the group (HS, LS) and group (HN, LN). This means that the friendliness factor depends on the body side-swaying. But, according to the results of the T-test, all pairs for the four motions are not significantly different.

(b) With respect to the quickness factor, the pairs HS-LS, HS-LN and HN-LN differ significantly; but the pair HN-LS is not significantly different. Hence, the motions are generally divided into the group (HS, HN) and group (LS, LN).This means that the quickness factor mainly depends on knee bending. Furthermore, walking with the knees stretched (H) has a higher quickness than walking with the knees bent (L).

(c) With respect to the activity factor, the pairs HS-LS, HS-LN, HN-LS and HN-LN differ significantly. Hence, the motions are divided into the group (HS, LS) and group (HN, LN). This means that the activity factor depends on body side-swaying. Walking with the body side-swaying (S) has a higher activity than walking in an upright posture (N).

Because significant differences cannot be seen directly in the friendliness factor, we calculate the mean value of the grade for each walking motion with respect to each of 10 adjective pairs related to the friendliness factor. The results are shown in Fig. 10, where the horizontal axis represents the mean value of the grade. With respect to "uchitoketa (relaxed)," the motion HN is different from the others. Thus, combining the knee stretched and the upright posture gives a relaxed impression. With respect to the pair "akarui (cheerful)," the motions seem divided into the group (HN, LN) and group (HS, LS). Hence, body side-swaying gives a more cheerful impression than does an upright posture. With respect to "kawairashi-i (lovable)," the motion LS is different from the others. Thus, combining the knees bent and body side-swaying gives a lovable impression.

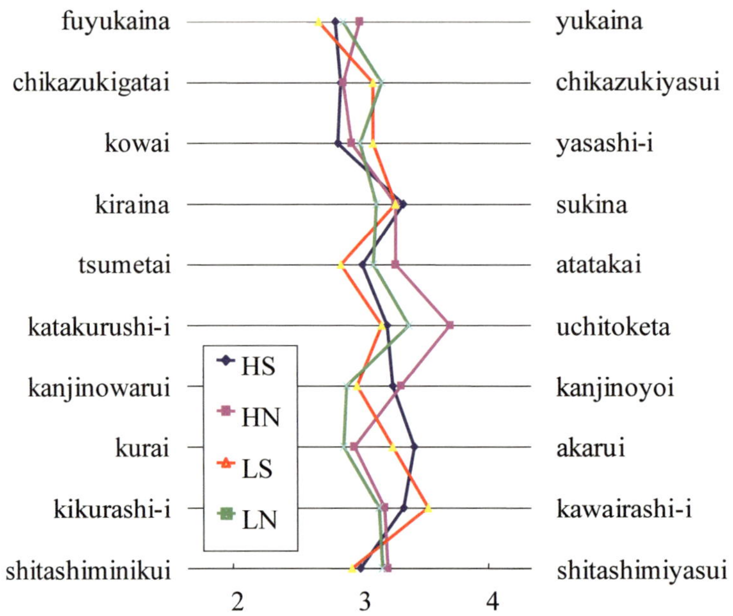

Fig. 10. Rating grades of four biped walking patterns on adjective pairs related to friendliness

5. Summary

Human impressions of humanoid robots are investigated using immersive 3D visualization system CAVE. In the first experiment, human impressions for rough shape parameters (thickness of head, body and legs) of humanoid robots are investigated. In the second experiment, human impressions for walking humanoid robots are evaluated, because robots and humans will pass each other frequently in future human-robot coexisting society.

Notice that factor analyses found the similar factors of the impressions for appearance (rough shape parameters) and motion (way of biped walking): friendliness, quickness and (leadership-)activity. Among them, the friendliness factor is an important property for robots coexisting with people. We obtain the following knowledge on the friendliness factor, which will be useful to design rough shape of humanoid robots as service robots and to generate their biped walking motion:

(a) Making the overall thickness of robot fat or slim (or keeping the ratio of the thickness of head, body and legs) gives little influences on the friendliness.

(b) Making the body or legs fat deceases the friendliness.

(c) Making the head fat or slim increases the friendliness. The reason why the friendliness increases is different for the fat head and slim head. The slim head gives good impression, and the fat head gives pleasant impression.

(d) Combining the knees stretched and the upright posture gives a relaxed impression, and combining the knees bent and the body side-swaying gives a lovable impression. The body side-swaying gives a more cheerful impression than does the upright posture.

These experiments limit the parameters of humanoid robots to two or three. The reason is to reduce the time of the experiment per subject and to make the difference between the robot shapes clearly understandable. In the future, we will continue with experiments on the influences of other parameters on human impressions.

6. References

Cruz-Neira, C., Sandin, D. J., DeFanti, T. A. (1993). Surround-screen Projection-based Virtual Reality: The Design and Implementation of the CAVE, Computer Graphics (Proc. SIGGRAPH93), pp. 135–142.

Goetz, J., Kiesler, S., Powers, A. (2003). Matching Robot Appearance and Behavior to Tasks to Improve Human-Robot Cooperation, Proc. RO-MAN2003, pp. 55-60.

Inoue, K., Nonaka, S., Ujiie, Y., Takubo, T., Arai, T. (2005a). Evaluation of Human Sense of Security for Coexisting Robots Using Virtual Reality (2nd Report: Evaluation of Humanoid Robots Passing by Humans in Corridors), Proc. of the 2nd International Conference on Systems and Human Science (CD-ROM).

Inoue, K., Ujiie, Y., Takubo, T., Arai, T. (2007). Psychological Evaluation for Rough Shape of Humanoid Robots Using Virtual Reality, Proc. ICAR2007

Inoue, K., Nonaka, S., Ujiie, Y., Takubo, T., Arai, T. (2005b). Comparison of Human Psychology for Real and Virtual Mobile Manipulators, Proc. RO-MAN 2005, pp. 73-78.

Kanda, T. & Ishiguro, H. (2004). Friendship Estimation Model for Social Robots to Understand Human Relationships, Proc. RO-MAN2004, pp. 539-544.

Kanda, T., Miyashita, T., Osada, T., Haikawa, Y., Ishiguro, H. (2005). Analysis of Humanoid Appearances in Human-Robot Interaction, Proc. IROS2005, pp. 62-69.

Kaneko, K., Kanehiro, F., Kajita, S., Hirukawa, H., Kawasaki, T., Hirata, M., Akachi, K., Isozumi, T. (2004). Humanoid Robot HRP-2, Proc. ICRA2004, pp. 1083-1090.

Kubota, N., Nojima, Y., Sulistijono, I. A., Kojima, F. (2003). Interactive Trajectory Generation using Evolutionary Programming for A Partner Robot, Proc. RO-MAN2003, pp. 335-340.

Nonaka, S., Inoue, K., Arai, T., Mae, Y. (2004). Evaluation of Human Sense of Security for Coexisting Robots using Virtual Reality (1st report: Evaluation of Pick and Place Motion of Humanoid Robots), Proc. ICRA2004, pp. 2770-2775.

Robins, B., Dautenhahn, K., Dubowski, J. (2004). Investigating Autistic Children's Attitudes Towards Strangers with the Theatrical Robot -A New Experimental Paradigm in Human-Robot Interaction Studies, Proc. RO-MAN2004, pp. 557-562.

Sakamoto, D., Kanda, T., Ono, T., Kamashima, M., Imai, M., Ishiguro, H. (2004). Cooperative Embodied Communication Emerged by Interactive Humanoid Robots, Proc. RO-MAN2004, pp. 443-448.

Shibata, T., Wada, K., Tanie, K. (2003). Subjective Evaluation of a Seal Robot at the National Museum of Science and Technology in Stockholm, Proc. RO-MAN2003, pp. 397-402.

Shibata, T., Wada, K., Tanie, K. (2004). Subjective Evaluation of Seal Robot in Brunei, Proc. RO-MAN2004, pp. 135-140.

Ujiie, Y., Inoue, K., Takubo, T., Arai, T. (2006). Evaluation of Human Sense of Security for Coexisting Robots Using Virtual Reality (3rd Report: Evaluation of Biped Walking of Humanoid Robot), Proc. of the 3rd International Symposium on Systems and Human Science, 2006.

Woods, S., Dautenhahn, K., Schulz, J. (2004). The Design Space of Robots: Investigating Children's View, Proc. RO-MAN2004, pp. 47-52.

16

A Novel Arm-Wrestling Robot Using Motion Dependant Force Control

Chul-Goo Kang
School of Mechanical Engineering, Konkuk University, Seoul 143-701, Korea

1. Introduction

In recent decades, aging population of most advanced countries has increased sharply, and elderly health care has become economically a big social issue. One way to cope with this problem is to introduce various service robots such as an arm-wrestling robot proposed in the article into our daily lives, which are able to promote elderly health and to save elderly welfare cost.

For successful execution of many tasks of service robots (Engelberger, 1989) such as robotic arm-wrestling, control of interacting force between a robot and the environment is crucial. Several interacting force control strategies have been developed during the past two decades (Whitney, 1987; Gorinevsky et. al., 1996; Siciliano & Villani, 1999; Natale, 2003), which can be classified into two groups; indirect force control without explicit closure of a force feedback loop, and direct force control with explicit closure of a force feedback loop (Natale, 2003).

Indirect force control includes compliance (or stiffness) control (Salisbury & Craig, 1980; Mason, 1981) and impedance control (Hogan, 1985; Anderson & Spong, 1988), in which the position error is related to the contact force through a mechanical compliance or mechanical impedance between the robot and the environment. On the other hand, direct force control includes hybrid position/force control (Raibert & Craig, 1981) and inner/outer motion/force control (Caccavale et al., 2005), in which actual force is measured using a force sensor and then is controlled via direct force feedback. If a detailed model of the environment is available, the hybrid position/force control can control position along a constrained direction and force along an unconstrained direction using a selection matrix. If a detailed model of the environment is not available, inner/outer motion/force control can control force and motion, in which outer force control loop is closed around the inner motion control loop.

These force control strategies generally assume that the environment is fixed in space. However, there are applications that the environment being contacted by the robot is no longer fixed in space, but is moving. In this case, contacting force control is much more challenging problem compared to the fixed environment case. For example, in robotic arm-wrestling considered in this article, the environment should be modeled as a moving object with compliance, viscous friction and inertia.

In the reference (Gorinevsky et al., 1996), the problem maintaining contact force with a moving object was considered, and the force regulation control of manipulator motion was treated as a motion control with imposed mechanical constraints. However, some application such as robotic arm-wrestling doesn't require motion control but pure force control, and require different force generation scenario according to the present position and velocity situation.

Recently, there have been other efforts in the emerging field of human-centered robotics, which focuses on safety-critical applications such as medical robots, service robots and entertainment robots. These human-centered robotic systems closely interacting between robots and humans require an additional issue, dependability in addition to conventional force control performance (Zinn et al., 2004; Bicchi & Tonietti, 2004).

In this article, we propose a force control logic for the moving environment (i.e., moving object), called motion-dependent force control. Position and velocity information of the moving object as well as interacting force information are fed back and affect actual force generation in the proposed control logic. Differently from the hybrid force control, force, position and velocity are measured and driven at the same direction. The proposed force control logic is applied to an arm-wrestling robot recently developed in our laboratory. Also, we present a user safety issue at the arm-wrestling robot.

Section 2 introduces an arm-wrestling robot developed in our laboratory. In section 3, the proposed force control logic is presented, and then application results of the proposed force control logic to the arm-wrestling robot are described in section 4. Discussions and conclusive remarks are in the final section 5.

2. A novel arm-wrestling robot

An arm-wrestling robot called Robo Armwrestler (Fig. 1) has been recently developed in our Intelligent Control and Robotics Laboratory, Konkuk University, in order for the elderly to help to keep their muscular strength, and to improve the quality of life of them physically and mentally. This project has been supported financially by the Korean government.

The Robo Armwrestler has salient features. Namely, it generates automatically the force level appropriate to each person after sensing the human arm-force at the early stage of the match, and therefore any person with either a strong or weak arm-force can enjoy arm wrestling without any parameter changes. Its generated force profile varies with each match, so one person can enjoy arm wrestling with the robot for a long time without becoming bored. The winning average of the robot is determined randomly at the start of the match, but the robot measures the person's will to win during the match, and the result influences the winning average of the robot. For example, the robot's arm-force becomes weaker and weaker, and the winning probability of the human increases if the human tries to stand out to win the match for a long time. The robot recognizes a human's approach and begins to talk, such as "Hello, welcome to the Robo Armwrestler! If you want to try arm wrestling, please sit down on the chair." and also automatically recognizes the human's sitting on the chair, and guides them on how to play and enjoy arm wrestling. The facial expression of the avatar changes synchronously according to arm wrestling situations.

The arm-wrestling robot is composed of an arm-force generation mechanism, a force control system, and an intelligent game scenario generation engine.

The arm-force generation mechanism comprises a servo motor, a position/velocity sensor, a speed reducer, a torque sensor, three inclinometers, a mechanical arm, and an adapter with a mechanical stopper. The electric servo motor provides necessary torque according to motor control input signals, and the position/velocity sensor detects angular position and angular velocity of the motor. An incremental encoder is selected as the position/velocity sensor for high resolution. The speed reducer decreases the speed and increases the torque of the motor. A harmonic drive is selected as the speed reducer instead of conventional gears that have large backlash, and thus cause trouble in torque control performance. The torque sensor detects the torque acting on the mechanical arm, which is installed between the speed reducer and the mechanical arm. The torque sensor should have reasonable resolution in order to get a reasonable force control performance. The adapter with a mechanical stopper is utilized to restrict the angle range of motion of the mechanical arm in order to guarantee safety of the user. The adaptor is further utilized to set the initial absolute angle of the mechanical arm by means of low speed control of the motor. The initial setting of the absolute angle of the arm can also be achieved redundantly by means of using multiple inclinometers (Kang et. al., 2004).

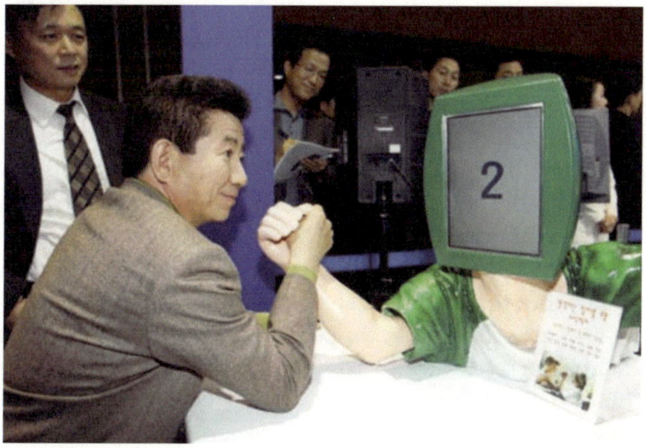

Fig. 1. The developed arm-wrestling robot. The President of Korea, Roh Moo-hyun arm-wrestled against Robo Armwrestler at the 2005 Future Tech Korea Exhibition (COEX in Seoul).

Two ultrasonic sensors are attached at the right and left sides on the front of the table, and detect a human's approach within a prescribed range of angles near the arm-wrestling robot. Ultrasonic sensors have an advantage of high noise immunity compared to other types of sensors, and can easily measure the distance of an approaching human under any circumstances. One photoelectric sensor using infrared rays detects a human sitting on a chair. In order to guide the player, the display monitor and speakers are prepared at an appropriate position of the table.

Control hardware is comprised of a CPU, a memory, an amplifier, a logic circuit, a pulse generation part, and output ports. The CPU produces motor control input using the control program and the feedback signals, and produces voice and image signals. The memory stores the control program including control logic and scenarios. The amplifier amplifies the low level voltage signal coming from the torque sensor, and achieves signal conditioning. The logic circuit conditions the feedback signal from the position/velocity sensor, and the pulse generation part produces a pulse signal for the ultrasonic sensors. Voice speakers and a display monitor are driven by the CPU through output ports.

Torque sensor, inclinometer, photoelectric sensor, and ultrasonic sensor signals are converted to digital signals through A/D converters, and transmitted to CPU. Encoder signals are transmitted to CPU directly through digital input channels. Motor driving signals are converted to analog voltage through a D/A converter, and transmitted to the motor driver.

When the CPU is down, the D/A converter can still output the last signal of the motor control input, and thus a dangerous situation can occur if the electric power is applied to the motor again. To resolve this problem, the CPU transmits an initialization completion signal to the motor power control part through a D/A converter, and sends 0 value to the motor driver through the D/A converter when the initialization procedure is completed (the initialization procedure starts when the main switch is pressed on). The motor power control part turns on the mechanical relay (MR) to supply the electric power to the motor according to the output signal of the sold state relay (SSR), which in turn is actuated by the initialization completion signal.

User safety is thus guaranteed even if the motor power switch is turned on before completing the initialization procedure or at abnormal conditions of the control system since the electric power is not transmitted to the motor.

When using the incremental encoder as the position/velocity sensor, we initially set the absolute zero degree angle of the mechanical arm using the mechanical stopper and velocity feedback control. More specifically, the control part slowly drives the motor clockwise or counterclockwise using position feedback control, and measures torque value of the torque sensor. If the measured torque is bigger than the threshold value, then the control part sets the present angular position as the specified degree of absolute angle, since the big measured torque implies that the mechanical stopper hit the stopper seat block.

The intelligent game scenario generation engine generates an appropriate force profile for each match, using a human's maximum force, a human's force pattern, time duration, a human's will to win, and randomness. The winning or losing is not predetermined, but varied online according to the pattern of the game progression.

For a few seconds in the early stage of the game, the engine detects the human's maximum force according to the procedure. The system increases force up to the specified value with a parabolic fashion for a short time, and then the robot measures arm velocity at regular intervals for the next few seconds. If the velocity is positive, then the robot increases the force until the velocity becomes negative, and memorizes the force value.

To realize unpredictable and intelligent game patterns, we adopt a random number and a value called *will point* that quantifies the will of the arm-wrestler to win the match. If the will point is near 100, the user is considered to have a strong desire to win the match. If the will point is near 0, the user is considered to have a weak desire to win the match. The will point is calculated by

$$\text{will point} = (\text{average arm-force during one sub-scenario})/(\text{maximum arm-force of the user}) \times 100 \quad (1)$$

The game scenario is composed of several sub-scenarios. If a sub-scenario is selected by means of random number, the robot decreases or increases the force during randomly determined intervals of time. During these intervals of time, human force is measured and averaged in order to calculate the will point. As soon as the execution of the sub-scenario is completed, the next sub-scenario is immediately generated online at that instant.

Arm wrestling progression is affected by the will point and the pre-specified probability. For example, if the obtained will point is 86, the class of win, draw, or defeat scenario is determined according to winning probability with 8 %, drawing probability with 90 %, and defeat probability with 2 %. This class determination is conducted using a random number 0 ~ 99, that is, the generated random numbers 98 and 99 imply the defeat class, random numbers 8 to 97 imply the drawing class, and random numbers 0 to 7 imply the winning class.

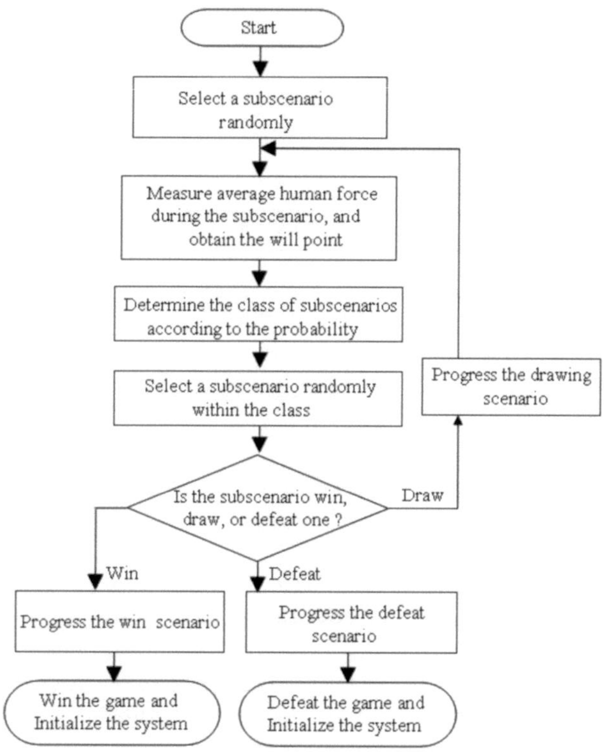

Fig. 2. Flowchart of the progression of robotic arm wrestling. The winning average is determined by random numbers and will points.

Using another random number, we select a sub-scenario randomly within sub-scenarios of the determined class. If the selected sub-scenario is a drawing one, then the will point is recalculated after the sub-scenario ends, and the above procedure is repeated. If the selected sub-scenario is a win or a defeat one, then the win or defeat sub-scenario is progressed, and the human wins or defeats, and the arm wrestling ends. Finally, the arm-wrestling system is initialized for the next match. Fig. 2 shows a simplified flowchart for the progression of robotic arm wrestling.

In an idle situation, the robot plays music, and waits for a person. As a person approaches the robot, the system automatically detects his approach, greets him, and encourages him to arm-wrestle. If the person sits down, the robot guides him to start the game

The actuation of the robotic arm-wrestling is composed of four steps. The first step is to initialize the arm-force generation mechanism and the control system, and to achieve the setting of an initial angle of the mechanical arm. The second step is to detect a human's approach to the arm-wrestling robot using ultrasonic sensors, and the human's sitting on the chair using a photoelectric sensor. The third step is to measure the maximum arm-force of

the user during a specified time interval based on the feedback signal coming from the torque sensor. The fourth step is to generate an intelligent arm-wrestling scenario, and to control the robot arm according to the proposed force control logic.
The winning sub-scenario implies a significant decrease of torque command value, the losing sub-scenario implies a significant increase of torque command value, and the drawing sub-scenario implies a small increase, a small decrease, or no change of torque command value.
The sub-scenario generated online is characterized by force increment, rising time, and maintaining time. These three values are determined using random numbers and the will point. The force increasing or decreasing in the sub-scenario is achieved by smooth polynomial curves.

3. Motion-dependant force control

For realizing robotic arm-wrestling, the arm-wrestling robot should increase arm-force rapidly in order to prevent from being lost easily when the human arm-wrestler tries to turn over the robot arm abruptly. For this purpose, the existing force control strategies such as compliance control, impedance control, hybrid position/force control, or inner/outer motion/force control, are not appropriate.
Conventional force control strategies for robot manipulators generally consider the situation that the environment is fixed in space as shown in Fig. 3(a). But there are situations that the environment being contacted by the robot is moving, and can be modeled as a moving object as shown in Fig. 3(b).
In the case of the moving environment, the motion of the moving object affects the contact force between the robot and the environment, and thus it should be considered quantitatively for the design of force control system. There could be two different situations. One situation is to regulate the contacting force constantly even if the environmental object is moving, and the other situation is to generate and control different interaction forces depending on current motions (that is, positions and velocities).

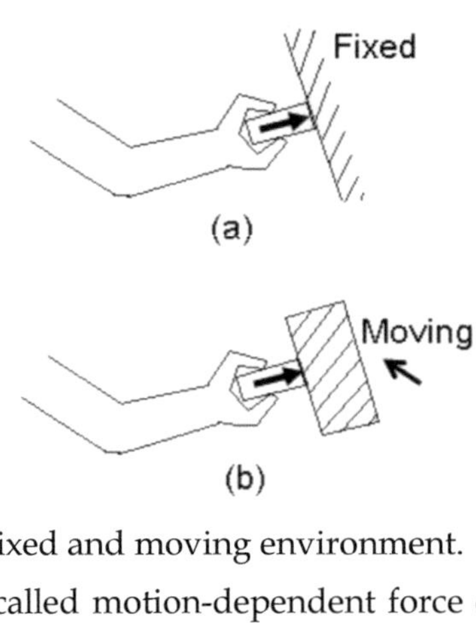

Fig. 3. Schematic models for fixed and moving environment.

We propose a control logic, called motion-dependent force control, for the latter situation. Being different from stiffness control, impedance control, hybrid position/force control, and inner/outer motion/force control, the motion-dependent force control logic does not possess explicit position or velocity feedback loops and does not control the position or

velocity of the robot arm directly. Instead, the measured position and velocity information are used only for generation of force commands of the force control loop.
The block diagram of the proposed force control scheme is shown in Fig. 4. The control system receives feedback signals of actual position, velocity, and force, and calculates force command using feedback signals and other operation requirements. The system then controls the mechanical arm by means of generating the motor control input according to the proposed force control law. The force control law in the block diagram could be a PID type control law or other advanced control laws.

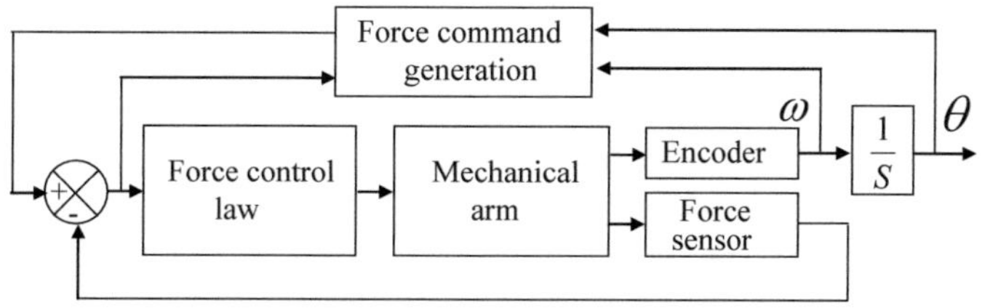

Fig. 4. Simplified block diagram of the motion-dependent force control system.

Force command $\tau_{command}$ is generated by the following formula in the proposed force control logic.

$$\begin{aligned} \tau_{command} &= \tau_{basic\_operation} + \Delta\tau \\ \Delta\tau &= \Delta\tau_v + \Delta\tau_p \\ \Delta\tau_v &= K\omega \end{aligned} \tag{2}$$

where $\tau_{basic\_operation}$ is a part of the force command demanded from the basic operation requirement, for example, game scenario in the robotic arm-wrestling, and $\Delta\tau$ is an added term demanded from the current motion. $\Delta\tau$ is composed of a correction term $\Delta\tau_v$ affected by the current velocity of the robot arm, and a correction term $\Delta\tau_p$ affected by the current position of it. In robotic arm-wrestling, for example, $\Delta\tau_v = K\omega$, and the value K is given from the gain scheduling table according to the actual angular velocity ω of the robot arm. $\Delta\tau_p$ affects $\tau_{command}$ to be zero whenever the robot arm reaches the limit angles set up initially in either direction.
In the arm-wrestling robot considered in the article, $\tau_{basic\_operation}$ is given by the intelligent game scenario described in the previous section. More specifically, $\tau_{basic\_operation}$ is determined on-line by the random combination of force magnitude, force-sustaining time, human's reaction pattern after detecting human's maximum force, and then by smoothing stepwise force command generated. One sample force command generated in this procedure is shown in Fig. 5.

4. Experimental results

To order to demonstrate the validity of the proposed force control logic, we applied it to the arm-wrestling robot and conducted force control experiments. Force control performance is mainly dependent on the accuracy of feedback signals, and real-time control capability, including the accuracy of sampling time, and the force feedback control logic itself.

Linux and RTAI (real-time kernel) have been adopted for the operating system of the arm-wrestling robot. When Linux and RTAI were implemented at desktop PC with Pentium IV 2.8GHz, timing errors less than ±0.02 ms occurred for generating 5 ms timer interrupts. However, Window XP had roughly ±1 ms latency when we executed the same program with the same PC platform.

Fig. 5 shows experimental results of the proposed motion-dependent force control when a 72-year-old woman played against the arm-wrestling robot. In the figures, blue solid lines imply torque commands, grey solid lines imply actual torques, red solid lines do angular velocities, and grey dotted lines do arm angles. From these graphs, we see that force command is generated intelligently using actual velocity and position information, and the generated scenario information. Furthermore, actual force reasonably tracks the force command by the proposed force control logic. Fig. 4 also shows that force pattern and elapsed time for each match are different from match to match even if the same person plays.

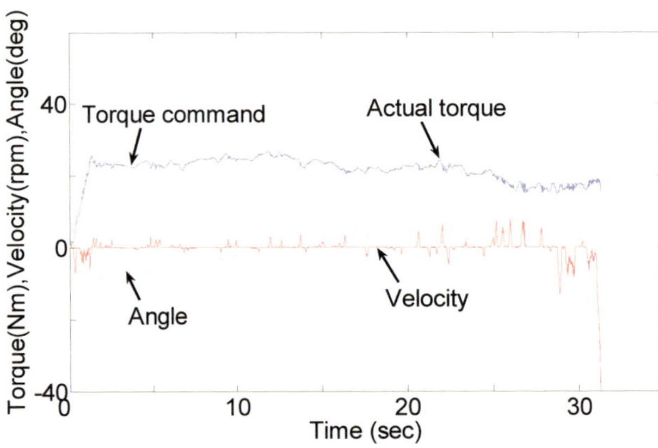

(a)

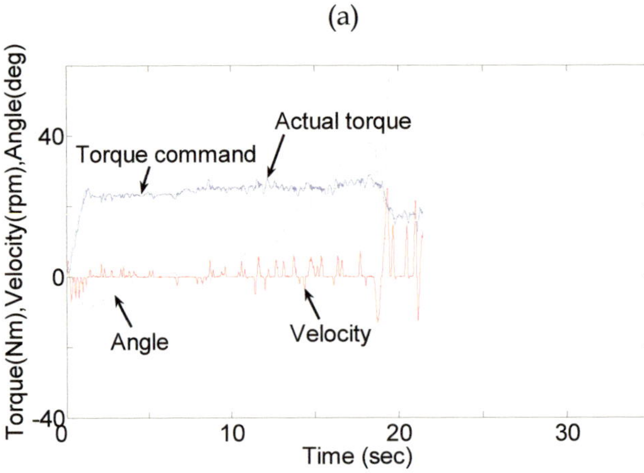

(b)

Fig. 5. Robotic arm-wrestling results of a 72-year-old woman. (a) a case where the woman won the game after the elapse of 31 seconds, (b) a case where the woman lost the game after the elapse of 21 seconds.

Fig. 6 shows a result of the match between a strong 25-year-old youth and the arm-wrestling robot, and Fig. 7 illustrates a match between a 10-year-old child and the arm-wrestling robot. As soon as the youth finished the match of Fig. 6, the child began the match shown in Fig. 7. Although the youth produced roughly 50 N·m, and the child roughly 20 N·m, the arm-wrestlings proceeded smoothly without changing any parameters of the robot.

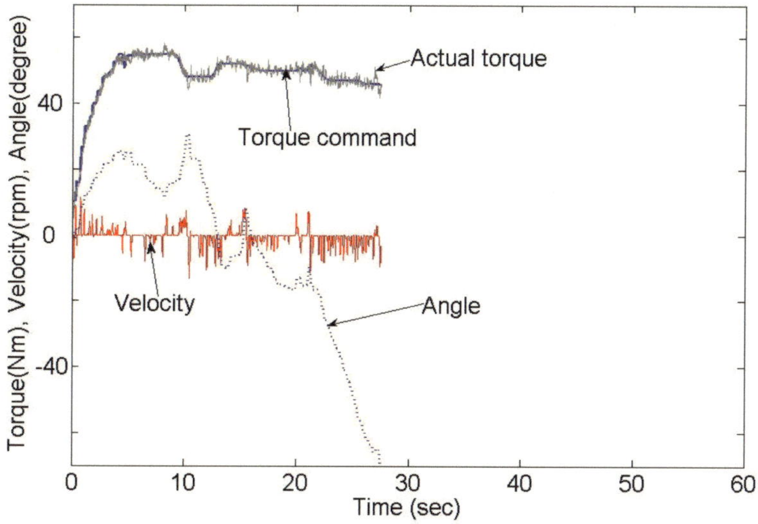

Fig. 6. Robotic arm-wrestling result of a 25-year-old youth who lost the game after the elapse of 28 seconds.

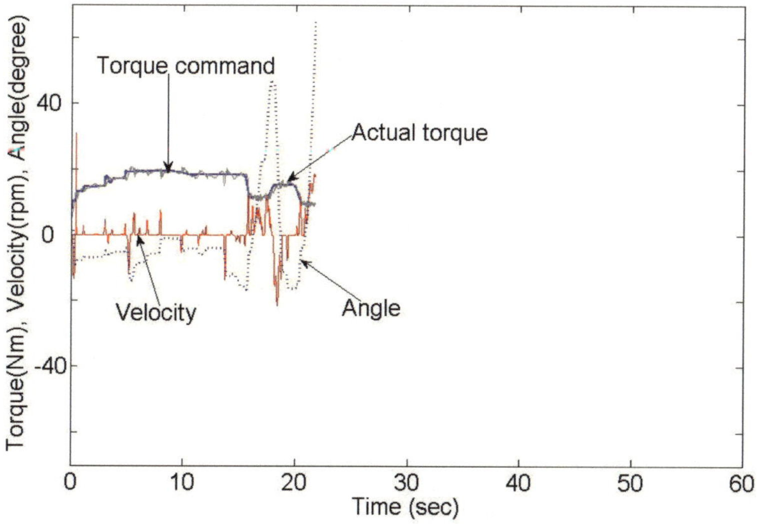

Fig. 7. Robotic arm-wrestling result of a 10-year-old boy who won the game after the elapse of 23 seconds.

Table 1 shows the elapsed time and winning/losing probability of each match when one user played arm wrestling 26 times with the arm-wrestling robot. From Table 1, we see that the elapsed time for a match varies each time, and the result of the match also varies each time, so that the enjoyment of the arm wrestling can be maintained for a long time without being bored. Table 2 summarizes results when two users played arm wrestling 26 times each with the Robo Armwrestler. Table 2 shows that the first user has 63 % of human's winning probability and the second user has 75 % of human's winning probability.
Operation experiences at the laboratory and Exhibitions have revealed that the proposed force control logic is appropriate for controlling contact force between the robot and the moving object, especially for controlling contact force between the arm-wrestling robot and the human arm-wrestler. The arm-wrestling robot with the proposed motion-dependent force control logic has generated intelligent scenarios unpredictably and reliably, and controlled the robot arm-force properly so that the human arm-wrestler feels as if he is arm-wrestling against a human. When the Robo Armwrestler was introduced to the public at the Future Tech Korea Exhibitions, it has been spotlighted in the mass media.

5. Conclusion

In this article, a novel arm-wrestling robot and a motion-dependent force control logic applicable to controlling contact force between a robot and a moving environment (i.e., moving object) have been proposed. In the motion-dependent force control logic, position and velocity information of the moving object as well as interacting force information are fed back and affect actual force generation and control. However, being different from stiffness control, impedance control, hybrid position/force control, and inner/outer motion/force control, the motion-dependent force control does not possess explicit position or velocity feedback loops, and does not control the position or velocity of the robot arm directly. Instead, the measured position and velocity information are used only for generation of force commands of the force control loop.
To demonstrate the validity of the proposed force control logic, we have applied it to the arm-wrestling robot recently developed in our laboratory, and showed the force control performance experimentally.
The experimental results and operational experiences at the Exhibitions have verified that the proposed force control logic is well suited for controlling contact force between the robot and the moving object, especially for controlling contact force between the arm-wrestling robot and the human arm-wrestler.
In this article, we have presented the system design, implementation, force feedback control, and generation of intelligent arm-wrestling scenarios of an arm-wrestling robot. Although the robot works as expected with the designed autonomy and reasonable control performance, we plan to further pursue research in order to build the arm-wrestling robot capable of recognizing facial expressions of the human using a webcam, and thus emotionally communicating with the human player. Moreover, we plan to add more degree-of-freedom for more human-like motion, and to eventually integrate arm-wrestling function into a humanoid robot.

Trial	Elapsed time (sec)	Result	Trial	Elapsed time (sec)	Result
1	11.0	Win	14	17.6	Lose
2	14.0	Win	15	14.5	Lose
3	20.0	Win	16	7.8	Lose
4	47.0	Lose	17	12.0	Lose
5	20.0	Win	18	11.5	Win
6	23.2	Win	19	23.4	Win
7	13.6	Win	20	8.4	Lose
8	11.7	Win	21	14.7	Win
9	14.5	Win	22	6.4	Win
10	22.2	Lose	23	18.8	Lose
11	12.1	Win	24	15.1	Lose
12	9.5	Win	25	13.6	Win
13	12.1	Lose	26	19.5	Win

Table 1. Elapsed time and match results when one user played 26 times with the arm-wrestling robot.

	Elapsed time average	Percentage of human wins
Person 1	17.0 sec	63 %
Person 2	14.0 sec	75 %

Table 2. Results when two users played 26 times each with the Robo Armwrestler

6. Acknowledgement

The author gratefully acknowledges the financial support from Seoul R&BD Program (e-Printing Cluster; Project Number: 10848). He also acknowledges Mr. Ki-Seon Ryu, Young-Wo Kim, Ik-Xu Son, Eun-Jun Park, Hae-Lyong Jung, Han-Seung Lee and Ho-Yeon Kim for their helps.

7. References

Anderson, R. & Spong, M.W. (1988). Hybrid impedance control of robot manipulators, *IEEE Trans. on Robotics and Automation*, Vol. 5, No. 6, pp. 723-739.

Bicchi, A. & Tonietti, G. (2004). Dealing with the safety-performance tradeoff in robot arms design and control: Fast and "soft-arm" tactics, *IEEE Robotics and Automation Magazine*, pp. 22-33, June.

Caccavale, F.; Natale, C.; Siciliano, B. & Villani, L. (2005). Integration for the next generation: Imbedding force control into industrial robots, *IEEE Robotics and Automation Magazine*, pp. 53-64, September.

Engelberger, J.F. (1989). *Robotics in Service*, ISBN 0-262-05042-0, MIT Press.

Gorinevsky, D.M.; Formalsky, A.M. & Schneider, A.Y. (1996). *Force Control of Robotics Systems*, CRC Press.

Hogan, N. (1985). Impedance control: An approach to manipulation, Part 1, II, and III, *ASME J. of Dynamic Systems, Measurement and Control*, Vol. 107, pp. 1-24.

Kang, C.-G.; Park, E.-J.; Son, I.-X.; Kim, Y.-W. & Ryu, K.-S. (2004). Conceptual Design for Robotic Arm Wrestling, *Proc. of IEEE Conf. on Robotics, Automation and Mechatronics*, pp. 1072-1076, Dec. 1 – 3, Traders Hotel, Singapore.

Mason, M.T. (1981). Compliance and force control for computer controlled manipulators, *IEEE Trans. on Systems, Man, and Cybernetics*, Vol. 14, pp. 418-432.

Natale, C. (2003). *Interaction Control of Robot Manipulators*, ISBN 3-540-00159-X, Springer.

Raibert, M.H. & Craig, J.J. (1981). Hybrid position/force control of manipulators, *ASME J. of Dynamic Systems, Measurement and Control*, Vol. 103, pp. 126-133.

Salisbury, J.K. & Craig, J.J. (1980). Active stiffness control of a manipulator in Cartesian coordinates, *Proc. 19th IEEE Conf. on Decision and Control*, pp. 95-100, Albuquerque, NM.

Siciliano, B. & Villani, L. (1999). *Robot Force Control*, ISBN 0-7923-7733-8, Kluwer Academic Publishers, Massachusetts 02061, USA.

Whitney, D.E. (1987). Historical perspective and state of the art in robot force control, *Int. J. of Robotics Research*, Vol. 6, No. 1, pp. 3-14.

Zinn, M.; Khatib, O.; Roth, B. & Salisbury, J.K. (2004). A new actuation concept for human-friendly robot design: Playing it safe, *IEEE Robotics and Automation Magazine*, pp. 12-21, June.

17

Sewer Robotics

Alireza Ahrary
Graduate School of Information, Production and Systems, Waseda University
Japan

1. Introduction

The usage of tethered, remotely controlled robot platforms in sewer pipes has become widely accepted. Research and development in engineering have moved the original application field of merely visual inspection to manipulative tasks such as the actual repair of pipe sections, installation of cable, e.g. for communication.

Despite all engineering advances in the mechanical design of the platforms and improvements of the video equipment, these efforts are directed to increase the quality of the data provided to the human operator. Yet the potential for automatic processing and data analysis in combination with hardware platform, as well as the application of IT technology for the development of an integrated sewer information system is neglected.

In this chapter, we being with the introduction of the legal framework together with the present inspection method, in which environment sewer inspection is embedded. Further, we will point out critical issues involved with the present inspection method.

Sewer systems are prone to damages due to aging, traffic, geological change, to name a few. Due to these damages, the groundwater is increasingly contaminated. Furthermore, heavy rainfall events may lead to inroad of the systems, resulting in overflow. In the case of separate sewer system as widely present in Japan (Table 1), this results in the undesired mixture of wastewater and rainwater. Thus, in order to ensure an optimal functioning sewer system, extensive inspection is necessary. Preceding repair is extensive qualified diagnostic measurement. Tethered mobile robots used for visual inspection in municipal and industrial sewer networks are one example. Beyond this classic task of application of the tethered robot as an extended tool of the human operator, new fields have emerged. Due to ongoing research and development of the mechanical design, tethered robots are now able to perform repair tasks with varying degree of complexity. Other emerging fields of application are the installation of communication cable in the sewer system as well as the encasement of inner pipe walls during renovation or repair.

(Most of the materials presented in this chapter have been selected from the material provided by "Steinbeis Japan Inc., Kitakyushu Foundation for the Advancement on Industry, Science and Technology," March 2002.)

separate sewer (sanitary sewer)	separate sewer (rainwater)	combined sewer
73%	5%	22%

Table 1. Division of sewer system in Japan

In Japan, matters regarding sewer pipes are regulated by the Sewer Water Law. The municipal governments as owners of the facilities are responsible for installation and maintenance of the public sewer system. In practice, authorities such as the construction bureau will authorize local companies to conduct inspections and after deciding upon the necessity also outsource the repair.
Although inspection cycles are clearly defined and also governed in the sewer system book, in reality, inspections are done after the occurrence of noticeable damages. Due to limitations in budget and equipment, the given inspection intervals are not strictly adhered to. This can also be clearly seen by comparison of actual inspected distance, 77km (Table 2) with total length of applicable sewer pipes, 3,220,053m (Table 3), for the financial year 2000. This gives us a ration of 0.0017%.

Year	1996	1997	1998	1999	2000	2001	2002
Length (m)	46,374	45,103	59,786	52,922	53,956	53,019	77,711

Table 2. Inspected sewer length per year at Kitakyushu city, Japan

We have to distinguish between two different types of inspection. For sewer pipe systems with an inner diameter larger than 800mm, human inspection is possible and conducted. The above mentioned tethered robot inspection platforms are used for pipes with a diameter less than 800mm.

Diameter (mm)	Length (m)
0	18,692
150	210,424
200	2,074,502
250	329,528
300	144,505
350	81,914
380	18,099
400	64,869
450	61,921
480	102
500	65,141
600	85,373
700	64,983
Sum (m)	3,220,053

Table 3. Present condition of sewer pipe length (diameter < 800mm) at Kitakyushu city, Japan (financial year 2000).

Both inspection procedures common is the acquisition of damages by the human inspector, either on loci or by means of video equipment. Based on the above information, manual inspection is done at maximum 77km/year which would take more than 40 years to complete all of the 3220km of pipes in Kitakyushu. In additional, reliability of the current

inspection method is depends on experience of an operator and it is also prone to human error. Accordingly, an effective autonomous robot capable of online identification and extraction of objects of interest such as cracks from sensory signals is of immediate concern.
To realize an autonomous system with capability of the pipe inspection, several companies and institutes in the world put efforts and researches at building of the robots with the different qualitative degrees of autonomy. In general the degree of autonomy in developed inspection pipe robots is classified as follow:

- **No autonomy:** The robot is completely tele-operated, usually via a tether cable, by a human operator. The pipe condition is assessed by the human operator who watches the sensor data (usually video) as the robot drives through the pipe. Mostly all the commercial sewer inspection robots are not autonomous system.
- **Semi-autonomy:** The tethered robot (in a few cases un-tethered) is partially controlled by automatic control programs or modules, or the assessment of the pipe condition is partially performed by sensor data interpretation programs.
- **Full autonomy:** The un-tethered robot carries all required resources onboard. Navigation is performed completely by control programs running on onboard computing equipment. Status messages may be communicated to a human inspector over a radio link. Assessment of the pipe condition may be performed partially onboard, or offline after retrieval of the recorded sensory data.

In general, we observe that fully autonomous robots are not yet marketable systems today. There are two main reasons for this:

- The degree of development of complete autonomous sewer robots does presently not warrant to use them safely and robustly in sewers. As mentioned in above, most of these robots have a complex moving mechanism and multi-sensor equipment for navigation and motion control. These complexities in mechanism and data processing make not easy to realize reliable commercial products especially for small range of the pipes up to 300mm in diameter.
- There seems to be some general skepticism about using fully autonomous systems without a possibility to interfere the robot control at any time, as would be the case for a tether-less sewer robot out of the range of some radio link.

However, semi-autonomy does make sense, implementing in a tele-operated sewer robot separate modules (like special sensors with data interpretation software, or special motion control modules) that operate autonomously within the overall control of a human operator. The purpose of such embedded autonomous modules would be to leverage the work load of the human operator in controlling the robot and interpreting sensor data, or to make certain measurements like in sewer inspection automatic, and therefore operational and reproducible. This is a serious issue in ensuring the quality of sewer maintenance, given that the pressure to provide service as cheap as possible imperils the quality of the inspection results.

In consequence, it makes sense in a mid-term market-oriented R&D strategy in sewer robotics to consider even those autonomous systems or components that do not seem to be sufficiently far developed for the market now, but may include separate or separable components that may have a mid-term market impact. The systems and approaches reviewed in the following are to be understood in that sense. On the long run, fully autonomous systems may become a reality in sewer maintenance, but it seems difficult now to predict a time line for this to happen.

2. Conventional sewer robots

2.1 PIRAT

PIRAT (**P**ipe **I**nspection **R**eal-Time **A**ssessment **T**echnique) is a semi autonomous tethered robot for the quantitative and automatic assessment of sewer condition (Kirkham et al., 1999). The PIRAT sewer inspection system has been developed between 1993 and 1996 within a joint project of the Manufacturing Systems and Automation Division of Australian CSIRO Manufacturing Science and Technology and Melbourne Water. Just like a conventional sewer inspection device, PIRAT (Fig. 1) is deployed to a sewer, and tele-operated from a surveillance unit via a cable by a human operator. The maximum cable length of 250m gives PIRAT a fair operating range. The added value of the PIRAT system is its ability to perform automatic evaluation of its sensory data.

PIRAT's innovative instrument system contains a video camera and a laser scanner. For flooded sewers, the latter can be substituted by a sonar scanner, but at the price of less resolution and inspection speed. In 600mm sewer pipes and at PIRAT's usual inspection speed, the laser scanner produced a resolution of about 1.5mm radially and 4mm axially and circumferentially.

The sensory data are evaluated by means of PIRAT's interpretation system, which is an expert system that runs on a Sun workstation in the mobile surveillance unit. From both types of scanner data, the interpretation system generated in a first step a three-dimensional model of the scanned sewer pipe section. In a second step, the interpretation system uses techniques from artificial intelligence to detect and classify damages on the basis of the three-dimensional model data. The result is a sewer condition assessment report that is readable for the human operator.

Fig. 1. The automated video inspection platform, PIRAT (http://vision.cmit.csiro.au/project/pirat/)

2.2 KARO

KARO (**KA**nal**RO**boter - German for sewer robot) is an experimental semi autonomous carrier for sewer inspection sensory equipment (Kuntze & Haffner, 1998), (Kuntze et al, 1995). It was developed by a group of research institutes and some industrial partners in Germany. The project was partly funded by the German Ministry for Research and Education (BMBF).

The monolithic KARO robot prototype (see Fig. 2) resembles much a standard non-autonomous sewer inspection robot, and it is tethered via a cable to a surveillance unit. Using inclinometers and an onboard control program, KARO is able to automatically correct for tilt in its pose and wheel slippage when driving inside a sewer pipe, thus freeing the human operator from this crucial control task.

The main innovation of KARO is its wide range of novel inspection and navigation sensors, namely, a microwave sensor and a 3D optical sensor, complementing the standard video camera and some ultrasound transducers.

The method for 3D measurement of hollow spaces using axial symmetric structured infrared light has been patented. It is applied for measuring pipe deformations, larger pipe cracks and obstacles inside the pipe, the latter being detected early by the US transducers. The microwave sensors are aimed at detecting leakages.

KARO has been continued as an internal research project within Fraunhofer IITB at least until 2000. Most recent research deals with fuzzy methods for data fusion of inspection sensors.

Fig. 2. The multi sensorial sewer inspection robot prototype, KARO

2.3 KURT

The development of the experimental sewer robot test platform KURT (**K**anal-**U**ntersuchungs-**R**oboter-**T**estplattform) started at the former GMD - now Fraunhofer - institute AIS in 1995 (Kirchner & Hertzberg, 1997). KURT is a six-wheeled autonomous untethered robot of approximate dimensions $30 \times 45 \times 30$cm.

KURT1 has been successfully employed for navigating autonomously in a dry sewer test net at the premises of the Fraunhofer campus in Sankt Augustin. To achieve this, the robot is provided with a map of the test net, representing the topology of the 80m of sewer pipes and the nine manholes in between, with a start position (number of a manhole) and a goal manhole.

The robot can determine the sequence of manholes or pipe junction types, respectively that it should pass on its path from start to goal. Since all pipe junctions inside the sewer test net are ground level connections, the robot is mechanically able to perform turns at such junctions. With its pivoted ultrasound sensor, KURT1 is able to classify the type of a pipe junction, i.e. whether it is L-shaped, X-shaped, or T-shaped. A special patented method for navigation under uncertainty enables KURT1 to detect and correct errors due to false classification or due to errors while performing turns at pipe junctions (Hertzberg & Kirchner, 1996). This work has been complemented by a method for probabilistic mapping of objects, like landmarks in the sewer (Schonherr et al, 1999).

Since its very first version, the KURT type robot platform has been further developed for indoor applications. The current version KURT2 can be equipped with a variety of sensors. The standard configuration includes inclinometers and sensors for odometry, either infrared or ultrasound distance transducers for obstacle detection, and optional bumpers. Fig. 3 shows a KURT2 system with a custom mounted 2D laser distance scanner and a notebook computer "Tough book" as CPU. Alternatively, an industry standard PC/104+ CPU can be provided.

Fig. 3. The commercial research robot platform, KURT2

2.4 MAKRO

MAKRO (**M**ehrsegmentiger **A**utonomer **K**anal**RO**boter/multi-segmented autonomous sewer robot) is the prototype of a fully autonomous, un-tethered, self-steering articulated robot platform (Fig. 4). It is designed for autonomous navigation in roughly cleaned sewer pipes within a diameter range of 300 to 600mm at dry weather conditions.

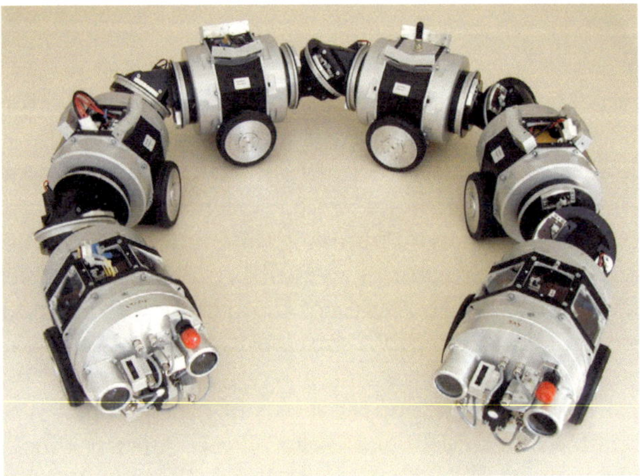

Fig. 4. Automated inspection platform, MAKRO

MAKRO's case design, consisting of six segments connected by five motor driven active joints, allows for simultaneously climbing a step and turning, e.g. at a junction consisting of a 600mm pipe and a branching 300mm pipe with equal top levels. MAKRO's autonomy and its kinematic abilities extend its potential mission range enormously, compared to conventional inspection equipment that is limited by the cable and poor kinematics (Rome et al, 1999).

MAKRO carries all the needed resources onboard. Standard NiCd batteries provide the power for its 21 motors, the sensors, and the electronics, including an industry standard

PC104 computer system and seven micro controllers (Kepplin et al, 1999), allowing for an autonomous uptime of about two hours.
The robot MAKRO has been developed by a German group of two research institutes and two industrial partners. The MAKRO project was funded partly by the German Ministry for Research and Education (BMBF) between 1997 and 2000. Since 2001, the MAKRO project is being continued as internal projects at Fraunhofer AIS and FZI (Scholl et al, 2000).

3. Sewer sensors

In this section, we review innovative sensors and sensor interpretation methods for sewer maintenance. The range of the depicted sensors ranges from widely applicable sensors such as compasses for navigation, to general usage sensor such as laser scanners up to especially for the application in sewer system developed ground penetrating sensors. Besides sensor systems, innovative methods of sewer maintenance, states of sewer pipes are of interest. The fusion of these innovative sensors and application methods will ensure that not only exhibited damage in the pipe in a narrow sense like a leak or crack or a blockage, but are potential heralds of a growing damage, like a bent or twisted pipe segment are successfully identified.

3.1 Laser scanners

Laser scanners are a technology that is in broad use in autonomous robots in general. Originally, they were developed as security sensors for surveying free space in production processes (such as protected areas around dangerous machines). For indoor applications, they are available in many different varieties at a highly developed technological level. The general principle is to measure by run-time of a laser beam the distance from a laser source to target objects. In scanners, the laser beam is deflected at different angles, measuring distances along a plane or half-plane, depending on the rotation and form of the deflecting mirror. The accuracy of the individual measurements is in the order of 1mm, depending on the measuring device and the distance.
The application principle for sewer inspection is to mount the scanner such that it scans radially the inner side of a pipe wall, see Fig. 5. If the inspection device moves along the pipe while measuring, the laser beam measures along a spiral on the pipe inside, where the resolution of the spiral depends on the turn rate of the reflection mirror and the speed of horizontal motion of the inspection device, typical turn rates being 1-2 turns per second.

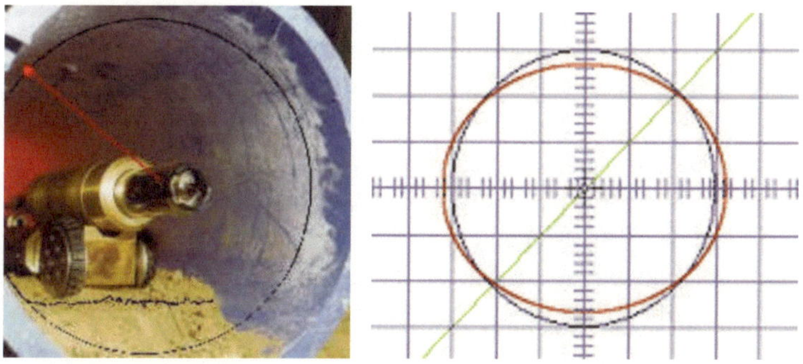

Fig. 5. OMC pipe profiler (City University of London) measured distances (red) overlayed with the circle (blue) that would represent the ideal pipe inside (profile image by Hytec)

As a result, a model of the pipe geometry can be directly acquired, which makes possible to detect deformations within the accuracy and resolution of the measurement principle.
Recently, prototypes of such sewer laser scanners have become available on the market. We mention two such systems here. Due to the precision and simplicity of the measurement principle and due to the simplicity of data interpretation, we expect laser scanners to become important sensors on fully autonomous or semi-autonomous sewer inspection robots.

-Hytec

The company Hytec (Montpellier, France) has developed a rotating laser, which performs exact measurement of the pipe's shape. It operates in pipelines of with 200 to 1000mm. The software for deflection calculation is included; the measured pipe profile can be viewed in real time on a PC color monitor. The VSLPC laser systems come as integrated with a color TV camera.

-Optimess

The company Optimess (Gera, Germany) offers the laser scanner OptiScan, originally developed by MFPA in Weimar, Germany (Kampfer et al, 1998). The variant OptiScan200 (Fig. 6) operates in pipes from 200mm, at a measuring distance of 5-100cm and an accuracy of ±1mm (according to vendor). Again, the software for calculating and visualizing the measured profile are available.

Fig. 6. OptiScan

3.2 Radar-like sensors

Ground Penetrating Radar (GPR) systems are in general heavy and power consuming. The antenna systems are large with respect to the dimension of a pipe robot.
A first use of GPR with a sewer system was announced by the developers of the KARO. They developed a GPR small enough for a sewer system. They also used a micro wave device to inspect anomalies behind the pipe walls. They report that it is possible to distinguish from type of sources of different anomalies based on bubbles of air or water, pipe couplings and house connections.
Commercially available system is presented by Oyo and Burn Am. In the case of Oyo, no data in regard to the applicability of this tool is available (i.e. do the number and severity of hidden cavities balance the expense of the radar).
In contrast, Burn Am has developed the Ground Penetrating Radar Sensor to marketable stage. The present model is integrated to a tethered, remotely controlled platform, Fig. 7.

Fig. 7. Ground penetrating radar by Burn Am

3.3 Chemical sensors

An analysis of chemical and physical parameters in the sewer as well as in the atmosphere in sewers may yield information about corroding and poisonous substances in the sewer as well as damages in the pipe system (ground water leaking in). Sensors for measuring isolated basic parameters of that kind (temperature, acidity) are widely available. They can be made small. Their energy requirements are normally negligible. Their handling may be non-trivial for a fully autonomous robot, as sensors measuring in fluids (such as sensors for the acidity level) can function for short periods of time only, i.e., they need to be dipped actively into the sewer. We report here on some sensor prototypes aimed at sensing a wider range of chemicals.

-"Nose on a chip"

The "nose on a chip", which could be incorporated into household gas appliances, consists of an array of tiny sensors on one integrated circuit and electronics on another. By selectively coating the micro cantilever arrays with appropriate chemicals, the chip can be customized to detect virtually any chemical or biological species. Developers say a single chip could detect thousands of chemicals.

-KAMINA

KAMINA is a micro system for detecting and characterization of corrosive and process gas or smell in the atmosphere. The "Karlsruher Mikronase (Karlsruhe micro nose)" is constructed as a micro chip array made out of SnO_2, WO_3 a.o. based on measuring the electric conductivity of each array element. In a cycle once per second, gases like CO, NO_2, NH_3, H_2S or organic gas and steam are detectable and quantifiable. The on-line evaluation is based on modern pattern recognition techniques and is customizable dependent on the purpose.

More sensors for navigation and motion control Autonomous sewer robots must include sensors for their own control, navigation and localization, not only those for sewer state assessment and damage detection. These sensors may overlap with the inspection sensors (e.g., a camera may be used for both navigation and damage detection). Localization is an issue not only for the proper robot control, so that the robot knows where it is, but also for inspection, as detected damages have to be reported with their location. Sensors in this direction are available in great varieties, and they are used as standard sensors in autonomous and semi-autonomous robots in other areas than sewer robotics. Note that sewer robots have the requirements in addition driver-less indoor transportation vehicles that their sensors be physically small and energy efficient.

-Compass

For navigation purposes it is helpful to know the northern direction. Accordingly, a compass may be of help. Electronic compasses are widely available on the market. Note that compass readings may be noisy in a sewer robot due to the many objects of city infrastructure that can typically be found close to sewers (steel constructions, cables, etc.). Therefore, among available electronic compasses, only those should be used on a sewer robot that report possible disturbances of their readings.

-Inertial sensors

- **Inclination sensors** Essential for helping to control tilt in fully autonomous or semiautonomous platforms. They are in use, e.g., in the KURT, MAKRO and KARO systems reported elsewhere in this thesis. They are widely available on the market, fitting the space and energy requirements of a sewer robot.
- **Accelerometers** Essential for helping the localization and control of fully autonomous robots.
- **Gyroscopes** Essential for measuring turns in autonomous robots. Widely available, but highly precise sensors are costly and may be large in build. All gyroscopes suffer from a systematical error of drift over time caused by the earth rotation.

4. Conventional inspection methods

Over the past ten years, Closed Circuit Television (CCTV) is the most commonly used internal sewer pipe inspection method. This method is carried out using a cable-tethered robot with an onboard video camera system. An operator remotely controls the movement of the robot and the video system (Fig. 8). By this video-supported visual inspection, any notable damages or abnormalities are recorded in video stream (Fig. 9). The reliability of this system depends on the experience of an operator.

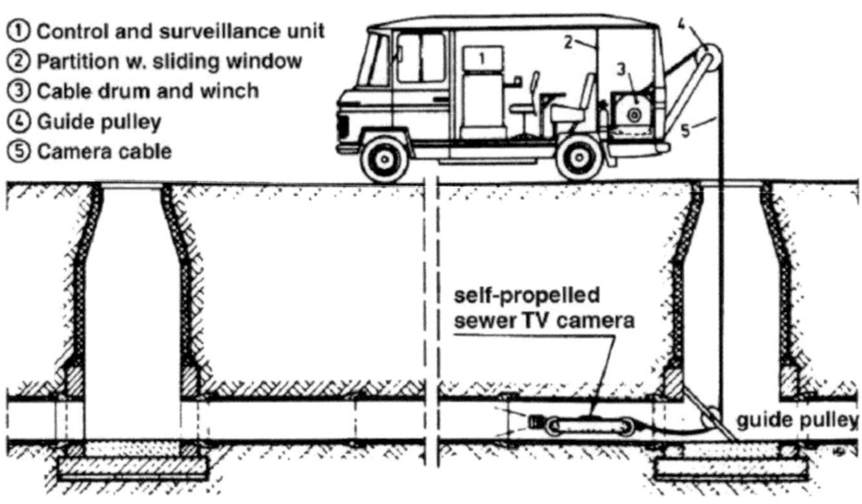

Fig. 8. Diagram of visual inspection

Fig. 9. Operator with video monitor

Recently, new and more accurate techniques have been fielded to conduct internal sewer pipe assessments. The methods used for inspecting pipe condition, fall into three broad categories:

- Inspection of inner pipe surface
- Inspection of pipe structure and bedding conditions
- Inspection of the pipe bedding

The first category, with which most practicing engineers are most familiar, included CCTV (conventional, light line and computer assisted), laser scanning and ultrasound techniques. The second and third categories include micro deflection, impact echo and ground-penetrating techniques to assess the conditions of the pipe outer layer and surrounding soil. The cost and accuracy of these methods vary widely and some methods are only justified in very specific situations where other methods will not produce an acceptable quality of information. Other inspection methods may be combined with CCTV inspection to verify the pipe condition assessment (Thomas & Laszczynski, 1997). These methods may include soils testing, physical and chemical testing of removed pipe segments (Biery et al, 1998).

5. Methods of inspecting the inner sewer wall surface

5.1 CCTV inspection systems

Standard pipe inspection systems are based on **C**losed **C**ircuit **T**elevision (CCTV) cameras in a large range of application fields such as waste pipes and drains. The CCTV method consists of a mobile, remotely operated platform usually equipped with a color, high-resolution video camera and a lighting system. The camera platform is connected via a multi-core cable to a remote inspection station with video recording facilities situated over ground. An engineer then assesses the recorded images off-line.

There are two basic types of the CCTV system. Each uses a television camera in conjunction with a video monitor, video cassette recorders and possibly other recording devices. In one case the inspection is performed using a stationary or zoom camera mounted at a manhole so that it looks into the sewer, while in the other a mobile, robotic system is placed within the sewer itself. Either form of CCTV inspection may miss certain types of defects, especially those that are hidden from the camera by obstructions as it looks down the sewer. Slight deformations of the sewer may go unnoticed, and any defects hidden beneath water inside the sewer will definitely not be found.

This is a subjective and time-consuming task that considerably increases the inspection costs. Moreover, only gross anomalies are evident to the human eye, which reduces the detection of faults at early stages. Another drawback associated with those systems in these particular environments is the lack of visibility inside the pipes and the poor quality of the acquired images that hinders a complete assessment of the pipe condition and sometimes even the detection of large defects.

5.2 Stationary CCTV systems

Stationary video cameras mounted at a manhole are limited with respect to what they can see. Defects that are close to the manhole will be found, but the farther into the sewer a defect is, the harder it will be to identify and evaluate. Defects beyond the range of the camera would be missed entirely unless they cause secondary effects that can be identified at the manhole (such as changes in water flow within the pipe between two manholes). One vendor of this technology suggests that the equipment be used as part of a screening process to determine which sewer sections should be completely examined by mobile CCTV systems. Stationary CCTV's usefulness in this respect will depend on whether the damage that can be detected by this type of system near a manhole in a sewer line is representative of that throughout the entire section of sewer line.

A survey of defects by IRC (Makar, 1999) indicates that, based on structural factors alone, stationary CCTV can readily be used to inspect vitrified clay sewers. In brick sewers the most efficient use of stationary CCTV would be to restrict it to inspecting sewer lines that are shorter than 50 meters in length, although a slightly greater factor of safety in the inspections would be produced by using the technique only on pipe sections that are less than 40 meters in length. While the results suggest that concrete pipes may be inspected in the same manner as vitrified clay pipes, too few concrete pipes were examined for a definitive conclusion to be reached.

5.3 Mobile CCTV systems

Mobile CCTV systems are the most common means of inspecting sewer lines. This type of CCTV system uses a camera mounted on a robot that enters the sewer system. The camera generally looks forward as the robot system moves along the sewer axis, allowing the operator to examine and evaluate the entire length between a pair of manholes. It is possible to modify this type of CCTV system to overcome many of the limitations of CCTV inspection discussed above. Some CCTV systems have "pan and tilt" cameras attached to the robot, which can find defects hidden from a forward looking camera behind connections and other obstructions within the sewer line. Sonar or ultrasound systems are often attached to robots to examine the sewer below the waterline. It is also possible to obtain CCTV equipment with a"light line"attachment to assist in quantifying smaller sewer deformations. This system projects a line of light around the circumference of the sewer being examined in order to assist in assessing the shape of the sewer. Inspecting within the pipe wall and the bedding condition Although CCTV, laser and ultrasonic systems provide images of the inside surface of a pipe wall, they do not indicate what is happening within the pipe wall or behind it. While in some cases the observed damage to a pipe is due to internal problems such as erosion, in many others the damage is caused by external forces. The following inspection techniques allow the sewer owner to examine the overall condition of an entire pipe wall, the soil behind a pipe or the pipe-soil system. Their ability to look beyond a pipe wall surface gives sewer owners opportunities to evaluate sewer condition in ways that are not possible with CCTV and similar techniques.

5.4 Wall micro-deflections and natural frequency of vibration

Measurements of wall micro-deflections and the natural frequencies of vibration of sewer lines are being developed specifically as a means of diagnosing brick sewer condition. The methods give information on the overall mechanical condition of the sewer line, rather than identifying specific defects.

A micro deflection in a pipe wall surface is created by applying pressure to the inside surface of the wall to very slightly deform it. In this case the intent is to measure the change in position versus the increase in pressure applied to the wall in order to indicate how well the grout between the bricks has been applied or whether the walls of a concrete or brick pipe have been damaged. It would be expected that a well grouted brick wall would expand continuously (although not equally) in all directions as the pressure increases, provided the pressure is below that which would damage the grouting. A similar, equal increase would be seen in an undamaged concrete or vitrified clay pipe.

Increasing in micro deflection in one direction while decreasing in another or a sudden change in the slope of a graph of applied force versus micro deflection would suggest that the wall was damaged. The major difficulty with this technique is determining the maximum safe pressure for use on a brick wall so that the inspection method does not damage it. This pressure will depend on the pre-existing condition of the sewer. While these pressures can be readily calculated for an undamaged sewer, the accuracy of such calculations is dependent on knowledge of the strength of the mortar or concrete at any given point in the sewer. This will vary depending on the age and condition of each sewer section. Care must therefore be taken to avoid damaging sewer sections that have below normal strength but are still able to function properly. This safety consideration is not as important for concrete pipes, where the strength of the pipe material is more uniform around the pipe circumference. Micro deflections are restricted in use to rigid pipes where an entire pipe wall will be deflected by the applied force.

Plastics such as PVC and HDPE can not be inspected using this method as local deformation of the pipe wall would tend to provide a false indication of the pipe condition. The restriction of the technique to materials such as brick, concrete, metal and vitrified clay means that it is only sensitive to the wall condition, rather than that of the surrounding bedding. Measuring the natural frequency of vibration also gives information about the mechanical behavior of a pipe wall, but in this case the process involves vibrating the wall at a range of frequencies and determining which frequency gives the largest vibrations (the natural frequencies). A section of good wall would be expected to have certain characteristic natural frequencies, while deviations from those frequencies would indicate that the wall or surrounding bedding was deficient in some manner. The application of this technique depends on the development of an understanding of exactly how the natural frequencies of different types of pipe wall would be expected to change with increasing damage. However, other factors can also affect the results of the natural frequency measurement, including changes in bedding material or quality, the amount of water in the pipe and the height of ground water around the pipe. Considerable research is needed to determine if these effects can be separated from those produced by actual damage to the pipe wall.

5.5 Impact echo/spectral analysis of surface waves

These closely related techniques they have been successfully applied to the inspection of large, empty concrete pipes and large brick water lines. The SEKISUI Company has made a sewer inspection robot with a hammer and microphone. The apparatus consists of a source of controlled impacts, such as a falling weight or a large pneumatic hammer and one or more geo phones that are mounted against the wall of the pipeline. Low frequency surface waves are produced when the wall of the pipe is struck by the hammer or weight. These waves are then detected by the geo phones. The major difference between the two techniques is that impact echo generally looks only at the actual waveform produced by the impact, while spectral analysis of surface waves (SASW) uses more geo phones and separates the waves into different frequency components (Opara et al, 1996). These different components travel at different speeds and penetrate to different depths in the soil beyond the pipe, allowing more information to be gathered about the condition of the pipe and surrounding soil.

Although the two techniques are similar, the use of the additional sensors and analysis in SASW means that it is possible to easily separate effects produced by soil conditions from those produced by problems in the pipe wall. SASW therefore is the most flexible of all the techniques discussed in this section since it is capable of investigating both pipe wall and soil condition at the same time. A drawback of Impact Echo and SASW inspection is that they are currently only available for manual use in large diameter tunnels that are easily accessible by human operators. Both techniques need to be automated to increase their inspection rate and allow deployment in smaller diameter pipelines. A second drawback is that cleaning of the pipe walls is likely to be required before they can be used.

Detecting conditions behind the pipe wall although some of the inspection techniques described previously can give information about conditions behind a pipe wall, their primary use is likely to be in determining the structural soundness of the wall itself. By contrast, ground penetrating radar may occasionally give information about delaminating in concrete sewers, but its major use in sewer lines is in detected potential problems behind the sewer walls.

5.6 Ground penetrating radar

Ground Penetrating Radar (GPR) is a geophysical method that has been developed over the past thirty years for shallow, high-resolution, subsurface investigations of the earth. GPR uses high frequency pulsed electromagnetic waves (generally 10 MHz to 1,000 MHz) to acquire subsurface information.

Energy is propagated downward into the ground and is reflected back to the surface from boundaries at which there are electrical property contrasts, Fig. 10 shows a diagram of the process. GPR is a method that is commonly used for environmental, engineering, archaeological, and other shallow investigations.

Radar is well known for its ability to detect airplanes or other flying objects, but with significant adaptations it can also penetrate rocks, sand and other solid materials to detect buried objects and voids. GPR is widely used in locating lost utilities, environmental site characterization and monitoring, agriculture, archaeological and forensic investigation, unexploded ordnance and land mine detection, groundwater, pavement and infrastructure characterization, mining, ice sounding, permafrost, void, cave and tunnel detection, sinkholes, subsidence, etc. It may be deployed from the surface by hand or vehicle. It has the highest resolution of any geophysical method for imaging the subsurface, with centimeter scale resolution sometimes possible. Resolution is controlled by wavelength of the propagating electromagnetic wave in the ground. Resolution increases with increasing frequency (shorter wavelength). Depth of investigation varies from less than one meter in mineralogical clay soils like montmorillonite to more than 5,400 meters in polar ice. Depth of investigation increases with decreasing frequency but with decreasing resolution. Typical depths of investigation in fresh-water saturated, clay-free sands are about 30 meters. Depths of investigation (and resolution) are controlled by electrical properties through conduction losses, dielectric relaxation in water, electrochemical reactions at the mineralogical clay-water interface, scattering losses, and (rarely) magnetic relaxation losses in iron bearing minerals. Scattering losses are the result of spatial scales of heterogeneity approaching the size of the wavelength in the ground (like the difference between an ice cube and a snowball in scattering visible light).

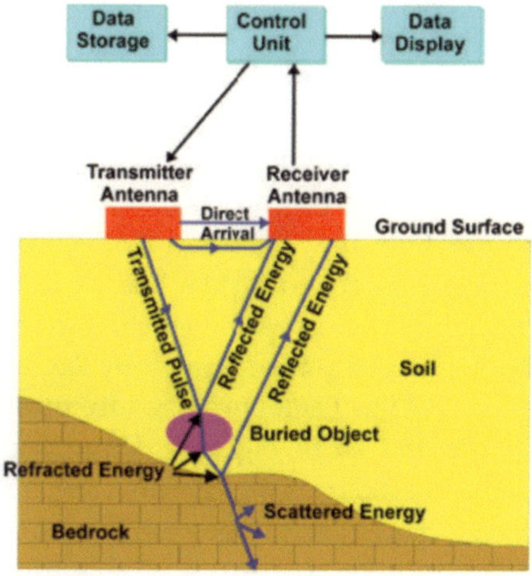

Fig. 10. Ground penetrating radar schema

The ability of GPR to detect subsurface voids has lead to an interest in using it to evaluate the condition of sewers and other pipes. While delaminating in concrete sewers could be detected by GPR systems, much of the interest in the technique is due to its ability to examine the bedding behind the pipe wall. Voids, rocks and regions of water saturation produced by ex-filtration should all be readily detectable by the technique. Recent research on this application has investigated its use in brick sewers (Price, 1995), transport tunnels and small diameter sewer lines (Daniels & Schidt, 1995).

Radar systems work by emitting a coherent beam of radio waves (Peters et al, 1994). These waves travel through space, air or the ground until they reach an object with differing conductivity and dielectric constant, such as an airplane, a void in the ground or an area saturated with water. Part of the radar wave is reflected off the interface between the two objects and propagated back to the transmitter. The rest of the wave passes into the new object and continues to travel in the original direction. Radar beams can also be attenuated by the nature of the material through which they travel. Materials that are highly conductive, have high dielectric constants, or are magnetic will rapidly attenuate the radar

beam. As a result radar is attenuated very rapidly in metals, giving essentially zero penetration, but can travel very long distances in air and space.
Sand, asphalt, clay and ice fall between these two extremes, with the degree of attenuation dependant on amount of liquid water and salts present in the material. Ice is essentially transparent to GPR, allowing the technique to be used to map the bottoms of glaciers. It can also penetrate deeply in dry sand. However, the depth of penetration in wet sand is much less, and in clays the penetration is further reduced (Benson, 1995). In these materials the presence of water increases the conductivity, while clays can also have significant dielectric constants. The presence of salt in the ground increases the soil conductivity and therefore further decreases the maximum penetration depth of a GPR system.
Radar can be used to locate leaks in buried water pipes either by detecting voids in the soil created by leaking water as it circulates near the pipe, or by detecting segments of pipe which appear deeper than they are because of the increase in the dielectric constant of adjacent soil saturated by leaking water. Ground penetrating radar waves are partially reflected back to the ground surface when they encounter an anomaly in dielectric properties, for example, a void or pipe. An image of the size and shape of the object is formed by radar time-traces obtained by scanning the ground surface. The time lag between transmitted and reflected radar waves determines the depth of the reflecting object.
In sewer pipes application, a radar antenna moves inside the sewer and is cable connected to a recording unit above ground. GPR may be used to detect water leaks in two ways:

- Identifying soil cavities created by the turbulent flow of leaking water.
- Identifying pipe segments, which appear deeper than expected because of the increase of the dielectric constant of adjacent soil, saturated by leaking water.

Despite a number of problems GPR has been employed in the condition assessment process for sewer lines in France since the early eighties. Data inspection requires substantial experience and training because the radar output is very difficult to interpret.
These techniques may be used from either the ground surface or from within the pipe but does not provide specific information on the pipe wall condition. For above ground use the technique is used to either identify the presence of pipe segments or the presence of voids surrounding the pipe, both by measuring the changing dielectric of the soil, cavities and pipes. For within pipe use the technique is used to identify cavities in the soil behind the pipe wall. In traditional above ground penetrating radar techniques, both transmitting and recording devices are mounted on the ground surface above the pipe.

6. Difficulties in conventional systems

The conventional inspection systems used for inspecting pipe condition in three broad categories are described in section 4. Among them, we focus on inspection of inner pipe surface and discuss about the difficulties in conventional systems.
Each of the systems for inspecting the inner surface of the pipe wall provides similar types of information to the manager of a sewer system. Conventional CCTV's long history of use means that new systems inspecting the same area must provide substantial advantages in the quality of information provided or in lower costs before they will be adopted for common use. Table 4 summarizes the advantages and disadvantages of each inspection methods.

Methods	Advantage	Disadvantage
CCTV inspection systems	• Standard technique. • Considerable body of knowledge. • Available to aid in interpreting results. • Relatively cheap. • Evaluates the entire length of sewer.	• Substantial operator interpretation of results. • Difficult to accurately compare two evaluations of the same sewer conducted at different times. • May miss defects hidden behind obstructions or under water.
Stationary CCTV systems	• Cheaper than CCTV. • Possibly useful as a screening mechanism for other techniques.	in addition to those listed for CCTV; • Only examines the sewer near manholes. • Long brick sewers are likely to be incorrectly. • Classified as undamaged.
Mobile CCTV systems	As for conventional CCTV except; • Better estimation of sewer deformation.	As for conventional CCTV except; • Greater expense than conventional CCTV.

Table 4. Summarizes the advantages and disadvantages of conventional inspection methods

The described modifications to conventional CCTV can assist in interpreting the results of a CCTV inspection, but are unlikely to be widely adopted unless their cost is not significantly higher than that of conventional CCTV alone. Stationary CCTV cameras are therefore the most likely application to enter common use, as they offer the opportunity to do preliminary examinations of the shorter pipes in a city's sewer system without cleaning.

Mobile CCTV systems offer noticeable advantages in identifying the presence of deformation, but are essentially an evolutionary enhancement of the standard CCTV system, rather than a revolutionary improvement.

7. Fully autonomous sewer robot, "KANTARO"

Hereafter, we describe an innovative method for realization of the fully autonomous sewer pipe inspection to cover the above mentioned problem. We improved the complexities of navigation for the motion control based on multi-sensor equipment by definition and implementation of intelligent module architecture for KANTARO (Ahrary et al, 2007), (Nassiraei et al, 2007). KANTARO is the prototype of a passive-active intelligent, fully autonomous, un-tethered robot which has an intelligent architecture in its sensor and mechanism. KANTARO, including a novel passive-active intelligent moving mechanism, is able to move into the straight pipe and passes different kinds of pipe bends without need to any intelligence of the controller or sensor reading. In addition to realize a fully autonomous inspection robot, we developed a small and intelligent 2D laser scanner for detecting of the navigational landmarks such as manholes and inlets (Ahrary et al, 2006) as stand-alone system and fusion with a fish eye camera to assist the pipe state and fault detection..

Consequently, we purpose a fully autonomous sewer inspection system but with a possibility to interfere the KANTARO control as it drives through the pipe either via an optical underground wireless communication module, developed in this project, or via a life optic un-tethered cable.

7.1 KANTARO architecture

KANTARO is designed as a fully autonomous, un-tethered robot, which fits to the pipes within a diameter range of 200-300mm. KANTARO's autonomy and its kinematic abilities extend its potential mission range enormously, comparing to conventional inspection equipment that is limited by the cable and poor kinematics. KANTARO carries all the necessary resources onboard. Standard lithium polymer batteries provide the power for its 4 motors, the sensors, the light and electronics, including a developed computer system and an optical underground wireless communication module, allowing for an autonomous uptime of about one hour.

To realize a reliable, robust robot and an easy maintenance system, KANTARO is designed to have a complete modular architecture in its mechanic, hardware and software. KANTARO, as shown in (Fig. 11), consists of two main modules: Bottom and Upper box modules. Bottom box, including a passive-active intelligent mechanism and a battery pack, can be presented as a robot platform. Electronic boards, the sensors and light are installed in upper box which be connected via the main connector to the bottom box.

In addition, KANTARO has IP67 waterproof standard that it achieved by waterproof design of KANTARO's modules including the upper and bottom box, motor boxes and battery pack.

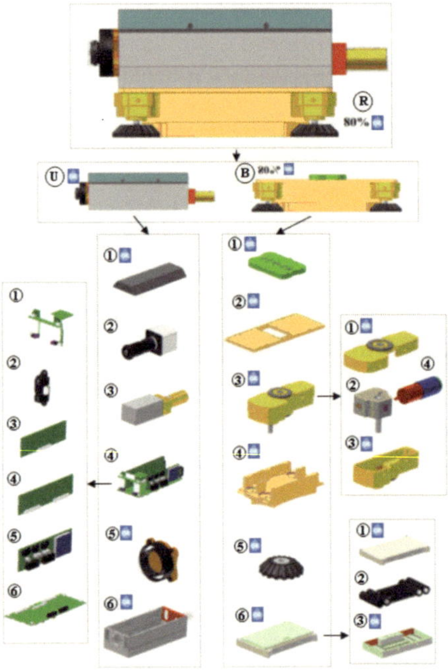

Fig. 11. KANTARO has a modular architecture consist of two main modules: Bottom box and Upper box. Modularity of bottom and upper boxes showed in right and left of the figure, consequently.

7.1.1 KANTARO platform

With regard to the current sewer pipe inspection technology, all commercial robots are capable to move in straight pipes but not any kind of pipe-bends. In addition, current sewer inspection robots are not able to pass different size of pipes. However, because of the large number of pipe-bends in the sewer pipe network, especially in Japan, pipe-bends pose one of the biggest issues for these kinds of robots. Design a robot that can move into the straight pipes and pass pipe bends will be a great industrial progress in sewer inspection industry. There have been a number of efforts and researches to develop a robot with capability of passing the pipe bends. Most of these robots, such as MAKRO or Pipe Rover/Pear Rover mentioned in introduction, have a complex mechanism and several sensors for detecting the pipe bends and pass them. These complexities in mechanism and data processing makes not easy to realize reliable commercial products.

KANTARO has a special patented mechanism, which is a passive active intelligent, compact and novel moving mechanism that can move into the straight pipes and passes different kinds of pipe bends without need to any intelligence of the controller or sensor reading. This passive active mechanism as a KANTARO platform (bottom box module, Fig. 12) itself can move into the pipe and passes wide variety of the pipe bends even without controller for the wheels actuators (without upper box module). In addition, this moving mechanism has capability to pass the different size of pipes in diameter even from a bigger diameter pipe to smaller diameter and also can pass a small obstacle and go down a small step. The design is based on the concept "passive adaptation of robot wheels to the bends in the pipe". This is accomplished by proper wheels orientation and passive damping of springs. KANTARO's mechanism, where as most of inspection robots have a moving mechanism with four, six or more vertical wheels, has four "horizontal" wheels connected to the four arms. Each arm-wheel is jointed to the platform base independently and they connected to each other by using four springs. Because of its special morphology, has an intelligence in selecting the correct direction of its movement same as water-flow direction with no control or sensor reading. This mechanism was improved by changing the wheels orientation from 0 degree to 5 degrees with respect to the horizontal level. In this approach, KANTARO has a more smooth motion in T-junctions and also it is capable to move on the flat surfaces.

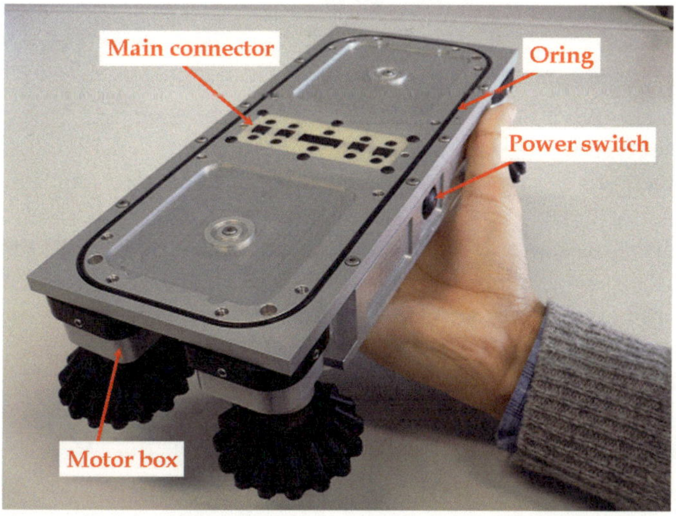

Fig. 12. KANTARO platform (Bottom box module)

7.1.2 KANTARO hardware

We designed all necessary electronic boards used for control, saving sensory data and wireless communication as a module, fit to the upper box of KANTARO (Fig. 13). This module contains a mother board, power and motor controller and CPU boards, hard disk and an optical underground wireless communication board. All boards are designed as separated modules to communicate to each other via their own connectors through the mother board. In the newest version of electronic board module, power board and hard disk are combined with mother board to be a more compact and wiring fewer modules. Wiring less was one of our targets in this project to realize a compact, reliable and robust hardware for KANATRO. To achieve the wiring less electronics, KANTARO hardware is designed base on modular architecture.

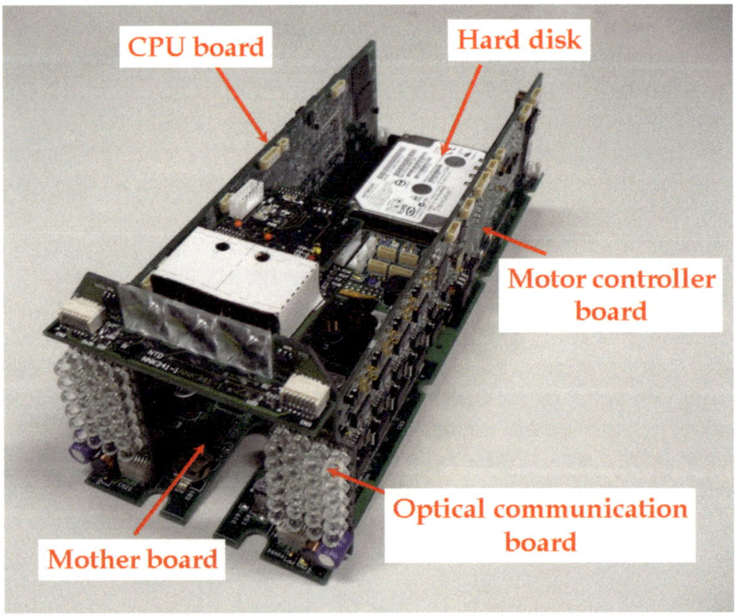

Fig. 13. The newest version of KANTARO electronic board module

7.1.3 KANTARO sensors

Autonomous sewer robots must include sensors for their own control, navigation and localization, not only those for sewer state assessment and fault detection. In addition, localization is an issue not only for the proper robot control, so that the robot knows where it is, but also for inspection, as detected faults have to be reported with their location. Therefore navigation and localization for the sewer inspection robots, as a whole, can be classified in two main groups;

- Motion-navigation, used for control and robot motion and,
- Fault-navigation, applied for the location of faults in the pipe.

The sensors used for motion navigation may overlap with the inspection sensors (e.g. a camera may be used for both motion navigation and fault detection). Motion-navigation sensors, mostly, mounted in front of the robot that it makes complex design to avoid the overlap of their workspaces.

KANTARO's sensor system includes an intelligent laser scanner, one fish eye camera, two IR sensors and a tilt sensor. The first two sensors are only used for fault detection and navigation, because KANTARO, dose not requests any kind of sensor for motion-navigation. In this approach we had no constraint for mounting the sensors, mentioned above, into the upper box of KANTARO. Using a tilt sensor mounted on the mother board, and onboard control program, KANTARO is able to automatically correct for tilt in its pose when driving inside a sewer pipe. To avoid driving the robot in a step down more than 10cm and protection from obstacle, two IR sensors are installed in front of robot in two side of the camera. Fig. 14 illustrates the arrangement of KANTARO sensors. Rotating laser scanner has been developed as a small and intelligent sensor installed in the robot rear to scan radially the inner side of a pipe wall. If the KANTARO moves along the pipe while measuring, the laser beam measures along a spiral on the pipe inside, where the resolution of the spiral depends on the turn rate of the reflection and the speed of robot. To reduce the data processing on the electronic board module, the developed laser scanner is improved to detect the navigational landmark such as manholes, inlets and pipe joints used for fault-navigation, independently, by using a powerful microprocessor (SH-2) as well as measuring distance, scanning angle, filtering and modeling data. For realization of a 360 degrees infinitive rotation in laser scanner, we developed a small and special magnet coupling to supply power to rotary part and transmit the laser signal to the microprocessor box.

Fig. 14. KANTARO's sensors is consists of a camera, one laser scanner, a tilt sensor mounted on the mother board and two IR sensors

7.2 An automated intelligent fault detection system

Basically, maintenance or inspection process starts by collecting information about the utility. It highlights useful information about conditions of the utility such as the number and the location of faults.

Conventional inspection of a sewer pipe system is carried out using a cable-tethered robot with an onboard video camera system. An operator remotely controls the movement of the robot and the video system. By this video-supported visual inspection, any notable damages or abnormalities are recorded in video stream. The reliability of this system depends on the experience of an operator. The system is also prone to human error, and tends to be time consuming and expensive. Consequently, effective automated online techniques to identify and extract objects of interest such as cracks are of immediate concern.

All previous works focused on specific types of faults in pipes and none of them proposes a method for detecting various types of faults. Accordingly, an automated fault detection system is not available in the real world. In this project, we propose a method for detecting faulty areas based on images, and propose an automated intelligent system designed to facilitate diagnosis of faulty areas in a sewer pipes system (Ahrary et al, 2007). The system utilizes image analysis and efficient techniques for providing the location and the number of faults in a sewer pipe system.

An overview of the automated intelligent fault detection system is shown in Fig.15. Digital images of sewer pipes taken by the camera system on the inspection robot are given to the fault detection system. The system, then, extracts a global ring ROI image to which edge enhancement is applied as preprocessing. Next, a newly defined measure of horizontal similarity is computed in order to extract candidates for visible faulty area in the ring ROI areas. Conjecture here is that the measure of similarity between images without faulty area is large. Hence, we focus on the area where the horizontal similarity value is smaller than a horizontal threshold, th_h, ranged between 0 and 1. Next, we extract a rectangular ROI and compute the vertical similarity value in the candidate faulty areas. Here, the area with vertical similarity value smaller than a vertical threshold, th_v, is defined as a faulty area. The proposed approach can detect even faint faults in this rectangular ROI area. Finally the detected faults and its locations are compiled as a report. Here the location information on faults is provided by sensors on the robot such as an encoder, IR and a laser scanner.

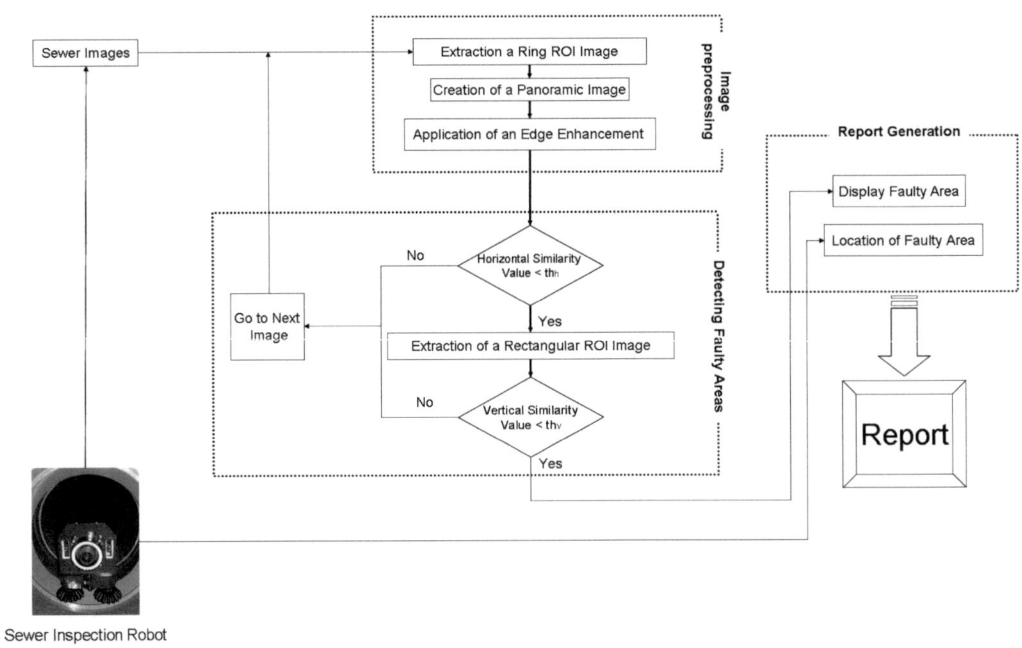

Fig. 15. An overview of the automated intelligent fault detection system

7.3 Experimental results

We evaluate the proposed system and KANTARO with designed GUI in the real sewer pipe net. The general information of inspection by using proposed system in real sewer pipe net is shown in Table 5.

Inspection date	2006.1.25	2006.9.20	2006.10.13
Inspection length	32(m)	88(m)	132(m)
Inspection time	10(min)	28(min)	45(min)
Numbers of existing faults	2 cracks	2 cracks	2 cracks+2 root invasions
Number of manholes	2	4	6

Table 5. The general information of inspection by using proposed system in real sewer pipe net

Table 5 shows sewer pipes net of length of 32(m), 88(m) and 132(m) with 2, 4 and 6 manholes is inspected. The inspection time shown here depends on the speed of the robot. The inspection results are shown in Table 6. Note that the inspected pipe net has been checked by sewer Inspection Company using the conventional system, therefore the numbers and locations of existing faults in each area were known in advance. The inspection results show that all existing faults in sewer pipe net are detected by the proposed system. Also, for generating a final report, we only need to print the report folder images (each image has the location, time and marked faulty areas information). We can say that the proposed system succeeds in attaining high detection performance, time and cost reduction in sewer inspection. Fig.16 shows scenery of inspection by proposed system in real sewer pipe net.

Inspection date	2006.1.25	2006.9.20	2006.10.13
Numbers of detected faults	2	2	4

Table 6. The inspection results by using proposed system in real sewer pipe net

Fig. 17. Scenery of inspection in real sewer pipe net by proposed system

8. Conclusions and future works

In this chapter we proposed an innovative, fast and robust sewer inspection method by using a passive active intelligent, fully autonomous, un-tethered robot, called "KANTARO", which fits to the pipes within a diameter range of 200-300mm. KANTARO prototype robot, including a novel passive active intelligent moving mechanism, has a robust movement into the straight pipe and smooth and rapid motion while passes different kinds of pipe bends without need of any intelligence of the controller or sensor reading. In this approach KANTARO does not request any kind of sensor for its motion inside of the pipe.
In addition, we developed a small and intelligent 2D laser scanner for detecting of the navigational landmarks such as manholes and pipe joints independently with main computer system and fusion with a fish eye camera, mounted on the KANTARO, used for assessing the pipe state and fault detection. Realization of KANTARO as a fully autonomous, reliable and robust pipe inspection robot has been achieved by designing an architecture based on intelligence at its modules and definition and implementation of a life optic un-tethered cable to make the inspection process easily and safely. To prove that KANTARO is able to use as a commercial product, we still have to perform further experiments in real world sewer pipe network with different in size and state pipe condition. Improving the offline fault detection software to get high accuracy is on the future work.

9. References

Ahrary, A.; Kawamura, Y. & Ishikawa, M. (2007). An automated intelligent fault detection system for inspection of sewer pipes, *IEEJ Transactions on Electronics, Information and Systems,* Vol.127, No.6, pp.943-950, June 2007.

Ahrary, A.; Nassiraei, A.F. & Ishikawa, M. (2007). A study of an autonomous mobile robot for a sewer inspection system, *Journal of Artificial Life and Robotics,* Vol.11, No.1, pp.23-27, January 2007.

Ahrary, A.; Kawamura, Y. & Ishikawa, M. (2006). A laser scanner for landmark detection with the sewer inspection robot KANTARO, *Proceedings of the IEEE International Conference on System of Systems Engineering (ICSoSE'06),* pp.310-315, Los Angeles, USA, April 2006.

Benson, A. (1995). Applications of ground penetrating radar in assessing some geological hazards: examples of groundwater contamination, faults, cavities, *Journal of Applied Geophysics,* Vol.33, No.1, pp.177-193, January 1995.

Biery, J.; Ratliff, A. & Hall, S. (1998). Evaluation of 70-year-old non-reinforced concrete sewers, *Proceedings of Pipeline Division Conference,* pp.318-327, San Diego, CA, USA, August 1998.

Daniels, D. & Schidt, D. (1995). The use of ground probing radar technologies for non-destructive testing of tubes, *Proceedings of the International Symposium of Nondestructive Testing in Civil Engineering,* pp. 429-436, Berlin, Germany, September 1995.

Hertzberg, J. & Kirchner, F. (1996). Landmark-based autonomous navigation in sewerage pipes, *Proceedings of First Euromicro Workshop on Advanced Mobile Robots (EUROBOT '96)*, pp. 68-73, Kaiserslautern, Germany, October 1996.

Kampfer, W.; Bartzke, R. & Ziehl, W. (1998). Flexible mobile robot system for smart optical pipe inspection, *Proceedings of SPIE Conference Nondestructive Evaluation of Utilities and Pipelines II*, pp. 75-83, San Antonio, Texas, March 1998.

Kepplin, V.; Scholl, K.U. & Berns, K. (1999). A mechatronic concept for a sewer inspection robot, *Proceedings of IEEE/ASME International Conference on Advanced Intelligent Mechatronics (AIM '99)*, pp. 724-729, Piscataway, NJ, September 1999.

Kirchner, F. & Hertzberg, J. (1997). A prototype study of an autonomous robot platform for sewerage system maintenance, *Autonomous Robots*, Vol.4, Issue 4, pp.319-331, October 1997.

Kirkham, R.; Kearney, P.D. & Rogers, K.J. (2000). PIRAT - A system for quantitative sewer assessment, *The International Journal of Robotics Research*, Vol.19, No.11, pp. 1033-1053, November 2000.

Kuntze, H.B. & Haffner, H. (1998). Experiences with the development of a robot for smart multisensoricpipe inspection, *Proceedings of IEEE International Conference on Robotics and Automation (ICRA '98)*, pp. 1773-1778, Leuven, Belgium, May 1998.

Kuntze, H.B.; Schmidt, D.; Haffner, H. & Loh, M. (1995). KARO - A flexible robot for smart sensor-based sewer inspection, *Proceedings of 12th International No-Dig Conference*, pp. 367-374, Hamburg, September 1995.

Makar, J. M. (1999). Diagnostic techniques for sewer systems, *Journal of Infrastructure Systems*, Vol.5, No.2, pp. 69-78, June 1999.

Nassiraei, A.F.; Kawamura, Y.; Ahrary, A.; Mikuriya, Y. & Ishii, K. (2007). Concept and design of a fully autonomous sewer pipe inspection mobile robot "KANTARO", *Proceedings of the IEEE International Conference on Robotics and Automation (ICRA'07)*, pp.136-143, Roma, Italy, April 2007.

Opara, N.K.; Wooda, R.D. & Shayea, N. (1996). Nondestructive testing of concrete structures using the Rayleigh wave dispersion method, *Journal of American Concrete Institute Materials*, Vol.93, No.1, pp. 75-86, January 1996.

Peters, L.; Daniels, J. & Young, J. (1994). Ground penetrating radar as a subsurface environmental sensing tool, *Proceedings of the IEEE International Conference on Robotics and Automation*, Vol.82, pp. 1802-1822, December 1994.

Price, T. (1995). Inspecting buried plastic pipe using a rotating sonic caliper, *Proceedings of the Second International Conference on Advances in Underground Pipeline Engineering*, pp.126-137, Bellevue, Washington, June 1995.

Rome, E.; Hertzberg, j.; Kirchner, F.; Licht, U.; Streich, H. & Christaller, TH. (1999). Towards autonomous sewer robots: The MAKRO project, *Journal of Urban Water*, Vol.1, No.1, pp. 57-70, March 1999.

Schonherr, F.; Hertzberg, J. & Burgard, W. (1999). Probabilistic mapping of unexpected objects by a mobile robot, *Proceedings of the IEEE/RSJ International Conference on Intelligent Robots and Systems (IROS '99)*, vol.1, pp. 474-481, 1999.

Scholl, K.U.; Kepplin, V.; Berns, K. & Dillmann, R. (2000). Controlling a multi-joint robot for autonomous sewer inspection, *Proceedings of IEEE/RAS International Conference on*

Robotics and Automation (ICRA 2000), pp. 1701-1706, Piscataway (NJ), San Francisco, April 2000.

Thomas, A. & Laszczynski, G. (1997). New technology for enhanced pipeline evaluation, *Proceedings of the ASCE Conference on Trenchless Pipeline Projects*, pp. 291-297, Boston, MA, USA, June 1997.

18

Visual Novelty Detection for Mobile Inspection Robots

Hugo Vieira Neto[1] and Ulrich Nehmzow[2]
[1]*Federal University of Technology – Paraná*
[2]*University of Ulster*
[1]*Brazil*
[2]*UK*

1. Introduction

Novelty detection – the ability to identify perceptions that were never experienced before – and, more generally, selective attention are extremely important mechanisms to autonomous mobile robots with limited computational resources. From the operational point of view, the robot's resources can be used more efficiently by selecting those aspects of the surroundings which are relevant to the task in hand or uncommon aspects which deserve closer analysis.

In fact, identification of new concepts is central to any learning process, especially if knowledge is to be acquired incrementally and without supervision. In order to learn concepts, it is necessary to determine first if they are not already part of the current knowledge base of the agent. Simultaneous learning and recognition (Artač et al., 2002) is a fundamental ability to robots aiming at true autonomy, adaptability to new situations and continuous operation.

From the application point of view, reliable novelty detection mechanisms facilitate automated environment inspection and surveillance by highlighting unusual situations. Novelty detection and incremental learning are also vital in applications that demand unsupervised environment exploration and mapping.

In this chapter we are particularly interested in environment inspection using mobile robots to assist in the detection of abnormalities. An example of a practical application is the automatic identification of cracks, tree roots or any other kinds of faults in sewer pipes. Sewer inspection is currently performed manually by human operators watching logged video footage, a time-consuming and error-prone approach, due to human fatigue. This task would benefit immensely from the assistance of an inspection robot that is able to pinpoint just the unusual features in the sewer which are likely to correspond to potential faults.

Fault detection tasks are different from the usual pattern recognition problems in which the features of interest are usually determined beforehand – the aim in fault detection is to detect something that is unknown *a priori*. Therefore, it is argued that the most feasible approach to be followed is to learn a model of *normality* of the environment, and then use it to filter out *any abnormal* sensory perceptions (Tarassenko et. al., 1995). Abnormal perceptions are thus defined as anything that does not fit the acquired model of normality.

Previous work has demonstrated that the approach of learning models of normality and later on using them to highlight abnormalities is very effective for mobile robots that use low-resolution sonar readings as perceptual input (Marsland et al., 2002a).

The sensor modality used as perceptual input obviously plays an important role in the robot's performance for a given task or behaviour. If relevant features of the surroundings can not be properly sensed and discriminated, it will be impossible for the robot to respond appropriately. Mobile robots are usually equipped with tactile, distance (infrared, sonar and laser range finders) and vision sensors. Of all range of sensors, vision is the most versatile modality because it can detect colour, texture, shape, size and even distance to a physical object. Moreover, vision also has the advantage to be able to generate high-resolution readings in two dimensions, making the detection of small details of the environment more feasible.

Vision is therefore a primary source of information for a mobile robot operating in real world scenarios, such as in sewer fault inspection. The main reason for this is that the environment needs to be sensed in high resolution and preferentially using a two-dimensional field of view, so that the chances of missing important details are minimised. Furthermore, vision is a sense shared with humans and therefore provides common ground for collaboration between robots and operators while performing the inspection task.

Much of the previous research done in novelty detection applied to environment inspection using real mobile robots was made using exclusively sonar sensing (Crook et al., 2002; Marsland et al., 2002a) and little work was done using monochrome visual input in very restricted ways (Crook & Hayes, 2001; Marsland et al., 2001). There is also work related to novelty detection using sonar readings in simulated robots (Linåker & Niklasson, 2000; Linåker & Niklasson, 2001). The main idea behind these approaches was to use on-line unsupervised learning mechanisms in order to acquire models of normality for the environment.

Marsland et al. (2002a) have developed the Grow-When-Required (GWR) neural network and used it to highlight novel patterns in sonar scans, while Crook & Hayes (2001) used a novelty detector based on the Hopfield neural network (Hopfield, 1982). These approaches were qualitatively compared in (Crook et al., 2002) during a novelty detection task using sonar readings in a corridor. Linåker & Niklasson (2000) developed the Adaptive Resource Allocating Vector Quantisation (ARAVQ) network and used it in simulations. All of these mechanisms have shown to work very well with low-resolution sonar data according to qualitative assessment criteria. However, none of them was employed using high-resolution visual data in real world application scenarios. Also, quantitative tools to assess and compare the performance of novelty filters objectively were missing. Therefore, qualitative evaluation of novelty filters is one of the issues addressed in this work.

Here we are mainly interested in investigating novelty detection using colour visual input in real robots. However, a major difficulty that comes with vision is how to select which aspects of the visual data are important to be encoded and processed. It is undesirable to process raw high-dimensional visual data directly due to restrictions in computational resources in mobile robots. Hence, a possible solution to cope with massive amounts of visual input (tens of thousands of pixels per image frame) is the use of a mechanism of attention to select aspects of interest and concentrate the available resources on those (Itti & Koch, 2001).

A mechanism of visual attention selects interest points within the input image according to some criteria (for instance, edges or corners). Interest points selected by such attention mechanisms are usually locations containing very descriptive information – they are visually *salient* in the scope of the image frame. A small region in the vicinity of an interest point can then be encoded to represent local visual features. This process not only localises salient features within the image, but also concentrates computational resources where they are necessary. Local encoding of a small image region also has the advantage of reducing data dimensionality while preserving details.

A particularly interesting attention mechanism is the saliency map model (Itti et al., 1998), which combines different visual features (such as intensity, colour and orientation in multiple scales) to obtain a general indication of saliency for each image location – saliency can be thought as the property to stand out from the background. This approach is very convenient for novelty detection and, more specifically, inspection tasks in which the identification of uncommon features is precisely what is desired. Also, the use of a model of visual attention is essential to localise *where* the unusual features are within the image.

The approach we follow in this work is to use the saliency map as mechanism of attention to select a number of salient regions in the input frame, which are encoded into feature vectors and then fed to an unsupervised learning mechanism, either a GWR neural network (Marsland et al., 2002b) or an incremental Principal Component Analysis (PCA) algorithm (Artač et al., 2002). These learning mechanisms are used to build a model of normality for the perceptions acquired in the operating environment and, after the learning process, are used as novelty filters to highlight arbitrary novel features that may be encountered. This approach, as well as some tools for qualitative and quantitative performance assessment of novelty detection systems, is described further in the next section.

2. An experimental framework for visual novelty detection

Although novelty detection using sonar sensing proved to be useful to detect open doors in corridors (Marsland et al., 2000) and even to identify in which corridor a mobile robot was operating (Marsland et al., 2002a), the very low resolution used – a robot's sonar ring is typically composed of a small number of sensors – and unreliable noisy readings pose serious limitations to more demanding real world applications. For example, it would be impossible to detect small cracks in a wall by using sonar sensors alone.

Vision, on the other hand, provides detailed information about the operating environment in high resolution. Of course, this comes at the expense of large amounts of data to be processed, which constitutes a serious difficulty when one desires real-time operation. Fortunately, the massive amount of information provided by a vision sensor is highly redundant and therefore can be compressed prior to higher levels of processing. Selecting which aspects of the visual data are the most relevant, however, is not a straightforward procedure and usually is dependant on the application.

Visual novelty depends on the multi-modal measures of the properties from the environment that the camera provides the robot with – some visual feature can be considered novel because of its colour, texture, shape, size, pose, motion or any combination of these and even other visual features – a much more complex scenario than the one of single mode sensors like sonars. Because multi-modal vision is very difficult to be accomplished in a mobile robot with limited computational resources, we had to decide which visual features were the most important to define novelty in our application domain.

In the context of an environment such as the inside of a sewer pipe, the visual novelties we are interested in are static, i.e. they do not move. Nevertheless, the sewer is a dynamic environment in the sense that new visual features that correspond to faults - cracks and tree roots, for instance - may appear at any time, hence the need of regular inspections. For this type of application, higher level visual interpretations - such as the concepts of size, pose or motion - are not as important as low-level features that characterise the essential appearance of visual objects. Therefore, we limited the visual features of interest in this work to colour, texture and shape.

Besides characterising novelty by visual appearance, spatial location of novel visual features in the environment is also important. Therefore, we are interested not only in detecting *which* features constitute potential faults but also *where* they are in the environment. Changes in location of visual features may also constitute novelty and are relevant in some application domains, such as automated surveillance. However, in the scope of this work we will not consider the location of a visual feature to be contextually important to determine novelty. In other words, it will not be possible to consider some visual feature as being novel based solely on its location in the environment.

Another difficulty related to visual novelty detection using a mobile robot concerns invariance to image transformations. Because the images are acquired from a moving platform, visual features are subject to several geometric transformations and it is undesirable that known visual features happen to be labelled as novel just because there were changes in appearance due to robot movement (e.g. changes in perspective). Hence, the image encoding procedure should be robust to small geometrical transformations that result from robot motion.

In order to localise novel features within the image frame, we decided to get inspiration from biological vision systems and use a mechanism of attention (Itti & Koch, 2001). Following this idea, smaller image regions selected by the visual attention mechanism from the input image can be encoded as feature vectors. Figure 1 depicts the block diagram of such an approach, in which the novelty filter is preceded by the attention mechanism. Instead of encoding the whole image frame in a single feature vector, several feature vectors are encoded per frame using the vicinity of salient image locations. Salient (or interest) points normally correspond to places with high information contents, i.e. strong peaks, edges or corners, depending on the criteria for their selection. In this work, we have used the saliency map (Itti et al., 1998) as our mechanism of selective attention.

By selecting interest regions in the form of image patches, we reduce the dimensionality of the data to be processed by the novelty filter and also gain robustness to some geometrical transformations, notably translations within the image frame due to robot movement. Furthermore, novel visual features can be immediately localised within the input image frame with the selection of local salient regions, as we will demonstrate experimentally.

The experiments that follow use raw image patches (24×24 pixels in size) extracted from salient locations within the input image (160×120 pixels in size) and compare performances of novelty filters based on the GWR neural network (Marsland et al., 2002b) and incremental PCA (Artač et al., 2002). There is hardly any other visual representation more specific than raw image patches and therefore generalisation in the experiments reported here is left to the learning mechanisms used as novelty filters. As a side-effect, the use of raw image patches allows visual feedback of the knowledge acquired during training (Vieira Neto & Nehmzow, 2005; Vieira Neto & Nehmzow, 2007).

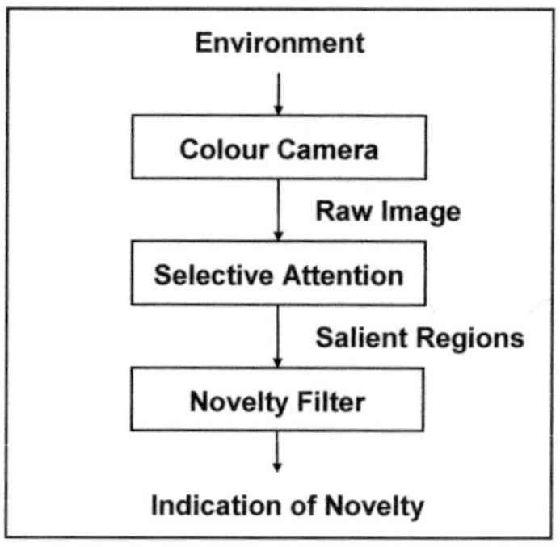

Figure 1. The framework for the investigation of visual novelty detection: an attention mechanism selects patches from the input image frame, which are then classified by the novelty filter as novel or non-novel.

2.1 Experimental setup

To evaluate the ability of our novelty detection framework to detect novel visual features that may appear in the robot's normal operating environment, we conducted experiments in controlled scenarios. Every experiment consisted of two stages: an exploration (learning) phase, in which the robot was used to acquire a model of normality of the environment, followed by an inspection (application) phase, in which the acquired model was then used to highlight any abnormal perception in the environment.

During the learning phase, images were acquired while the robot was navigating around a "baseline" environment (an empty arena or corridor, containing only "normal" features). These images were then processed by the attention mechanism to generate input feature vectors and train the novelty filter. After that, during the application phase, novel objects were placed in the environment so that a new sequence of images could be acquired and used to test the trained novelty filter.

The expected outcome of these experiments was that the amount of novelty detected would continuously decrease during exploration as a result of learning. At the beginning of the learning procedure everything is considered novel and, as the robot learns, less and less novelties should be found. During the inspection phase we expected that peaks in the novelty measure would appear only in areas where a new object had been placed. This hypothesis was tested using a real robot navigating in engineered (laboratory) and medium-scale real world environments. Figure 2 shows the experimental setup used for the laboratory experiments.

The colour vision system of *Radix*, the Magellan Pro robot shown in Figure 2a, was used to generate visual stimuli while navigating in the environment. The robot was equipped with standard sonar, infra-red and tactile sensors, and also with an additional laser range scanner whose readings were used for controlling the navigation behaviour. *Radix* operated completely autonomously in our experiments.

The robot's on-board computer was capable of processing on-line up to eight frames per second when running our control software, which was optimised for speed. Nevertheless, the images used in the experiments reported in this chapter were acquired at one frame per second (without stopping the robot) for off-line processing. This procedure was chosen in order to allow fair performance comparisons between different novelty detection mechanisms by using the same datasets.

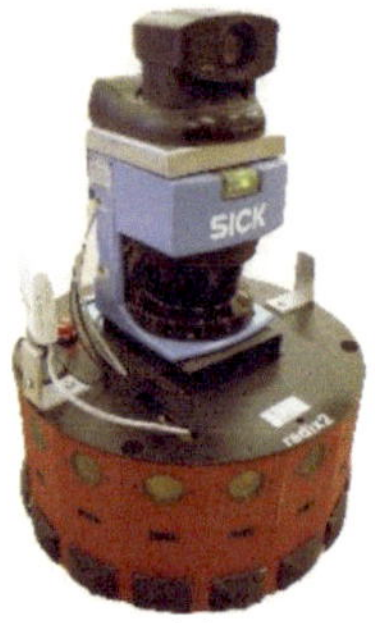

(a)

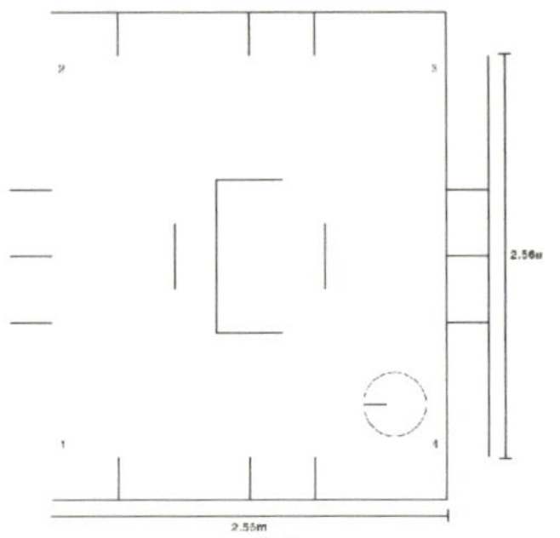

(b)

Figure 2. Experimental setup: (a) Magellan Pro mobile robot; (b) top view of a typical robot arena used as operating environment used in laboratory experiments. The robot arena was delimited by cardboard boxes, represented by rectangles, while the robot is represented by a circle with a line indicating its front, to where its camera is directed.

Figure 2b shows the top view of the engineered environment used in the laboratory experiments, a square arena delimited by cardboard boxes in whose corners (numbered from 1 to 4) novel objects were introduced. The cardboard boxes at the borders of the arena acted as walls (approximately 0.5m high) that limited the robot's trajectory and also its visual world. The images were acquired with the robot's camera tilted down to −25 degrees, so that the field of view was constrained to the arena's walls and floor, resulting in a completely controlled visual world for our experiments. A simple obstacle-avoidance algorithm using the robot's laser range scanner measurements was used as the navigation behaviour for the robot. In our experiments, this behaviour has shown to be very predictable and stable.

2.2 Assessment of results

Qualitative and quantitative assessment tools were devised to analyse our results. These assessment tools are very important to establish a reference for comparisons and therefore determine which of the studied methods perform better according to the desired application.

Qualitative assessment. In the following sections we use bar graphs in which novelty measurements provided by the novelty filter are plotted against time. They are used in order to obtain a qualitative indication of performance, in a similar fashion to (Marsland et al., 2002a). In these graphs, time essentially corresponds to a certain position and orientation of the robot in the environment because the navigation behaviour used in the experiments was highly repeatable.

The efficiency of learning during the exploration phase can be graphically assessed through inspection of the qualitative novelty graphs in multiple rounds. By looking at the novelty graphs for the exploration phase, one can determine how fast learning occurred and also assess if the amount of learning was adequate for the acquisition of an environmental model of normality.

In the inspection phase, a new object was introduced in the normal environment in order to test the system's ability to highlight abnormal perceptions. The measure of novelty was expected to be high only in places where the new object could be sensed, an expectation that should be reflected in the novelty graphs obtained. The inspection phase of experiments was also carried out in multiple rounds with the learning mechanism disabled, so that unusual features in the environment were highlighted every time that they were perceived. Hence, the consistency of a novel feature being detected in a particular location of the environment but in different inspection rounds can also be evaluated using this qualitative assessment scheme.

Quantitative assessment. The off-line processing of image frames acquired with the robot in exploration and inspection phases also allows a quantitative assessment and direct performance comparison between different approaches through the use of identical datasets. For that, we manually generated ground truth in the form of a binary image for each input image where the novel object was present. In these binary images, the pixels corresponding to the novel object were highlighted (see examples in Figure 3).

(a)

(b)

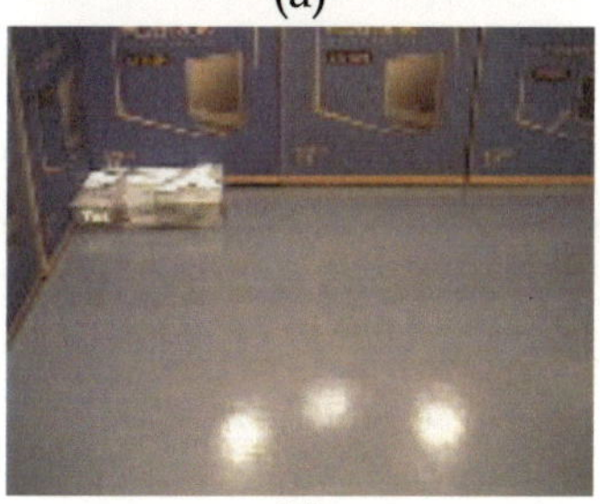

(c)

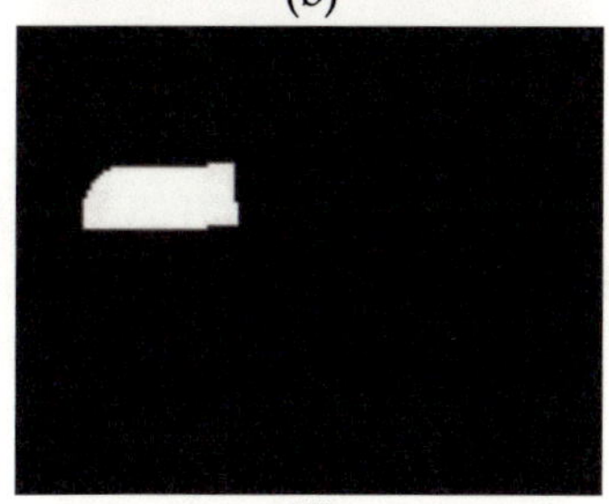

(d)

Figure 3. Example of typical input images containing a novel object: an orange football in (a) and a grey box in (c), and their corresponding ground truth novelty templates in (b) and (d), respectively.

Using the manually generated ground truth information, contingency tables were built relating system response to actual novelty status, as shown in Table 1. If a given region of the input image had more than 10% of highlighted pixels in the corresponding ground truth template, then this region's novelty status was considered as "novelty present". In this case, if the system response was "novelty detected" this configured true novelty and therefore entry *A* in the contingency table shown in Table 1 was incremented; otherwise, if the system response was "novelty not detected", this configured a missed novelty with entry *B* being incremented. On the other hand, if the novelty status of a given image region was considered as "novelty not present" (less than 10% of highlighted pixels in the corresponding ground truth region) and nevertheless the system responded as "novelty detected", this configured a false novelty with entry *C* being incremented. Finally, if the system response agreed with the actual novelty status by attributing "novelty not detected" to a region whose novelty status was "novelty not present", this represented true non-novelty and entry D was incremented.

	Novelty Detected	Novelty Not Detected
Novelty Present	A	B
Novelty Not Present	C	D

Table 1. Contingency table for the quantitative assessment of novelty filters.

An ideal association between system response and actual novelty status would have a contingency table in which values B and C in Table 1 are zero, while values A and D have non-zero values (in practice, A will be small in comparison to D as usually there are few examples of novel features in the inspected environment). The statistical significance of the association between the actual novelty status (ground truth) and the novelty filter response can be tested using χ^2 analysis (Nehmzow, 2003; Sachs, 2004). For the 2×2 contingency table shown in Table 1, the χ^2 statistic is computed using:

$$\chi^2 = \frac{N(AD-BC)^2}{(A+C)(C+D)(A+B)(B+D)}, \tag{1}$$

where $N = A+B+C+D$ is the total number of samples in the table.
If $\chi^2 > 3.84$ there is a significant correlation between novelty status and novelty filter response, with a probability $p \leq 0.05$ of this statement being wrong. If $\chi^2 > 6.64$ the significance level of the correlation is higher and the probability of being wrong decreases to $p \leq 0.01$. It is also important to mention that the χ^2 test is valid only for well-conditioned contingency tables – this entails the computation of a table of expected values, which must have no entries with expected values below 5 (Nehmzow, 2003).
The strength of the association was assessed by Cramer's V, which is directly based on the χ^2 statistic (Nehmzow, 2003):

$$V = \sqrt{\frac{\chi^2}{N}} = \sqrt{\frac{(AD-BC)^2}{(A+C)(C+D)(A+B)(B+D)}}. \tag{2}$$

The uncertainty coefficient U, an entropy-based measure, was also used to estimate the strength of the association. Computation of the uncertainty coefficient relies on the fact that each sample in the contingency table shown in Table 1 has two attributes, the actual novelty status S and the novelty filter response R. The entropy of S, $H(S)$, the entropy of R, $H(R)$, and the mutual entropy of S and R, $H(S, R)$, are given by the following equations:

$$H(S) = -\frac{A+B}{N}\ln\left(\frac{A+B}{N}\right) - \frac{C+D}{N}\ln\left(\frac{C+D}{N}\right), \tag{3}$$

$$H(R) = -\frac{A+C}{N}\ln\left(\frac{A+C}{N}\right) - \frac{B+D}{N}\ln\left(\frac{B+D}{N}\right), \tag{4}$$

$$H(S,R) = -\frac{A}{N}\ln\left(\frac{A}{N}\right) - \frac{B}{N}\ln\left(\frac{B}{N}\right) - \frac{C}{N}\ln\left(\frac{C}{N}\right) - \frac{D}{N}\ln\left(\frac{D}{N}\right). \tag{5}$$

When applying equations 3, 4 and 5, one must remember that $\lim_{p\to 0} p\ln p = 0$. The uncertainty coefficient U of S given R, $U(S|R)$, is finally computed using (Nehmzow, 2003):

$$U(S|R) = \frac{H(S) - H(S,R) + H(R)}{H(S)}. \tag{6}$$

Both V and U provide normalised measures of strength ranging from zero to one. Good associations result in V and U having values close to one, while poor associations result in values close to zero. Therefore, the values of V and U can be used to determine which among two or more novelty systems perform better in a given situation.
A further statistic that was used is the κ index of agreement, which is computed for 2×2 contingency tables as follows (Sachs, 2004):

$$\kappa = \frac{2(AD - BC)}{(A+C)(C+D) + (A+B)(B+D)}. \tag{7}$$

This statistic is used to assess the agreement between ground truth data and novelty filter response, in a similar way to what is done with V and U. However, it has the advantage of having an established semantic meaning associated with some intervals, as shown in Table 2 (Sachs, 2004).

Interval	Level of Agreement
$\kappa \leq 0.10$	No
$0.10 < \kappa \leq 0.40$	Weak
$0.40 < \kappa \leq 0.60$	Clear
$0.60 < \kappa \leq 0.80$	Strong
$0.80 < \kappa \leq 1.00$	Almost complete

Table 2. κ intervals and corresponding levels of agreement between ground truth and novelty filter response.

Unlike V and U, the κ statistic may yield negative values. If this happens, the level of *disagreement* between system response and manually generated ground truth can be assessed. Negative values result when the entries B and C in the resulting contingency table are larger than the entries A and D. In such a case, both U and V would still result in positive values because they are designed to measure the strength of the association (be it positive or negative) rather than the level of agreement (positive association) or disagreement (negative association).

3. Experiments in a laboratory environment

3.1 The GWR neural network as novelty filter

The GWR network (Marsland et al., 2002a; Marsland et al., 2002b) is a self-organising neural network based on the same principles as Kohonen's Self-Organising Map (Kohonen, 1984). Its structure is composed of nodes that represent the centres of clusters (model weight

vectors) in input space - every time that an input is presented, each network node will respond with higher or lower activity depending on how good its weight vector matches the input vector.
A novelty filter based on the GWR network basically consists of a clustering layer of nodes and a single output node. The connecting synapses to the output layer are subject to a model of habituation, which is a reduction in behavioural response to inputs that are repeatedly presented. In other words, the more a node in the clustering layer fires, the less efficient its output synapse becomes.
What makes the GWR network superior to the SOM is its ability to add nodes to its structure - hence the name Grow-When-Required - by identifying new input stimuli through the habituation model. Given an input vector, both the firing node's activity and habituation are used to determine if a new node should be allocated in order to represent the input space better.
The habituation rule of a clustering node's output synaptic efficacy is given by the following first-order differential equation (Marsland et al., 2002b):

$$\tau \frac{dh(t)}{dt} = \alpha[h_0 - h(t)] - S(t) \, . \tag{8}$$

where h_0 is the initial value of the efficacy $h(t)$, $S(t)$ is the external stimulus, τ and α are time constants that control the habituation rate and the recovery rate, respectively.
$S(t)=1$ causes habituation (reduction in efficacy) and $S(t)=0$ causes dishabituation (recovery of efficacy). It is important to mention that only habituation was modelled in our implementation - dishabituation was disabled by setting $S(t)=1$. Using $\alpha = 1.05$ and $h_0 = 1$ results in efficacy values ranging from approximately 0.05 (meaning complete habituation) to 1 (meaning complete dishabituation). As synaptic efficacy has a bounded output, it can be used neatly as a measure of the degree of novelty for any particular input: higher efficacy values correspond to higher degrees of novelty. More detail is given in (Vieira Neto & Nehmzow, 2005).
Normality model acquisition. Exploration was conducted in five consecutive loops around the empty arena, with the robot being stopped and repositioned at the starting point in every loop. This procedure was used in order to ensure that the robot's trajectory would be as similar as possible for every loop, resulting in consistent novelty graphs for qualitative assessment. Images were acquired at the rate of one frame per second, resulting in a total of 50 images per loop around the arena. We used normalised raw image patches, selected by the saliency map from the acquired images, as input to the GWR network.
During the exploration phase, learning of the GWR network was enabled to allow the acquisition of a model of normality. As expected, the amount of novelty measured - the efficacy of the habituable synapse of the firing node - decreased as the network habituated on repeated stimuli. Only four nodes were acquired by the GWR network by the end of the fifth exploration loop.
Novelty detection. Having trained the GWR network during the exploration phase, we then used the acquired model to highlight any unusual visual features introduced in the empty arena: an orange football was placed as novel object in one of the corners and the robot was used to inspect the arena. The ball was selected not only because it contrasted well with the

arena's colour features, but also because it did not interfere with the robot's trajectory around the arena (it could not be sensed by the laser range scanner).

Learning of the GWR network was disabled during inspection, so that consistency in novelty indications could be verified over different loops around the arena. The novelty graphs obtained for the inspection phase of the arena containing the ball are shown in Figure 4.

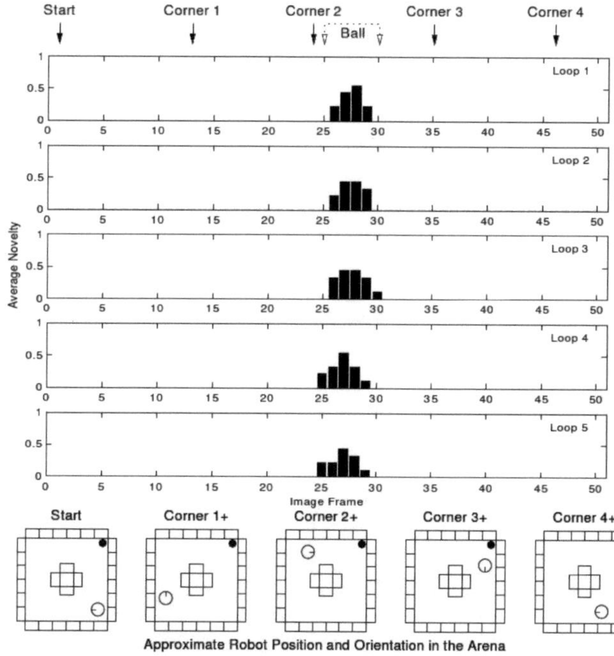

Figure 4. Inspection of the arena with the orange ball (novel stimulus) using the GWR network. The orange ball is clearly and consistently highlighted as novelty. The pictograms below the graphs indicate the approximate position and orientation of the robot in the arena while performing the inspection loops and also the location of the novel stimulus.

Because the images were acquired at the rate of one frame per second, the horizontal axis of the graphs in Figure 4 can also be interpreted as time in seconds. Pictograms indicating the approximate position and orientation of the robot in the arena are also shown (we use the notation "Corner 1+" to indicate position and orientation immediately after the robot has completely turned the first corner).

The set of frames where the orange football appeared in the camera's field of view are indicated by dotted arrows on the top of Figure 4. These frames correspond to locations where high values for the novelty measure were expected to happen (the ball appeared always in the same frames in every loop because the navigation behaviour was very stable). As one can notice in Figure 4, the ball was always correctly detected as the novel feature in the environment (see also figure 8a for the visual output of the system). Contingency table analysis through the χ^2 test revealed statistical significance between the novelty filter

response and actual novelty status ($p \leq 0.01$). The strength of the association revealed almost complete agreement between system response and actual novelty status.
A new inspection phase was then performed in the arena with another novel stimulus, a grey box instead of the orange ball. The grey box is much less conspicuous than the orange ball and the idea behind its use was to check the system's ability to detect a novel object similar in colour to the environmental background and therefore not very salient (the arena had predominantly grey and dark blue colours).
The frames in which the grey box appeared in the camera's field of view are indicated with dotted arrows in Figure 5, where the results of the new inspection round are also given.

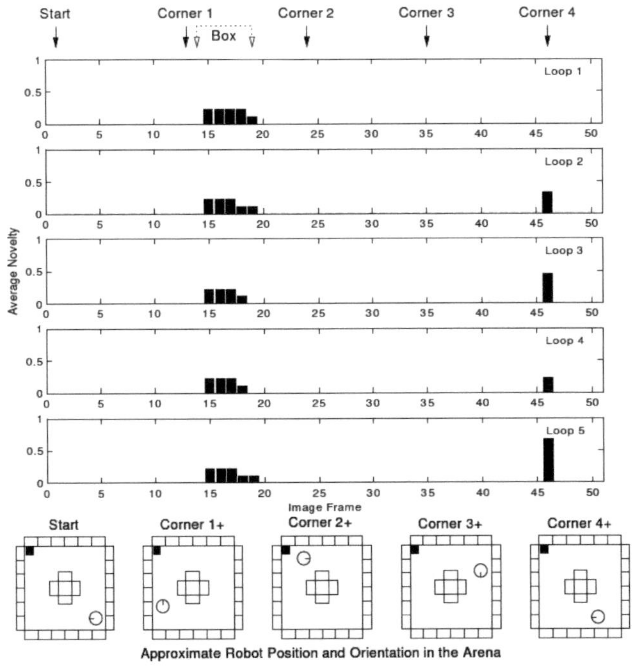

Figure 5. Inspection of the arena with the grey box (novel stimulus) using the GWR network. The grey box is clearly and consistently highlighted as novelty, but unexpected novelty indications also appeared consistently for image frame 46. The pictograms below the graphs indicate the approximate position and orientation of the robot in the arena while performing the inspection loops and also the location of the novel stimulus.

The GWR network correctly identified the grey box as novel, as shown in Figure 5 (see also figure 8b for the visual output of the system). However, unexpected novelty peaks also appeared consistently for image frame 46, when the robot turned a corner very close to the arena's wall. In this particular image frame, two thirds of the most salient regions correspond to a large edge between two of the cardboard boxes that constitute the wall. Although the robot was exposed before to edges between the cardboard boxes, it had never before been as close as happened in this case – this resulted in image patches containing edges larger in scale than the GWR network was habituated to, resulting in their classification as novel.

Nevertheless, contingency table analysis using the χ^2 test revealed statistical significance between system response and ground truth ($p \leq 0.01$). The strength of the association was also measured and revealing strong agreement between novelty filter response and actual novelty status. As one can notice from these results, the false novelty indications due to the features present in image frame 46 depressed results slightly, but not statistically significantly.

3.2 Incremental PCA as novelty filter

PCA is a very useful tool for dimensionality reduction that allows optimal reconstruction of the original data, i.e. the squared reconstruction error is minimised. It consists of projecting the input data onto its principal axes - the axes along which variance is maximised - and is usually computed off-line because the standard algorithm requires that all data samples are available a priori, making it unsuitable for applications that demand on-line learning.

A method for the incremental computation of PCA recently introduced by Artač et al. (2002) , which is an improvement to the algorithm originally proposed by Hall et al. (1998), makes simultaneous learning and recognition possible. Using this method, it is possible to discard the original input data immediately after the eigenspace is updated, storing only projected data with reduced dimensions.

We use incremental PCA as an alternative method to the GWR network to perform on-line novelty detection. The magnitude of the residual vector - the RMS error between the original input and the reconstruction of its projection onto the current eigenspace - is used to decide if a given input is novel and therefore should be added to the model. If the magnitude of the residual vector is above some threshold r_T, the corresponding input vector is not well represented by the current model and therefore must be a novel input. Complete implementation details of the incremental PCA algorithm used in this work can be found in (Vieira Neto & Nehmzow, 2005).

The previous laboratory experiments were repeated using the incremental PCA approach for comparison purposes. During the exploration phase it was verified that most of the eigenspace updates happened in the beginning of the first loop around the arena, becoming less frequent as the environment was explored. By the end of the fifth exploration loop, the incremental PCA algorithm acquired 35 model vectors.

As before, the model learnt during the exploration phase was used to highlight novel visual features in the arena during the inspection phase, while the learning mechanism was disabled. The results obtained for the inspection of the arena with the orange ball are given in Figure 6.

The orange ball was correctly identified as novel, as can be seen in Figure 6. Also, there were very few false indications of novelty. Inspection was repeated for the arena containing the grey box and the results obtained are given in Figure 7.

The grey box was also correctly highlighted by the incremental PCA approach with only a few spurious novelty indications. Incremental PCA coped better with the robot getting closer to the arena's walls (e.g. the large scale edge present in frame 46) because of our choice of parameters which influence generalisation - but this does not necessarily mean that incremental PCA always generalises better than the GWR network.

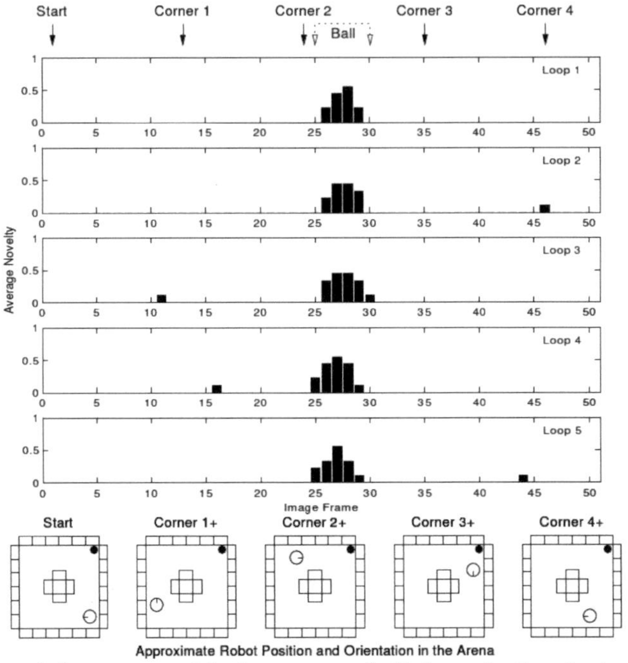

Figure 6. Inspection of the arena with the orange ball (novel stimulus) using incremental PCA. The orange ball is clearly and consistently highlighted with very few unexpected novelty indications.

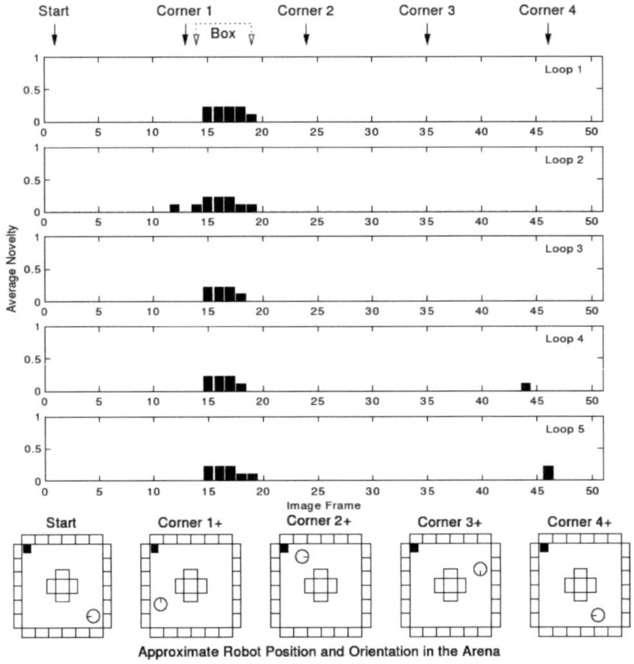

Figure 7. Inspection of the arena with the grey box (novel stimulus) using incremental PCA. The grey box is clearly and consistently highlighted with very few unexpected novelty indications.

3.3 Results

Table 3 shows a quantitative comparison of the results obtained with the GWR network and the incremental PCA approach. All cases presented statistically significant correlation between novelty filter response and actual novelty status according to the χ^2 analysis ($p \leq 0.01$).

Overall performances (combined performances for the orange ball and the grey box) of both approaches are quantitatively very similar (almost complete agreement between novelty filter response and ground truth), although the incremental PCA algorithm yielded slightly better and more consistent overall results.

	GWR network	Incremental PCA
Orange ball	$V = 0.91$ $U = 0.74$ $\kappa = 0.91$	$V = 0.86$ $U = 0.68$ $\kappa = 0.86$
Grey box	$V = 0.70$ $U = 0.44$ $\kappa = 0.70$	$V = 0.83$ $U = 0.60$ $\kappa = 0.83$
Overall	$V = 0.82$ $U = 0.58$ $\kappa = 0.82$	$V = 0.85$ $U = 0.64$ $\kappa = 0.85$

Table 3. Performance comparison for the laboratory experiments: all results correspond to statistically significant correlation between system response and actual novelty status (χ^2 analysis, $p \leq 0.01$).

Figure 8 gives examples of output images in which the novel object was present in the field of view of the robot's camera. In these output images the numbers indicate the location of the interest points identified by the saliency map in order of relevance (0 corresponding to the most salient), while the presence of white circles indicate that the surrounding region was classified as novel by the filter.

One can notice in Figure 8 that in both cases, the system was able to indicate *where* the regions containing part of the novel object were within the image frame. For the particular output images shown, the results yielded by both novelty filters - GWR network and incremental PCA - were identical.

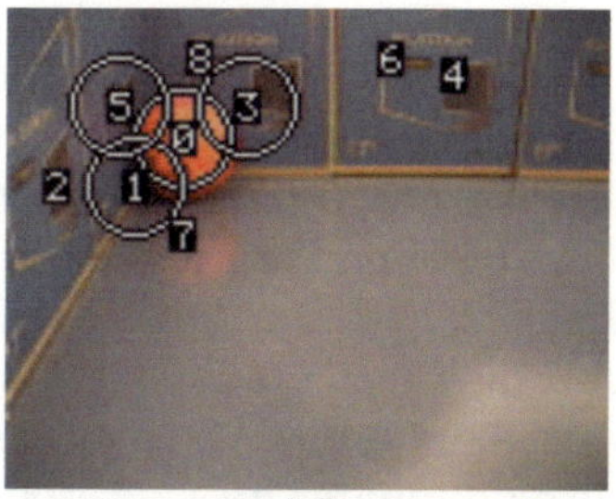

(a)

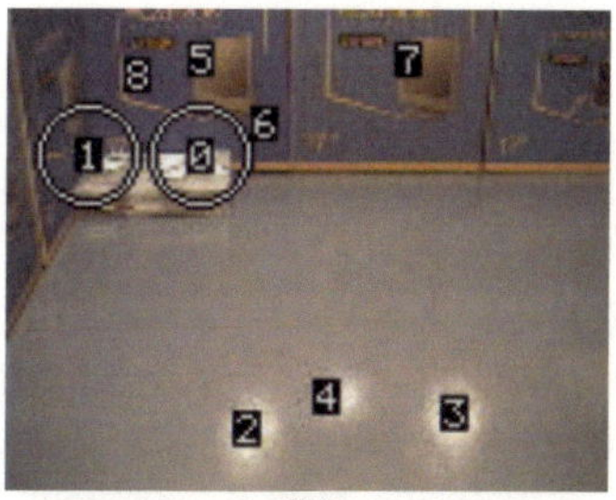

(b)

Figure 8. Examples of output images for the laboratory experiments: (a) orange ball (GWR and PCA); (b) grey box (GWR and PCA). The numbers indicate the location of salient points in order of importance (0 corresponds to the most salient) and the white circles indicate that the region corresponding to a particular salient point was labelled as novel.

4. Experiments in a real world environment

After the successful results obtained in experiments conducted in a laboratory environment, it was time to test the proposed visual novelty detection approach in a medium-scale real world environment.

The ideal scenario would be to send the robot down a sewer pipe to inspect for cracks, tree roots and other types of faults. However, our research robot was too large and also was not suitable to operate in such an environment. Furthermore, rigorous analysis and assessment of the system's behaviour in this kind of situation would be very difficult to perform due to the lack of knowledge and control of environmental characteristics - construction of the novelty ground truth, for instance, would constitute a difficult task.

Experimental setup. Hence, we decided to conduct experiments in one of the corridors at the Network Centre building at the University of Essex. The robot navigated along the corridor using the same navigation behaviour previously used in the laboratory experiments, acquiring one image frame per second, which resulted in the acquisition of 50 images per journey along the corridor. Differently from the previous laboratory experiments, the camera's pan-tilt unit was driven to its home position (facing straight towards the forward direction of the robot) for the experiments in the corridor.

Exploration was performed in the "empty" corridor to acquire a model of normality, as in the previous laboratory experiments, but limited to three journeys along the corridor. Finally, the learnt model of normality was used to inspect the corridor for unusual visual features that were manually inserted *a posteriori*.

We placed three different novel objects in the corridor at different times: a black rubbish bag, a dark brown bin and a yellow wooden board. These objects appeared in the robot's field of view immediately after the traversal of a door, which was present in the corridor.

Results. After three exploration journeys along the empty corridor, the GWR network acquired 48 nodes, while the incremental PCA acquired 80 model vectors. Although the GWR network acquired fewer concepts, the incremental PCA algorithm has a more efficient and compact representation.

Apart from the wooden board, the chosen novel objects are dark and therefore to some extent similar to the dark areas of the normal environment. Contrast in the images acquired in the corridor was generally poor because no extra illumination was used, just the weak lighting already present. In spite of this fact, both GWR network and incremental PCA algorithm were able to correctly highlight the novel objects in the corridor during inspection, as shown in Figure 9.

However, both novelty filters also responded with false novelty indications for a pair of fire extinguishers that were present in the corridor. These novelty indications were unexpected because the fire extinguishers were already present during the exploration phase and therefore should have been part of the acquired model of normality. We attribute such false novelty indications to changes in the scale of the already known visual features of the fire extinguishers and believe that use of an image encoding method that is robust to changes in scale would contribute to reduce false novelty indications and enhance general performance of the visual novelty filter. This hypothesis is currently under investigation (Vieira Neto & Nehmzow, 2007).

Table 4 shows a performance comparison in terms of Cramer's V, uncertainty coefficient U and κ index of agreement. All results showed statistically significant correlation between system response and actual novelty status (χ^2 analysis, $p \leq 0.01$). Overall performance (combined results for all three novel objects) indicates strong agreement between system response and actual novelty status. The GWR network presented the most consistent results.

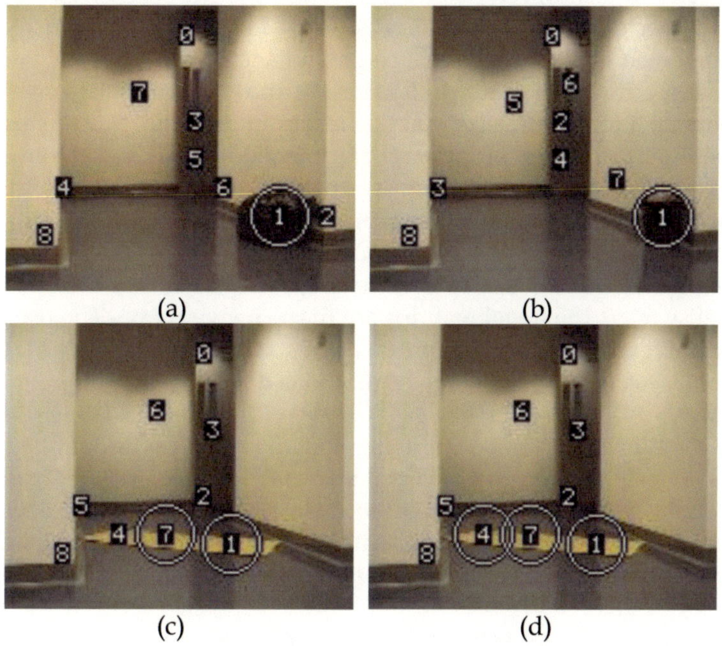

Figure 9. Examples of output images for the real world experiments: (a) black rubbish bag (GWR and PCA); (b) dark brown bin (GWR and PCA); (c) yellow wooden board (GWR only); (d) yellow wooden board (PCA only). The white circles correctly indicate regions containing novel features.

	GWR network	Incremental PCA
Black bag	$V=0.63$ $U=0.35$ $\kappa=0.63$	$V=0.65$ $U=0.37$ $\kappa=0.65$
Brown bin	$V=0.64$ $U=0.38$ $\kappa=0.64$	$V=0.50$ $U=0.23$ $\kappa=0.50$
Yellow board	$V=0.67$ $U=0.37$ $\kappa=0.67$	$V=0.84$ $U=0.69$ $\kappa=0.84$
Overall	$V=0.65$ $U=0.36$ $\kappa=0.65$	$V=0.70$ $U=0.44$ $\kappa=0.70$

Table 4. Performance comparison for the real world experiments: all results correspond to statistically significant correlation between system response and actual novelty status (χ^2 analysis, $p \leq 0.01$).

5. Conclusion

To achieve novelty detection on visual images, for the purpose of automated inspection tasks, we used a mechanism of visual attention that selects candidate image patches in the input frame, combined with methods that classify image patches as novel or non-novel.
For real world inspection applications vision is the most appropriate sensor modality, as it provides colour, texture, shape, size, and distance information. All of this information is useful for robots operating in complex environments, but of course comes at the cost of processing large amounts of data, which is a particular challenge on autonomous mobile robots with limited computing power available.
We demonstrated that the use of the saliency map (Itti et al., 1998) as selective attention mechanism minimises the amount of data to be processed and at the same time makes it possible to localise *where* the novel features are in the image frame. The use of this attention mechanism avoids explicit segmentation of the input image (Singh & Markou, 2004) and makes the overall system more robust to translations due to robot motion.
Because novelty is of contextual nature and therefore can not be easily modelled, the approach that we follow is to first acquire a model of normality through robot learning and then use it as a means to highlight any abnormal features that are introduced in the environment. For this purpose, we have used unsupervised clustering mechanisms such as the GWR neural network (Marsland et al., 2002b) and the incremental PCA algorithm (Artač et al., 2002), which were both able to learn aspects of the environment incrementally and yielded very good results.
We proposed an experimental setup to evaluate performance and functionality of visual novelty filters, dividing the experimental procedure in two stages: an exploration phase, in which the learning mechanism was enabled to allow the robot to build a model of normality while experiencing the environment; and an inspection phase, in which the acquired model of normality was used as a novelty filter. Novel objects were inserted in the robot's environment during the inspection phase of experiments with the expected outcome that the visual novelty filter would produce indications of novelty and localise these new objects in the corresponding input image frame.
As the precise location and nature of the novelty introduced during the inspection phase is known by the experimenter, it is possible to generate ground truth data to be compared with the responses given by the novelty filter. In order to assess the performance of a novelty filter objectively, we used 2×2 contingency tables relating actual novelty status (ground truth) to system response, followed by the computation of statistical tests to quantify the association or agreement between them. Here we used the χ^2 test in order to check the statistical significance of the association between ground truth and novelty filter response, followed by the computation of Cramer's V, the uncertainty coefficient U and the κ index of agreement (Sachs, 2004).
Extensive experimental data was logged to evaluate and compare the efficiency of the visual novelty filter. The χ^2 analysis of the generated contingency tables revealed statistical significance in the associations between system response and actual novelty status in all of the reported experiments. Typical quantitative analyses resulted in strong agreement with the ground truth data.
Qualitative assessment of the learning procedure during exploration, as well as consistent identification of novel features during inspection was made through the use of novelty bar

graphs. In these graphs, a measure of the degree of novelty in each image frame is plotted against time/position. Novelty graphs are particularly useful to identify novelty indications in unexpected locations of the environment and investigate their reasons, leading to improvements in overall system robustness and ability to generalise.

We consider the results obtained to be very good and likely to succeed in real world applications that involve exploration and inspection of environments using vision. An example of such an application is the automated inspection of sewer pipes. However, more elaborate processing to become more robust to general affine transformations is likely to be necessary for applications in which the environment is not as structured as the arena and corridor that were used as operating environments in this work.

5.1 Contributions

One of our main contributions was to implement and experiment with visual novelty detection mechanisms for applications in automated inspection using autonomous mobile robots. Previous work done in novelty detection used only low-resolution sonar readings (Crook et al., 2002; Marsland et al., 2002a) or very restricted monochrome visual input (Crook & Hayes, 2001; Marsland et al., 2001). In contrast to this, the work presented here used colour visual stimuli with an unrestricted field of view. The selection of algorithms had emphasis on bottom-up and unsupervised learning approaches to allow exploitation of relevant characteristics of the acquired data from the ground-up.

Quantitative performance assessment tools based on contingency table analysis and statistical tests were developed in order to support objective comparisons between different visual novelty filters. For comparison purposes, novelty ground truth maps were generated in the form of binary images, in which novel visual features are highlighted manually. Because vision is a sensor modality shared between robots and humans, generation of novelty ground truth maps occurs in a natural and relatively easy way (although it demands time because of the volume of images involved).

Another main contribution was the demonstration that attention mechanisms extend the functionality of visual novelty filters, enabling them to localise where the novel regions are in the input frame and improving image encoding robustness to translations due to robot motion. Also, the use of an attention mechanism avoids explicit segmentation of the input image frame.

5.2 Future research

The results and conclusions drawn from the experiments in visual novelty detection reported in this work open a series of avenues for future investigations and improvements.

It would be interesting, for instance, to conduct more experiments using alternative attention mechanisms, especially those which can determine affine transformation parameters for the selected regions of interest. Possible options are the Harris-affine detector (Mikolajczyk & Schmid, 2002; Mikolajczyk & Schmid, 2004) and the interest point detector developed by Shi & Tomasi (1994). The use of such algorithms is expected to result in image encoding with extra robustness to affine transformations, improving the ability to generalise and reducing the number of stored vectors or nodes by the novelty filter. Experiments are needed to compare performances with the attention mechanism already studied here to confirm or reject this hypothesis.

There are also some alternative methods of interest for the image encoding, which are likely to improve robustness to changes in scale and orientation of visual features. One possibility is the use of space-variant (log-polar) foveation (Bernardino et al., 2002) and the Fourier-Mellin transform (Derrode, 2001; Reddy & Chatterji, 1996) in order to encode visual features. For applications that demand a systematic exploration of complex large-scale environments, such as a whole floor in a building, the integration of the proposed visual novelty detection framework with the environment exploration scheme developed by Prestes e Silva Jr. et al. (2002; 2004) is of particular interest. This approach uses potential fields to generate a dynamic exploration path that systematically covers the entire free area of the robot's environment, while generating a grid map of the obstacles that are present. Later on, the generated grid map can be used to produce arbitrary inspection paths or even paths towards specific goals.
If a novelty detection algorithm is used to learn and associate the local visual appearance of the environment to the grids of the environmental map, it is possible to determine novelty not only in terms of uncommon features that may appear in the environment, but also to establish if known features appear in unusual locations. A potential application of such ability is the automated organisation of a room, in which an autonomous mobile robot would be able to identify which objects are not in the places they were supposed to be and then take actions to correct the situation, "tidying up" the environment.

6. Acknowledgements

The research and experimental work reported in this chapter was conducted at the Robotics Research Laboratory of the University of Essex (UK) with the financial support of CAPES Foundation (Brazil), which is gratefully acknowledged

7. References

Artač, M.; Jogan, M. & Leonardis, A. (2002). Incremental PCA for on-line visual learning and recognition. *Proceedings of the 16th International Conference on Pattern Recognition*, Vol. 3, pp. 781-784, ISBN 0-7695-1695-X, Quebec, Canada, August 2002.

Bernardino, A.; Santos-Victor, J. & Sandini, G. (2002). Model-based attention fixation using log-polar images, In: *Visual Attention Mechanisms*, Cantoni, V.; Petrosino, A. & Marinaro, M. (Ed.), pp. 79-92, Plenum Press, ISBN 978-0306474279, New York, USA.

Crook, P. & Hayes, G. (2001). A robot implementation of a biologically inspired method for novelty detection. *Proceedings of TIMR 2001 - Towards Intelligent Mobile Robots*, Manchester, UK, April 2001.

Crook, P.; Marsland, S.; Hayes, G. & Nehmzow, U. (2002). A tale of two filters - on-line novelty detection. *Proceedings of the 2002 IEEE International Conference on Robotics and Automation*, Vol. 4, pp. 3894-3899, ISBN 0-7803-7272-7 Washington DC, May 2002.

Derrode, S. (2001). Robust and efficient Fourier-Mellin Transform approximations for gray-level image reconstruction and complete invariant description. *Computer Vision and Image Understanding*, Vol. 83, No. 1, pp. 57-78, ISSN 1077-3142.

Hall, P.; Marshall, D. & Martin, R. (1998). Incremental eigenanalysis for classification. *Proceedings of the 9th British Machine Vision Conference*, Vol. 1, pp. 286-295, ISBN 1-901725-04-9, Southampton, UK, September 1998.

Hopfield, J. J. (1982). Neural networks and physical systems with emergent collective computational abilities. *Proceedings of the National Academy of Sciences,* Vol. 79, No. 8, pp. 2554-2558, ISSN 0027-8424.

Itti, L. & Koch, C. (2001). Computational modelling of visual attention. *Nature Reviews Neuroscience,* Vol. 2, No. 3, pp. 194-203, ISSN 1471-0048.

Itti, L.; Koch, C. & Niebur, E. (1998). A model of saliency-based visual attention for rapid scene analysis. *IEEE Trans. on Pattern Analysis and Machine Intelligence,* Vol. 20, No. 11, pp. 1254-1259, ISSN 0162-8828.

Kohonen, T. (1984). *Self-organization and Associative Memory,* Springer-Verlag, ISBN 978-3540513872, New York, USA.

Linåker, F. & Niklasson, L. (2000). Time series segmentation using an adaptive resource allocating vector quantization network based on change detection, *Proceedings of the 2000 International Joint Conference on Neural Networks,* pp. 323-328, ISBN 978-0769506197, Como, Italy, July 2000, IEEE Computer Society Press.

Linåker, F. & Niklasson, L. (2001). Environment identification by alignment of abstract sensory flow representations, In: *Advances in Neural Networks and Applications,* Mastorakis, N.; Mladenov, V.; Suter, B. & Wang, L. (Ed.), 79-116, WSES Press, ISBN 960-8052-26-2.

Marsland, S.; Nehmzow, U. and Shapiro, J. (2000). Detecting novel features of an environment using habituation, *From Animals to Animats 6: Proceedings of the 6th International Conference on Simulation of Adaptive Behavior,* pp. 189-198, ISBN 978-0262632003, Paris, France, September 2000, MIT Press.

Marsland, S.; Nehmzow, U. & Shapiro, J. (2001). Vision-based environmental novelty detection on a mobile robot, *Proceedings of the International Conference on Neural Information Processing*, ISBN 978-7-309-03012-9, Shanghai, China, November 2001.

Marsland, S.; Nehmzow, U. & Shapiro, J. (2002a). Environment-specific novelty detection. *From Animals to Animats 7: Proceedings of the 7th International Conference on the Simulation of Adaptive Behaviour,* pp. 36-45, ISBN 0-262-58217-1, Edinburgh, UK, August 2002, MIT Press, Cambridge, USA.

Marsland, S.; Shapiro, J. & Nehmzow, U. (2002b). A self-organising network that grows when required. *Neural Networks,* Vol. 15, No. 8-9, pp. 1041-1058, ISSN 0893-6080.

Mikolajczyk, K. & Schmid, C. (2002). An affine invariant interest point detector, *Computer Vision - ECCV 2002: Proceedings of the 7th European Conference on Computer Vision,* pp. 128-142, ISBN 978-3540437444, Copenhagen, Denmark, May 2002, Springer-Verlag, Berlin, Germany.

Mikolajczyk, K. & Schmid, C. (2004). Scale and affine invariant interest point detectors. *International Journal of Computer Vision,* Vol. 60, No. 1, pp. 63-86, ISSN 0920-5691.

Nehmzow, U. (2003). *Mobile Robotics: A Practical Introduction,* 2nd ed., Springer-Verlag, ISBN 978-1852337261, London, UK.

Prestes e Silva Jr., E.; Engel, P. M.; Trevisan, M. & Idiart, M. A. P. (2002). Exploration method using harmonic functions. *Robotics and Autonomous Systems,* Vol. 40, No. 1, pp. 25-42, ISSN 0921-8890.

Prestes e Silva Jr., E.; Idiart, M. A. P.; Trevisan, M. & Engel, P. M. (2004). Autonomous learning architecture for environmental mapping. *Journal of Intelligent and Robotic Systems,* Vol. 39, No. 3, pp. 243-263, ISSN 0921-0296.

Reddy, B. S. & Chatterji, B. N. (1996). An FFT-based technique for translation, rotation, and scale-invariant image registration. *IEEE Transactions on Image Processing,* Vol. 5, No. 8, pp. 1266-1271, ISSN 1057-7149.

Sachs, L. (2004). *Angewandte Statistik: Anwendung statistischer Methoden,* Springer Verlag, ISBN 978-3540405559, Berlin, Germany.

Shi, J. & Tomasi, C. (1994). Good features to track, *Proceedings of the IEEE Conference on Computer Vision and Pattern Recognition,* pp. 593-600, ISBN 0-8186-5825-8, Seattle, USA, June 1994, IEEE Computer Society.

Singh, S. & Markou, M. (2004). An approach to novelty detection applied to the classification of image regions. *IEEE Transactions on Knowledge and Data Engineering,* Vol. 16, No. 4, pp. 396-407, ISSN 1041-4347.

Tarassenko, L; Hayton, P.; Cerneaz, N. & Brady, M. (1995). Novelty detection for the identification of masses in mammograms. *Proceedings of the 4th IEEE International Conference on Artificial Neural Networks,* pp. 442-447, ISBN 0-85296-641-5, Cambridge, UK, June 1995.

Vieira Neto, H. & Nehmzow, U. (2005). Automated exploration and inspection: comparing two visual novelty detectors. *International Journal of Advanced Robotic Systems,* Vol. 2, No. 4, pp. 355-362, ISSN 1729-8806.

Vieira Neto, H. & Nehmzow, U. (2007). Visual novelty detection with automatic scale selection. *Robotics and Autonomous Systems,* Vol. 55, No. 9, pp. 693-701, ISSN 0921-8890.

19

An Inspection Robot for High Voltage Power Transmission Line and Its Dynamics Study

Xiaohui Xiao, Gongping Wu, Hua Xiao and Jinchun Dai,
Wuhan University
P. R. China

1. Introduction

Inspection robot for power transmission line is a new type of technical carrier for detection and inspection of mechanical/ electrical failures in high-voltage (110kV, 220kV) and extra-high voltage (500kV or above) power transmission system. The robot equipped with sensors, detection instruments and data communication apparatus, is capable of inspecting transmission system without suspending power supply. As shown in Fig. 1, an inspection robot developed by Wuhan University can autonomous move along the 220kV phase line and overcome all kinds of obstacles to carry out the inspection tasks (Wu et al., 2006: Xiao et al., 2005). Comparing with such current inspection approaches as inspectors and unmanned aerial vehicles (UAV), the robot is more efficiency and safer to assure higher detection quality, especially in severe conditions (mountain areas, river-crossing, grasslands, etc). Thus it has broad engineering application prospects.

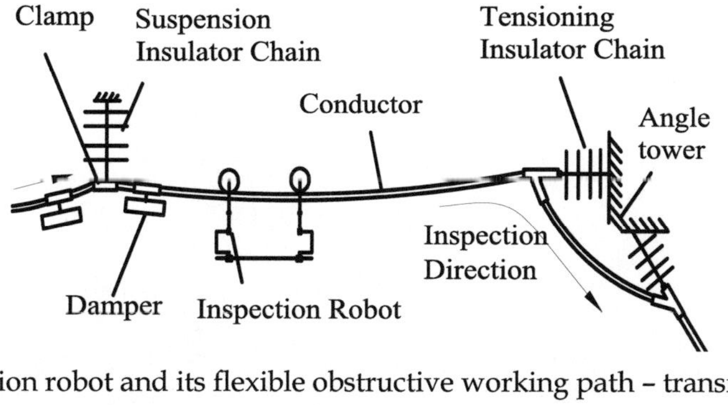

Fig. 1. An inspection robot and its flexible obstructive working path – transmission line

Since the end of 1980s, a number of relative achievements have been carried out about inspection robot system (Montambault, 2003: Peungsungwal, 2001: Sawada, 1991: Tang & Fang, 2004: Wu et al., 2006), mechanism schematic (Wu et al., 2006: Zhou et al., 2004: Zhou et al., 2004), control system (Tang, 2004: Wang et al., 2006: Zhu et al., 2006), obstacles recognition and navigation (Fu, 2005: Li, 2007: Zhang et al., 2007).

The multi-rigid-body dynamics modeling, simulation and experimental tests were performed with an phase-line inspection robot prototype (Xiao et al., 2005). Considering the flexible obstructive working environment, the performances of the inspection robot,

especially the precision of obstacles' location and failure signals' detection, are affected by the coupling vibration of the robot and overhead transmission line. Thus, the rigid-flexbile coupling dynamics modeling and simulations were studied in typical working conditions (Xiao et al., 2006, 2007 & 2008).
This chapter will introduce three generations prototypes of the inspection robot for 220kV phase line developed by Wuhan University and analyze its dynamic performances, including multi-rigid-body dymanics of the robot, coupling dynamics of the robot and flexible line.

2. Inspection robot prototypes

Since 1997, Wu et al. in Wuhan University have developed three generations of inspection robot prototypes, namely, remotely operated vehicle - ROV, auto-crawling robot - ACR, and auto-rolling/crawling robot - ARCR.
(a) ROV: The first generation prototype ROV is composed of three extension arms, three wheels and one translation rail (Wu et al., 1999). There is one rotation degree of freedom (DOF) between each extension arm and wheel, and one rotation DOF between fore/rear extension arm and robot body as well as one translation DOF between middle arm and robot body. By remotely operation, the ROV is able to travel along no-obstacle phase line by means of synchronization drive of three wheels, and overcome insulator chains, dampers and suspension clamps in manner of three arms' stepping in turn. However, it is incapable of climbing overhead line's sag and spanning tensioning tower.
(b) ACR: Since the performance limitaions of ROV, an auto-crawling robot (ACR), was developed in 2000 (Wu, et al, 2006). As shown in Fig. 2, the ACR prototype is composed of two umbrella-shaped suspension wheels, two clamping jaw mechanisms, two stroke amplification mechanisms, and hydraulic servo/drive system. The three wheels, of which the angle between centerlines is 120°, can rotate around the tongue wheel together with bearing shaft. Hydraulic servo is adopted for motion controlling, including the clamps' adaptive constant force grasping, amplification mechanism's stretching motion, and coordinated crawling. A single-action cylinder is used to drive clamping jaw mechanism, while a double-action cylinder is for stroke amplification mechanism. However, the slow crawl speed and inability to span tension towers limite ACR's application.

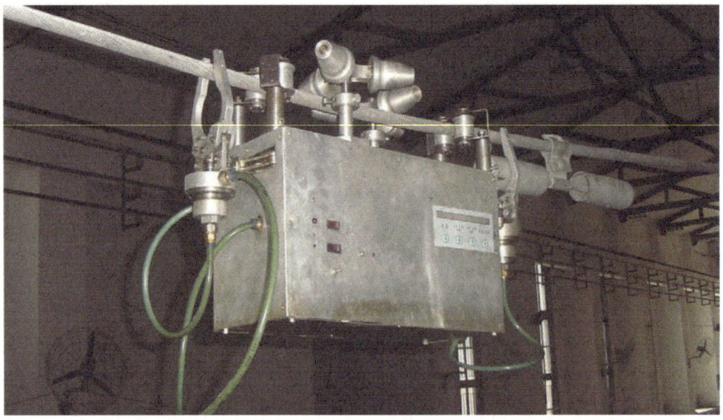

Fig. 2. ACR prototype (by Wuhan University, 2000)

(c) ACRC: Based on the above two generaions prototypes, an auto-rolling/crawling robot prototype(ARCR) was developed for autonomous online full-path inspection of 220kV transmission phase line (Wu et al., 2006). ARCR is composed of three sub-sysytems including the inspection robot, remotely control ground station, and signal analysis/ diagnosis/inspection management system software. The remotely control ground station is avalable for wireless data transceiver and picture capturing. The diagnosis/inspection management system software is for visual light and infrared image analysis, failure diagnosis, and inspection database management.

Fig. 3. ARCR prototype (by Wuhan University, 2005)

As shown in Fig. 3, the inspection robot is composed of mechanism, power on-line supply, sensor and obstacles detection, navigation, image scanning and detection, data wireless transceiver, and control system. The self-governing on robot's obstacle-overcoming is realized by means of autonomous navigation of multiple electromagnetic sensors and machine visual hybrid servo. Magnetic energy of transmission conductor is converted into electric energy for power supply. Therefore, the robot can fulfill six basic functions as follows: (1) full path moving along 220kV phase line with obstacles, (2) online power supply and monitoring, (3) navigation including obstales' detecting, identifying and location, (4) visible light/infrared image scanning and detection, (5) wireless communication, (6) robot self-postion detection, grasping force detection, and motions programming.

The performance tests of ARCR was conducted on 220kV field live lines of Wuhan Power Company. The main performances parameters are listed as following: weight: 30kg; dimensions (length×width×height): 650mm×300mm×600mm; valid wireless communication distance: 4km; average power consumption: 40W; available power supply: 40W (as load current of phase line = 220A); rolling speed: 5km/h; maximum climbing grade: 15°; crawling speed: 200m/h; crawl grade: 75° (Wu et al., 2006).

In consideration of the obstructive working environment and the requirement on inspection tasks, ARCR prototype mechanism is designed into double-arms symmetrical and suspending structure (Fig. 4). There are nice DOF in total. There is one driving wheel on the end of each arm enabling the robot to roll along non-obstacle section of line, and a pair of claws enabling the robot to grasp/loose the line when it encounters obstacles. Each arm has two rotation degrees of freedom (DOF) to realize rotation of robot arms on two vertical axes. Between the two arms, there is an interactive translation DOF available for their interactive sliding and transposition along the slide rail.

3. Working environment analysis and obstacles-overcoming programming

3.1 Kinematics tasks

The typical structure of the transmission phase line, as shown in Fig. 1, includes suspension and tensioning angle towers, phase lines and accessories (dampers, suspension or tensioning line clamp, insulator chains, etc.). Taking the phase line as its moving path, the ACRC has to carry out three kinematics tasks as follows:

a) Moving along the no-obstacle segment of the phase line:

b) Overcoming the obstacles along the phase line including the suspension/tensioning tower, dampers, clamps, and insulator chains, etc.:

c) Varying moving paths between phase line and jumper line.

3.2 Flexible obstructive inspection moving path

The flexibility of the transmission line is very high, because the span between two adjacent towers is usually as much as hundreds of even more than one thousand meters, and the sag is scores of meters as while.

Moreover, the environmental wind loads may excite Aeolian vibration, or galloping in the the winter (Guo et al., 2002), of which the vibration and force can be transferred to the robot. On the other hand, when the robot overcomes obstacles or change moving paths, it has to adjust postures and thus produces unbalanced force. The coupling of the robot and overhead line will force the robot to vibrate and thus decreases its performance.

3.3 Obstacle-overcoming programming

In kinematic and dynamics modeling, we only consider 6 degrees of freedom, namely, rotation Joint 2 and 3 of Arm I, and rotation Joint 5 and 6 of Arm II, translation Joint 1 and the horizontal translation Joint 4 between two arms. The axis of Joint 2 and Joint 6 are horizontal, intersecting vertical with that of Joint 3 and Joint 5, respectilvely.

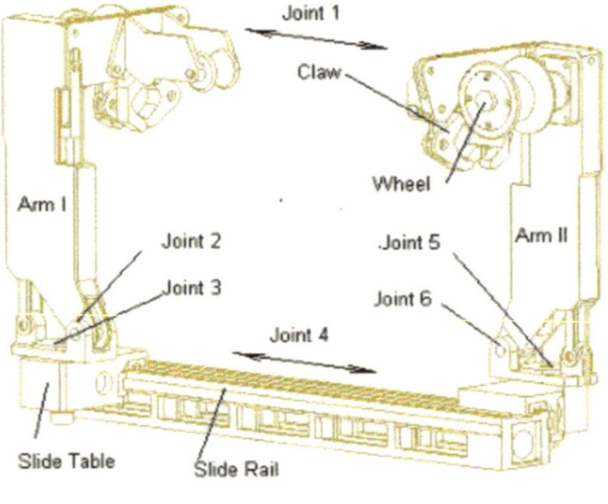

Fig. 4. Symmetrical mechanism structure of the ACRC

As the symmetrical structure, the motion of six DOF can be abstracted into four basic sub-actions, with which the robot is able to carry out all the three required kinematics tasks. Taking damper-overcoming as an example, the four sub-actions are programmed in Fig. 5 (a)-(d). Sub-action 2, 3, and 4 are basic for obstacles-overcoming.

(a) Sub-action 1: Two wheels roll along the transmission line with two arms parallelly suspending on the line.

(b) Sub-action 2: Arm I (or Arm II) end manipulator clamps the line, while the robot rotates with Joint 2 (or Joint 6) to lift/descend the robot body by 30°.
(c) Sub-action 3: Arm I (or Arm II) end manipulator clamps the line, while another arm rotates with the axis of Joint 5 (or Joint 3) by 180°.
(d) Sub-action 4: Arm I (or Arm II) end manipulator clamps line, while another arm translates along Joint 4 the slide rail to transpose two arms.

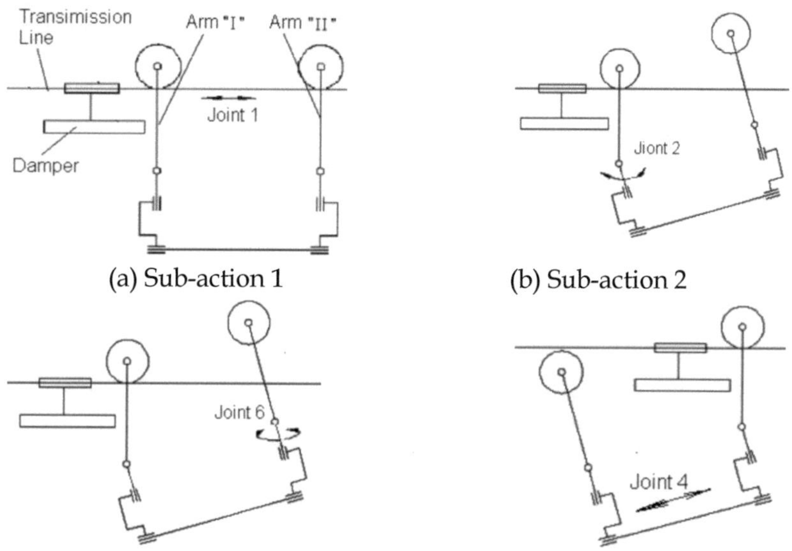

(a) Sub-action 1 (b) Sub-action 2

(c) Sub-action 3 (d) Sub-action 4

Fig. 5. Action programming for damper-overcoming

4. Multi-rigid-body dynamics of the robot

4.1 Dynamics modeling

The dynamics model of the robot is derived with Lagrange method. Taking E_k as the kinetic energy and E_p the potential energy of the system, the Lagrange function is defined as:

$$L = E_k - E_p \tag{1}$$

Then, the Lagrange eqaution of the system is :

$$\tau_i = \frac{d}{dt}\left(\frac{\partial L}{\partial \dot{q}_i}\right) - \frac{\partial L}{\partial q_i}, \; i = 1, 2, \cdots n \tag{2}$$

where, q_i is the generalized displacement at Joint i (*m*, or *rad*): $\dot{q}_i$ the generalized velocity at Joint I (*m/s*, or *rad/s*): τ_i the generalized force at Joint I (*N*, or *N·m*).

With E_k and E_p represented by homogeneous coordinate, the general dynamics equation of multi-rigid body system is:

$$\tau_i = \sum_{j=1}^{n} D_{ij}\ddot{q}_j + I_{ai}\ddot{q}_i + \sum_{j=1}^{n}\sum_{k=1}^{n} D_{ijk}\dot{q}_j\dot{q}_k + D_i \tag{3}$$

where, n is the number of robot links; D_{ik} is the acceleration item; While $i = k$, D_{ii} is the effective inertia; while $i \neq k$, D_{ij} is the coupled inertia of Joint i and Joint j; D_{ikm} is the inverse torque item imposing on joint i generated by the acceleration of joint k; while $k = m$, D_{ikk} is the Centripetal Force coefficient caused by velocity of Joint j at Joint i; while $k \neq m$, D_{ikm} is the Coriolis Force coefficient caused by velocity of Joint j and k at Joint i; D_i is the gravity at Joint i.

$$D_{ik} = \sum_{j=\max i,k}^{n} Trace\left(\frac{\partial T_j}{\partial q_i} J_j \frac{\partial T_j^T}{\partial q_k}\right) + I_{ai}\delta_{ik} \text{ where } \delta_{ik} = \begin{cases} 1, i = k \\ 0, i \neq k \end{cases} \tag{4}$$

$$D_{ijk} = \sum_{p=\max i,j,k}^{n} Trace\left(\frac{\partial^2 T_p}{\partial q_j \partial q_k} J_p \frac{\partial T_p^T}{\partial q_i}\right) \tag{5}$$

$$D_i = \sum_{i=1}^{n} -m_i \vec{g}^T \frac{\partial T_p}{\partial q_i} {}^p\vec{r}_p \tag{6}$$

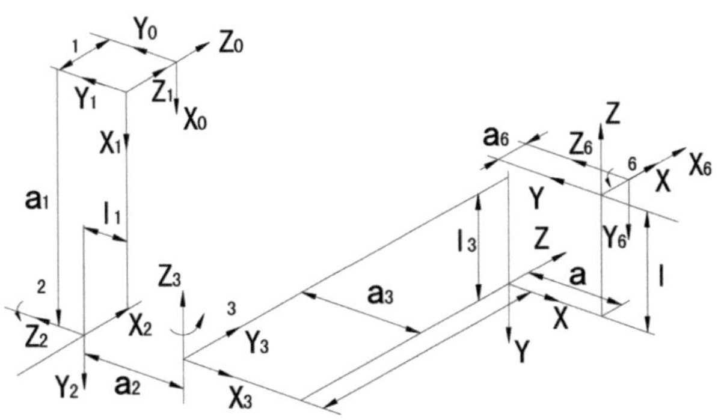

Fig. 6. The links coordinates setting of the ACRC Prototype

Link i	θ_i	α_{i-1}	a_{i-1}	d_i
1	0°	0°	0	d_1 (variable)
2	θ_2 (variable)	-90°	a_1	$d_2(-l_1)$
3	θ_3 (variable)	90°	a_2	$d_3\ (-l_2)$
4	180°	90°	a_3	d_4 (variable)
5	θ_5 (variable)	-90°	a_4	$d_5(l_5)$
6	θ_6 (variable)	-90°	a_5	0

Table 1. Link Parameters of ACRC prototype

Considering the six joints defined in section 3, the coordinates of each link were formed in Fig. 6. Based on D-H method robot link parameters were obtained (Table 1). α_i, a_i, d_i stand for the link twist angle, link length, and the link offset, respectively.

The initial value of the six variables are listed as follows:

$$d_1 = 0; d_4 = l_4; \theta_2 = -90^\circ; \theta_3 = 90^\circ; \theta_5 = -90^\circ; \theta_6 = 0^\circ$$

Then we can obtain link transformation matrix T_i, and derive D_{ik}, D_{ikm}, D_i and pseud-inertia matrix J_i from Equ. (4)-(6), which were detailed in paper (Xiao , 2005). The effective inertia items are listed in Equ.(7) - (12),

$$D_{11} = \sum_{p=1}^{6} Trace\left(\frac{\partial T_p}{\partial q_1} J_p \frac{\partial T_p^T}{\partial q_1}\right) = m_1 + m_2 + m_3 + m_4 + m_5 + m_6 \quad (7)$$

$$\begin{aligned} D_{22} = \sum_{p=2}^{6} Trace\left(\frac{\partial T_p}{\partial \theta_2} J_p \frac{\partial T_p^T}{\partial \theta_2}\right) &= I_{2xx} + c_3^2 I_{3xx} + c_3^2 I_{4xx} + c_{35}^2 I_{5xx} + \left(c_6^2 c_{35}^2 + s_6\right) I_{6xx} + I_{2yy} \\ &+ s_3^2 I_{3yy} + I_{4yy} + s_{35}^2 I_{5yy} + \left(s_6^2 c_{35}^2 + c_6^2\right) I_{6yy} + I_{3ZZ} + s_3^2 I_{4ZZ} + I_{5ZZ} + s_{35}^2 I_{6ZZ} + m_3 l^2 + m_4 l^2 \\ &+ m_4 d_4^2 s_3^2 + m_5 d_4^2 s_3^2 + m_6 d_4^2 s_3^2 + 2 m_4 d_4 Z_4 s_3^2 - 2 m_6 \bar{y}_6 d_4 s_3 s_6 c_{35} \end{aligned} \quad (8)$$

$$\begin{aligned} D_{33} = \sum_{p=3}^{6} Trace\left(\frac{\partial T_p}{\partial \theta_3} J_p \frac{\partial T_p^T}{\partial \theta_3}\right) &= I_{3xx} + I_{4xx} + I_{5xx} c_6^2 I_{6xx} + I_{3yy} + I_{5yy} + s_6^2 I_{6yy} + I_{4ZZ} + I_{6ZZ} \\ &+ 2 d_4 \left(m_4 \bar{Z}_4 + m_6 \bar{y}_6 S_5 S_6 \right) + d_4^2 \left(m_4 + m_5 + m_6\right) \end{aligned} \quad (9)$$

$$D_{44} = \sum_{p=4}^{6} Trace\left(\frac{\partial T_p}{\partial d_4} J_p \frac{\partial T_p^T}{\partial d}\right) = m_4 + m_5 + m_6 \quad (10)$$

$$D_{55} = \sum_{p=5}^{6} Trace\left(\frac{\partial T_p}{\partial \theta_5} J_p \frac{\partial T_p^T}{\partial \theta_5}\right) = I_{5xx} + I_{5yy} + c_6^2 I_{6xx} + s_6^2 I_{6yy} + I_{6ZZ} \quad (11)$$

$$D_{66} = \sum_{p=6}^{6} Trace\left(\frac{\partial T_p}{\partial \theta_6} J_p \frac{\partial T_p^T}{\partial \theta_6}\right) = I_{6xx} + I_{6yy} \quad (12)$$

Wherein, c_i and s_i stand for cosine θ_i and sine θ_i respectively, c_{jk} and s_{jk} stand for $\cos(\theta_j + \theta_k)$ and $\sin(\theta_j + \theta_k)$, respectively.

4.2 Experimental tests and simulation

The experimental tests were performed with the III ACRC prototype in the simulative 220 kV 1: 1 overhead transmission line laboratory, Wuhan University. The experimental test system is detailed in Fig. 7. A 20-meters model with three spans and two towers was set up for full-path inspection tests. The test variables include motor dirve current, angular displacement, velocity and acceleratopm of each joint. The angular motion sensors are embedded in the robot control system. The motor current test system was composed of 6 hall-effect current transducers, a 6-channel amplifier, INV360D data acquisition instrument, and a computer with data analysis software. The tests were performed under 4 sub-actions. The experimantal results were listed in the paper (Xiao et al., 2005).

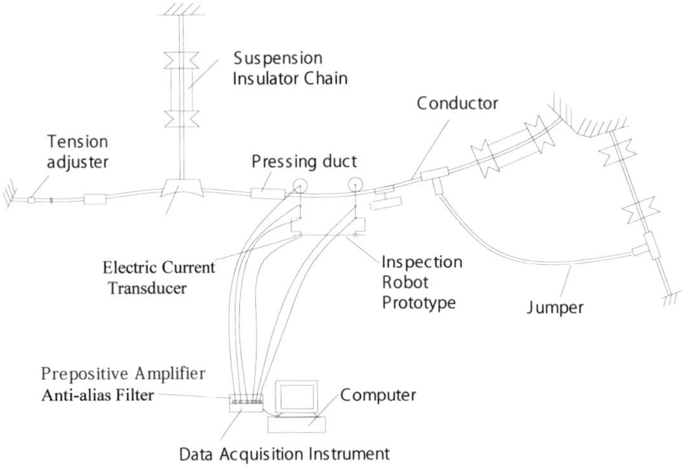

Fig. 7. Experimental test system scheme (Xiao et al., 2005)

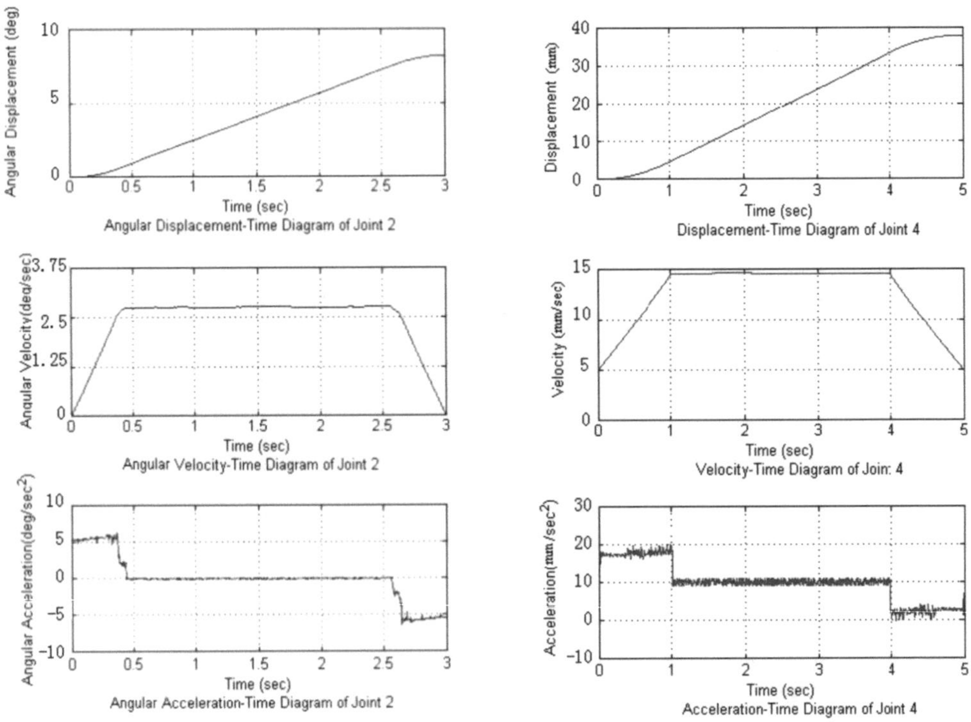

(a) Sub-action 2 - Joint 2 (b) Sub-action 4 - Joint 4

Fig. 8. Simulation resutlts (Xiao et al., 2005)

Based on the dynamic model in section 4.1, we performed forward dynamics simulations of 4 sub-actions in MATLAB. Fig. 8 shows the simulation results of Sub-action 2 and Sub-action 4. Comparing with the experimental results, the angular displacement and velocity in simulation are more stable, because we didn't consider the flexibility of the transmission line in dynamics modeling.

5. Rigid - flexible coupling dynamics of robot and transmission line

To explore the influences of the flexible path on the robot's dynamic performance, coupling modeling and simulation were conducted based on multi-flexible body dynamics theories. First, a finite element model (FEM) of one span of line was built to obtain its dominant modals for spatial configuration. Second, a multi-flexible-body dynamics model of the line was obtained with Lagrange method. Third, the multi-rigid-body model of the robot and the multi-flexible-body model of the line was coupled to conduct coupling dynamics simulation.

5.1 Multi-flexible modeling of the transmission line

For the rigidity of the large span of flexible line has little impact on its spatial configuration, we can assume that the line takes on "Catenary state" to calculate the coordinates of the key points(Li, 2001). Considering the general condition in 220 kV hight-voltege transmission system, we chose the conductor's type LGJ-185/30: diameter = 18.88 mm, density = 3.473×103 kg/m^3 elastic modulo = 7,600Mpa, tensile force of the line = 500 N.

A FEA model was built in ANSYS with the key poits data. The modal frequencies and modal shape are obtained with subspace method. Then, the spatial configuration of overhead line can be described with selected modal vectors and corresponding modal coordinates, namely, the physical coordinate vectors of the line can be indicated by superposition of the selected dominant models (Xiao et al., 2007).

5.2 Coupling contact model under sub-action 1

In ADMAS, the contact model of flexible line and the rigid robot wheel was built via discretizing the actual continuous contact modeling. We simplified the model of the robot and line, and equalized their contact force to two dimensional contact between central node group of flexible line FEA model and rigid edge circle of the robot wheel. Dynamics model for inspection robot rolling on non-barrier segment of transmission line contained 300 contract force units in total as shown in Fig.9, where 1 is transmission line finite element model; 2, dumb object; 3, fixing pair; 4, contract force unit; 5, wheel of Arm "I"; 6, two-dimensional circle; 7, kinematical input; 8, robot body; 9, rotating pair; 10, co-planer restraint:

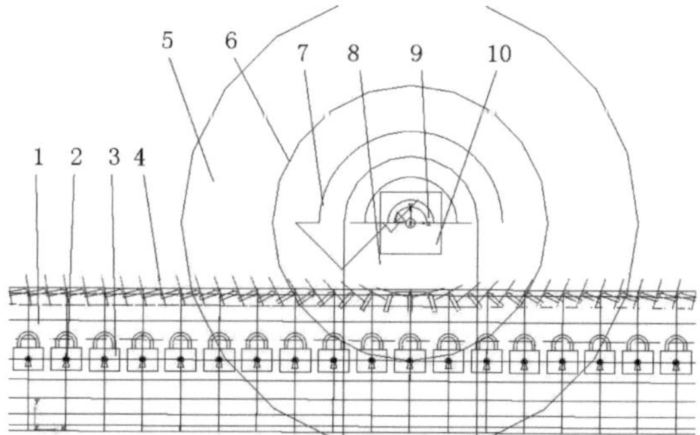

Fig. 9. Contact mode of one robot wheel and flexible transmission line

5.2 Simulation results

The joint's kinematical function was defined with *STEP* function in ADMAS. The form of STEP is:

$$STEP(t,t_0,x_0,t_1,x_1) \tag{13}$$

where, t is the independent variable ; t_0 , t_1 the initial and final value of t , respectively; x_0 , x_1 the initial and final function value of *STEP* , respectively.

According to the parameters of the robot prototype, the joint *STEP* functions are set as follows.

Taking 5s for simulation time, and 1s, 0.5s, and 0.3s for accelerate/decelerate time, respectively, the simulation of sub-action 1 was conducted with three different *STEP* functions.

STEP 1: $3\times360\times(STEP(t,0,0,1,1)-STEP(t,4,0,5,1))$ (deg)

STEP 2: $3\times360\times(STEP(t,0,0,0.5,1)-STEP(t,4.5,0,5,1))$ (deg)

STEP 3: $3\times360\times(STEP(t,0,0,0.3,1)-STEP(t,4.7,0,5,1))$ (deg)

The dynamics simulation results of the robot rolling along a 30-meters-span of overhead transmission line are shown in Fig. 10, where x-axis is horizontal direction between two adjacent towers, and y is the vertical direction. Fig. 10 shows that the vibration amplitude in XY plane is much higher than that in Z direction, which is corresponding with the overhead line's wind-deduced vibration characteristics. The robot can carry out the preset kinematic target in flexible working environment. And, the coupling between the robot and line forces the robot vibrate, thus the fluctuation of the robot body with flexible moving path are larger than that with rigid path(Fig. 8).

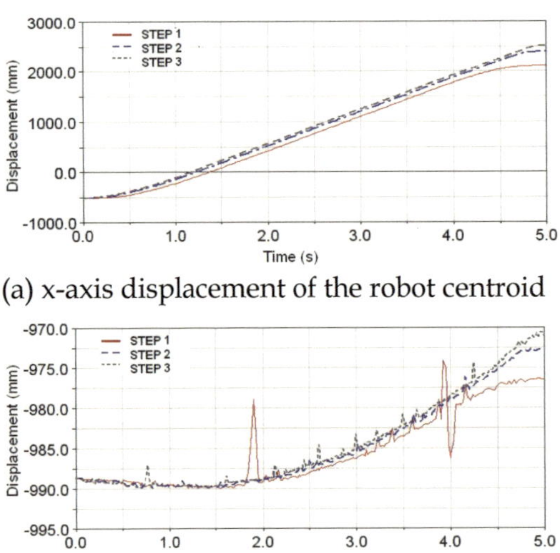

(a) x-axis displacement of the robot centroid

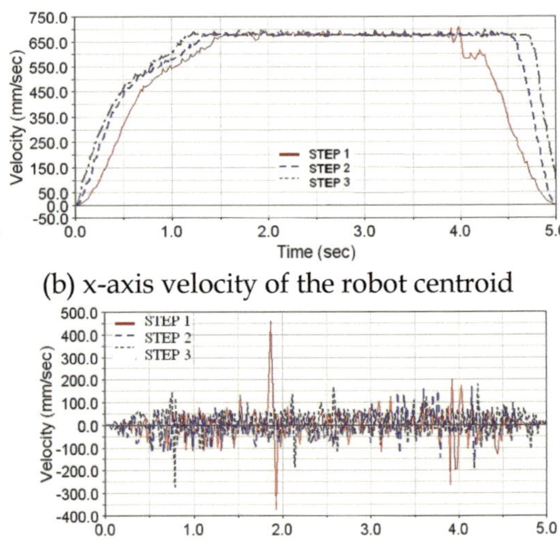

(b) x-axis velocity of the robot centroid

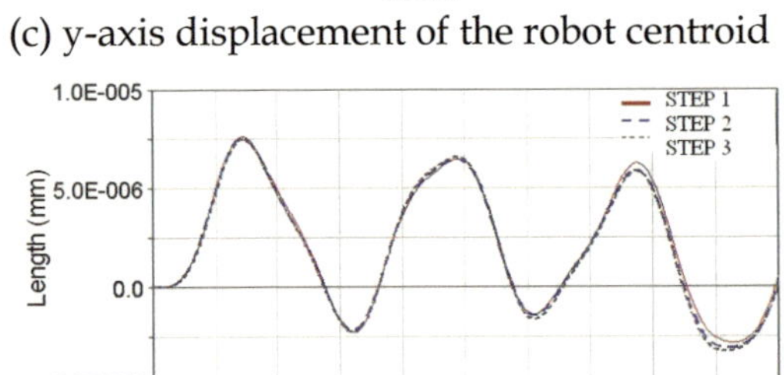

(c) y-axis displacement of the robot centroid

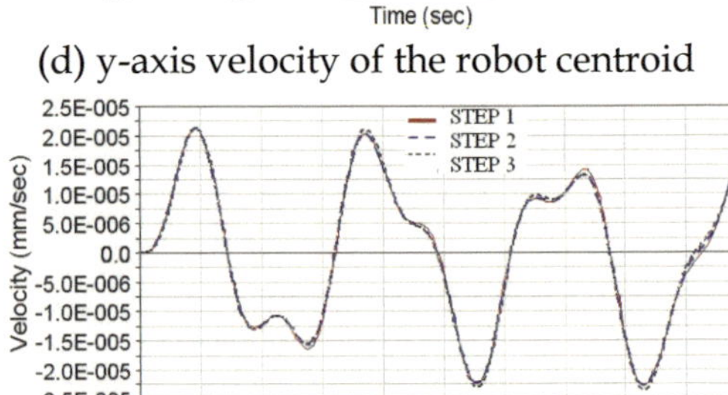

(d) y-axis velocity of the robot centroid

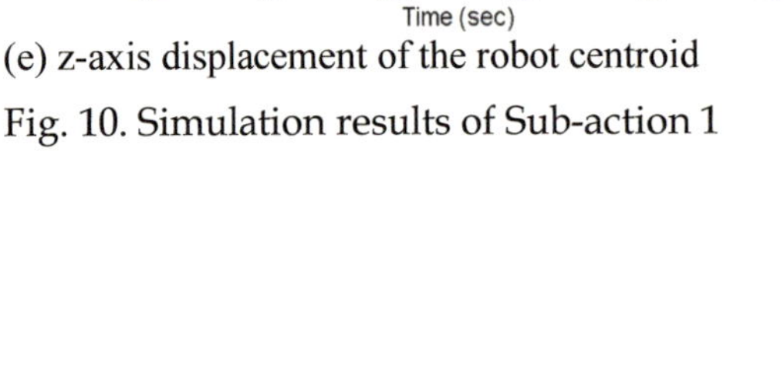

(e) z-axis displacement of the robot centroid

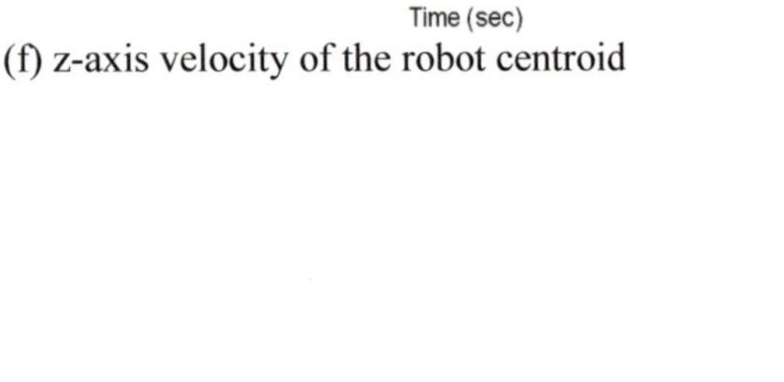

(f) z-axis velocity of the robot centroid

Fig. 10. Simulation results of Sub-action 1

6. Conclusions and future plan

Through kinematic analysis, dynamics modelling, simulation and tests, we can conclude as follows:
1) The proposed double-arms inspction robot prototype can fulfill full-path kinematic target, including moving along the no-obstacle segment, overcoming the obstacles, and varying moving paths.
2)The flexible working path decreases the perfomance of the robot, but the robot is capable of carrying out the preset kinematic target along flexible path.
More detailed dynamics analysis can refer to other papers (Xiao et al., 2005, 2006, 2007 & 2008). The model proposed in this chapter are far from fully demonstrating the actuality and those nonlinear factors in flexible obstructive inspection work environment. Further research is conducting to improve the robot's dynamic performance, such as: considering the flexibility of the joints and robot arm on dynamic model improvement, simulation for obstacle-overcoming in flexible working environment, and the effects of natural wind loading, etc.
The chapter proposed an inspection robot for 220kV phase line, and detailed the three generation prototypes developed in the past decade. Under the support of "863 Plan" and NSF in China, the research is now performing in further perfect of the robot prototype and reliability for feild application. The futre plan is to expand mobile robot technical platform in inspection robot to the application of icebreaking and repairing on transmission.

7. Acknowledgement

The authors gratefully acknowledge the financial support provided by the National Natural Science Foundation of China under Grant No: 50575165, the National High Technology Research and Development Program of China under Grant No: 2002AA420110, 2005AA2006-1, and 2006AA04Z202, the Natural Science Foundation of Hubei Province in China (2006NS-402).

8. References

Fu S. F.; Wang H. G.; Fang L. J. & Jiang Y. (2005). On obstacle-navigation control of inspection robot for the extra-high voltage power transmission line. *Robot*, Vol.27 No. 4, pp. 341-345+366, 1002-0446

Guo Y. L.; Li G. X; You C. Y. (2002). *Transmission line galloping*, Electric Power Press, 7508312317, Beijing

Lu Y. F. (1996). *Dynamics of Flexible Multi-Body System*, High Education Press, 7-04-005711-5 Beijing

Li Q. M.; Zhang Y. C.; Li J. C. (2007). Visual navigation for power transmission line inspection robot. *Journal of Computer Engineering and Applications*, Vol.12, No.19, pp: 526-530, 1002-8331

Montambault S. & Pouliot N. (2003). The HQ lineROVer: Contributing to innovation in transmission line maintenance. *Proceedings of IEEE 10th Int Conf in Trans and Dist Construction*, pp. 33-44, 0-7803-7917-9, Orlando, Florida, April, 2003, Institute of Electrical and Electronics Engineers, INC., Orlando

Peungsungwal S.; Pungsiri B. & Chamnongthai K. (2001). Autonomous robot for a power transmission line inspection. *Proceedings of 2001 IEEE International Symposium on*

Circuits and Systems, pp. 121-124, 0-7803-6685-9, Sydney, NSW, Australia, May 2001, Institute of Electrical and Electronics Engineers, Sydney

Sawada J.; Kusumoto K. & Maikawa Y. (1991). A mobile robot for inspection for power transmission lines. *IEEE transaction of Power Delivery,* Jun, 1991, pp. 309-315, 1000-6446

Tang L.; Fang L. J. & Wang H. G. (2004). Development of an inspection robot control system for 500kV extra-high voltage power transmission lines. *The SICE Annual Conference,* pp. 1819-1824, 4-907764-22-7, Sapporo, August, 2004, Society of Instrument and Control Engineers, Sapporo, Tokyo, Mar, 2004, Sapporo

Wu G. P.; Dai J. C. & Guo Y. L.; (1999). Small Running Vehicle with automatic surmount obstacles on high voltage transmission line. *Water Conservancy & Electric Power Machinery,* Vol.21, No.1, pp. 46-49+54, 1000-6446

Wu G. P.; Xiao X. H.; Guo Y. L. & Hu J. C. (2006). Development of a Crawling Robot for Overhead High-Voltage Transmission Line. *China Mechanical Engineering,* Vol.17, No.2, pp. 237-240, 1004-132X

Wu G. P.; Xiao X. H.; Xiao H.; Dai J. C.; Bao W. J. & Hu J. (2006). Development of a Mobile Inspection Robot for High Voltage Transmission Lines. *Automation of Electric Power System,* Vol.30, No.13, pp. 91-93+107, 1004-1026

Wang L. D.; Fang L. J.; Wang H. G.; et al. (2006). Development and control of an autonomously obstacle-navigation inspection robot for extra-high voltage power transmission line, *Proceedings of the International Joint Conference of SICE-ICASE*, pp. 5400-5405, 89-950038-5-5, Busan, March, 2006, Publisher, Busan

Xiao X. H.; Wu G. P.; Du E. & Shi T. L. (2005). Dynamics simulation and experimental study of inspection robot for high-voltage transmission-line. *Journal of Central South University of Technology (English Edition),* Vol.12, No.6, pp: 726-731, 1005-9784

Xiao X. H.; Wu G. P. & Li S. P. (2006). The rigid-flexible coupled dynamics characteristic between mobile robot along overhang flexible cable and its moving path, *WSEAS Transaction on Computer,* Vol.5, No.3, pp: 521-527, 1109-2750

Xiao X. H.; Wu G. P. & Li S. P. (2007). The coupling simulation of a power transmission line inspection robot with its flexible moving path when overcoming obstacles, *Proceedings of International Conference on Automation Science and Engineering*, pp. 326-331, 978-1-4244-1154-2, Scottsdale, AZ, September, 2007, Scottsdale

Xiao X. H.; Wu G. P.; Du E.; & Li S. P. (2008). The impacts of flexible obstructive working environment on the dynamic performances of an inspection robot for power transmission line. *Journal of Central South University of Technology,* Vol.15, No.3, pp: 525-530, 1005-9784

Zang Y. C.; Ling Z. Z; Fu S. Y.; Tan M. & Wu G. P. (2007). Structure-constrained obstacles recognition for power transmission line inspection robot, *Robot,* Vol.29 No. 1, pp. 1-6, 1002-0446

Zhou F. Y.; Wu A. G.; Li Y. B.; Wang J. D. & Ling Z. Z. (2004). Development of a Mobile Robot for Inspection of High Voltage Overhead Transmission Lines. *Automation of Electric Power System,* Vol.8, No.23, pp. 89-91, 1000-1026

Zhu, X. L.; Zhou J. P.; Wang H. G.; Fang L. J. & Zhao M. Y. (2006). Single arm running control method of inspection robot based on obliquitous sensor, *Proceedings of International Conference on Robotics and Biomimetics,* pp. 187-192, 1-4244-0571-8, Kunming, China, Dec, 2006, Kunming

Zhu, X. L.; Wang H. G.; Fang L. J.; Zhao M. Y. & Zhou J. P. (2006). Dual arms running control method of inspection robot based on obliquitous sensor, *Proceedings of IEEE/RSJ International Conference on Intelligent Robot and Systems.,* pp. 5273-5278, 1-4244-0259-X, Beijing, China Oct, 2006, Beijing

Teaching Introductory Programming Concepts with Lego MindStorms in Greek High Schools: A Two-Year Experience

Maya Sartatzemi[1], Vassilios Dagdilelis[2] and Katerina Kagani[1]
University of Macedonia
[1]Department of Applied Informatics,
[2]Department of Educational and Social Policy
Thessaloniki,
Greece

1. Introduction

Even today, teaching the basic principles of programming still presents numerous difficulties for students, which in the last several decades have comprised the subject of extensive research (Mayer, 1989; Mendelsohn et al., 1990; Du Boulay, 1989; Du Boulay et al. 1989; Brusilovsky et al., 1994). On the one hand attempts have been made to investigate the factors that render the learning of programming principles difficult, and on the other, to examine the effectiveness of programming languages and integrated environments, as well as the actual methods applied to the teaching of programming.
The difficulties, according to (Du Boulay, 1989; Du Boulay et al. 1989) fall into one of the following categories:

1. Orientation: what programming is; the types of problems that can be solved; the positive effects of learning programming.
2. Notional machine: difficulties in understanding the general properties of the "machine" which the student learns to control and its relation to the physical machine.
3. Rules of the programming language (notation): syntax and semantics as an extension of the properties and behaviour of the notional machine.
4. Learning the established structures.
5. Pragmatics: acquiring the ability to determine, develop, control and debug a program with the available tools.

Many researchers agree with the view that one of the most significant factors which contributes or even gives rise to the above-mentioned difficulties is the use of the classic approach (Brusilovsky et al., 1997; Xinogalos, 2002) to the teaching of programming. By this we mean teaching based on:

- The use of a general purpose language (such as Pascal, C, etc);
- The use of a professional environment for this language; and
- The proposed problems that require the processing of numbers and/or symbols.

Much research has shown that the problems arising from general-purpose languages are due to the fact that they are complex and novices need to learn a lot in order to be able to perform relatively simple tasks (Brusilovsky et al., 1994; Brusilovsky et al., 1997). Many such languages require the user to key in a very "dark" code in order to achieve insignificant and commonplace outcomes. Frequently, the students are either taught the syntax before writing their first program, which can be rather discouraging, or they develop programs without having any interest in them whatsoever.

In order to overcome these obstacles, various approaches have been proposed. One application, which appears to be particularly popular with students in secondary level education, is based on the use of computer-controlled models, such as the Logo turtle or Karel's Robot (Pattis et al., 1995; Bergin et al., 1997; Becker, 2001). Besides these simulated models, an attempt has lately been made to apply physical models in the teaching of programming. Generally speaking, the use of models, and especially robots, seems to exert an irresistible allure on young aged students. Many educators believe that robots are a powerful way to motivate learning (Hirts et al., 2003) thus they are used to increase student interest, interaction and retention.

The teaching of programming with the application of robots and appropriate environments has numerous benefits, as many researchers point out (Kaskalis et al, 2001; Hirst et al, 2003; Lawhead & Bland, 2003;). More specifically, the advantages of this approach are:

- It is concrete since students program things they can handle, to behave in ways they can observe in the physical world.
- It is incremental.
- It is creative.
- It allows manipulation within a constrained context.
- It provides immediate feedback.
- Children's constructions have behaviour (i.e. anthropomorphic behaviour which is very interesting.
- It uses a variety of skills.
- It allows complete novices to create interesting outcomes (e.g. "go collect a tennis ball" rather than "Hello world").

The above view appears to be internationally widespread, which also corresponds to the accepted perception in the Greek educational system. Of course, a successful (or otherwise) introduction to programming as a subject in any educational system cannot but be influenced by local factors. In the Greek educational system, for instance, and more specifically at secondary level, teaching programming is quite restricted time-wise: both third year junior and first year senior high school students are taught the basic concepts of programming in 10 one-hourly lessons. Within this short time period they are taught the role of a programming language, the basic commands for input/output data, the selection commands, as well as learning to use a programming environment in order to develop, edit, debug and run their programs. Usually, the teachers apply a general-purpose language and unavoidably the students meet with difficulties in their attempt to understand the program's behaviour and associate it to their activities. So far, the results of these endevours have not been particularly promising, which appears to be the case in other countries as well.

In this research we present the findings of the studies conducted in the last two years on teaching the basic concepts of programming to third year junior and first year senior high school students with the application of robots, and more specifically LEGO MINDSTORMS (LM). LM have been available on the market since 1996 through the LEGO Company and enable the construction of a large range of physical models, which can be programmed with the help of a RCX processor integrated into them.

Our research team conducted 3 series of lessons to junior and senior high school students. The results from the 1st series of lessons were presented in a previous paper (Satratzemi et al., 2005). Our latest findings are included in this chapter, whose structure is as follows: In section 2 we give a description of LM, and the programming languages and environments used. Then, we give a brief description of the lessons where LM are used worldwide. In the 3rd section an analytical description is given of the lessons taught by our research team to the two different groups of junior and senior high school students. Following this, the comparative results of the students' assignments of both school years are presented. Lastly, the chapter ends with a discussion of the findings.

2. Description of Lego Mindstorms, programming languages and environments

The first products marketed by LEGO consisted of simple small bricks, which children assembled in order to build houses, castles, towers, bridges and so on. In this way they first came into contact with the basic ideas of architecture and civil engineering, in the same way that the renowned "Meccano" first brought young people into contact with mechanical engineering. Later, the LEGO kits were enriched with the addition of new pieces such as gears, axes, motors and batteries, with whose help the children were able to create mechanical constructions, like automobiles, lorries, cranes and ferris wheels. Within the framework of these constructions the children were gradually initiated in the science of engineering. The first-generation kits, therefore, allowed children to form structures, while the second-generation kits enabled them to create mechanisms.

In the last few years, new LEGO kits have appeared, LM, third-generation kits, which permit children to produce "behaviours". These kits are used to make a wide range of machines, which present specific behaviours, such as a robot, a machine that moves in reaction to a person's or other machine's movements. Now by attaching sensors onto the electronic LEGO bricks, children can transform a simple door into a "talking" door that "welcomes" those who enter it; or a door which counts how many people enter each day (Resnick, 1993). LM embodies the constructionism learning theory described in the book Mindstorms: Children, Computers, and Powerful Ideas by Seymour Papert.

The first retail version of LM was released in 1998 and marketed commercially as the Robotics Invention System (RIS). The current version was released in 2006 as Lego Mindstorms NXT.

The original Mindstorms Robotics Invention System kit comprises a large range of bricks, sensors, motors and other components, which can be used to make physical models. These extra components are attached onto a larger LEGO brick, on which the RCX processor is integrated. With the appropriate programming of RCX, one can create a wide variety of LEGO robots.

The three-steps involved in the creation of an autonomous robot are as follows:

1. Construct the robot according to the instructions given in the kit or by using one's imagination.
2. Program development using the iconic programming environment, which accompanies the kit or another programming language.
3. Load the program onto the robot, using the Infrared transmitter.
4. Program execution.

In order for the robot to express a particular behaviour, it is usual for steps 2 to 4 to be repeated several times.
RCX (Robotic Control X) is an autonomous microcomputer integrated onto a LEGO brick, which can be programmed through a computer and allows the construction to interact with the environment autonomously, independently of the computer.
This processor has:

- Three input ports (1, 2, 3);
- Three output ports (A, B, C);
- A small one line, LCD screen;
- A microphone; and
- Four buttons.

At the input ports sensors are connected (touch, light, temperature), while at the output ports motors and lamps are connected. The buttons are to turn the RCX on/off, to choose the program that will be executed and to start/stop the execution.
Due to RCX's ability to control motors or lights and to collect data with the help of sensors, both adults and children can easily make creations that move, react and carry out various tasks.
Inside the RCX there is a microcontroller Hitachi H8 with 32K external RAM. Its function is to control three motors, three sensors and an infrared serial communication port. Also, in the interior of RCX there is a ROM 16K chip, where a program driver is stored which is executed as soon as RCX is activated (http://graphics.stanford.edu/~kekoa/rcx/).
A firmware program is also loaded which requires 16K storage space. Both these programs - driver and firmware -undertake the translation and execution of the programs loaded onto RCX by the user to control the robot (http://emhain.wit.ie/~p98ac25/).

2.1 Programming of RCX

In order to program RCX it is necessary to have two software on the computer (http://mapageweb.umontreal.ca/cousined/lego/4-RCX/Survey/):

- An operating system (OS): without this operational system RCX cannot carry out any activities.
- A series of communication processes (libraries): these are necessary only for the computer to be able to send information to or receive data from RCX. These libraries can be a set of .exe or .dll or .ocx (ActiveX) files. The libraries should assist the program to transmit/save/retrieve data to and from the microcomputer.

Despite the fact that the two elements mentioned above were intended for different platforms (the operating system for RCX, the libraries for the computer), they must be compatible so that communication between computer and RCX is possible (http://mapageweb.umontreal.ca/cousined/lego/4-RCX/Survey/).
The first time RCX is activated, the code stored in its memory (ROM) is executed. Following, the software "firmware" must be loaded. This concerns RCX's operational system, which is responsible for the management of the system's resources and the execution of the programs that are loaded. As soon as firmware is loaded onto RCX, the user can load their program. This software receives the programs from the infrared transmitter and responds to the buttons found on RCX, which are used to start or stop the program's execution. It facilitates communication between the motors and sensors (robot I/O), assists the basic types of data and operations (e.g., a math library), it supports tools in order to make interaction with the

robot feasible. There are, however, particular circumstances, where firmware is not enough since greater control of RCX is required, e.g. in the use of more variables. In order to deal with such situations, alternative type of software have been developed that can be used in LM RCX Controller; these are legOS, leJOS and PbFORTH.

Along with the LM kit the company also provides an Active-X called "Spirit.ocx". This operates as an interface between the program created by the user and RCX. It includes a series of predefined procedures and variables, which support communication with RCX. With the help, therefore, of this file one can create a form in Access, for instance, which is comprised of buttons that control the robot's movement. Moreover, theoretically one can program RCX using any programming language whatsoever (http://emhain.wit.ie/~p98ac25/).

The software which comes with LM to create the programs is relatively simple as it was mainly designed for children and for adults who do not have programming experience.

Program loading - execution:

Communication between the computer and RCX occurs with infrared rays. An infrared transmitter (IR- transmitter - tower) is connected to the serial port of the computer (http://emhain.wit.ie/~p98ac25/). With the help of IR, programs are loaded onto RCX. 6KB of memory are available for the storage of each program the user loads onto RCX (http://graphics.stanford.edu/~kekoa/rcx/). From the moment that a program is loaded onto RCX, the robot is ready to operate autonomously. The absence of wires facilitates the robot's free movement. The only thing the user needs to do is to press the correct button on RCX in order to start program execution. The execution can be repeated. In addition, there is the possibility to load many programs on the computer and each time to activate and execute whichever the user wishes.

Below are presented the programming languages and environments used in the lessons we conducted. More specifically, these include the programming environment and language Robolab, the language NQC, and the environment Bricx Command Center.

2.2 Robolab

ROBOLAB was created with the aim to meet the needs of the educational community for software that could be used by students of different ages and varying programming abilities.

This programming environment is the result of the cooperation among the following institutions: Tufts University, LEGO Dacta (the LEGO department which is involved with educational material) and National Instruments.

It is a graphical programming environment based on LabView; it is appropriate for PC and MAC; and it has numerous levels that require varying degrees of programming ability.

The environment's programming abilities are divided into two sections so that they can meet the needs of students of different levels. In both sections, icons are used for the representation of commands and structures.

The first and basic section, called Pilot can be seen in Fig. 1. Its limitations concern the number and the sequence of icons that the student uses for their attempt to be successful. It is suitable for young ages, as it allows for the quick and easy syntax of programs, it uses icons such as lights, arrows and clocks, which can help students during program use.

Programming in this section mainly involves choosing from the pull-down menu so that a simple program is in sequential form.

In Pilot there are 4 programming levels, which gradually increase in degree of difficulty going from the simpler programs to more complex ones with more selections for the sensors and more parameters. At the highest level, Pilot can be used to create programs which are implemented in 2 or more steps depending on the program's needs. Even though the ROBOLAB software includes a very detailed manual, the Pilot section, allows first-time users to create programs quickly and to gradually progress to more complicated programming. Especially in regards to children, teachers have ascertained that they can proceed through the 4 levels of Pilot with very little guidance.

Pilot does not enable users to fully exploit RCX's capabilities; it simply provides an environment where one can observe the connection between programming and the actions executed by RCX.

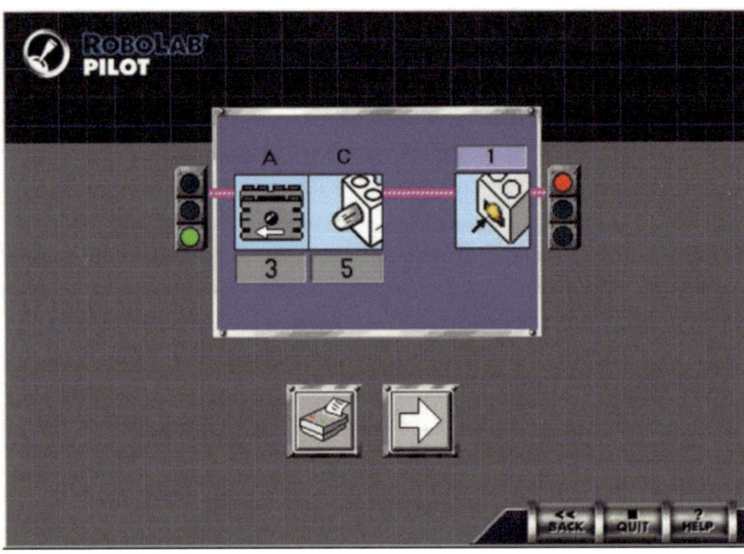

Fig. 1. A screenshot of Pilot section

These capabilities are sufficient for primary school teachers and pupils, who do not need to reach a higher level of programming. On the other hand, more experienced programmers can use Pilot to test their creations or their programming ideas quickly and easily.

The second and more advanced section is called Inventor (Fig. 2) and was created to satisfy the needs of users at higher levels who have more programming ability. The environment in this section is likewise graphical, however, it offers the user more flexibility. Here too there are 4 levels related to complexity.

In Inventor the commands that form a program are chosen from a function pallet. This gives users an unlimited number of choices, as well as a high level of responsibility for the program's success.

Inventor provides a higher level of programming structures, such as loops, variables, selection structures, multiple selections and the possibility to create subroutines. For this reason, programmers can take advantage of all of RCX's properties.

The transition from Pilot to Inventor presupposes that the user understands the rationale with which the commands-icons must be selected and connected in order to develop a program.

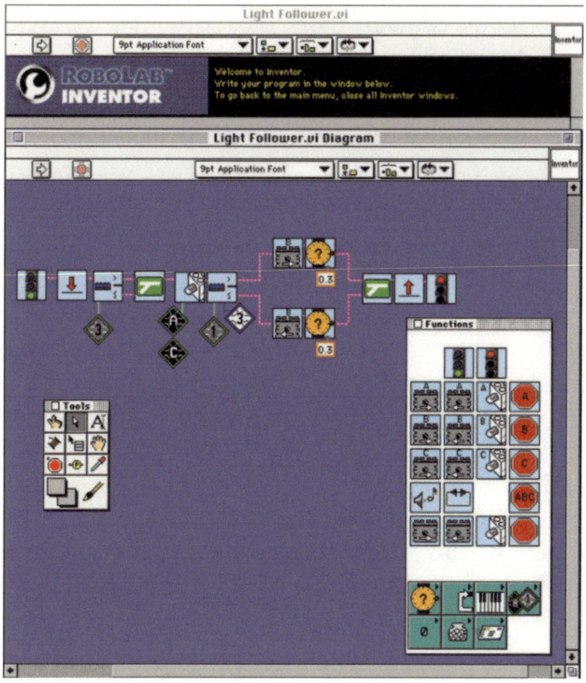

Fig. 2. A screenshot of Inventor section

The main advantage of ROBOLAB is that it applies a learning model where the large number of available commands is dependent on the level that the user has reached (http://www.umcs.maine.edu/~pbrick/).

2.3 Not Quite C (NQC)

Not Quite C was developed by David Baum and is available freely on the Internet for Windows, Unix and MacOS. It is a simple programming language with syntax similar to C; it is based on firmware, which accompanies the Robotics Invention System and it does not use Spirit.ocx. It is not suitable for teaching programming to novices. On the contrary, it is a good solution for someone who knows to program in C (http://www.baumfamily.org/nqc/). It assists the use of variables, a counter, an array and subroutines, a fact that makes it an effective programming tool.

Initially NQC was designed to operate on a command-line. Later, quite a few users created Integrated Development Environments (IDEs) whose aim was to provide a user-friendlier environment for programming with NQC. They developed easy-to-use graphical environments that incorporate the NQC compiler and are enriched with further features, such as an RCX remote control, highlighting the syntax among others (Kaskalis et al., 2001).

RCX Command Center (RCXCC), WinNQC, NQCedit, Bricx Command Center and Visual NQC 2001 are such programming environments.

In the lessons we conducted, Bricx Command Center (BricxCC) was used. It is built around NQC Compiler that makes it possible to program RCX. The foundation of BricxCC is a full-featured programmer's editor in which you can write your programs with the integrated ability to compile the programs and download them to the brick. In case of errors, they are reported at the bottom of the editor window so that they can be corrected easily. In addition to providing an editor, BricxCC contains tools to control your robot directly, to watch what is happening in the brick, to download (new) firmware, etc.

3. Using mechanical models in educational institutions

From their initial appearance in 1998 LM have attracted the attention of numerous educators and researchers. They started being used at all levels of education in order to teach a variety of subjects and fields, such as: basic programming concepts, artificial intelligence, robotics etc (Avanzato, 1998; Barnes, 2002; Genalo, 1997; Goldweber et al., 2001; Lawhead, P. & Bland, 2001; Fagin, 2000; Wolz, 2001). In an attempt to get the most detailed literature review possible regarding the use of LM as a teaching tool, we sought those institutions which have integrated LM in their syllabus. This was perceived to be necessary in order for us to be fully informed which would aid us in the preparation of the series of lessons to be conducted for the teaching of programming to 3rd year junior and 1st year senior high school students in the Greek educational system.

Since it would require a lot of space to give a detailed description of the study programs where LM have been used, we have organized the findings of this literature search into table form and present them the table below. Table 1 includes the type of lesson (incorporated in the existing curriculum, seminar etc), length-content of the lessons and educational level.

Program - Educational Institution	Type of lesson			Lengthα				Educational Level			
	Incorporated in the existing curriculum curriculum	Seminar	Extra-curricula Activities	Less than 6 months	Six months/	12 months	Over 12 months	Primary Education	Secondary Education	Tertiary Education	Independent of educational Level
Wellesley College, "Robotic Design Studio"	X				X					X	
University of Mississippi, "Java, Event and Robots"	X				X					X	
Indiana University, "Learning with LEGO Robots"		X		X						X	
Penn State Abington College, CMPSC 201	X				X					X	
University of New Jersey, CS1	X				X					X	
Texas A & M University Corpus Christi, "Systems Analysis and Design"	X				X					X	
Colby College, "Exploration with Robots" (CS117)	X				X					X	
US Air Force Academy	X				X					X	
Xavier University	X				X					X	

University of Evansville, "LEGO 101 Project"			X		X					X	
Villanova University, "Introduction to Artificial Intelligence"(CSC4500)	X				X					X	
Case Western Reserve University, "Autonomous LEGO Robotics", BIOL/EECS 375/475	X				X					X	
Uppsala University, "Real time systems & Real time programming"	X				X					X	
National University of Singapore, "CS 3243- Foundations of Artificial Intelligence"	X			X						X	
University of Edinburgh, "Intelligent Autonomous Robotics"	X				X					X	
University of Salzburg, "Embedded Software Engineering Course, CCN 25754"	X				X					X	
Waterford Institute of Technology, "Robotics course for Industrial Computing"	X				X					X	
University of Southern California, CSCI 445 "Introduction to Robotics"	X				X					X	
University of Pennsylvania, "MEAM 247 Mechanical Engineering Laboratory"	X				X					X	
Trinity College," Computer Technology"	X				X					X	
Universität des Saarlandes, "Talking Robots with LEGO MindStorms", 2004	X				X					X	
University of New Mexico	X				X					X	
University of Applied Sciences Hamburg	X				X					X	
Iowa State University, "Toying with Technology"		X		X							X
Carnegie Mellon University, Robotics Academy		X					X				X
Bethany Christian High School	X			X					X		
Brooks-Smith School			X	X				X			
Shady Hill School			X			X			X		
Blake School			X				X	X			
S.P. Waltrip High School			X				X		X		
Davis Creek Elementary			X	X				X			

Table 1. Educational Institutions which use LM

By examining the subjects of the institutions, briefly presented in the above table, we arrived at the following conclusions:

Scientific Field - Level

- The technology of physical models has been integrated more into subjects on robotics. In a number of cases it is considered a basic precondition that students have programming experience (mandatory subject at University), in most cases however, the teaching of robot programming comprises a separate study unit.
- Most applications are implemented in primary or tertiary education.

Length – Content of Lessons

- In tertiary education, the lessons are usually incorporated into the department's curriculum that last for one semester. In contrast, the use of physical models at younger ages usually takes place within the context of extracurricular activities, which can last from a few weeks to up to a year.
- From a study of the curricula, the findings show that at tertiary level some theoretical lessons are first given, whereas for younger ages learning is accomplished exclusively through the students' activities with Legos.
- At all levels of education, lessons in computer laboratories are backed up by related class assignments. University students, were called on to design extra assignments, which lead to a consolidation and expansion of what had been discussed during the lesson, giving stimulation for study and personal involvement in the subject. During the design of assignments, university students were quite active and showed particular interest. In addition, they were more concerned about issues related to the robot's behaviour and the program philosophy rather than the syntax of a particular programming language.
- University student assignments were usually accompanied by written documentation.
- The selection of a physical model, in almost all cases, regardless of educational level, is made by the teachers.
- In addition, in almost all the university institutions, at the end of the semester there is an exhibition of all the physical models which the participants constructed and programmed; a fact which no doubt is an incentive to try harder as well as providing consistency.

Programming environment

- At younger ages preference was for one of the graphical environments (Robolab) for programming robot behaviour, while university students usually program their robots on environments based on more general-purpose languages, such as Java, Basic, C (NQC), Ada.
- The selection of programming environments in each case depends on the level of programming abilities which the student has or should attain. The needs of users who are experienced in programming are met by programming environments that are based on some general-purpose language. In contrast, novice programmers avoid such environments and turn to graphical based ones. In this way, they do not focus their attention on learning the syntax of the language, but rather on basic programming principles.

General comments

- The interest, enthusiasm, and activity shown by students were much greater than in previous years when lessons were conducted in a more traditional manner.

- It is imperative that the teachers who are likely to apply physical mechanical models in their lessons are trained. In order for these training needs to be met quite a few seminars are organized geared to high school teachers to become acquainted with robot technology, learn how it can be incorporated into the existing curriculum and to develop it within the context of the educational process.
- One fairly serious drawback was that in many cases there was a lack of sufficient numbers of robots at the informatics laboratories, which can cause difficulties in conducting the lesson.

4. Description of lessons in the Greek educational system

As stated above, LM are a contemporary tool applied in Informatics lessons in various institutions all over the world. However, few studies have been conducted regarding the effectiveness of these models in teaching in related subjects. The first general impressions in the application of LM in the classroom were encouraging. In spite of this, one of the first formal assessments of Mindstorms as a teaching tool by Fagin and Merkle (2004) showed that LM not only did not have a positive effect on students' performance but actually had a negative one. In a recent research, (Clinburn, 2006) presented projects and courses where LM were effective teaching tools and listed the lessons where LM could be applied.

Taking into consideration the findings related to the cases of LM use presented in the previous section, the research on the effectiveness of LM, the curriculum and the time constraints placed on the teaching of programming for third year junior and first year senior high school students in the Greek educational system, we designed a series of lessons for the subject of Informatics. For the teaching of programming with LM we conducted 3 series of lessons to high school students as already stated The results from the 1st series of lessons were presented in a previous paper (Satratzemi et al., 2005), where we referred to our first attempt to teach programming with the aid of LM. Our latest findings are presented in this section.

Here an analytical description is given of the lessons taught by our research team to two different groups of junior and senior high school students.

In the school year 2005-2006 the environment Robolab was used to teach 84 3rd year junior high school students. The students worked in pairs, each pair having its own PC and RCX processor. Nine lessons were conducted on the following topics:

1st Introduction to programming with LEGO. Presentation of Lego (RCX, infrared transceiver); Introduction to the programming environment Robolab; Loading-Execution of available programs on Robolab.

2nd Presentation of programming with Legos, presentation of Pilot programs levels 1, 2, 3.

3rd Input - Output command, Wait command. Presentation of Inventor programs level 1.

4th Parameters - Modifiers, Inventor level 3.

5th -6th Repeat structures Jump/land, loop

7th -8th Selection structure.

9th Containers - Modifiers

In the school year 2006-2007 the language NQC and the Bricx Command Center environment were used to teach 48 3rd year junior high school students. Likewise, the students did pair-work, and each pair had its own PC and RCX processor. There were 6 lessons on the following topics:

1st. Introduction to programming with LEGO; Presentation of Lego (RCX, infrared transceiver); Introduction to the programming environment BricCC; Loading-Execution of available programs on NQC.

2nd. Output command (Description - motor, lamp, sound); Usefulness of Wait command; Examples of how to apply commands presented.

3rd-4th. Input command (Sensors) - Repeat command; Usefulness of Repeat structure; Repeat command - until command; Examples of how to apply commands presented.

5th-6th. Usefulness of selection structure; If command; Combination of if command with repeat commands; Nested if commands; Examples of how to apply commands presented.

For both series of lessons presentations of the concepts of each lesson were developed along with worksheets, check sheets, brief notes on the basic icons of Robolab and the basic commands of NCQ.

The comments made by the teacher in 2005-2006 were taken into account for the design of the lessons in the following year. Specifically, the alterations made are the following:

1. In 2005-2006 new concepts were introduced by the teacher with the help of digital presentations, following this, the students had to answer questions in writing and they implemented their own programs in order to control the robot's behaviour. According to the teacher's observation, this approach was not particularly effective. Teaching with the use of presentations excluded the students from participating and as a consequence they did not fully comprehend the concepts taught. This resulted in producing misconceptions in terms of the significance of the commands. The commands were learnt at the stage of program development with the students often requesting the teacher's assistance.

   Hence, in 2006-2007 a problem-based approach was applied; that is, the teacher posed the problem for which the students had to develop a code. Through discussion the teacher guided the students to understanding the necessity of the existence of the related commands which would solve the problem. Following this, the teacher presented the commands to the students. The teacher used the board to present the new concepts - commands. Then, exercises were proposed to the students which were based on the implementation of the code. At the end of the lesson students had to answer to True/False questions and the recapitulation of the concepts taught in the current lesson was made through digital presentations.

2. Further, it was observed that in 2005-2006 there were too many exercises set and that they were rather similar as regards the control of the robot's behaviour. This had resulted in the students losing interest in programming as the lessons progressed since there were no challenging new situations of operation to encounter of the robot's behaviour. In addition, the students were forced to work under pressure in terms of time due to the large number of exercises they were required to get through.

   In 2006-2007 in order for student interest to be maintained, fewer exercises were set and thus the time available to solve each problem increased, allowing each group to work at their own pace. In order for the results to be comparable the problems presented below are those that were given both years.

After each exercise tables are given showing the percentages of students who solved the problem correctly, who did not solve it correctly, and who did not attempt the exercise at all. This is followed by another table with data on particular commands of the code which are considered to be significant in formulating conclusions regarding the degree of understanding and use of the particular data of each programming language by each group of students. In the tables, G1 is the students of 2006-2007 and G2 the students of 2005-2006.

Exercise 1: Program the vehicle you constructed to move forward using the two motors for 6 seconds and then stop.

Groups	correct	Error/ misconception	No answer
G1	75.0%	21.0%	4.0%
G2	57.0%	0.0%	43.0%

Table 2a. Results of 1st Exercise

	Correct answers			
Groups	motors	time	stop	syntax
G1	83.0%	92.0%	83.0%	83.0%
G2	57.0%	57.0%	57.0%	

Table 2b. Data details of the correct answers of the 1st Exercise

Exercise 2: Program your vehicle to move forward using both motors for 4 seconds, to stop for 2 seconds and then to move in the opposite direction for 4 seconds.

Groups	correct	Error/ misconception	No answer
G1	83%	8%	8%
G2	50%	7%	43%

Table 3a. Results of 2nd Exercise

	Correct answers				
Groups	motors	direction	time	stop	syntax
G1	88%	92%	92%	83%	92%
G2	57%	57%	57%	50%	57%

Table 3b. Data details of the correct answers of the 2nd Exercise

Exercise 3: Program the vehicle you constructed to produce an audio signal and then to wait for 1 second. This procedure will be repeated 10 times.

Groups	correct	Error/ misconception	No answer
G1	92%	4%	4%
G2	93%	7%	0%

Table 4a. Results of 3rd Exercise

	Correct answers			
Groups	Repetition structure	Number of repetitions	sec	syntax
G1	96%	96%	96%	92%
G2	100%	100%	86%	

Table 4b. Data details of the correct answers of the 3rd Exercise

Exercise 4: Program the vehicle you constructed to move straight-ahead (using both its motors) until it hits an obstacle. Then to move in the opposite direction and to stop as soon as it hits another obstacle.

Groups	correct	Error/ misconception	No answer
G1	67%	33%	0%
G2	50%	43%	7%

Table 5a. Results of 4th Exercise

	Correct answers				
Groups	Repetition structure	sensor	Stop after until	Condition until	syntax
G1	100%	92%	79%	92%	92%
G2	100%		86%	50%	86%

Table 5b. Data details of the correct answers of the 4th Exercise

Exercise 5: Program the vehicle you constructed to move (using both its motors) with the lamp on. When the light level of the room increases by 5 the following should occur for one second: the lamp turns off and the vehicle stops moving. This procedure will be repeated 4 times. At the end a sound is heard.

Groups	correct	Error/ misconception	No answer
G1	46%	25%	29%
G2	14%	79%	7%

Table 6a. Results of 5th Exercise

	Correct answers					
Groups	Repetition structure	Number of repetitions	Light, motor	Wait for	Stop	sound
G1	67%	71%	58%	63%	58%	67%
G2	79%	79%	71%	21%	64%	71%

Table 6b. Data details of the correct answers of the 5th Exercise

Exercise 6 Program the vehicle you constructed to produce a sound SOUND_DOWN if the touch sensor in the front of the vehicle has been pressed, otherwise, if it has not been pressed, it must move for 3 seconds.

Groups	correct	Error/ misconception	No answer
G1	67%	29%	4%
G2	57%	43%	0%

Table 7a. Results of 6th Exercise

	Correct answers				
Groups	If/Fork	Condition (if/fork)	sound	motor	merge
G1	96%	92%	92%	79%	
G2	57%	57%	100%	-	100%

Table 7b. Data details of the correct answers of the 6th Exercise

Exercise 7: Make a program which will light up the lamp of the vehicle you constructed for 6 seconds, if the touch sensor at the back of the vehicle is pressed. Otherwise it will move the vehicle for 2 seconds in a forward direction.

Groups	correct	Error/ misconception	No answer
G1	63%	25%	13%
G2	36%	57%	7%

Table 8a. Results of 7th Exercise

	Correct answers					
Groups	if	condition	Touch sensor	lamp	motor	syntax
G1	83%	71%	75%	71%	71%	75%
G2	79%	79%	79%	36%	71%	93%

Table 8b. Data details of the correct answers of the 7th Exercise

Exercise 8: Make a program that will produce a sound 4 times, if the light sensor detects a value smaller than 60. Otherwise, the vehicle must move in a forward direction for 4 seconds.

Groups	correct	Error/ misconception	No answer
G1	17%	42%	42%
G2	14%	50%	36%

Table 9a. Results of 8th Exercise

	Correct answers					
Groups	if	condition	Touch sensor	repeat	motor	syntax
G1	86%	50%	93%	79%	79%	100%
G2	14%	57%	57%	36%	43%	64%

Table 9b. Data details of the correct answers of the 8th Exercise

Below we present a sample of the questions given to students, which refer to issues of understanding the commands and locating program errors.
Question 1
List the categories of sensors.

Groups	correct	Error/ misconception	No answer
G1	71%	29%	0%
G2	22%	64%	14%

Table 10a. Results of 1st Question

Groups	touch	light	temperature	rotation
G1	100%	92%	92%	75%
G2	64%	79%	64%	21%

Table 10b. Data details of the correct answers of the 1st Question

Question 2
Where is the mistake? The program must activate motor A for 4 seconds and then stop it.
Find the error in the program below:
After loading the program the robot must move backwards and stop as soon as it comes up against an obstacle with the help of the touch sensor in port 3. The program must activate motor A for 4 seconds and then stop it.

```
task main()
{
      SetDirection(OUT_A+OUT_C, OUT_REV);
      SetPower(OUT_A+OUT_C, 5);
      SetOutput(OUT_A+OUT_C, OUT_ON);
      until (SENSOR_3==1);
}
```

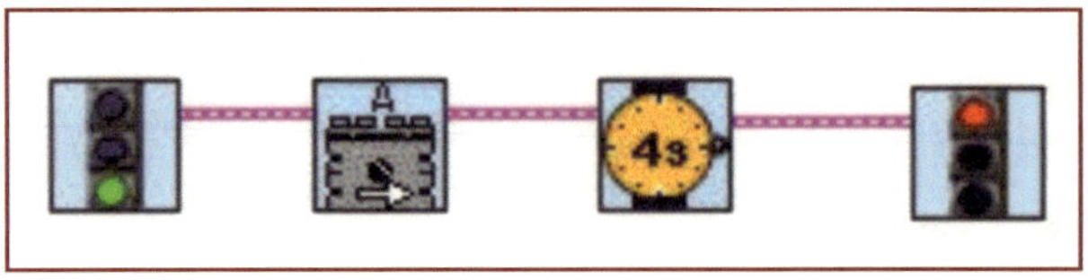

Groups	correct	Error/ misconception	No answer
G1	79%	0%	21%
G2	86%	0%	14%

Table 11. Results of 2nd Question

Question 3

Find the error in the program below :

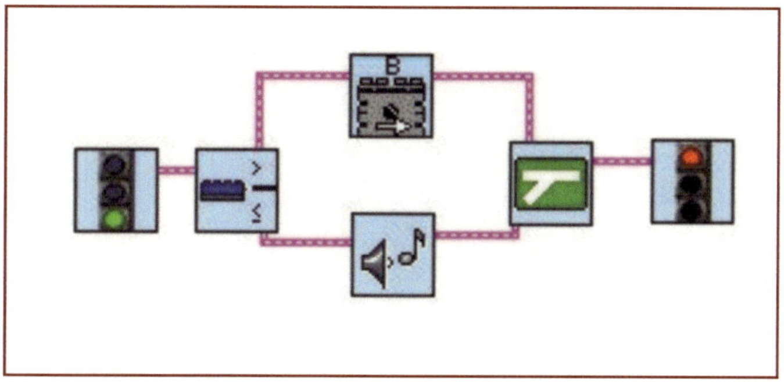

```
task main()
{
 SetSensor(SENSOR_1, SENSOR_TOUCH);
 if (SENSOR_1==1)
 {
    SetDirection(OUT_A+OUT_C,OUT_FWD);
    SetOutput(OUT_A+OUT_C, OUT_ON);
    Wait(200);
    SetOutput(OUT_A+OUT_C, OUT_OFF);
 }
 else
 {
    PlaySound(SOUND_CLICK);
 }
}
```

This program must produce a sound when the light sensor detects a value below 50. Otherwise, it must activate motors A and C for 2 seconds in a forward direction. This program must activate motor B when the light sensor detects the light level above 80, otherwise it will produce a sound signal.

Groups	correct	Error/ misconception	No answer
G1	100%	0%	0%
G2	36%	57%	7%

Table 12a. Results of 3rd Question

Groups	SetSensor / sensor	Condition / expression for 50
G1	100%	100%
G2	36%	93%

Table 12b. Data details of the correct answers of the 3rd Question

Question 4
A) What behaviour do you expect your robot to have?
B) Change your program so that it produces the same result (only for students in NQC).

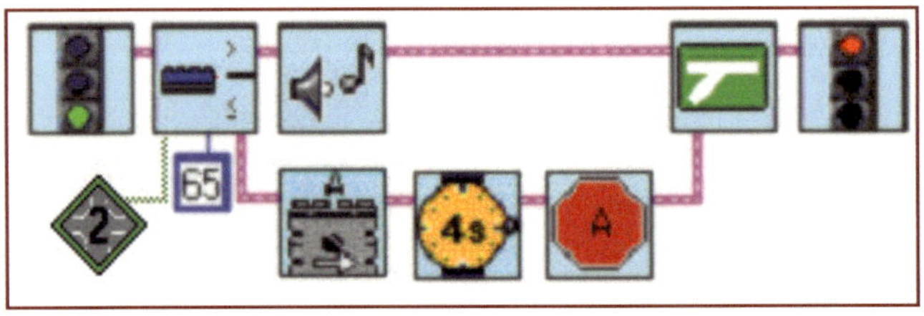


```
task main()
{
 SetSensor(SENSOR_2, SENSOR_LIGHT);
 SetSensor(SENSOR_1, SENSOR_TOUCH);
 if (SENSOR_2>=20)
 {
    SetDirection(OUT_A+OUT_C,OUT_FWD);
    SetOutput(OUT_A+OUT_C, OUT_ON);
    until (SENSOR_1==1)
    SetOutput(OUT_A+OUT_C, OUT_OFF);
    PlaySound(SOUND_CLICK);
 }
 else
 {
    SetOutput(OUT_B, OUT_ON);
    Wait(300);
    SetOutput(OUT_B, OUT_OFF);
    PlaySound(SOUND_CLICK);
 }
}
```

Groups	correct	Error/ misconception	No answer
G1	100%	0%	0%
G2	57%	43%	0%

Table 13a. Results of 4th Question

Groups	Use of if	condition	change
G1	100%	100%	100%
G2	79%	64%	

Table 13b. Data details of the correct answers of the 4th Question

Question 5

A) What behaviour do you expect your robot to have?

```
task main()
{
        repeat (3)
        {
                    SetDirection(OUT_A+OUT_C, OUT_REV);
                    SetPower(OUT_A+OUT_C, 5);
                    SetOutput(OUT_A+OUT_C, OUT_ON);
                    Wait(200);
                    SetOutput(OUT_A+OUT_C, OUT_OFF);
                    Wait(100);
                    PlaySound(SOUND_CLICK);
        }
}
```

Groups	correct	Error/ misconception	No answer
G1	58%	25%	17%
G2	100%	0%	0%

Table 14. Results of 5th(A) Question

(B) Change your program so that the sound is produced only once, before the motor is activated.

Groups	correct	Error/ misconception	No answer
G1	46%	21%	33%
G2	57%	36%	0%

Table 15. Results of 5th(B) Question

5. Evaluation of the findings

An analysis of the data of the previous section is made here. A comparative analysis and evaluation of the results for both years of teaching will be attempted. The use of a different programming language and environment give the opportunity to formulate hypotheses in relation to the source of the difficulties, such as problems on account of the environment and the language, problems owing to the mechanical models and lastly to intrinsic problems of programming concepts which appear independently from the language or the environment.

Analysis of results of exercises:

In the majority of the exercises solved with Robolab a large percentage of students forgot to use the stop icon in order to stop the motors or the lamp.

Exercises 1 & 2: The use of the icon which determines waiting for a particular period of time appears to present some difficulty for the students in Robolab where only 57% used the correct icon, in contrast to NQC where 92% used the related command correctly.

Exercise 3: Robolab significantly favours the understanding of the repeat structures since all students with Robolab solved the exercise while a small percentage with NQC developed an incorrect code.

Exercise 4: The use of the repeat structure with Robolab is indirect since the icon that refers to the touch sensor (on/off) is used. It appears that this icon is used effectively, however, almost half the students did not use the modifier to determine the input port between the sensor located in the front and back parts of the robot. The percentage of programs with incorrect syntax was larger with the use of Robolab.

Exercise 5: More students used the repeat structure correctly in Robolab than in NQC (79% and 67% respectively). In Robolab students confused the icon related to the light sensor with the touch sensor and usually did not use the correct icons for the modifiers to control the level of lighting or the equivalent port which the light sensor is connected to. In NQC, the command is wait while in Robolab students must use the icon From the results, this seems to have created difficulties for the students since a mere 21% used the correct icon in conjunction with the correct Modifiers.

Exercise 6: The percentage of students who solved this exercise with Robalab is smaller than with NQC. This exercise is related to the select structure If. It must be noted that although more time was needed in order to explain the if structure in NQC than in Robolab, in the end students made more mistakes in the code with Robolab since they forgot to include in the if/fork icon the modifier that determines which input port the touch sensor is connected to. The icon which represents sound appears to be used more effectively in Robolab than in NQC. As in exercise 5, the majority of students also had problems in this exercise with the syntax of the sound command in NQC than in Robolab.

Exercise 7: In this exercise only 36% of students with Robolab used the correct icon for the operation of the lamp. Often they got the lamp icon confused with that of the sensor. On the contrary with NQC, the percentage of students which used the lamp correctly is practically double (71%), since in their code they did not require a particular command for the lamp. In NQC the commands for the operation of the components which are connected to the output ports are the same, the activation of the motor or the lamp and the determining of their power is defined by the same commands, which are SetOutput & SetPower respectively, as long as, of course, in each one of these commands is included the respective port where each mechanism is connected. Unfortunately, in Robolab different icons are needed for the lamp and the sensor, and in fact the two icons are next to each other, with the result that the students often confused them by selecting the one instead of the other.

Exercise 8: A small percentage of students solved this exercise in both environments. We observe that also in this exercise which refers to the use of if with a nested repeat structure, although the students understood both structures much faster in

Robolab than in NQC in the end they made more mistakes in Robolab. A mere 14% used the correct if icon to control whether the light sensor detected a value smaller than 60. In this case also they confused the light sensor icon of the select structure with that of the lamp. Further, in the repeat command in Robolab 36% used the modifier to state the number of repetitions, whereas 79% in NQC correctly developed the repeat structure. If we compare the percentage of students who developed the repeat structure correctly in exercises 3, 5 and 8, we see that it is 96%, 71% and 79% respectively in NQC, while in Robolab it is 100%, 79%, and 36% respectively. An explanation for this significant reduction in the percentage of correct responses in Robolab, in contrast to the relative consistency which the responses in NQC present, is that when the complexity of the code increases, then the students' code in Robolab presents more errors even for commands which appear to have been comprehended faster and easier in Robolab.

Analysis of the question results:

Question 3: For both years we see that the percentage of incorrect answers is similar and in fact in 2006-2007 slightly higher. It appears that students do not consider the icon or the command related to stopping the operation of the components found at the output ports of the mechanism to be necessary. In other words, the necessity of the stop icon in Robolab or the command SetOutput with the parameter OUT_OFF in NQC, is probably understood by the students as a feature of the mechanism and they do not recognize the equivalent icon or command when these are missing from their programs, but also often they do not include them in the code they develop.

Question 4: In this question, which concerns locating the error in the port that the light sensor is connected to, we observed that the students who worked with Robolab had a very low percentage of correct responses (36%) in relation to those who worked with NQC (100%). In the incorrect Robolab program the students had to add the modifier to the if icon, which determines the port which the light sensor is connected to.

Question 5: 64% of students with Robolab could correctly describe the condition in the control structure. In NQC all students recognized the duplication of the code not only in the if structure but also in the else structure.

Question 6: In the questions regarding the repeat structures students with Robolab achieved a higher percentage than students with NQC. The majority of students with NQC described the body of commands included in the repeat structure correctly but they did not describe the actual command repeat. They probably did not have enough time as it was one of the last questions they were given, or the icon in Robolab which describes the repeat command is quite effective and the students understand the operation of this command quickly without misconceptions, something which does not appear to happen to the same effect in NQC.

General Observations recorded by the teacher:

- The graphical environment of Robolab did not particularly facilitate the syntax of the code. In Robolab students had a problem with connecting the icons together, whereas the same was not observed for NQC.

- A graphical environment helps in the first lessons in terms of time needed to comprehend and develop the code but the use of modifiers counteracts this initial positive start.
- With NQC more time was needed to explain if in relation to Robolab where the graphical representation makes it easier and faster to for this structure to be understood.
- Modifiers are the components of Robolab which produce problems since they involve a large number of different icons, such as regulating the input/output ports, sound, light level, the number of repetitions in the repeat structures etc, which not only did students not apply correctly, but very often they confused them.
- It was ascertained that how the general properties of the "machine" the student learns to control and its relationship to the physical machine are understood, did not produce the same type of ambiguities in both environments. In other words, physical models help students to understand the "mechanism" of program execution and to associate commands of the code with the robot's activities.

6. Conclusions

Below, conclusions are drawn of the findings from the 2 series of lessons; they are related to other studies, and proposals for further research are made.

The analysis of the findings of the two-year study allowed us not only to make a comparison of the teaching methods applied each but also to formulate certain general conclusions in relation to the application of LegoMindstorms robots within the framework of an introduction to programming in secondary education.

Our teaching approaches with the use of physical models present certain features relative to earlier studies:

- Our lessons took place within a specified curriculum at the secondary educational level. This is a significant factor in evaluating the representational value of our results. One consequence of this conditions was, for example, that the participation in the lesson was mandatory thus not all students were highly motivated, as would have been the case in an elective subject. In addition, the lessons occurred within a context, which to some extent determined certain teaching parameters, such as imposed teaching objectives as part of the school syllabus. Furthermore, a relatively large number of students were obliged to participate in the lessons and generally the whole framework within which the lessons were conducted was a typical classroom setting. In some way, therefore, we consider that our lessons correspond more to the average student at secondary education level and subsequently the related findings are fairly representative.
- Our lessons were in the context of an introduction to programming (not robotics or technology). This means that certain aspects of the students' activities were more important than others: for instance, the learning and correct use of common programming structures (such as selection and repeat) were more significant than the constructional features of the robotic systems applied. Thus, while the usual implementation of robotic systems in teaching focuses on construction and "behaviour", we used robots as a means of teaching the basic concepts of programming.
- Our lessons were conducted during two consecutive school years using 2 different programming environments. As previously mentioned, the change of programming

environment in the second year occurred in an attempt to improve the effectiveness of the lessons. For the same reason certain lessons were also altered in that same year. After finalizing the second series of lessons, we made a new set of proposals intended to further improve the teaching methodology. However, these findings will not be discussed here as they are relatively extensive and cannot be incorporated into the scope of the present discussion.

Our general finding could, therefore, be summarized as follows:

Generally, the application of physical models, such as robots, undoubtedly present advantages since they are attractive and thus constitute a strong incentive for students. Nevertheless, as observed by other researchers, their use does not automatically mean that all students fully understand all the new concepts and techniques presented, nor of course to the same degree. Therefore, the use of these environments, in order for them to have teaching value, require a careful design of activities, time etc. In reality, the evaluation of the pros and cons of the use of these environments in teaching is fairly complex, since, as our findings show, they demand explicitly specified teaching conditions in order to achieve positive results. In addition, student results seldom have an "absolute" character; for example, the comparative findings between the two environments used in our study did not clearly favour one or the other environment, since at the initial stages it appeared that one prevailed, while at more advanced stages the other seemed to do so (see below for a detailed analysis). In other circumstances, the students appear to understand certain basic concepts or mechanisms but they make important mistakes when answering the related questions. Consequently, we are not yet in a position to formulate general rules that describe student behaviour in relation to the difficulties they encounter. It is characteristic, therefore that, as referred to in the comparative study in the above section, even the incentives for students must, in some way, be "pre-designed, as often their interest rapidly wanes when progress in the lesson is "slow" or when the proposed activities are not "interesting". In certain cases, it appears that the teaching demands are incompatible or even conflicting with students' incentives.

At another level, as was anticipated, the understanding of the "machine's" general properties that the student learns to control and its relation to the physical model did not produce problems in either environment.

Nevertheless, many of the difficulties encountered in teaching an introduction to programming were present in our lessons as well. In many cases, for example, the requirements for programming the robots comprised a source of difficulty for the students when it is it is different to their daily experience. A characteristic example of this is the command or symbol STOP (or Wait respectively) which is needed to stop the operation of different parts of the mechanism often appears to be ignored by students, as is the use of modifiers for the selection of the correct input port. The students do not include the related symbols in their programs, nor do they indicate the absence of these in pre-prepared programs they have been given to check for mistakes, even though they show that they have an understanding of the role these play and the way they are used. A possible explanation is that students are not familiar with the need for an accurate and explicit formulation of the respective commands, which in their daily experience occur in an indirect manner (for example, finalizing an operation of the mechanism or after some predetermined time span, usually for students means that the mechanism stops without this requiring a specific command in order to do so). Difficulties of this nature are observed in introductory teaching to programming almost irrespective of the environment used.

Our comparative study also enabled us to examine the relative teaching value of the environments which use iconic or "textual" programming languages, an issue which has often preoccupied the international research community.
The use of an iconic language naturally requires the selection of icons and symbols which refer to semantics while at the same time these icons and symbols must be distinctly different to avoid getting them confused (as in the case with the light and touch sensors, or the lamp and light sensor which the students often mix up). The characteristic icons, however, are not always easy to create, since there are elements in the programming of robots which are represented better with words rather than with icons. A typical example of this are modifiers: it is difficult for someone to "guess" the purpose or how they are used by simply looking at the icon. In addition, the large number of icons required even for common programs, could create confusion since the students must choose the appropriate icon among a range of icons that are alike. The weakness of parameterizing the iconic environments seems to produce yet another problem, as for every operation an appropriate icon must be created ad hoc, which refers to it. In contrast, in textual languages the simple existence of parentheses in the syntax of a command refers directly to the need to give a value to the particular parameter. In this way, while in the initial stages in the negotiation of a concept, an iconic environment appears to be better, later, however, this advantage is lost.
A similar thing seems to happen with the repeat structures, since the repeat structure is obvious in the flow chart created by the icons. Students' mistakes, however, are many when the problem requires nested structures and generally a more concise structure and a more complex code than that required for a simple repetition. Lastly, a similar phenomenon was located in the use of the selection commands if.
In some way although written commands are less direct than icons, they permit the creation of relatively complex structures with greater ease.
It is likewise characteristic that while naturally in iconic environments syntactic errors are not an issue, nevertheless, students appear to have difficulty in the "syntax" of their programs, meaning that they found it difficult to connect and correctly locate the icon – commands.
The nature of the problems encountered within the context of the lessons, did not allow for further investigation of students' perception to the lessons and generally to the evaluation of the teaching benefits of these types of environments. It was not possible to examine student strategies on problems which were more complex and algorithmically more interesting. Our findings, however, allow us to conclude that taking into account the conditions described above, teaching with physical models such as robots appears to be enticing to students and helps them to learn elementary programming concepts.
Two adjacent results arising from the lessons conducted, which did not, however, constitute set questions in our research, must nevertheless be mentioned.
The first is perhaps interesting for those who intend to apply physical models in their teaching, like the ones we describe. The use of these models requires a much longer preparation period than that required by using only software. The reasons are quite obvious: since the physical models operate as mechanisms, they must be regularly checked to ensure that they are operating properly (for example, if there is a satisfactory level of energy in the batteries, if the connections are good, etc).
The second is related more to teaching as such rather than the preparation of the material. Since these systems operate under real conditions, they are not governed by the absolute determinism of the software. So, an operational command of a motor, for instance, does not

always cause the vehicle to move for the same distance (it depends on the energy of the battery, ground friction etc). Moreover, certain difficulties encountered by users came precisely from the fact that they were dealing with physical systems, which the students must implement with precision.

Although neither of these points is a serious drawback to teaching, they, nevertheless, do exist and they do affect teaching, even if to a small degree.

The findings so far allow us to formulate the conclusion that physical models comprise a good alternative solution for an introduction to programming, but further research must be conducted, which will determine those conditions with greater accuracy (e.g. selection of programming environment in accordance with teaching aims) which will enable teaching to be more effective.

8. References

Avanzato, R. (1998). Collaborative Mobile Robot Design in an Introductory Programming Course for Engineers, *Proceedings of the ASEE Annual Conference* ", June 1998, Seattle, Washington.

Barnes, D. (2002). Teaching Introductory Java through LEGO Mindstorms Models, *Proceedings of the 33rd SIGCSE Technical Symposium on Computer Science Education,* pp. 50 – 54, Northern Kentucky, Cincinnati, February 2002, ACM.

Becker, B. (2001). Teaching CS1 with Karel the Robot in Java, *Proceedings of the Thirty Second SIGCSE Technical Symposium on Computer Science Education*, pp. 50 – 54, Charlotte, NC USA, February 2001, ACM.

Bergin, J.; Stehlik, M.; Roberts & Pattis J. (1997). *Karel++ - A Gentle Introduction to the Art of Object-Oriented Programming, 2nd edn. New York, Wiley.*

Brusilovsky, P.; Calabrese, E.; Hvorecky, J.; Kouchnirenko, A. & Miller, P. (1997). Mini-languages: a way to learn programming principles, *Education and Information Technologies. 2*, pp. 65-83.

Brusilovsky, P.; Kouchnirenko, A.; Miller, P. & Tomek, I. (1994). Teaching programming to novices: a review of approaches and tools. *Proceedings of ED-MEDIA '94*, pp. 103-110, Vancouver, British Columbia, Canada, June 1994.

Cliburn, D. (2006). Experiences with the LEGO MindstormsTM throughout the Undergraduate Computer Science Curriculum, *36th ASEE/IEEE Frontiers in Education Conference*, pp. T2F1-6, San Diego, October 2006, CA, IEEE.

Du Boulay, B.; (1989). Some Difficulties Of Learning To Program, *In Studying The Novice Programmer*, Soloway, E., Sprohrer, J. (Ed.), Lawrence Erlbaum Associates, pp. 283-300.

Du Boulay, B.; O'Shea, T. & Monk, J. (1989). The Black Box Inside the Glass Box: Presenting Computing Concepts to Novices, In: *Studying The Novice Programmer*, Soloway, E., Sprohrer, J. (Ed.), pp. 431-446, Lawrence Erlbaum Associates.

Fagin, B.; (2000). Using Ada-Based Robotics to Teach Computer Science, *Proceedings of the 5th Annual Conference on Innovation and Technology in Computer Science Education*, pp. 148-151, June 2000, Helsinki, Finland, ACM.

Fagin, B.; Merkle, L. (2003). Measuring the Effectiveness of Robots for Teaching Computer Science, *Proceedings of the 34th SIGCSE Technical Symposium on Computer Science Education (SIGCSE 2003)*, pp. 307-311, February 2003, Reno, Nevada, ACM.

Genalo, L. (1997). Toying With Technology: Mobile Robots and High School Interns, *Proceedings of the ASEE Annual Conference.*

Goldweber, M.; Congdon, C.; Fagin, B.; Hwang, D.; Klassner, F. (2001). The Use of Robots in the Undergraduate Curriculum: Experience Reports, *Proceedings of the Thirty Second SIGCSE Technical Symposium on Computer Science Education*, pp.404-405, Charlotte, North Carolina, February 2001, ACM.

Hirst, A.; Johnson, J. ; Petre, M.; Price, B. & Richards, M. (2003). What is the best programming environment / language for teaching robotics using Lego Mindstorms?, *Artificial Life Robotics*, Vol. 7, pp.124-131.

Kaskalis T.; Dagdilelis V.; Evangelidis G. & Margaritis K. (2001). Implementing Applications on Small Robots for Educational Purposes: Programming the LEGO Mindstorms, *Proceedings of the 5th Hellenic – European Conference on Computer Mathematics and its Applications (HERCMA 2001)*, pp. 337 – 341, Athens, Greece, September 2001.

Lawhead, P. & Bland C. (2001). Events, Robots and Programming Using Legos in CS1, SIGCSE Bulletin, Proceedings of the 6th Annual Conference on Innovation and Technology *in Computer Science Education*, Vol. 33, No 3, June 2001, ACM.

Mayer, E. (1989). The Psychology of How Novices Learn Computer Programming. *In Studying The Novice Programmer*, Soloway, E. and Sprohrer, J. (Eds.), Lawrence Erlbaum Associates, pp. 129-160.

Mendelsohn, P.; Green, T.R.G. & Brna, P. (1990). Programming Languages in Education: The Search for an Easy Start. *In Psychology of Programming*, Green H., T., Samurcay, R. & Gilmore, D. (Ed.), Academic Press, pp. 175-200.

Pattis, R. E.; Roberts, J. & Stehlik, M. (1995). *Karel - The Robot, A Gentle Introduction to the Art of Programming*. New York, John Wiley & Sons.

Resnick M. (1993). Behavior Construction Kits, *Communications of the ACM*, Vol. 36, No. 7, pp. 64-71.

Satratzemi M.; Dagdilelis V.; Kagani K. (2005). Teaching Porgramming with robots: A case Study on Greek Secondary Education, P. Bozanis, E.N. Houstis, (Ed.), pp. 502-512, *Lecture Notes in Computer Science (LNCS)*, Vol. 3746.

Wolz U. (2001). Teaching Design and Project Management with Lego RCX Robots, *Proceedings of the Thirty Second SIGCSE Technical Symposium on Computer Science Education*, pp. 95-99, Charlotte, North Carolina, February 2001, ACM.

Xinogalos, S. (2002). *Educational Technology: A Didactic Microworld for an Introduction to Object-Oriented Programming* (in Greek), Ph.D. Thesis, Dept. of Applied Informatics, University of Macedonia.

21

Agricultural Robots – Applications and Economic Perspectives

Pedersen S. M[1], Fountas S[2] and Blackmore S[2]
[1]University of Copenhagen,Institute of Food and Resource Economics,
[2] University of Thessaly
[1]Denmark
[2] Greece

1. Introduction

1.1 General aspects

For many years robotic systems have been widely used for industrial production and in warehouses, where a controlled environment can be guaranteed. In agriculture and forestry, research into driverless vehicles has been a vision initiated in the early 1960's with basic research on projects on automatic steered systems and autonomous tractors (Wilson, 2000).
Recently, the development of robotic systems in agriculture has experienced an increased interest, which has led many experts to explore the possibilities to develop more rational and adaptable vehicles based on a behavioural approach. A combined application of new sensor systems, communication technologies, positioning systems (GPS) and geographical information systems (GIS) have enabled researchers to develop new autonomous vehicles for high value crops in the agriculture and horticulture soctor, as well as for landscape management.
Several autonomous prototypes have been described for orchards and horticultural crops, such as oranges (Hannan and Burks, 2004), strawberries (Kondo et al., 2005) and tomatoes (Chi & Ling, 2004). Moreover, automated systems for site specific irrigation based on real time climatic conditions have been described for high value crops (e.g. Miranda et al, 2005)
For field crops there are also a number of systems, such as the Demeter system for automated harvesting equipped with a camera and GPS for navigation (Pilarski et al., 2002), and the autonomous Christmas tree weeder (Have et al., 2005) and the API platform for patch spraying (Bak & Jacobsen, 2004). In addition we have seen automated systems for animal production in indoor environments such as automated feeding and cleaning.
In the open and outdoor environment, which will be the focus here, robotic and autonomous systems are more complex to develop - mainly because of safety issues. The robots safety system would have to be reliable enough for it to operate autonomously and unattended. It is relatively costly to develop safety systems if the vehicle has to be completely autonomous. In principle, they can work 24 hours a day but if a robot has to be attended then the time is limited by the person. In this matter different scenarios and degrees for autonomy have been investigated depending on the task to be carried out. Concepts have been initiated to investigate if small autonomous machines would be more

efficient to replace the traditional large tractors. These vehicles should be able to carry out useful tasks all year round, unattended and able to behave sensibly in a semi-natural environment over long periods of time. The small vehicles may also have less environmental impact replacing the over-application of chemicals and fertilizers, requiring lower usage of energy with better control matched to requirements, as well as causing less soil compaction due to lighter weight.

Fig 1. MF-Scamp robots for scouting, weeding and harvesting (Designed by Blackmore. Copyright©2008 AGCO Ltd)

1.2 Economics

So far, only a few studies have been published on the economic consequences by introducing autonomous field machinery to achieve more sustainable production systems. Goense (2003) compared autonomous with conventional vehicles, equipped with implements having working widths from 50 to 120 cm. He showed that if an autonomous vehicles can be utilised 23 hours a day, it would be economic feasible with slight reductions in prices of navigation systems or with slight increases in labour costs. Goense also discussed a number of other changes that will affect the final result, such as the fraction of labour time needed out of the total machine time and the machine tracking system, which provides better utilisation of machine working width and there is no need for operators rest allowance. On the other hand, there may be negative effects in the form of higher costs in travelling distances for service personal.

Additionally, Have (2004) analysed the effects of automation on machinery sizes and costs for soil tillage and crop establishment. He assumed that the ratios between an autonomous tractor and a manned tractor, in terms of price, labour requirement and daily working hours would be 1.2, 0.2, and 2 times, respectively. The analysis, which included all direct machinery and timeliness costs showed that the shift to automatic control would decrease the tractor size, implement sizes and implement investments to about half; decrease the tractor investment to about 60% and decrease the sum of annual tractor and machinery costs to approximately 65%. Pedersen et. al. (2006 & 2007) outlined a comparison between 3 different applications of robot systems with conventional systems for crop production and landscape treatment. The economic feasibility study in thepresent chapter relies on some of these scenarios and descriptions.

1.3 Objectives

The aim of this chapter is to present the status of the current trends and implementation of agricultural and horticultural robots and autonomous systems and outline the potential for future applications. Different applications of autonomous vehicles in agriculture have been examined and compared with conventional systems, where three main groups of field operations have been identified to be the first potential practical applications: crop establishment, plant care and selective harvesting.

Moreover we will give examples of the economic potential of applying autonomous robotic vehicles compared to conventional systems in two different applications: robotic weeding in high value crops, particularly sugar beet, and crop scouting in cereals. The comparison was based on a systems analysis and an individual economic feasibility study for each of the applications. Focus will be put on potential labour cost savings, farm structure implications and sizes for operation, daily working hours, potential environmental impact, energy costs and safety issues.

2. Autonomous vehicles scenarios

2.1 Crop establishment

Several concepts of crop production and cultivation are now being revisited and reconsidered in the light of developing smarter machines.

For traditional crop establishment and seed bed preparation it has been common that the whole topsoil of a field is inverted with a plough to create a suitable seed bed. This is a well known operation that suits many circumstances but it also uses a lot of energy. If this approach is turned around to consider the actual seed requirements, other methods with reduced tillage may become an option. The seed requires contact with soil moisture to allow uptake of water and nutrients; it requires a structure that can hold the plant upright but also allow the roots to develop and the shoots to grow. Moreover it requires space with little competition from other plant seeds. If the same seed environment can be achieved by only mixing the soil within a few centimetres of the actual seed then the rest of the soil does not need to be disturbed as it can be well conditioned by natural soil flora and fauna.

Another traditional concept is to grow crops in rows. It would seem that the only explanation as to why this is done is that it requires the simplest type of machines. Seeds are placed relatively densely along each row. The problem is that in principle, each plant requires equal access to light, air, water and nutrients, which are often spatially related. Intra crop competition can be reduced by giving a more even or equal spacing and seed distribution with accurate placement of seeds in a more uniform pattern (see figure 2).

In figure 2 the main crop is illustrated with green spots and the weeds are red spots. On the left of figure 2 the crop is seeded in traditional rows. Here, it is possible for the weed to grow fast and overtake the main crop at an early development stage. On the right side the crop is distributed uniformly with seed placement. In principle, the main crop will cover the weed before the weed develops and thereby reduce the weed pressure. Initial trials indicate that it may be possible to reduce the weed biomass with up to 60% by using uniform patterns with accurate seed placement (Weiner and Olsen 2007).

If the location of each seed is known and the position of each emerged crop plant is estimated, it will be possible to identify each plant by its spatial location. Improved information about plant characteristics allows improved management and decision making and allows a number of improved, more targeted operations that can improve the overall crop growing efficiency. As only a small volume of soil is needed to be cultivated there are a number of different methods that could be used. Rotary mechanical tillage in two dimensions on a vertical or horizontal axis (as indicated in the seed map in Figure 2) could be used instead of just one dimension or water-jetting or the injection of a hygroscopic polymer gel could be used to create a micro climate for the seed.

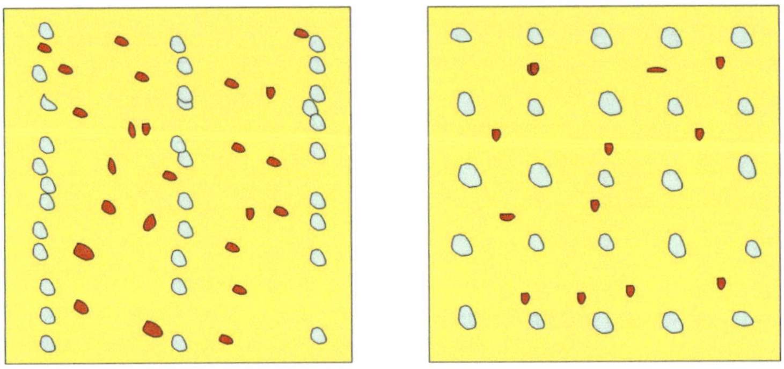

Fig. 2 Precise seed placement in uniform patterns (Source. Based on Weiner J and Olsen J, 2007. Momentum 2007)

Seed mapping is relatively simple in practice as a tilt adjusted RTK-GPS could be fitted to the high precision seeder and infra red sensors mounted below the seed chute. As the seed drops, it cuts the infrared beam and triggers a data logger that records the indicated position and orientation of the seeder. A simple kinematic model can then calculate the actual seed position (Griepentrog et al., 2005a). However, ultra high precision placement of seed is difficult from a moving machine. Even the most sophisticated commercial machines have problems with consistently separating individual seeds and will often damage the seeds as they pass through the metering mechanisms. Some mechanisms do ensure that each seed has zero ground velocity which is important to stop the seed bouncing after impact with the soil (Griepentrog et al., 2005b).

2.2 Crop scouting

An important part of good management is the ability to collect timely and accurate information about the crop development. Quantified data has tended to be expensive and sampling costs can quickly out weigh the benefits of spatially variable management. (Godwin et al., 2003).

Data collection would be less expensive and timelier if an automated system could remain within the crop canopy for continual monitoring that can be used for assessing crop status. This could be achieved by either embedding cheap wireless sensors at strategic positions within the crop, or placing more expensive sensors onto a moving platform.

Crop characteristics would include Leaf Area Index, crop height (giving growth rates), growth stage, biomass, senescence, etc. Crop nutrient status is difficult to assess in the early stages and assessing it independently from moisture stress become very complex (Christensen et al., 2005).

With the advent of biosensors, a whole new set of opportunities will become available to monitor growing crops for pest and disease attack (Tothill, 2001). As the robotic/autonomous vehicle could patrol the fields continually looking for weeds and other threats, real-time alerts could be sent to the manager whenever certain conditions were encountered. These could take the form of noting actual pest or disease attack or by monitoring environmental conditions where they are likely to occur or that the risk of attack is significant. Differing growth rates could also be used to identify potential problems.

2.3 Selective harvesting

At present, crops are usually harvested when the average of the whole field is ready as this simplifies the harvest process. Selective harvesting involves the concept of only harvesting those parts of the crop that meet certain quantity or quality thresholds. It can be considered to be a type of pre sorting based on sensory perception. Selective harvesting has been well known in forestry for many years where certain trees are harvested according to quality and size or to improve the quality of the remaining trees.

In agriculture, examples could be to only harvest barley below a fixed protein content or combine grain that is dry enough (and leave the rest to dry out) or to select and harvest fruits and vegetables that meet a size criteria.

As these criteria often attract quality premiums, increased economic returns could justify the additional sensing. Benefits of multiple-pass system than a single-pass for corn harvested for biomass has been reported by Shineers et al. (2003), as well as the advantages for selective harvesting of asparagus (Cembali et al., 2005) and dates in Iran (Abounajmi, 2004). To be able to carry out selective harvesting effectively, two criteria have to be met; the ability to sense the quality factor before harvest and the ability to harvest the product of interest without damaging the remaining crop.

Most agricultural vehicles are getting bigger and hence not suited for this approach. Therefore, smaller and more versatile selective harvesting equipment is needed for this purpose. Either the crop can be surveyed before harvest so that the information needed about where the crop of interest is located, or that the harvester may have sensors mounted that can ascertain the crop condition. The selective harvester can then harvest that crop that is ready, while leaving the rest to mature, dry, or ripen etc.

Alternatively, small autonomous whole crop harvesters could be used to selectively gather the entire crop from a selected area and transport it to a stationary processing system that could clean, sort and maybe pack the produce. This is not a new idea, but updating a system that used stationary threshing machines from many years ago. Alternatively, a stripper header could be used to only gather the cereal heads and send them for threshing. As selective harvesting only harvests products of desired quality the harvesting process will be phased over a longer periods of time, negating the need for large equipment. As the products are already graded or sorted, it also adds value to the products before they leave the farm.

3. Economic scenarios

In this section, we have compared the costs and benefits of the potential commercial use of robotic vehicles. We based the calculations on partial budgeting, where the cost change is compared to conventional practices. In this model, we included changes in initial investments, labour costs, change in speed, daily working hours, energy consumption, control and surveillance costs. We compared the saved labour and spraying costs and additional costs for electronic devices, GPS-system and platforms for the robotic systems with conventional manned systems.

Technical parameters such as dimensions, capacities, speed and related costs were based on recommendations from other research groups and experts. However the economic figures such as period of depreciation, real interest rate (5 %) and maintenance costs were based on the authors' assumptions.

For the two scenarios: **field scouting for weed detection** and **weeding**, the model was built to project potential applications in conventional farms with average field size and crop

rotations. These two application technologies are fairly mature currently at a pre-commercial development stage.
The data for the conventional applications, were taken from general economic statistics for the Danish farm management standards, regarding specific costs for contracting. Moreover, we have received data from researchers at Danish Institute of Agricultural Sciences (DIAS), Bygholm, that are working with these specific applications on an autonomous platform. The autonomous platform that they are using is the Autonomous Plant Inspection (API) research platform (Bak and Jakobsen, 2004).
We assumed that the period of depreciation is 10 years with linear depreciation. The real interest rate was 5% and we assumed that it would be necessary to add some additional labour time (two weeks) for testing and making the necessary calibrations for the systems.

3.1 Field scouting for weed detection

Weed mapping is the process of recording the position, density (biomass) and preferably species of different weeds using machine vision. For automatic weed detection, several studies have been performed applying different discrimination and classification techniques. Manh et al. (2001) used parametric deformable templates to segment individual weed leaves, Sokefeld et al. (2000) identified more than 20 weed species using Fourier descriptors and shape parameters, while Sogaard (2005) used Active Shape Models to identify 19 weed species. Artificial neural networks have also been used by many researchers to discriminate weeds (e.g. Burks et al., 2005; Granito et al., 2005) with machine vision. Other researchers have used image analysis techniques for weed discrimination using remote sensing from ground sensors (e.g. Vrindts et al., 2002) and airborne photography (e.g. Godwin and Miller, 2003).
Weed mapping is a hypothetical scenario in the sense that most farmers do not conduct systematic field scouting in their cereal fields today. Farmers either practice conventional farming with conventional spraying or they conduct organic farming with mechanical weeding. In this comparison, we assume that the alternative to autonomous weed mapping is manual weed mapping, which implies that the farmer has to register and map the weeds in the field manually with a handheld GPS.
In the weed mapping scenario, we compared autonomous field scouting for weeds in cereals with the manual detection of weeds. The autonomous system requires an API vehicle and cameras for weed detection and mapping. The Danish Institute of Agricultural Sciences (DIAS) has performed tests using such a vehicle for weed recognition and data from personal communications with the researchers have been used for our calculations. The API platform, (Fig.2) was initially developed by Madsen and Jakobsen (2001). Now, there is the third generation of API vehicle, further developed by Aalborg University in Denmark. This prototype has four wheel-drive, four-wheel steering with two motors per wheel, one providing propulsion and the other steering to achieve higher mobility (Bisgaard et al., 2004).
The platform has a height clearance of 60 cm and track width of 1 m. It is equipped with an Real Time Kinematic Global Positioning System (RTK-GPS) and on the top of the frame there is an operating console and an implement for the agricultural operation such as spraying devices, sensors or weeding tools. The vehicle communicates with the farm management PC for navigation, according to the computed route plan, as well as collision avoidance (Bak and Jakobsen, 2004). Based on shape recognition of weeds it is possible to create a weed map of the entire field.

Fig. 3. The API platform.

In this scenario, the area for field scouting and weed mapping is limited to 500 ha to match large production units with the necessary flexibility. We have focused on cereal crops but it may also be relevant for other crops such as high value crops like sugar beet, potatoes or other horticultural crops. However, the shorter the time for carrying out the activity, the lower the overall capacity required.

In weed mapping/field scouting, the robotic system is compared with manual detection of weeds. Most of the time for manual weed scouting will take place in the first year, which is followed by shorter update scouting in the following years. Manual weed scouting is assumed to require about 0.7 man h/yr/ha (Pedersen, 2003). The weed patches are registered by using GPS and GIS systems to create weed maps of the individual fields. Autonomous field scouting using the API platform has a speed of 3.6 km/h and a capacity of 4.32 ha/h, which adds up to 116 h/y for autonomous weed scouting on a 500 ha area giving 0.232 h/yr/ha.

An automated weed mapping system will enable the farmer to produce weed maps, which will be useful to carry out precise patch spraying with the right mixture and dose rates of herbicides where it is needed. The technology and decision support systems to apply patch spraying is available and herbicide savings can vary from 30% and up to 75% (Heisel et al., 1999; Sogaard, 2005).

The autonomous field scouting system in cereals reduces the costs by about 20 % compared to manual weed scouting but it should be possible to increase the capacity to 1000-2000 ha since the system, as presented here, is only used 116 h/y (Pedersen et. al 2006)

Since the costs of the autonomous platform is based on estimated costs of producing the platform it might be the case that a commercial selling price will be significantly higher. An 100 % increase of the cost of the API-platform to 30,300 € will imply that the overall costs of the autonomous field scouting system will increase to 20.3 €/ha/year, which is slightly above the labour costs for manual weed scouting.

3.2 Weeding

3.2.1 Weeding methods

As outlined above it could be possible to automate weed mapping by using active shape models and GPS systems in cereals. However, one issue is to collect data about weeds; another issue is to physically remove the weeds.

There are several methods that can remove or kill unwanted plants without using chemicals (Nørremark and Griepentrog, 2004). These can range from total removal down to simple retardation. A classic example would be to promote the wilting of the weed plants by breaking the soil and root interface by tilling the soil within the root zone.

In principle, there are three main areas within the crop environment that require different types of treatment: The inter-row area (the space between the crop rows), the intra-row area (the space between the plants within the row), and the close-to-crop area that is within the leaf and root envelope. The closer to the crop that a machine should operate, increased care and accuracy is needed so as not to damage the crop plant tissue.

The inter-row area is relatively easy to keep free of weeds as it is an open strip in line with the direction of travel. The intra-row area is more difficult to manage as it is an intermittent space delineated by irregular spacing of crop plants. The close-to-crop area should not have any soil disturbance as this would lead to possible wilting. Weeds within this area are likely to incur the most competition of all as they are so close to the crop plant's resources. Retardation of weeds in this area must rely on intelligent sensing and highly targeted energy inputs such as micro spraying or laser weeding.

Laser weeding holds great promise as it uses a highly mobile focused beam of infra red light to thermally disrupt cell membranes in the weed leaves and shoots. The beam can be controlled easily by computer and can significantly reduce the energy needed for thermal weed control (Griepentrog et al., 2006). Location of weed leaves can be achieved by machine vision (as above)

3.2.2 Micro spraying

Micro spraying takes the concept of a spray boom down to the centimetre level (Søgaard et al., 2006). It applies highly targeted chemicals and can treat small areas by selectively switching the jets on and off. It is part of a larger system that can recognise individual weed plants and locate their leaves for treatment (see weed mapping).

Trials have shown that when herbicide is targeted in the right way at the right time, the usage can be drastically reduced. Tests were carried out by a human operator to identify and treat individual weed plants that resulted in reducing the application of glyphosphate from 720 grams per hectare down to about 1 gram per hectare for an infestation of 100 weeds per square meter and maintain acceptable efficacy (Graglia, 2004). If this same approach can be carried out by an autonomous micro sprayer then there will be significant economic and environmental advantages.

Within the close-to-crop area, great care must be taken not to damage the crop nor disturb the soil. The use of a micro spray that delivers very small amounts directly on to the weed leaf has been fitted to the crop scouting robot described above. Machine vision can be used to identify the position of an individual weed plant and a set of nozzles mounted close together can squirt a herbicide on to the weed. Tests have shown that splashing can be reduced when a gel is used as a carrier rather than water (Søgaard & Lund, 2005).

3.2.3 Robotic weeding

The financial analysis here focuses on the potential application of micro spraying in sugar beets in four regions that differ in terms of farm structure, farm sizes and land topography, namely, Denmark, Greece, UK and USA (Red River Valley). Sugar production in the four regions are: Denmark (48700 ha), Greece (35973 ha), US (528890 ha) and UK (154000 ha). FAO statistics 2004.

We have focused on four case areas that all produce sugar beet but differs in terms of labour cost and average farm sizes - with the highest in the US. However, in the US, most crops are produced less intensively with less input factors. Sugar beet production occurs in 12 US states and about 48% of the US sugar beet acreage is located in the Red River Valley (Minnesota and Eastern North Dakota). The average US sugar beet farm area is about 80 ha (Ali 2004) and the average field size for sugar beet is 6.5 ha in UK although some fields may be as large as 40-50 ha. Currently in the UK, there are a little under 9,000 growers and about 150,000-180,000 ha of sugar beet is grown annually.

In the ***robotic weeding*** scenario, we compared an autonomous vehicle equipped with a micro spraying system with a conventional sprayer for sugar beet. The micro spraying system would be mounted on an API platform as illustrated in figure 3.

The micro sprayer consists of a set of eight micro valves with a driver circuit for each nozzle. It is assumed that this system can reduce the application of herbicides by 90% compared with standard doses in sugar beet (Lund and Sogaard, 2004). The working speed is 1.8 km/h with recharging of batteries every 5 h. The vehicle has a width of 2 m with a capacity to treat 4 rows simultaneously.

Inter-row weeding (between the rows) is envisaged to be carried out conventionally. We only focus on the intra-row weeding close to the crop. The season for operation is limited from April to July. The robotic weeding is compared with the costs of conventional weeding in sugar beet. The costs for these operations are based on average prices for contracting. The API platform, as designed for this scenario, is equipped with 4 micro spraying systems. The autonomous platform is able to cover 4 rows at a time. The speed is 1.8 km/h and the capacity is 0.4 ha/h, which adds up to between 417 h/y in Greece and 883 h/y in UK and US for autonomous weeding. For the autonomous micro spray system, inter-row hoeing has to be conducted twice whereas, for conventional spraying, we assume one treatment. For comparison, it might be relevant to inter-row hoe 3 times when conducting band spraying.

Based on the assumptions above, the potential economic viability of operating this system compared to a similar treatment with conventional practices were determined. The total investments for the systems are indicated as depreciation and capital costs.

The RTK-GPS system is still fairly expensive for these practices although the price is expected to be reduced as the technology becomes more widespread. The cost of receiving a GPS reference signal accounts for a significant share of the yearly costs for both systems. For both systems, we assume a depreciation period of 10 years. Maintenance of investments was assumed to be an additional 3 %.

For weeding in sugar beet, the primary savings were related to the reduced application of herbicides and additional costs were related to investments in micro-sprayers. A comparison between the costs in Red River Valley and Denmark indicated that the pesticide costs on conventional average farms are about 200-220 € in both regions.

	Robotic weeding in sugar beet			
	DK	GR	US*	UK
Platform with RTK-GPS	API system	API system	API system	API system
Total area treated with autonomous system , ha	80	50	100	100
Operation time per day, h/day	16	12	16	16
Operation hours, h/yr	667	417	883	883
Days for operation, days	42	35	52	52
Wages, unskilled labour, €/hour	14	6	9	10
Electricity costs 100kwh, in €	9.30	6.68	8.18	9.05

Table 1. Technical and financial assumptions for the system analysis
Source: Pedersen et al 2007.
* Red River Valley (Minnesota and Eastern North Dakota)
Exchange rates: 1 € = 7.45 DKK, 0,68£, 1,34 $

With the autonomous system, it is possible to handle 4 rows with the API platform and possible more (6-10 rows) in the future. Moreover, the costs of each spraying system are likely to be reduced with larger systems.

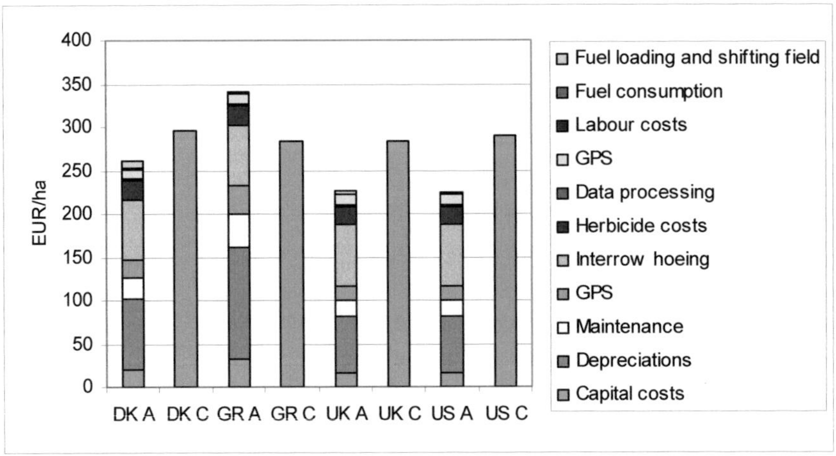

Fig. 4. Cost structure for autonomous (A) and conventional (C) weeding in sugar beet

The calculations were based on fairly conservative economic figures and prices in the 4 case regions. However, a further reduction in the price of RTK-GPS and other electronic systems in line with increased supply of these systems are expected in the future. Software costs were not explicitly included in this study, apart from the RTK software system. In this matter, some additional costs should be expected depending on the diffusion of the systems. Moreover, safety regulations for using these systems may be a further financial burden. To implement a complete independent and autonomous system, it will be necessary to include additional control and safety regulation and insurance costs depending on the task and location.

4. Discussion and future trends in crop production

Most new machines brought to the market are bigger than the previous model. When discussing this issue with equipment manufactures, this trend is likely to continue into the future. The driving force for this growth would seem to be to take advantage of the economies of scale that larger machines bring with them. This is easily demonstrated if the cost of the operator is taken into account. As most operators are paid by the hour, a larger machine that can increase the work rate over a smaller one can have a significant economic advantage.

The size increase does not only bring benefits. Large machines are only viable when working in large fields as turning, positioning and transport are all non productive activities. Although many farms have removed field boundaries to take advantage of the larger machines, many smaller farms cannot follow suite due to environmental concerns and suffer economically because of it.

As this equipment becomes larger, it also becomes very capital intensive with new tractors and combines becoming prohibitively expensive for the small and medium sized farm. Reliability also becomes an issue as all processes are carried out in series. If one part of the mechanisation system breaks down then all field operations stop.

An alternative approach would be to use available information technologies to automate these processes to the point where they do not need a human operator. By removing the person from the immediate control of the system, it offers new opportunities but also creates new problems.

Once the person is outside the control loop, then the economies of scale that applied to the larger, manned tractors does not apply and alternative smaller smarter systems can be developed. Work rates (per day) can be kept high by working longer hours and using multiple machines.

By taking a systems approach to designing robotic systems, consideration can be given to a system in terms of its action, interactions and implications. The result should be a new mechanisation system that collectively deals with the crop's agronomic needs in a better way than is done now. Most people define agronomic processes in terms of how they are currently carried out and a break from this mentality, or paradigm shift, is needed to define the processes in terms of the fundamental plant needs. When the plant requirements are defined independently of the machine that carries out the corresponding operations, this improved specification can be used in conjunction with mechatronic principles to help design smarter and more efficient machines.

In this study we have analysed the economic viability of two hypothetical autonomous robotic systems. In both scenarios we have replaced trivial labour intensive tasks for specific areas with autonomous systems based on highly accurate GPS-systems. These concepts and applications could be expanded to other field cultivation systems, tillage systems and grass cutting tasks at sport facilities and at public recreation areas. The autonomous weeding system with micro spraying in sugar beet may reduce the overall herbicide application with 90% and thereby improve the socio economic benefit. The autonomous field scouting system opens up the possibility for easier weed mapping, which again may give an incentive to conduct patch spraying in cereals and other crops. In addition, these robotic systems may further improve flexibility and expand the daily time constrains to night operations in the field and thereby improve the efficiency in modern crop production.

Based on the various systems and technical assumptions above, we have provided the potential economic viability of operating these systems compared to a similar treatment with conventional practices. The RTK-GPS system is still fairly expensive for these practices,

although the price is expected to be reduced as the technology becomes more widespread. The cost of receiving a GPS reference signal accounts for a significant share of the yearly costs for both systems - although both systems seem to be economically viable given the technical and economic assumptions above.
For both systems, we assume a depreciation period of 10 years. However, given the intensive utilisation of the robotic weeding system, it may be necessary to reduce the period of depreciation to about 5-8 y. In contrast, the autonomous field scouting system might have a longer lifetime than outlined above. It should also be possible to reduce field scouting costs by nearly 20 % in cereals and for the autonomous weeding in sugar beet, it might be possible to reduce costs by 12 %. For the latter however, it might be possible to reduce costs by 24 % compared to conventional treatment if inter-row hoeing could be reduced to only one treatment as for conventional weeding. In these calculations, we have used fairly conservative economic figures based on current prices. However, we may expect a further reduction in the price of RTK-systems and other electronic systems in line with increased supply of these systems. Software costs are not explicitly included in this study, apart from the RTK software system. Sensors for safety, such as ultrasonic, laser scanner, bumpers, were also not included in this study for the API platform not to further complicate the analysis, while they are very important for safety reasons In this matter, some additional costs should be expected depending on the diffusion of the systems.

5. Conclusions

An initial outcome from this study indicates that most of these autonomous systems are more flexible than conventional systems and may reduce labour costs and restrictions on the number of daily working hours significantly. Moreover, it is possible to substitute the most trivial working routines with autonomous systems although some routines are nearly impossible to automate due to the required accuracy of the specific tasks. In addition, at this stage of development, the initial investments and annual costs for expensive GPS systems are still relatively high but it seems possible to design economic viable robotic systems for grass cutting, crop scouting and autonomous weeding.
Findings show that there is a significant potential for applying these systems if it is possible to impose adequate control and safety regulations systems at reasonable costs. Moreover, a comparison between different European countries indicates that labour costs, crop rotation and farm structure may have a tremendous impact on the potential use of these systems.

6. References

Abounajmi, M., 2004. Mechanization of Dates Fruit Harvesting. ASAE Paper No. 041028. St. Joseph Mich.: ASAE.

Bak, T. and Jakobsen, H. 2004. Agricultural Robotic Platform with Four Wheel Steering for Weed Detection. Biosystems Engineering 87(2), 125–136.

Bisgaard, M. Vinther, D., and Østergaard, K. Z. 2004. Modelling and Fault-Tolerant Control of an Autonomous Wheeled Robot. Group Report 04gr1030a. Institute of Control Engineering, Aalborg University, Denmark.

Blackmore, S., Griepentrog, H.W., Fountas, S., Gemtos, T.A., 2007. Specifications for an autonomous crop mechanization system. Agricultural Engineering International: the CIGR Ejournal. Manuscript PM 06 032. Vol. IX. September, 2007.

Burks, T. F., Shearer, S. A., Heath, J. R. and Donohue, K. D. 2005. Evaluation of Neural-network Classifiers for Weed Species Discrimination. Biosystgems Engineering 91(3), 293-304.

Cembali, T., J.R., Folwell, , T. Ball and D.D. Clary. 2005. Economic Comparison of Selective and Non-Selective Mechanical Harvesting of Asparagus. ASAE Paper No. 053003. St. Joseph Mich.: ASAE.

Chi, Y. T. and P. P. Ling. 2004. Fast Fruit identification for Robotic Tomato Picker. ASAE Paper No: 043083. St. Joseph Mich.: ASAE.

Christensen, L. K., S. K. Upadhyaya, B. Jahn, D. C. Slaughter, E. Tan, and D. Hills. 2005. Determining the influence of water deficiency on NPK stress discrimination in maize using spectral and spatial information. Precision Agriculture Journal 6: 539-550.

Godwin, R. J., R. Earl, J. C. Taylor, G. A. Wood, R. I. Bradley, J. P. Welsh, T. Richards, B. S. Blackmore, M. C. Carver, S. M. Knight and B. Welti. 2001. Precision Farming of cereal crops: A five-year experiment to develop management guidelines. Project report 264e, - 328pp. London, UK: Home Grown Cereals Authority.

FAO-statistics 2005, www.fao.org

Godwin, R. J., Miller, P. C. H. 2003. A Review of the Technologies for Mapping Within-field Variability. Biosystems Engineering 84(4), 393-407.

Goense, D. 2003. The economics of autonomous vehicles. In Proceedings of the VDI-MEG Conference on Agricultural Engineering, VDI-Tatung Landtechnic,Hannover, Germany, 7-8 November, 2003.

Graglia, E. 2004. Importance of herbicide concentration, number of droplets and droplet size on growth of Solanum nigrum L, using droplet application of Glyphosphate. XIIeme Colloque International sur la Biologie des Mauvaises Herbes.

Granitto, P. M., Verdes, P. F. and Ceccatto, H. A. 2005. Large-scale investigation of weed seed identification by machine vision. Computers and Electronics in Agriculture 47 (2005), 15–24.

Griepentrog, H. W., M. Nørremark, H. Nielsen, and B. S. Blackmore. 2005a. Seed Mapping of Sugar Beet. Precision Agriculture Journal 6: 2157-165.

Griepentrog, H. W., P. T. Skou, J. F. Soriano and B. S. Blackmore. 2005b. Design of a seeder to achieve highly uniform sowing patterns. In Proc. 5th European conference on Precision Agriculture, ed. J. Stafford, 675-682, 8-11 June. Uppsala, Sweden: Wageningen Academic Publishers.

Griepentrog, H. W., M. Nørremark, J. Nielsen and J. F. Soriano. 2006. Close-to-crop thermal weed control using a CO2 laser. In Proc. Proceedings of CIGR World Congress, 3- 7 September, Bonn Germany: EurAgEng, CIGR.

Hannan , W. M. and F. T. Burks. 2004. Current Developments in Automated Citrus Harvesting.ASAE Paper No. 043087. St. Joseph Mich.: ASAE.

Kondo, N., K. Ninomiya, S. Hayashi, T. Ohta and K. Kubota. 2005. A New Challenge of Robot for Harvesting Strawberry Grown on Table Top Culture. ASAE Paper No: 043083. St Joseph Mich.: ASAE.

Have, H. 2004. Effects of automation on sizes and costs of tractor and machinery. In: EurAgEng04: Proceedings of the European Conference of Agricultural Engineering. Leuven, Belgium, 12-16 September 2004, European Society of Agricultural Engineers, paper 285.

Have, H., Nielsen, J. Blackmore, S. and Theilby, F. 2005. Development and test of an autonomous Christmas tree weeder. In: ECPA 05: Proceedings of the of 5th

European Conference of Precision Agriculture, Uppsala July 9-12, edited by. J. V. Stafford, Wageningen Academic Publishers, The Netherlands, 629-635.

Heisel, T., Christensen, S., Walter, A. M. 1999. Whole-field experiments with site-specific weed management. In: ECPA 02: Proceedings of the Second European Conference on Precision Agriculture, Odense, Denmark, 11–16 July, 1999 (Stafford J V, ed), Part 2, pp 759–768

Madsen, T. E. and Jakobsen, H. L. 2001. Mobile Robot for Weeding. MSc. Thesis. Danish Technical University, Lynby, Denmark.

Manh, A.G., Rabatel, G., Assemat, L., Aldon, M.J. 2001. Weed leaf image segmentation by deformable templates. Journal of Agricultural Engineering Research, 80(2), 139–146.

Miranda, F.R. Yoder, R.E, Wilkerson, J.B., Odhiambo, L.O., 2005. An autonomous controller for site-specific management of fixed irrigation systems. Computers and Electronics in Agriculture 48 (2005) 183–197.

Norremark, M. and Griepentrog, H. W. 2004. Physical methods to control weeds within crop rows. In: EurAgEng04: Proceedings of the European Conference of Agricultural Engineering. Leuven, Belgium, 12-16 September 2004, European Society of Agricultural Engineers, paper 405.

Pedersen, S. M. 2003. Precision farming - Technology assessment of site-specific input application in cereals. Ph.D. dissertation., IPL, Danish Technical University, Lyngby, Denmark.

Pedersen, S. M., S. Fountas, H. Have and B.S. Blackmore. 2006. Agricultural robots - system analysis and economic feasibility. Precision Agriculture 7(4): 295-308., Springer

Pedersen S.M, Fountas S and Blackmore B.S (2007): Economic potential of robots for high value crops and landscape treatment, Precision Agriculture 08457-465, Wageningen Academic Publishers.

Pilarski, T., Happold, M., Pangels, H., Ollis, M., Fitzpatrick, K. and Stentz, A. 2002. The Demeter system for automated harvesting. Autonomous Robots 13, 9-20.

Shinners, J. K., N. B. Binversie and P. Savoie. 2003. Whole-plant corn harvesting for biomass:comparison of single-pass and multiple-pass harvest systems. ASAE Paper No: 036089. St. Joseph, Mich.: ASAE.

Søgaard, H.T., 2005. Weed classification by active shape models. Biosystems Engineering 91 (3), 271–281.

Søgaard, H. T. and I. Lund. 2005. Investigation of the accuracy of a machine vision based robotic micro spray system. In Proc. 5th European conference on Precision Agriculture, 8-11 June., ed. J.

Sökefeld, M.,Gerhards, R., Kühbauch, W. 2000. Site-specific weed control—from weed recording to herbicide application. In: Proceedings of the 20th German Conference on Weed Biology and Weed Control, Stuttgart-Hohenheim, Germany, 14–16 March, 2000

Tothill, I. E., 2001. Biosensors developments and potential applications in the agricultural diagnosis sector. Computers and Electronics in Agriculture, Volume 30 (1-3) 205-218.

Vrindts, E., De Baerdemaeker, J., Ramon, H., 2002. Weed Detection Using Canopy Reflection. Precision Agriculture, 3, 63–80.

Weiner J and Olsen J (2007): Konkurrencevenen kan udnyttes, Momentum nr. 12/07.p 28-30.

Wilson, J. N. 2000. Guidance of agricultural vehicles - a historical perspective. Computers and Electronics in Agriculture 25, 1-9.

22

Service Robotics in Construction

Bock Thomas
Technical University Munich
Germany

1. Introduction

The first service robots for construction had been developed end of seventies beginning of eighties of the last century in Japan. These early service robots were used in the construction sector for inspection of nuclear power plants, exterior walls of high rise buildings and cleaning of high rise facades or glass roofs.

Fig.1 Kajima façade diagnostic robot

In the nineties service robots were applied to civil engineering projects such as inspection of tunnels, railroad tracks and bridges.

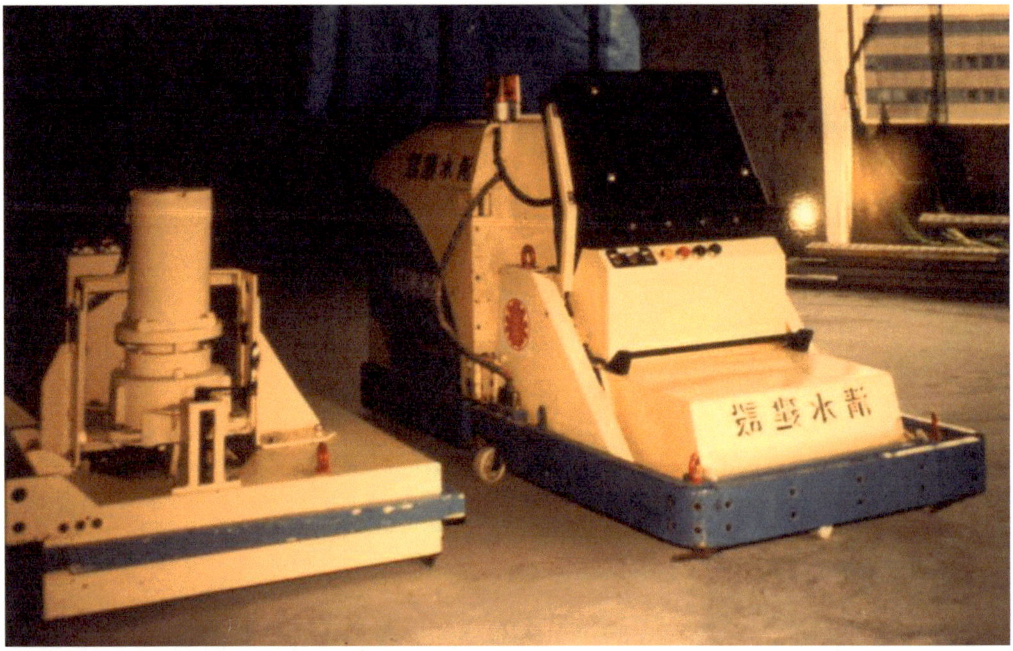

Fig. 2 MTV Cleaning robot

Fig. 3 Comatec robosoft glass cleaning robot

Advances in the autonomous robotics research resulted in service robots for office logistics of files and security against intruders, gas leaks or fire hazards etc.

Fig. 4 Tron office service robot

Due to the demographic changes especially in Japan service robots are being developed as partner, pet or personal robots for day care centers, hospitals and apartments.

2. Information and communication technology for service robotics in buildings

2.1 Home automation

Home automation systems are offered by all major electronic companies, such as Hitachi (home control system), Nihon Denki (Jema Home Bus), KEC (Kansai home bus system), Matsushita Denki Sangyo (Stavtec, Hals, HA system), Matsushita Denko (Napic Remote Control, Homen Mast, HA Installation System), MITI (MITI Home Bus), Mitsubishi Denki (House Keeping System, Home Automation System), MPT (HBS), NEC (Home Media Bus), NHK (Home Bus System), NTT (HAN), Sanyo Denki (Shains), Sharp (home Terminal System, House Info System, Home Bus System), Sony (Access 3), Toshiba (Nexhome Media System, Home Control System). In Europe Siemens Building Systems and Philips Home Lab and in USA Honeywell, Johnson Controls, Microsoft's "cyber home" in Medina and the Georgia Tech's Aware Home's Living room and MIT House conduct research by building real living environments

Major functions of standard installation cover fire alarm, gas leakage alarm (detection and notification), oxygen shortage alarm (detection and notification), electric lock (unlocking function, when a visitor is checked), burglar alarm (sensor notifies when a door or window is opened or broken), emergency alarm (notification to the public emergency center), earthquake alarm (notification of an inherent earthquake while switching appliances to appropriate mode), guest monitor at door, interphone between rooms, doorphone, power control, HVAC (heating-ventilation-air conditioning) control and control of water level and temperature of bathtub. As an example of home automation follow as a brief outline of Mitsubishi Melon home automation system.

Mitsubishi Electric thinks of home automation as information oriented home management. Center piece is the "in house total information system." The next level is formed by "household control systems" and "lifestyle information utilization system". On the third level we find "house keeping", "home management", and "life-culture".

Last level consists of "fire and crime prevention", "Energy control", "environmental and equipment control", "Data communications", "health care", "education and training ", "bussiness and culture" and "hobbies and entertainment".

This means that home medical care, home banking, home shopping, at home ticket reservation, at home employment, electronic mail, home nursing, home study, tele-control-monitoring-metering can be done without leaving home.

2.2 TRON house projects and Toyota personal robotics business model

This project supported by the Ministry of international trade and industry developped a new kind of computer chip. For international communication an automatic translation system was developed. With the help of TRON (Tron realtime operating nucleus) by Prof. Ken Sakamura of University of Tokyo, information and communication technology, intelligent robots should enable aged people and handicapped people in independant daily life.

The Tron project was started in 1979. In 1981, a committee was set up to envision the information society of the 90's and the requirements of the future computer technology needed for the social development on all levels and with regard to all aspects.

This project supported by the Ministry of international trade and industry aimed at developing a new kind of computer. With the help of the fifth generation computer technology, intelligent robots should support elderly and handicapped people in their daily life.

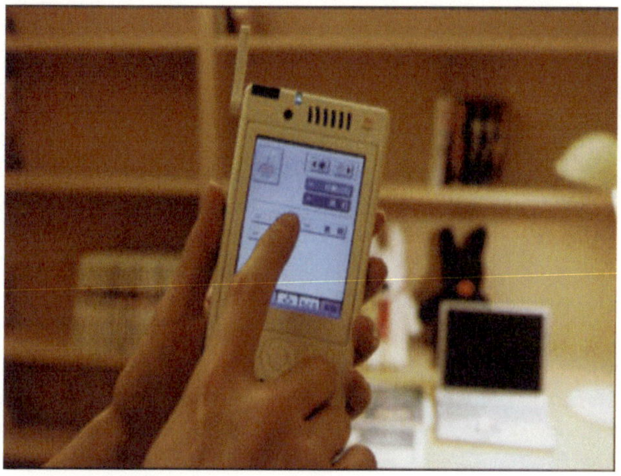

Fig. 5 Tron ubiquitous controller

The goal of the TRON Project has always been to create of a total "computer architecture", so it would be possible to establish a computer-based society in the 21st century. The "TRON Intelligent house" was one of the many TRON application projects that was actually built. The core of the concept was the symbiotic fusion of nature, humans and computers. It was completed in 1989 in Nishi Azabu for a price ticket of 1 billion yen or roughly 7 million €.
At that time, the intelligent house was the most computerized building, with 380 computers interconnected via the TRON architecture. While these computers were able to control every window or door, television, security system, and even the kitchen appliances remotely and automatically, the building itself and the major amenities were made of traditional Japanese wood material - thus combining Japanese computer technology with the traditional aspects of domestic architecture. Nowadays, many of these appliances are considered standard or helpful - especially considering the demographic changes in some countries.

Fig. 6 Tron House Nishi Azabu

In 2005 Toyota Home K.K. has developed a new intelligent home, based on TRON and other leading edge technologies. The house is called "Toyota Dream House PAPI." It is a smart or intelligent home and was created to reflect the all-purpose computing technologies that will be available for intelligent home construction in the year 2010.

Toyota Home creates modules to construct homes in factories that are similar to automobile businesses. With this application, Toyota demands that Prof. Sakamura learns about modular home construction techniques and adapts his design accordingly. That is why the company fully intends to bring the technologies on the market after they become commercially feasible. The main objective of this project was to create and realize an environmentally friendly, energy saving, intelligent house design in which the latest network computing technologies developed by the T-Engine project could be tested and further developed.
The total area of the house is 689 square meters. The house is mainly made of recyclable materials such as glass and aluminium. There is no need to clean the large glass windows regularly because of a special self-cleaning coating. In addition, Toyota Dream House PAPI possesses the technology along the lines of TRON to secure communication between the nodes. Nano T-Engine and pico T-Engine are used to make the construction of the nodes easy. The respective network is fairly sophisticated and the Toyota personal robot uses this network for navigation.
This personal robot among I-Foot, I-Unit and I-Real indicates the future business model of Toyota motor corporation for 2020 when the core business will shift from cars to personal robots. The Ubiquitous Communicator (UC) identifies people moving through the computerized living space and their personal preferences can be recognized. Though, the UC is only one of many HMIs, which can be used in the Toyota

3. Health care technologies for service robots in day care centers, hospitals and at home

3.1 Patient robots

While industrial robots perform the role of producing something, medical robots are simulators that play a passive role, and are used in educational programs for physicians and nurses. These medical robots are called patient simulators.

Fig. 7 Japan Science Society 1986

There is another type of simulator robot, called a dummy robot, which is placed in a car to collect data on the possible damage which human passengers might suffer in an accident of automobile collision. The patient robot shares the same concept.
The first patient robot in USA was an education robot for anaesthesia developed by the University of Southern California. Called SIM-1 (meaning simulator No. 1), this robot was designed for use as an educational model to show young physicians and nurses how anaesthesia should be administered to patients before surgery.
First, the condition of a patient and the surgery to be performed are shown to the students, who then determine the kind of anaesthesia to be administered to the patient. When an action corresponding to administering anaesthesia is taken, the robot that is connected to a computer reacts to the anaesthesia, showing a rise in the palpitation of the heart or a rough breath. The computer outputs the data which show whether the action that has just been taken is correct. The simulator robot also shows change in blood pressure, reaction of the pupil, etc. and, if an intravenous drip injection is administered, allows monitoring of the reaction of a living body. Patient simulators for the blood circulatory system were also developed in the United States.
In Japan, a training robot for bringing drowned persons to life again was developed by a group of Tokyo Women's Medical College. The robot has reactive mechanisms responding to a heart massage or mouth-to-mouth breath, showing restoration to the normal pulse and normal blood pressure, or to normal reaction of the pupil, or to a regular heart beat. This robot is also connected to a computer to give data in numerical values.
The monitor displays the condition of the patient and the instructions to the students, such as "Blood not circulating enough," "Massage more correctly," and "Breath not enough." The training results can be confirmed again with recordings on paper. Other robots train dentists etc.

3.2 Medical diagnostic robots in buildings

Robots are also being used for medical diagnostic purposes. Apart from many electromechanical devices, such as CT scanners and ultrasonic appliances, that are used as auxiliary tools of diagnosis, diagnostic robots and wearable medical diagnostic devices scanning pulse, oxygen etc. using UMTS , WLAN or blue tooth are emerging.

Fig. 8 Paro therapeutic robot

The present robots for industrial use perform work only on hardware, that is materials and machines. But robots will then be able to work on not only hardware but also human beings. One of the fields where such robots will be usable is medicine. Another is rehabilitation. Service robots will also be used at home to help sick or old people. The technique of the breast cancer diagnostic robot will be basic and important, opening the way to application to one field after another. It will provide basic know-how for developing what may be called a home robot or a personal robot which will be used at home to help sick or old people.
This diagnostic robot has nearly reached the level of practical usefulness. The technique of this robot that has been developed so far carries an important meaning as a basic technique to develop artificial senses close to those of the human being, which future robots must have.

3.3 Pet robots

Pet robots which will directly help the sick are being developed.
Patients who have just been operated on and old sick persons who have to stay in bed need the help of someone else. The illness may cause any change in their physical condition or raise physiological needs anytime. So, someone must always be there to help the patient. The fact, however, is that this kind of service is getting increasingly difficult to expect. There are anticipations, therefore, that tender robots will be available to help patients and sick old persons.

Fig.9 Pet Robot

Medical pet robots may be required to have functions similar to those of ordinary industrial robots in some cases. To help a sick person lying in bed drinking something, for example, the robot needs to handle only hardware, that is, a glass. That kind of technique has already been developed for industrial robots.

When it comes to do something more complex, however, problems may occur. For example, a person lying in bed may have to be taken up in the arms from the bed and moved to a wheelchair. Or, a sick person may need help in taking a bath. In such cases, a human helper carefully holds and helps the sick one making full use of the senses of the skin. If a robot tries to do it with its hard mechanical hands, the sick person could be injured.

Therefore, robots for these purposes must have senses similar to those of the human skin, which the breast cancer diagnostic robot has in the form of artificial fingers.

Another system uses a cushion and extends two arms under it to hold a patient up.

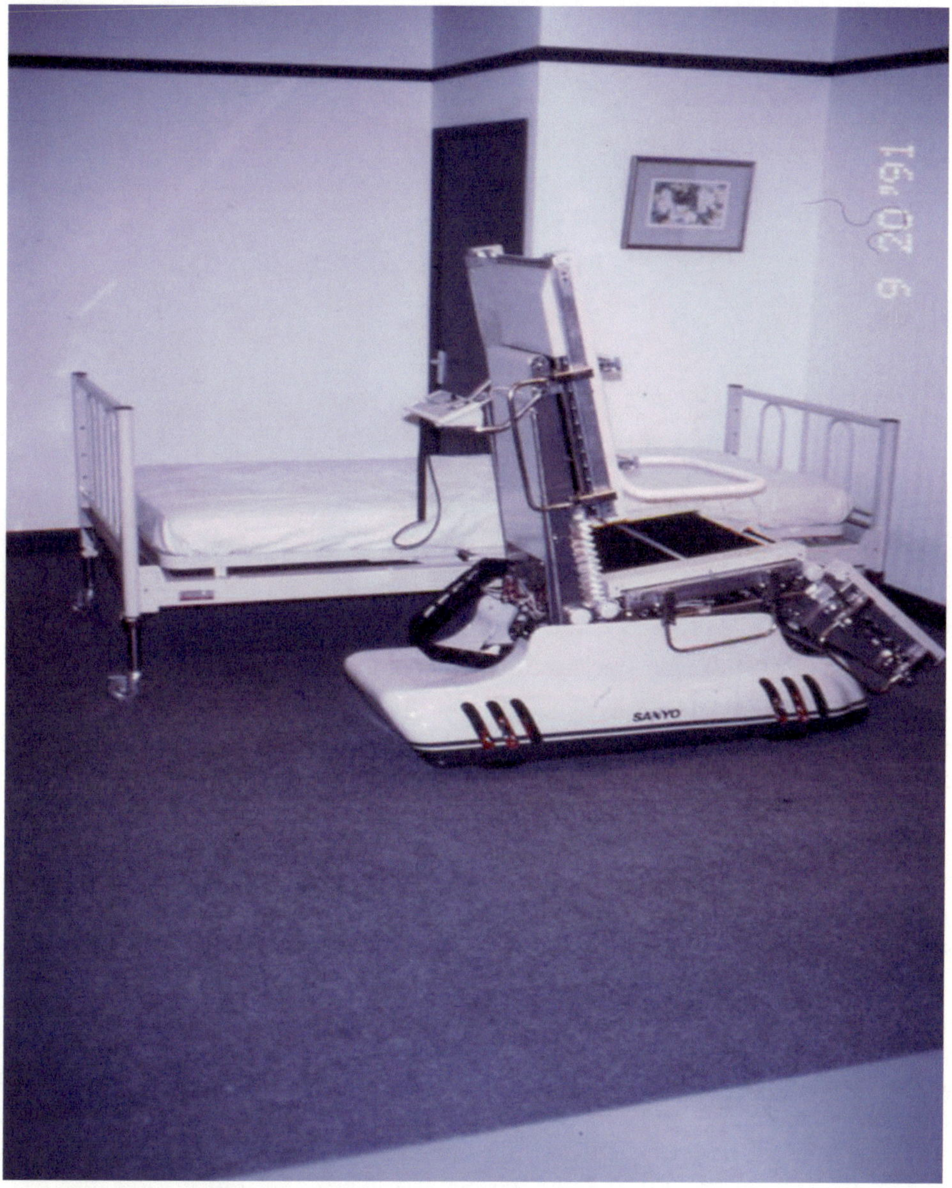

Fig. 10 Sanyo robotic patient transfer and wheelchair System

3.4 Assistive robots

Attention is paid to robots which will help physically handicapped people in many ways. Guide dogs, for example, help those who cannot see. A guide robot to substitute guide dogs is being developed. It is a system which transmits information as signals by electric stimulation from an electrode box, to be hand-held by a visually handicapped person, to guide him or her.

Fig. 11 Mel Dog , Prof. S. Tachi, Mechanical Engineering Laboratory , MITI, Japan

A mechatronic walking device was developed for disabled persons. Later manipulators will be added to enhance performance and future robot technology is also applied to artificial hands and legs for those who lost them. Artificial hands and legs of an electromyogram controlled system now available can be used in the same way as human hands and legs. As a command is transmitted from the brains via the motor nerves to the muscles, the muscles contract to move the hands or legs. Artificial hands and legs of the electromyogram controlled system are based on the same theory.

An electrode is placed on the skin of a limb extremity to receive a command transmitted via the motor nerves. The command is interpreted by a small electronic circuit - a kind of computer - and the artificial hands or legs are driven according to the meaning of the electronic signal deciphered.

Some robots have sensing device in an artificial hand. The human hands and legs have coetaneous sensation, and feed back information on the objects they touch via the sensory nerves to the brains. These robots work in a similar way. When the artificial hand touches something, the sensing device built in the hand converts the information into electrical signals to be transmitted via the electrode placed on the skin at the extremity of the hand to the sensory nerves and to the brains, where a decision is made according to the electrical signals received. Different from the conventional type of artificial hand which opens or closes the fingers using the force of body muscles, this type of artificial hand can be directly used almost in a natural way without long training.

Fig. 12 Hitachi walking device

If this electromyogram controlled system is applied to an artificial leg, walking speed can be automatically changed, or a stairway may be walked up or down using the artificial leg and the natural leg alternately - in the same way non-handicapped persons do.
This kind of system requires large power. One system uses a device of generating and storing power, implanted in the ankle of an artificial leg. When the wearer walks, the device generates energy from the rotary motion of the ankle in the walking direction, and stores that energy to drive the knee.

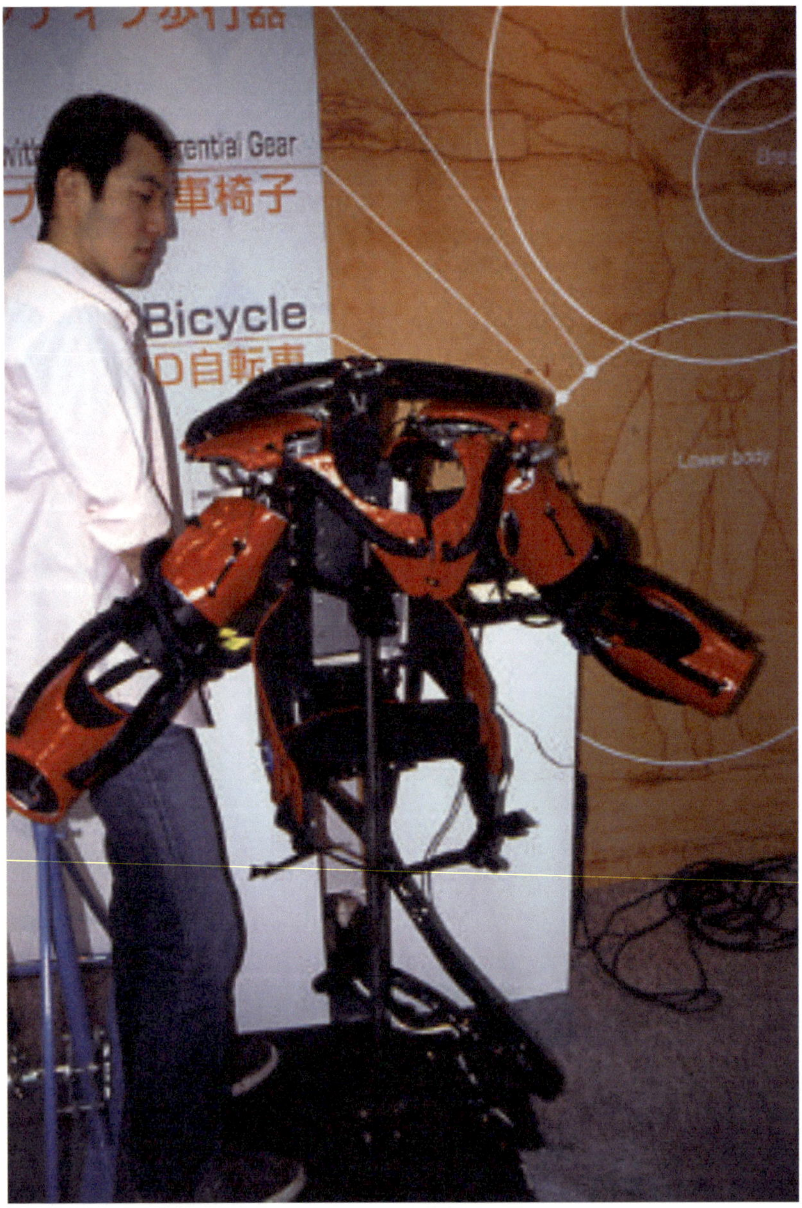

Fig. 13 Rika University, Tokyo

Fig. 14 HAL, Prof. Sankai, Tsukuba University

3.5 Service robots in hospitals and health care centers

Robotic techniques, computers and sensors assist and handle patients with sensory handicaps and support doctors with up-to-date computer-aided diagnosis. The pioneering work in this area was conducted by MITI in Japan at the Mechanical Engineering Laboratory (MEL). Here, the feasibility of a patient-handling robot, called "Nursy" was put to test. In addition, a multidirectional platform was developed with the potential use as a wheelchair for the disabled. Nursing costs and wages are steadily increasing. There is also a constant lack of nurses. Robots could perform some of these tasks in the future.

Repetitive hospital functions, such as delivering linens, preparing beds, giving medication, transporting patients, monitoring vital signs, supplying the patients with water and food are being automated.

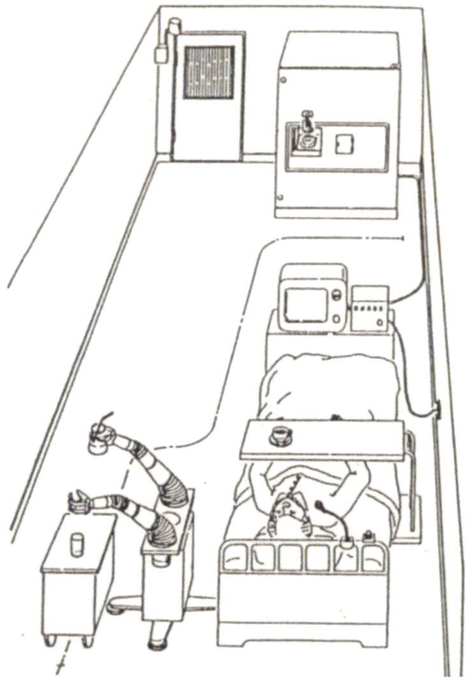

Fig. 15 Japan Science Society 1986

4. LIESE: LIfE Support systEm: mechatronic renovation of building stock for independent life style of elderly

To realize the "LIESE" concept in existing building stock, components were newly developed. Wooden wall finish component with handrail-shaped edge it functions as a handrail at a 75 cm level with incorporated RFID tags to detect position.

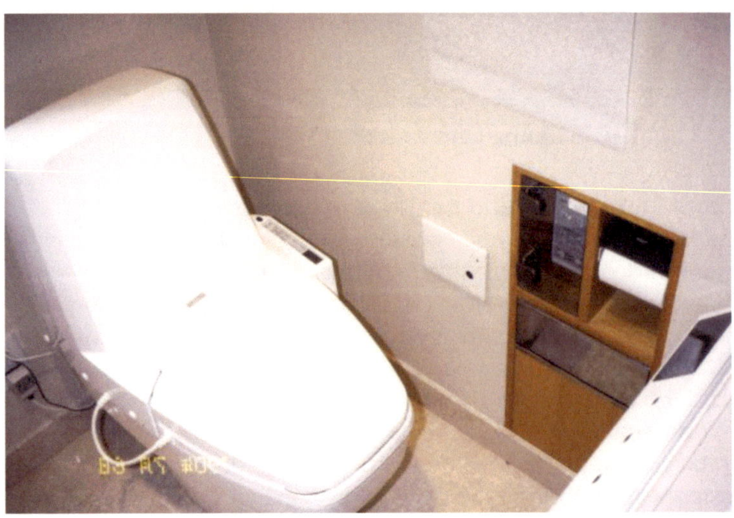

Fig. 16 Diagnostic toilet

Flexible wall with a handrail-shaped edge: this is the wall that separates a room and a bathroom and can easily be disassembled.

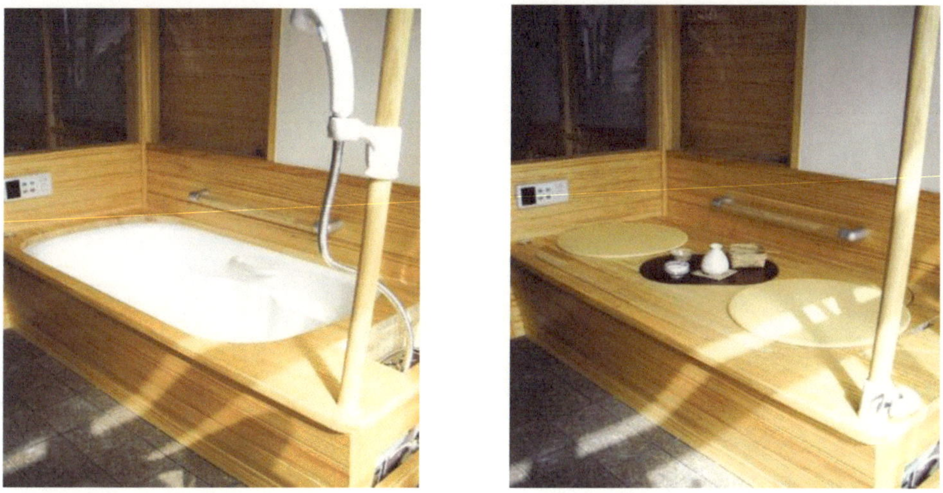

Fig. 17, 18 transformable bath tub

One direction movable bed with inbuilt servo motors which can easily move to support necessary care activities.

Fig. 19 Servo driven bed

Sliding door with a handrail-shaped edge: it has the function of stopping the door when one pushes the handrail-shaped edge downwards by installing the wheels equipped with the stop function below the door and easily supports the opening due to incorporated servo motor that detects the horizontal motion and assists the sliding of the door which is especially helpful while sitting in a wheelchair

Fig. 22 Servo driven sliding door

Pumping toilet covered with table: it is covered with a movable table to conceal the toilet, and the pressure pump is used for its drainage. It is supposed that one lays a pipe for water supply and drainage using a bathroom that is in the center of the house, and a method involving a pressure pump is used in order to minimize the height of the raised floor at 15 cm.

Fig. 20 , 21 Multi purpose toilet

The sensor floor panel features an inbuilt senor mat that detects fallen person. A waterproof sheet with synthetic rubber and plastic sensor sheet is placed on the floor with a waterproof panel on top of it.

Fig. 23 Multipurpose sink

Multipurpose kitchen sink and washbowl can be installed in a narrow space and used as a small kitchen sink in some cases and as a washbowl in other cases. For this purpose, the shape of the sink is uniquely designed to correspond to both functions. The sink bottom is flat for kitchen use and it is deep enough for washing one's face. The sink has an asymmetric shape and allows comfortable approach by wheelchair. A new process of robotic metal shaping was successfully used to produce individually adjusted multipurpose sinks.
The bathtub can be transformed into a bench, when the tub is not in use and t he shower can also be hidden.

5. Future robotic technology for service applications in buildings

Existing robotic technologies are based on industrial robotics application mostly for the production of cars and other mass production items. These industrial applications of robots require high speed motion, high accuracy and extended mean time between failure among other requirements. In the case of service robots for construction and building sector we need light weight robots with great payload and intuitive programming. Some of these technologies are available for example in space robotics such as the rotex light weight robotic arm developed by Professor Hirzinger.

6. Future trends towards urban service robotics

The next step according to Prof. Hasegawa in Japan will be the development of urban service robotics in which advanced robotic technology merge with architecture, buildings and urban infrastructure. The environment of human daily life is ill defined and dynamically changing in three spatial dimensions and time. Persons walk and drive around, layout of floor plans in buildings and furniture change so that any map, parametric world model facility management plans or building information models become obsolete.

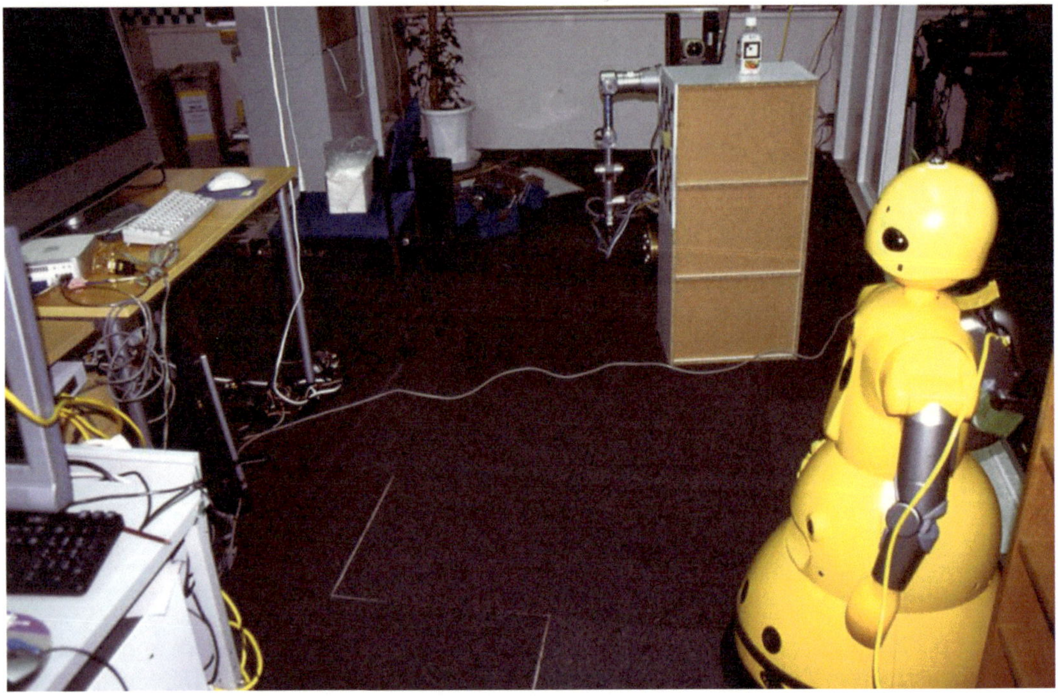

Fig. 24 Service robotics room at University of Tokyo, Prof. T. Sato and "Wakamaru" MHI

What is easy to recognize for humans is difficult to be realized by even advanced autonomous robots. Therefore the next step towards will be a structured environment using sensor network, vision systems, microwave and embedded distributed RFID tags and additional GPS for outdoors navigation.

7. Conclusion

Service robots in construction move from cleaning of the construction site and diagnosing of the building façade as security, logistic and assistive robots into the building and apartment and will in the future share urban environment with humans.

8. References

Matsumura S. ; Ando M, Ishizuka K., Kobata S., Seki E., Kumagai R., Development of the Renovation Technology of the Existing Residence for Elderly People, Journal of Architecture and Building Science, Vol.117 No.1498, pp.231-234, 2002

Matsumura S. ; "RAKU-INKYO"Infill System-upgrading existing dwellings for the elderly, Open House International, Vol.27, No.2, pp.64-68, 2002

Hasegawa T, ; Murakami K ; Robot Town Project : Supporting Robots in an Environment with its Structured Information, pp.119-123, 3rd URAI, 2006

Bock T., Rojin Oriented Design for Life Support System, 6th BAS, Gits, Belgium, 2006